Springer Series in Operations Research

Editor:
Peter Glynn

Springer
New York
Berlin
Heidelberg
Barcelona
Budapest
Hong Kong
London
Milan
Paris
Tokyo

Springer Series in Operations Research

Zvi Drezner
Editor

Facility Location

A Survey of Applications and Methods

With 57 Illustrations

 Springer

Zvi Drezner
Department of Management
  Science/Information Systems
California State University, Fullerton
Fullerton, CA 92634-9480
USA

Series Editor:
Peter Glynn
Department of Operations Research
Stanford University
Stanford, CA 94305
USA

Library of Congress Cataloging-in-Publication Data
Drezner, Zvi.
    Facility location : a survey of applications and methods / Zvi
  Drezner.
       p.   cm. — (Springer series in operations research)
    Includes bibliographical references.
    ISBN 0-387-94545-8 (hc : alk. paper)
    1. Industrial location — Mathematical models.   2. Store location —
  Mathematical models.   3. Offices — Location — Mathematical models.
  4. Public utilities — Location — Mathematical models.   I. Title.
  II. Series.
  HC79.D5D73   1995
  658.2′1 — dc20   95-19691

Printed on acid-free paper.

Production managed by Bill Imbornoni; manufacturing supervised by Jacqui Ashri.
Photocomposed pages prepared from the editor's LaTeX files.
Printed and bound by Edwards Brothers, Inc., Ann Arbor, MI.
Printed in the United States of America.

9 8 7 6 5 4 3 2 1

ISBN 0-387-94545-8 Springer-Verlag New York Berlin Heidelberg

Dedicated to Tammy and Taly

Preface

Facility location applications are concerned with the location of one or more facilities in a way that optimizes a certain objective such as minimizing transportation cost, providing equitable service to customers, capturing the largest market share, etc. Facility location problems give rise to challenging geometrical and combinatorial problems. The research on facility location problems spans many research fields such as operations research/management science, industrial engineering, geography, economics, computer science, mathematics, marketing, electrical engineering, urban planning, and related fields.

Applications to facility location models abound. Location of warehouses, plants, hospitals, retail outlets are classical examples. Applications are also found in the location of electronical components, warning sirens, sprinklers, radar beams, exploratory oil wells. These are less obvious "facilities". One should consider applying a location model to any scenario that involves finding a best location (or locations) for any object(s).

This book *Facility Location: A Survey of Applications and Methods* provides a state of the art review along with references to important contemporary topics in locational analysis. The book includes twenty chapters divided into four parts, each dealing with a different aspect of location modeling and implementation. The book concludes with over 1,200 references.

The first part *Methodology and Analysis* consists of a survey of methodological and technical aspects of location analysis. Issues such as estimating distances, the error introduced by assuming discrete demand rather than continuous one, global optimization techniques, inference of weights, conjugate duality, and using Voronoi diagrams are reviewed.

The second part *Various Location Objectives* concentrates on the wide variety of objectives and applications found in facility location models such as market externalities, distribution system design, global manufacturing strategy as well as review of various objectives, and the location of emergency facilities.

The third part *Competitive Facility Location* involves the location of facilities in a competitive environment. The most important decision for a retailer is choosing an outlet location. Chapters reviewing the state of the art in retail location and economic modeling are presented.

The last part *Routing and Location* discusses the interface between routing and location as well as capturing customers who drive on freeways, hazardous material location and its impact on the environment.

The book is an essential tool for marketers who need to find the best locations for new retail outlets, industrialists who wish to design a distribution system, planners who need to establish emergency facilities, and other practitioners, as well as researchers in the area of location who wish to design new algorithms and new approaches to solving location problems.

I am thankful to all the contributors, who come from 11 countries, each a leading experet in his subject matter, for their outstanding and comprehensive work. They made this book such a great reference guide to "everything you ever wanted to know" about location. The book would not have been possible without their contribution.

Zvi Drezner
Fullerton, California

Acknowledgements

Most of the chapters were reviewed by fellow contributors and we are grateful to the following reviewers who are not contributors: Emilio Carrizosa, George List, and Christian Michelot. I am also thankful to my graduate student Eddy Lin for his help in editing and translating many of the chapters to LaTeX.

The first two authors of Chapter 3 have been supported by Office of Naval Research grant #N00014-92-J-1194 and FCAR (Fonds pour la Formation de Chercheurs et l'Aide à la Recherche) grant 92EQ1048. The first author of Chapter 3 has also been supported by NSERC (Natural Sciences and Engineering Council of Canada) grant to HEC and NSERC grant GP0105574. The second author of Chapter 3 has also been supported by NSERC grant GP0036426 and a FRSQ (Fonds pour la Recherche en Santé du Québec) fellowship. The work of the third author of Chapter 3 was done during a visit to GERAD.

Chapter 8 was in part supported by grants from the Natural Sciences and Engineering Council of Canada under grant numbers OGP0009160 and OGP0039682 and by a grant under the Québec/New Brunswick Cooperation Agreement. This support is gratefully acknowledged.

Chapter 11 is an extended and updated version of the survey paper [960] written for the ISOLDE VI meeting in Chios, Greece. The author acknowledges the courtesy of the editors of the Studies in Locational Analysis in letting him heavily relay upon that paper.

Chapter 16 was partially funded by DCYGIT grants numbers PB92-1036 and PB92-1037, Ministry of Education, Spain

Chapter 17 was supported by the NSERC grants of the three authors.

Chapter 18 was partially supported by the NSERC Grant of the first author.

Chapter 19 is a modified and updated version of (parts of) the survey article by Boffey and Karkazis in *Studies in Locational Analysis*, Issue 5.

Chapter 20 was partially funded by the Natural Sciences and Engineering Research Council of Canada (OGP 25481).

Contents

III   COMPETITIVE FACILITY LOCATION   283

IV   ROUTING AND LOCATION   387

Contributors

Oded Berman
University of Toronto
Faculty of Management
246 Bloor St. West
Toronto, Ontario M5S 1V4
Canada.
berman@fmgt.mgmt.utoronto.ca

Brian Boffey
University of Liverpool
Department of Statistics
and Computational Mathematics
Liverpool, L69 3BX
United Kingdom.

Margaret L. Brandeau
IE/EM Department
Stanford University
Stanford, CA 94305-4024
U.S.A.
Margaret.Brandeau@forsythe.
stanford.edu

Jack Brimberg
Royal Military College of Canada
Department of Engineering
Management
Kingston, Ont. K7K-5L0
Canada.
brimberg@hp.rmc.ca

Samuel S. Chiu
EES Department
Stanford University
Stanford, CA 94305-4025
U.S.A.
samchiu@leland.stanford.edu

C. Samuel Craig
Stern School of Business
New York University
Management Education Center
44 West 4th Street
New York, NY 10012
U.S.A.
scraig@stern.nyu.edu

M. Cemal Dincer
Department of Industrial
Engineering
Bilknet University
06533, Bilknet, Ankara
Turkey.
dincer@bcc.bilknet.edu.tr

Tammy Drezner
School of Management
California State University
Dominguez Hills
Carson, CA 90707.
U.S.A.

Zvi Drezner
Department of Management
Science/Information Systems
California State University-Fullerton
Fullerton, CA 92634.
U.S.A.
drezner@fullerton.edu

H.A. Eiselt
University of New Brunswick
Faculty of Administration
P.O. Box 4400
Fredericton, N.B. E3B 5A3
Canada.
haeiselt@unb.ca

Erhan Erkut
Faculty of Business
University of Alberta
Edmonton, Alberta, T6G 2R6
Canada.
Erhan_Erkut@ualberta.ca

Arthur M. Geoffrion
Anderson Graduate School of
Management
University of California,
Los Angeles
Los Angeles, CA 90024-1481
U.S.A.
ageoffri@agsm.ucla.edu

Avijit Ghosh
New York University
Stern School of Business
44 West 4th Street (Room 8-74)
New York, NY 10012-1126
U.S.A.
agosh@stern.nyu.edu

Thomas A. Grossman, Jr.
Faculty of Management
University of Calgary
Calgary, Alberta T2N 1N4
Canada.
grossman@acs.ucalgary.ca

Pierre Hansen
GERAD & Ecole des Hautes
Études Commerciales
Département des Méthodes Quan-
titatives et Systèmes d'Information
5255 avenue Decelles
Montréal (Québec) H3T 1V6
Canada.
pierreh@crt.umontreal.ca

M. John Hodgson
University of Alberta
Department of Geography
3-32 HM Tory Building
Edmonton T6G 2H4
Canada.
jhodgson@geog.ualberta.ca

Patrick Jaillet
MSIS Department
The University of Texas at Austin
Austin, TX 78712-1175
U.S.A.
and
Laboratoire de Mathématiques
Appliquées
ANPC/CNRS, Paris
France.
jaillet@emx.cc.utexas.edu

Brigitte Jaumard
GERAD & Ecole Polytechnique de
Montréal
Département de Mathématiques
Appliquées
P.O. Box 6079, Station "Centre-
ville"
Montréal (Québec) H3C 3A7
Canada.
Brigitt@crt.umontreal.ca

Thomas R. Jefferson
Anderson Graduate School of
Management
University of California,
Los Angeles
Los Angeles, CA 90024-1481
U.S.A.
tjeffers@agsm.ucla.edu

Soheila Jorjani
School of Business Administration
California State University
San Marcos, CA 92069
U.S.A.
soheila_jorjani@csusm.edu

John Karkazis
University of the Aegean
Michalon 8, 82100
Chios
Greece.

Dmitry Krass
University of Toronto
Faculty of Management
Toronto, Ontario M5S 1V4
Canada.
krass@fmgmt.mgmt.utoronto.ca

Shiv Kumar
Applied Insurance Research
137 Newbury Street
Boston, MA 02116
U.S.A.

Gilbert Laporte
Universite De Montreal
Centre de Recherche sur les
Transports
C.P. 6128 Succ. A
Montreal H3C 3J7
Canada.
gilbert@crt.umontreal.ca

Robert F. Love
McMaster University
Faculty of Business
Department of Management
Science
Hamilton, Ontario L8S 4M4
Canada.
RLove@facbus.business.
mcmaster.ca

Vladimir Marianov
Department of Ingeneria Electrica
Universidad Catolica De Chile
Casilla 6177, Correo 22
Santiago
Chile.
marianov@ing.puc.cl

Sara McLafferty
Department of Geology and
Geography
Hunter College
695 Park Avenue
New York, NY 10221
U.S.A.
slm@everest.hunter.cuny.edu

James G. Morris
School of Business
Grainger Hall
975 University Ave.
University of Wisconsin-Madison
Madison, WI 53706-1323
U.S.A.
jmorris@bus.wisc.edu

Atsuyuki Okabe
Atsuyuki Okabe
Department of Urban
Engineering
The University of Tokyo
Tokyo 113
Japan.
a34532@tansei.cc.u-tokyo.ac.jp

Morton E. O'Kelly
Ohio State University
Department of Geography
103 Bricker Hall
190 N. Oval Mall
Columbus, OH 43210
U.S.A.
mokelly@magnus.acs.
ohio-state.edu

Dominique Peeters
Universite Catholique de Louvain
Department of Geography
Place Louis Pasteur 3
B-1348 Louvain-La-Neuve
Belgium.
peeters@geog.ucl.ac.be

Frank Plastria
Vrije Universiteit Brussel
Center for Industrial Location
Pleinlaan 2
B-1050 Brussels
Belgium.
faplastr@vnet3.vub.ac.be

Charles ReVelle
The Johns Hopkins University
319 Ames Hall
Baltimore, MD 21218
U.S.A.
revelle@jhuvms.hcf.jhu.edu

Carlton H. Scott
Graduate School of Management
University of California-Irvine
Irvine, CA 92717
U.S.A.
scott@ccmail.gsm.uci.edu

Daniel Serra
Dept. of Economics
Universitat Pompeu Fabra
Balmes 134
Barcelona 08008
Spain.
serra@upf.es

David Simchi-Levi
Department of Industrial Engineer-
ing and Management Sciences
Northwestern University
MLSB 2225 North Campus Drive
Evanston, IL 60208-3119
U.S.A.
levi@iems.nwu.edu

Atsuo Suzuki
Department of Information Systems
and Quantitative Sciences
Nanzan University
18, Yamazato-cho, Showa-ku
Nagoya, 466
Japan.
atsuo@michael.nanzan-u.ac.jp

Jacques -F. Thisse
CERAS, Ecole Nationale des Ponts
et Chaussees
28, rue des Saints-Peres
75343-Paris
France.
thisse@enpc.enpc.fr

Hoang Tuy
Institute of Mathematics
P.O. Box 631, Bo Ho
Hanoi
Vietnam.
hotuy@math.liu.se

Vedat Verter
Faculty of Business
University of Alberta
Edmonton, Alberta, T6G 2R6
Canada.
vverter@gpu.srv.ualberta.ca

Scott T. Webster
School of Business
Grainger Hall
975 University Ave.
University of Wisconsin-Madison
Madison, WI 53706-1323
U.S.A
swebster@bus.wisc.edu

BOOK OVERVIEW

The book is divided into four parts, each dealing with a different aspect of location modeling and implementation.

The first part *Methodology and Analysis* consists of a survey of methodological and technical aspects of location analysis. The second part *Various Location Objectives* concentrates on the wide variety of objectives and applications employed in and by facility location analyses. The third part *Competitive Facility Location* involves the location of facilities in a competitive environment. The introduction of new facilities among existing competing facilities is discussed. The last part *Routing and Location* discusses the interface between routing and location.

A list of books, survey papers, and special journal issues on locational issues beginning in 1980 is provided at the end of this book overview.

Methodology and Analysis

In Chapter 1 *Estimating Distances* Jack Brimberg and Robert F. Love discuss various methods to approximate travel distances between two points by a measure dependent only on the coordinates of the two points. When objects in space such as different cities in a geographic region, activity centers in a production plant, or the junction boxes in an electrical system, can be represented as points, a distance predicting function (DPF) may be used to transform the point coordinates of two points into an estimate of the distance between them.

Many areas of application for distance prediction functions are just now evolving. A few of these are: verifying arc lengths between nodes in road network data, providing road distances for continuous location models, dynamic fire control models, truck scheduling models and recently, to provide distances in Geographical Information System (GIS) models.

In Chapter 2 *Replacing Discrete Demand with Continuous Demand* Zvi Drezner discusses the error in the calculation of the objective function and consequently in determining the optimal location for a facility when continuous demand is modelled as a set of demand points. The error caused by this approximation is investigated and a particular location problem is used for testing. A simple approximation to the objective function of any location problem is proposed. It is recommended that this approximation be used in modelling any location problem.

In Chapter 3 *Global Optimization in Location* Pierre Hansen, Brigitte

Jaumard, and Hoang Tuy review various global optimization methods and how they apply to facility location issues. Global optimization methods aim to find the global optimum of a nonlinear and nonconvex function subject to nonlinear and nonconvex constraints. The main approaches to global optimization, i.e., branch-and-bound, Lipschitz underestimation, outer approximation, polyhedral annexation, linearization and decomposition are briefly reviewed and illustrated through their application to a series of extensions of Weber's problem.

In Chapter 4 *Inferred Ideal Weights* Morton E. O'Kelly discusses the issue of finding the weights for facilities that will result in the given optimal location. He shows that if a set of facility locations is given, *and hypothesized to be optimal with respect to the data*, there are still many interesting analytical questions to be answered concerning the demand for the services.

The model turns conventional locational analysis around, and asks for the "demand" weights that support the optimality of a particular location. The discussion gives an opportunity to review several planar facility location problems. The chapter discusses four variants of the problem: (a) the multifacility location problem; (b) the location-allocation problem; (c) the spatial interaction based location model; and (d) the hub location problem.

In Chapter 5 *Conjugate Duality in Facility Location* Carlton Scott, Thomas Jefferson, and Soheila Jorjani show how conjugate duality theory for convex programs can be applied to location models. Using the well known minisum and minimax models under various unconstrained and constrained scenarios, dual programs are constructed, as well as the relationship between the primal formulation and its corresponding dual. Further cases may be developed using the theory and examples given in the chapter.

In Chapter 6 *Using Voronoi Diagrams* Atsuo Suzuki and Atsuyuki Okabe discuss the application of Voronoi diagrams to solving continuous location problems. The Voronoi diagram is a tessellation of the plane. Given a set of points in the plane, we associate locations in the plane with the closest member of the point set. The result is a tessellation of the plane into a set of polygons associated with the points. The chapter introduces the concept of the Voronoi diagram, shows how to solve the continuous p-median problem, the continuous p-center problem, the time-space p-median problem, the mobile facility location problem, and some other problems. The chapter concludes with a list of available free computer programs.

Various Location Objectives

In Chapter 7 *Facility Location with Market Externalities* Margaret L. Brandeau, Samuel S. Chiu, Shiv Kumar, and Thomas A. Grossman Jr. discuss location of facilities when market externalities affect the objective function. Research on location decisions has generally assumed away market externalities, overlooked the latitude available to customers in choosing a facility to use, and ignored the effects of transportation and congestion costs on consumer behavior. However, congestion externalities, user choice, and demand elasticities do exist in practice and, therefore, it is quite appropriate that they be incorporated rigorously into locational analysis. The chapter describes models for the efficient location of facilities and the allocation of a limited resource (e.g. service capacity) among facilities in an environment with negative congestion externalities. Models that incorporate user choice and models that assume a dictatorial environment, as well as models with elastic and inelastic demand are considered. Published and new models and results are presented.

In Chapter 8 *Objectives in Location Problems* H.A. Eiselt and Gilbert Laporte provide a framework for objective functions in location models based on a decision-making scenario involving users and planners. Well-known classes of models are put into this framework and the most important contributions in the respective fields are surveyed and discussed. Throughout the chapter they emphasize the forces that different objectives exert on facilities.

In Chapter 9 *Distribution System Design* Arthur M. Geoffrion, James G. Morris, and Scott T. Webster present a guide to fast analysis tools for distribution system design. The chapter is presented in a realistic environment establishing the procedures and the practical considerations involved in such a design.

In Chapter 10 *Siting Emergency Services* Vladimir Marianov and Charles ReVelle review the considerations and the process involved in siting emergency facilities. They first discuss the issues and then review the solution procedures.

In Chapter 11 *Continuous Location Problems* Frank Plastria gives a structured overview of the literature on location theory in continuous space, with special attention to the most recent results. Possible avenues of further research are suggested. The following domains are explored: distance measures, dominance and efficiency, voting topics, single facility location, single facility location-allocation, and multifacility location, multifacility location-allocation. The paper concludes with some related problems.

In Chapter 12 *Global Manufacturing Strategy* Vedat Verter and M. Cemal Dincer focus on facility location decisions in global firms. They suggest that the global manufacturing strategy of a firm provides the framework for facility location. The locational decisions interact with the capacity and technology selection decisions within the international context. Thus, global firms need to consider all configurational decisions simultaneously. The authors point out the differentiating features of global manufacturing strategy planning. Then, they present the analytical approaches to the multicommodity, multiechelon production-distribution system design problem as building blocks of comprehensive location models for global firms. Finally, they review the prevailing studies that make an explicit effort to incorporate review of the distinguishing features of global manufacturing.

Competitive Facility Location

In Chapter 13 *Competitive Facility Location in the Plane* Tammy Drezner surveys recent developments in competitive facility location with emphasis on the location of new facilities anywhere in the plane rather than selected from a prespecified set of alternative sites. The underlying assumption is that the best location for a facility is at a point at which its captured market share is maximized. There are two main approaches to estimating market share attracted by a facility. The first, based on Hotelling's formulation, assumes that customers patronize the facility closest to them. It implicitly implies that facilities are equally attractive. The second approach, based on Huff's formulation, assumes a probabilistic behavior of customers and takes into account different attractiveness of facilities.

The Hotelling model is extended by replacing the proximity assumption by the assumption that customers base their choice on a utility function that the distance is one of its components. The discrete utility model assumes that all customers located at a demand point apply the same utility function and therefore all of them select the same facility. This assumption is further relaxed in the random utility model where it is assumed that customers draw the utility parameters from a statistical distribution. In the random utility model the customers located at the same demand point may patronize different facilities.

Gravity models (Huff's assumption) are also analyzed and the best locations in the plane for one or more facilities are found.

In Chapter 14 *Multifacility Retail Networks* Avijit Ghosh, Sara McLafferty, and C. Samuel Craig review the application of location allocation models for designing multifacility retail networks. Location allocation models determine the best locations for new retail outlets based on stated corporate objectives through different allocations of consumers based on their expected travel patterns. These travel patterns are influenced by the location of competing and complementary stores and the road network in the area. The overall goal of the chapter is to discuss the application of location allocation models for systematically evaluating and planning store networks and to illustrate these applications in a variety of retail contexts. The chapter: (i) provides a brief historical review of the development of retail location models, (ii) presents a typology of different types of models, (iii) discusses the mathematical structure of location allocation models used in retail location planning, (iv) describes the types of data needed to implement the models and the sources of these data, and (v) discusses some of the difficulties and challenges in practical application of the models. The chapter concludes with an agenda for future research.

In Chapter 15 *Economic Models of Firm Location* Dominique Peeters and Jacques-François Thisse deal with the integration of pricing strategies in firm location models when requirements for the firm's output are variable. Five spatial price policies and two location models are considered. Some economic properties are first developed. Algorithms yielding the optimal prices are then proposed. Finally, using numerical examples, the impact of the various pricing policies on the firm's location and distribution is discussed in light of the economic properties mentioned above.

In Chapter 16 *Competitive Location in Discrete Space* Daniel Serra and Charles ReVelle focus on the location of facilities in a discrete space. The siting of facilities in a competitive environment has recently become a mainstream topic in the field of location-allocation modelling. An overview of the emerging literature on discrete competitive models is presented. Some formulations of the location of competitive facilities in an oligopolistic market represented by a discrete network are addressed. Ways in which entry barriers can be applied to newcomers to the market are presented. Uncertainty in the information used in formulating strategies to maximize market share is discussed. Finally, a new model based on early works of ReVelle and Serra is presented.

Routing and Location

In Chapter 17 *Flow-Interception Problems* Oded Berman, John Hodgson, and Dmitry Krass discuss the problem of attracting demand (generated by traffic flow) on a network. They focus on models where demand for service originates not from the nodes of the network but from flow (customers) travelling on various paths of the network. Examples of such customer behavior include in particular discretionary services (e.g. automatic teller machines, gasoline stations and convenience stores). Here a significant fraction of customers consuming the service do so on an otherwise pre-planned tour. The purpose of the travel is not to obtain service, but if there is a facility on the pre-planned tour (or substantially close) the customers may choose to obtain the service. Therefore, the objective here is not to minimize the average travel time or the worst case travel time but to locate the facilities so as to maximize the total flow of customers that are 'intercepted' during their travel. Other examples of applications where the objective is to maximize the intercepted flow of customers include locating police radar traps, highway traffic monitors, billboards, etc.

In Chapter 18 *Location-Routing Problems with Uncertainty* Oded Berman, Patrick Jaillet, and David Simchi-Levi review recent results on location-routing problems with uncertainty. In these problems, a salesman must visit, on any given day, a random subset of a given set of customers. The salesman, upon receiving the daily list, starts a service tour which includes all the waiting customers and which ends at the salesman's home location. The problem is to find the optimal home location that minimizes the expected distance traveled, using either (i) a *re-optimization strategy:* solve optimally (or near-optimally with a good heuristic) every potential instance of the original problem (this problem is called the traveling salesman location problem) or (ii) an *a priori optimization strategy*: find an a priori solution to the original problem and then update in a simple way this a priori solution to accommodate each particular instance/variation (this problem is called the probabilistic traveling salesman location problem).

In Chapter 19 *Location, Routing and the Environment* Brian Boffey and John Karkazis discuss the implications of routing on the environment. The location of obnoxious facilities and/or the routing of vehicles carrying hazardous materials give rise to problems which are inherently multiobjective in nature. One natural objective concerns the possible detrimental effects on the environment that may result from location - routing decisions. These are considered with particular attention focused on the spread of airborne pollutants. The mechanism of spread of pollutants by wind is first described, then a brief description given of relevant concepts relating to multicriteria analysis. This is followed by a review of the location / routing literature relating to airborne pollutants and finally some possible directions for future research are indicated.

In Chapter 20 *Hazardous Material Logistics* Erhan Erkut and Vedat Verter provide a review of Hazardous Materials (hazmats) Logistics. The major concerns in strategic management of "hazmats" are the risks associated with them, the spatial distribution of these risks, and the costs incurred during the management process. They first review the methodology for assessment of the risks imposed on the public by potentially hazardous activities. Then they present the prevailing approaches for modeling equity in the spatial distribution of risk. Then they cover the cost aspects of hazmats management. They also discuss analytical approaches for facility design and transportation planning, respectively. The current trend towards the development of integrated hazmats logistics models, and the previous studies that would serve as building blocks in this endeavor are then presented.

References

The following is a list of books, review articles, and special volumes that has been published since about 1980.

Books	Review Articles	Special Volumes
[11], [36], [263] [284], [424], [469], [470], [533], [546], [606], [786], [853], [873], [897], [1132]	[127], [209], [250], [374], [385], [438], [466], [585], [717], [732], [960], [1047], [1119],[1163], [1193], [1194]	[133], [248], [251], [253], [254], [311], [367], [486], [523], [627], [669], [670], [773], [904]

Part I

METHODOLOGY AND ANALYSIS OF FACILITY LOCATION

1

ESTIMATING DISTANCES

Jack Brimberg
Royal Military College of Canada

Robert F. Love
McMaster University

1.1   Introduction

When objects in space such as different cities in a geographic region, activity centres in a production plant, or the junction boxes in an electrical system, can be represented as points, a distance predicting function (DPF) may be used to transform the point coordinates of two points into an estimate of the distance between them.

Many areas of application for distance predicting functions are just now evolving. A few of these are: verifying arc lengths between nodes in road network data; providing road distances for continuous location models, dynamic fire control models, truck scheduling models and recently, to provide distances in Geographical Information System (GIS) models. It is interesting to note that the two most widely used (in North America) truck dispatching software packages, [842, 1008] both use round norm DPFs either completely or to augment network data inputs. Some case studies involving these packages and others are discussed by Golden and Wasil [487]. The great advantage of using a DPF rather than attempting to assemble a distance data bank is the speed of installation and absolute coverage of a geographical area by a DPF compared to the limited set of distances available from a data set (only distances available are the ones in the data set) and the (usually) high cost and large amount of time required to prepare a distance data file.

The concept of using multi-parameter "round norms" as distance predicting functions was suggested by Love and Morris [782, 784, 785]. Applications and comparisons of two round norms have been made by Berens [79] and Berens and Körling [80, 81]. Alternately, Ward and Wendell [1171, 1173] have developed the notion of "block norms" for the same uses. One of the simplest distance predicting functions is the weighted Euclidean norm or $\tau\ell_2$ function. The single parameter τ can be estimated by dividing the sum of the actual distances between the members of a set of individual locations (randomly chosen from the geographic region of interest) by the sum of the Euclidean distances between the same pairs of locations. However, if it is desirable to utilize a more accurate distance predicting model such as the weighted ℓ_p norm model, or $\tau\ell_p$ model, then it is necessary to access a specially written computer program to estimate the two parameters τ and p. An efficient version of such a program is described by Brimberg and Love [137]. Alternatively, accurate estimates of the parameters τ and p may be obtained with any standard linear regression software package. In order to do this, the area of interest is first modelled with the weighted one-two norm described by Brimberg and Love [138] using linear regression. The regression coefficients thus obtained are then used to estimate the parameters τ and p of the $\tau\ell_p$ distance model using equations developed by Brimberg, Dowling and Love [144].

1.2   Norms as Distance Estimating Functions

The basic properties which define a norm are well suited for distance estimation. Any function $k\colon \mathcal{R}^N \to \mathcal{R}^1$ which satisfies the following properties is called a norm [1012]:

1. $k(x) \geq 0, \forall x \in \mathcal{R}^N$, (positivity)

2. $k(x) = 0 \;\leftrightarrow\; x = 0$,

3. $k(cx) = |c|k(x), \forall x \in \mathcal{R}^N, c \in \mathcal{R}^1$, (homogeneity)

4. $k(x) + k(y) \geq k(x+y), \forall x, y \in \mathcal{R}^N$ (triangle inequality)

5. Property 3 implies $k(-x) = k(x)$ (symmetry property).

The above properties ensure that when norms are used as distance estimators, the estimates have characteristics similar to actual distances. These are:

1. Non-negativity of distances.

2. Symmetry: typically the travel distance from point A to point B is the same as the distance from point B to point A.

3. The triangle inequality implies that one always chooses the shortest route. For example, a trip from A to any intermediate point B (denote this by the vector $x = B-A$) and then from B to C (vector $y = C-B$) cannot be shorter than the direct route from A to C (vector $(x+y)$).

1.2.1   Classification of Round and Block Norms

It is useful to characterize norms in more detail. To this end, we define the unit ball B of a norm k acting on $\mathcal{R}^N$ as follows: $B = \{x | k(x) \leq 1\}$.

Thus, B is the closed set of points in $\mathcal{R}^N$ contained by the unit contour of k. The symmetry property, $k(y) = k(-y)$, implies that if $y \in B$, then so is $-y$. Hence B is a symmetric set of points containing the origin.

Suppose x_1 and x_2 belong to B, and consider a point $y = \lambda x_1 + (1 - \lambda)x_2$ where $\lambda \in [0, 1]$. That is, y can be any point along the line segment joining x_1 and x_2. Then, using the triangle inequality and homogeneity properties of norms, we have

$$k(y) = k(\lambda x_1 + (1 - \lambda)x_2) \leq \lambda k(x_1) + (1 - \lambda)k(x_2) \leq 1. \qquad (1.1)$$

Thus $y \in B$, and we conclude that the unit ball is a convex set. In summary, the unit ball of any norm is a symmetric closed bounded convex set. Furthermore, it can be shown [1012, Theorem 15.2] that a one-to-one correspondence exists between the norms k and the symmetric closed bounded convex sets B.

Thisse, Ward and Wendell [1131] use the unit ball to distinguish between block and round norms. They classify block norms as those whose contours are polytopes (polygons in $\mathcal{R}^2$), as distinct from round norms whose contours contain no flat spots. This feature is illustrated in Figure 1. Referring to Figure 1, let z_1 and z_2 denote two points on some contour C of the norm k. This contour, which is the boundary of a convex set, is given by $C = \{x | k(x) = b\}$, where b is a scalar greater than zero ($b > 0$). Noting that $k(z_1) = k(z_2) = b$, and proceeding as in (1.1), we obtain

$$k(\lambda z_1 + (1 - \lambda)z_2) \leq b, \quad \lambda \in [0, 1]. \qquad (1.2)$$

In particular, if C is the unit contour, so that $b = 1$, relation (1.2) becomes

$$k(\lambda z_1 + (1 - \lambda)z_2) \leq 1, \quad \lambda \in [0, 1]. \qquad (1.3)$$

We are now ready to give a formal definition of the round norm.

Definition 1 *A norm k is a round norm if, and only if,*

$$k(\lambda z_1 + (1 - \lambda)z_2) < 1, \qquad (1.4)$$

for all z_1, z_2 on the unit contour of k such that $z_1 \neq z_2$, and all λ in the open interval (0, 1).

Note that the strict inequality in (1.4) implies that the unit ball B of a round norm is a strictly convex set. This geometric interpretation is equivalent to the definition of S-norms given by Pelegrin, Michelot and Plastria [930]. The unit ball of a block norm is not strictly convex. If z_1 and z_2 are on the same facet of the unit polytope of some block norm k, then $k(\lambda z_1 + (1 - \lambda)z_2) = 1$. Thus, for $k(x)$ a block norm, the "<" sign in (1.4) must be replaced by a "$\leq$" sign. The set of round norms can be further divided as follows:

FIGURE 1. Unit Contours in $\mathcal{R}^2$

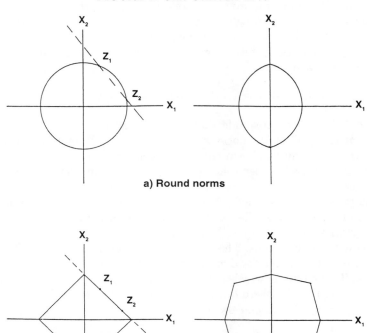

a) Round norms

b) Block norms

Definition 2 *A round norm $k(x)$ which is differentiable at all $x \in \mathcal{R}^N$, except $x = 0$ (the origin), is termed a differentiable round norm. Otherwise $k(x)$ is a nondifferentiable round norm.*

From the defining properties, it follows that all norms must be nondifferentiable at the origin (viz. an inverted cone-shape with apex at the origin). The differentiable round norm has the useful property that its first order derivatives exist everywhere else. However, this property does not hold for nondifferentiable round norms, or for block norms. For example, if $k(x)$ is a block norm, then clearly $k(x)$ is nondifferentiable at the edges of its polytope contours, while it is differentiable at all other points except $x = 0$. In $\mathcal{R}^2$, this means that the block norm is nondifferentiable at the corner points of its polygon contours, while it is differentiable everywhere else except at the origin.

Ward and Wendell [1173] have shown that a block norm can be defined in $\mathcal{R}^2$ as $k(x) = \sum\limits_{g=1}^{r} |y_g^T x|$ where $\{y_g : g = 1, 2, \ldots, r\}$ is a set of vectors which span $\mathcal{R}^2$ and $x = (x_1, x_2)^T \in \mathcal{R}^2$. Substituting y_g with $z_g(\sin \theta_g, \cos \theta_g)^T$ for $g = 1, 2, \ldots, r$, and noting that z_g is a nonnegative scalar, the two following block norm families have been investigated by Ward and Wendell. $B_1(x) = \sum\limits_{g=1}^{4} z_g |(\sin \theta_g, \cos \theta_g)x|$ where $\theta_g = \frac{180^0(g-1)}{4}$, and $B_2(x) = \sum\limits_{g=1}^{8} z_g |(\sin \theta_g, \cos \theta_g)x|$ where $\theta_g = \frac{180^0(g-1)}{8}$.

These two norms are called the block one and block two norms, respectively. The z_g's are the empirical parameters of the norms. Thus, the exact shape of the unit ball for the norm will be determined by these scalars. The block one norm is a specific subgroup of the block two norm in which some of the z_g's are set to zero. A subclass of the block one family with $z_1 = z_3$ and $z_2 = z_4$ is the family of one-infinity norms. A one-infinity norm can be denoted as $z_1 \ell_1(x) + z_2 \sqrt{2} \ell_\infty(x)$ where $\ell_1(x)$ and $\ell_\infty(x)$ are the rectangular and Tchebycheff norms, respectively (see section 1.3 for definitions).

By noting that a one-infinity norm can be written as $(z_1+z_2)[\frac{z_1}{z_1+z_2}\ell_1(x)+\frac{z_2\sqrt{2}}{z_1+z_2}\ell_\infty(x)]$ Ward and Wendell [1171] state that $(z_1 + z_2)$ is analogous to the weighting factor τ of the $\tau\ell_p$ norm (weighted ℓ_p norm in section 1.3).

The preceding discussion leads in summary to a practical classification of norms for use in location models. This classification scheme is shown in Figure 2. Sample contours of the various norms are illustrated in Figure 1.

The general class of block norms leads to location models which can be solved by linear programming and related techniques [786]. Location models which use round norms must be solved by nonlinear iterative methods such as the Weiszfeld procedure [786]. Iterative descent algorithms such as the Weiszfeld procedure require differentiability of the objective function, and hence, smoothing functions such as the hyperbolic approximation and hyperboloid approximation are often used in place of the norms (see [786] for details of these approximations).

1.2.2   Basic Properties

By means of the triangle inequality and homogeneity properties of any norm k, it follows immediately that $k(\lambda x_1 + (1 - \lambda)x_2) \leq \lambda k(x_1) + (1 - \lambda)k(x_2)$, for all $x_1, x_2 \in \mathcal{R}^N$ and all $\lambda \in [0, 1]$. Hence, k is a convex function of x on $\mathcal{R}^N$. This is a well known result [1012, p.131] which permits many location problems to be formulated as convex minimization models. In this sub section, we exploit the strict convexity of the unit ball B to provide a stronger convexity result for the case where k is a round norm.

FIGURE 2. A practical classification scheme for norms.

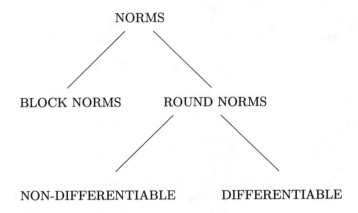

Property 1 *Let $k(x)$ be a round norm on $\mathcal{R}^N, N > 1$. Then $k(x)$ is strictly convex along any straight line which does not pass through the origin.*

Proof Choose any two points x_1, x_2, $x_1 \neq x_2$, such that the straight line through x_1 and x_2 does not pass through the origin. Clearly, $x_1 \neq 0$ and $x_2 \neq 0$. We can write $x_1 = \mu_1 z_1$ and $x_2 = \mu_2 z_2$, where μ_1 and μ_2 are positive scalars, and z_1 and z_2 are intersection points of the unit contour of k with the half lines from the origin through x_1 and x_2, respectively. Since the line through x_1 and x_2 does not pass through the origin, we must have $z_1 \neq z_2$. Without loss in generality, assume that $\mu_1 \leq \mu_2$. Then with $x_0 = \lambda x_1 + (1 - \lambda)x_2, \lambda \in (0,1), \mu' = \lambda\mu_1 + (1 - \lambda)\mu_2$, and $\lambda' = \lambda\frac{\mu_1}{\mu'}$, we obtain

$$
\begin{aligned}
k(x_0) &= k[\lambda\mu_1 z_1 + (1 - \lambda)\mu_2 z_2] = \mu' k[\lambda' z_1 + (1 - \lambda')z_2] \\
&< \mu' \quad (\text{relation}(1.4)) = \lambda\mu_1 + (1 - \lambda)\mu_2.
\end{aligned}
$$

But $\mu_1 = k(x_1)$ and $\mu_2 = k(x_2)$, so that $k(x_0) < \lambda k(x_1) + (1 - \lambda)k(x_2)$. $\square$

Let us now suppose that k is a block norm, and furthermore that x_1 and x_2 above are chosen so that z_1 and z_2 are on the same facet of the unit polytope of k. In this case, $k[\lambda' z_1 + (1 - \lambda')z_2] = 1$, and so, $k(x_0) = \mu' = \lambda k(x_1) + (1 - \lambda)k(x_2)$. We see that if k is a block norm, and the half lines from the origin through x_1 and x_2 intersect the unit polytope of k on the identical facet, then k varies linearly along the line segment connecting x_1 and x_2. It follows that k is convex piecewise linear along any straight line in $\mathcal{R}^N$

Property 1 is a fundamental result which distinguishes round norms from block norms. Consider, for example, the single facility minisum location problem in $\mathcal{R}^N$. It is readily shown using Property 1 that whenever the fixed points (demand points or customers) are not collinear and the distance function is a round norm, the objective function is strictly convex. Hence, the optimal facility location must be unique. This is not generally true when a block norm is used.

The ℓ_p distance function (section 1.3) is widely used in location models. The following result gives additional information concerning the classification of this important function. (The proof uses the Minkowski inequality. For details see [135]).

Property 2 $\ell_p(x)$ *is a differentiable round norm for* $1 < p < +\infty$.

The next property shows how a nondifferentiable round norm can be constructed [135].

Property 3 *Let* $k_1(x)$ *denote a block norm and* $k_2(x)$ *a differentiable round norm, where* $x \in \mathcal{R}^N$, $N > 1$. *Let* $k_3(x)$ *be a positive linear combination of* $k_1(x)$ *and* $k_2(x)$; *i.e.*,

$$k_3(x) \; = \; a_1 k_1(x) + a_2 k_2(x), \quad a_1, \; a_2 \; > \; 0. \tag{1.5}$$

Then $k_3(x)$ *is a nondifferentiable round norm.*

Property 3 implies that any block norm or differentiable round norm can be considered as a limiting case of a nondifferentiable round norm. For example, $k_1(x) = \dfrac{1}{a_1} \lim\limits_{a_2 \to 0} k_3(x)$, and $k_2(x) = \dfrac{1}{a_2} \lim\limits_{a_1 \to 0} k_3(x)$, where k_1, k_2 and k_3 satisfy (1.5). In this sense then, the family of nondifferentiable round norms contains the families of block and differentiable round norms.

1.3   The ℓ_p Norm

The ℓ_p norm is a popular distance measure used in continuous location theory. An extensive discussion of location models with ℓ_p distances can be found in [786]. The functional form of the ℓ_p norm in N-dimensional Euclidean space ($\mathcal{R}^N$) is given by:

$$\ell_p(x) \; = \; \sqrt[p]{\sum_{t=1}^{N} |x_t|^p}, \quad x = (x_1, \dots, x_N)^T \in \mathcal{R}^N, \; p \geq 1.$$

Typically in the literature it is assumed that $p = 1$ or 2, for which we obtain respectively the well known rectangular and Euclidean norms. When $p = +\infty$, we have the Tchebycheff norm, defined as $\ell_\infty(x) = \max\{|x_1|, \dots, |x_N|\}$.

When $p < 1$, $\ell_p(\cdot)$ no longer has the properties of a norm. Hyper-rectilinear distances $(0 < p < 1)$ are studied by Juel and Love [652].

From a practical viewpoint, the distance function chosen to model a transportation network should be as accurate as possible. Love and Morris [782, 784] present several measures which are, for the most part, norms multiplied (weighted) by an inflation factor that helps to account for hills, bends and other forms of noise in the transportation network. They carry out an empirical study in which the best fitting parameter values are obtained for sets of data from urban and rural regions. An important finding of their study is that an empirical distance function should be tailored to a given region whenever a premium is placed on accuracy. This conclusion resulted from the observed statistical superiority of the weighted ℓ_p norm over the weighted rectangular and Euclidean norms.

Berens and Körling [80, 81] examine the road network of the Federal Republic of Germany (FRG), and conclude that the weighted Euclidean norm is sufficiently accurate in this case. In a further study utilizing 25 of the largest cities in the FRG, Love and Morris [785] produced distinctly different results from those of Berens and Körling. These results demonstrated the superiority of the weighted ℓ_p function over the weighted Euclidean function. At about the same time, a further study of Berens [79] supported the earlier findings of Berens and Körling for the FRG. However, this latter study included 11 other countries and produced mixed results. The improvement Berens obtained with the weighted ℓ_p function varied from 11.27 per cent in the United States to zero per cent in Austria. In another study of 18 geographical areas, Brimberg, Love and Walker [145] found improvements as high as 64 per cent using the weighted ℓ_p norm with axis rotation.

Ward and Wendell [1171] fit the weighted one-infinity norm to two sets of data of intercity highway travel used by Love and Morris [782] for the weighted ℓ_p norm, and observe that the one-infinity norm is less accurate than the $\tau\ell_p$ norm. Further empirical work is done by Ward and Wendell [1173] using the general block norm and the data sets of Love and Morris [784]. Since block norms are linear in their parameters, Ward and Wendell are able to apply standard linear regression techniques to find the best fitting values. In comparative studies, the weighted ℓ_p norm has always given more accurate fits to empirical data than the comparable two parameter one-infinity block norm. Moreover, recent research by Love and Walker [788] shows that the weighted ℓ_p norm is generally more accurate even than a block norm with five parameters. Indeed, increasing the number of parameters to eight or more in a block norm does not guarantee that a more accurate distance predicting function than the weighted ℓ_p norm will be obtained. Block norms do allow the formulation of linear models but the number of variables and constraints increases rapidly with problem size.

1.3.1   Directional Bias

Since most location problems occur in the plane, we restrict our attention to predicting distances in 2 dimensional space ($\mathcal{R}^2$). For any norm k on $\mathcal{R}^2$, we have the following fundamental result.

Property 4 *The dimensionless ratio,*

$$r = \frac{k(x)}{\ell_2(x)}, \; x \neq 0, \tag{1.6}$$

is a function of θ alone, where $\theta = \tan^{-1}\left(\frac{x_2}{x_1}\right)$ is the angle specifying the vector $x = (x_1, x_2)^T \in \mathcal{R}^2$.

Proof: For the points $x \neq 0$ on any half line in $\mathcal{R}^2$ ending at the origin, we have $k(x) = c\ell_2(x)$, where c is a constant. It immediately follows that $r = r(\theta)$. $\square$

We shall call $r(\theta)$ the directional bias function of the norm k. In a qualitative sense, this function can be thought of as a measure of the relative difficulty of travel in any direction. If $r(\theta_1) > r(\theta_0)$, then one must travel a longer distance along a line at angle θ_1 with the x_1 axis than along a line at angle θ_0 with the x_1 axis, to cover the same Euclidean distance between pairs of points. In the physical world, the shortest possible path between two points is the straight line or Euclidean path. Hence, for norms used to approximate actual travel distances in $\mathcal{R}^2$, the directional bias function should satisfy the relation, $r(\theta) \geq 1$, $\forall \theta$. Otherwise, distances shorter than Euclidean are possible.

The traditional method of illustrating and comparing the directional bias of norms is by means of the unit circle $k(x) = 1$ (e.g., see [786, Figure 10.1]). The function $r(\theta)$ provides a formal definition of directional bias, and a new way of representing it graphically which we believe is more informative and easier to interpret than the unit circle, since the relevant information is now contained in a standard plot of a function of one independent variable (θ). From the symmetry property of norms, it follows that $r(\theta) = \frac{k(-x)}{\ell_2(-x)} = r(\theta + \pi)$. Thus, $r(\theta)$ has a periodicity of π/n where n must be some integer greater than or equal to one.

Let us consider now the directional bias of the ℓ_p norm, denoted by $r_p(\theta)$ where $p \geq 1$. Using the definition in (1.6), we obtain

$$r_p(\theta) = \frac{\ell_p(x)}{\ell_2(x)} = \frac{1}{\ell_2(x)} \sqrt[p]{|x_1|^p + |x_2|^p} = \sqrt[p]{|\cos\theta|^p + |\sin\theta|^p}. \tag{1.7}$$

Alternatively, we see that the ℓ_p norm can be expressed in terms of the Euclidean distance and the angle of travel (θ) as follows:

$$\ell_p(x) = \ell_2(x) \sqrt[p]{|\cos\theta|^p + |\sin\theta|^p}. \tag{1.8}$$

Examples of the directional bias function for different values of p include $r_1(\theta) = |\cos\theta| + |\sin\theta|$, and $r_2(\theta) = \sqrt{|\cos\theta|^2 + |\sin\theta|^2} = 1$, for the rectangular and Euclidean norms respectively.

Some useful properties of $r_p(\theta)$ are given below. The proofs can be found in [140].

Property 5 $r_p(\theta)$ *is periodic with period* $\frac{\pi}{2}(= 90^0)$.

Property 6 *For any real* Ω, $r_p(\frac{\pi}{4} + \Omega) = r_p(\frac{\pi}{4} - \Omega)$.

From the two preceding results, we see that $r_p(\theta)$ is the mirror image of itself about the line $\theta = \frac{\pi}{4}$, and that this function has a period of $\frac{\pi}{2}$. Hence, it is only necessary to consider θ in the interval $[0, \frac{\pi}{4}]$ to determine the behaviour of $r_p(\theta)$ over its entire range. Noting that $|\cos\theta| = \cos\theta$ and $|\sin\theta| = \sin\theta$ for $\theta \in [0, \frac{\pi}{2}]$, we obtain the following expressions for the first and second order derivatives of $r_p(\theta)$:

$$\frac{dr_p(\theta)}{d\theta} = \frac{\sin 2\theta}{2[r_p(\theta)]^{p-1}} \left(-\cos^{p-2}\theta + \sin^{p-2}\theta \right) \tag{1.9}$$

and

$$\frac{d^2 r_p(\theta)}{d\theta^2} = -r_p(\theta) + (p-1)[r_p(\theta)]^{1-2p} \left(\frac{\sin 2\theta}{2} \right)^{p-2}, \tag{1.10}$$

where $0 < \theta < \frac{\pi}{2}$.

Property 7 *In the interval* $0 \leq \theta \leq \frac{\pi}{4}, r_p$ *is a strictly increasing function of* θ *if* $0 < p < 2$, *while it is strictly decreasing in* θ *if* $p > 2$.

Property 8 *If* $p > 1$ *and* $p \neq 2$, *then* r_p *has a unique inflection point* (θ^*) *in the interval* $0 \leq \theta \leq \frac{\pi}{4}$.

The shape of $r_p(\theta)$ is illustrated in Figure 3 for various values of p, and for θ in the range $[0, \frac{\pi}{2}]$, i.e., one complete cycle. From Properties 6 and 7, it follows that r_p has its maximum value at $\theta = \frac{\pi}{4}$ and minimum value at $\theta = 0, \frac{\pi}{2}$, if $0 < p < 2$, while the converse holds if $p > 2$. Defining the direction of greatest (least) difficulty as the value of θ which maximizes (minimizes) r_p, we see that for $0 < p < 2$ the direction of greatest difficulty is at 45^0 to the axes ($\theta = \frac{\pi}{4}, 3\frac{\pi}{4}, 5\frac{\pi}{4}, 7\frac{\pi}{4}$), and the direction of least difficulty is parallel to the axes ($\theta = 0, \frac{\pi}{2}, \pi, 3\frac{\pi}{2}$). In other words, the distance $\ell_p(x - y)$ between any two points x and y separated by a straight line segment of fixed length $\ell_2(x - y)$, is maximized if this line segment is at 45^0 to the axes, and minimized if it is parallel to an axis. The converse holds when $p > 2$.

We note the following characteristics of the directional bias function at $\theta = 0, \frac{\pi}{4}$, and $\frac{\pi}{2}$:

$$r_p(0) = r_p\left(\frac{\pi}{2} \right) = 1, \quad p > 0; \tag{1.11}$$

FIGURE 3. Directional Bias Function of $\ell_p(x)$

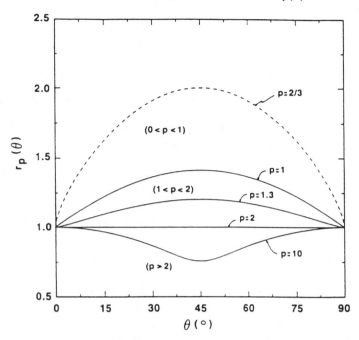

$$r_p\left(\frac{\pi}{4}\right) = 2^{\frac{2-p}{2p}}, \quad \forall p \neq 0; \tag{1.12}$$

$$\frac{dr_p(0)}{d\theta} = \frac{dr_p\left(\frac{\pi}{2}\right)}{d\theta} = 0, \quad 1 < p < +\infty; \tag{1.13}$$

$$\frac{dr_p\left(\frac{\pi}{4}\right)}{d\theta} = 0, \quad \forall p \neq 0. \tag{1.14}$$

1.3.2   General Considerations For Fitting the ℓ_p Norm

Three criteria mentioned in the literature for fitting the unknown parameters of a distance function to a set of data on travel distances in a transportation network [782, 141] will be discussed here. Applied to the weighted ℓ_p norm, these goodness-of-fit criteria are as follows:

Criterion 1: $\displaystyle \min_{\tau,p} AD_\ell = \sum_{i=1}^{n-1} \sum_{j=i+1}^{n} |d_\ell(a_i, a_j) - B_{ij}|,$

Criterion 2: $\displaystyle \min_{\tau,p} SD_\ell = \sum_{i=1}^{n-1} \sum_{j=i+1}^{n} \frac{[d_\ell(a_i, a_j) - B_{ij}]^2}{B_{ij}},$

Criterion 3: $\min_{\tau,p} SND_\ell = \sum_{i=1}^{n-1} \sum_{j=i+1}^{n} [\tau r_p(\theta_{ij}) - \eta_{ij}]^2,$

where

$n =$ the number of fixed points (cities, destinations, customers) in the road network, which are chosen for the data set;

$a_i = (a_{i1}, a_{i2})^T$ is the known location of the i'th destination, $i = 1, ..., n$;

$B_{ij} =$ the actual distance (road miles) between the fixed points a_i and a_j;

$d_\ell(a_i, a_j) = \tau \ell_p(a_i - a_j)$ is the weighted ℓ_p norm used to estimate the distance (road miles) between a_i and a_j;

$\theta_{ij} = \tan^{-1}\left(\frac{a_{i2}-a_{j2}}{a_{i1}-a_{j1}}\right), \forall i, j$;

$\eta_{ij} = \frac{B_{ij}}{\ell_2(a_i-a_j)}, \quad i, j = 1, \ldots, n, \; i < j,$ is referred to as the normalized travel distance (actual distance $\div$ Euclidean distance)

and $\tau, p > 0$ are the unknown parameters.

Note that an assumption of symmetric travel distances in the network is being made. Otherwise we should allow all permutations of the pairs (i, j) instead of combinations in the preceding summations.

As noted in [782], the empirical distance function tends to estimate long distances relatively more accurately than short ones when criterion 1 is used. This bias is partially removed in criterion 2 through the weights B_{ij}^{-1}, while criterion 3 has the least bias. Thus, we see that the three criteria measure goodness-of-fit in significantly different ways. The practitioner must choose a suitable one for his requirements.

We now proceed to develop a general methodology for fitting the weighted ℓ_p norm. First and foremost, note that implicit in our discussion of the directional bias function $r_p(\theta)$ is the fact that it pertains to a particular set of orthogonal reference axes. Thus, characteristics such as the directions of greatest and least difficulty of the distance function are measured relative to the given axes. Except for the weighted Euclidean norm (see (1.7 and the ensuing discussion), rotating the axes results in a different directional bias, or alternatively, the distance between any two points varies under this rotation. Consider, as an example, the hypothetical case where the roads in a transportation network form a perfect rectangular grid and the destinations a_i are all situated at intersection points of the roads. If the axes are chosen parallel to the grid, and we set $\tau = 1, p = 1$, then actual distances will be predicted exactly. On the other hand, if the axes are specified at 45^0 to the grid, $\tau = \sqrt{2}$ and $p = +\infty$, the actual distances will be predicted exactly once again. However, for any other choice of orthogonal axes, the parameters will take on intermediate values, and the predicted distances will not coincide with the actual.

Clearly, the specification of the reference axes is an important part of any empirical study. One must recognize the dual relation between the distance function and the reference axes. Both are required in order to obtain a specific form of the directional bias. The choice of axes and distance function should be made after a careful study of the road network. Based on the predominant pattern of the roads, one should ascertain the directions which are easiest and most difficult to travel in. The axes and the distance function should be chosen accordingly to coincide with this directional bias.

Specification of the reference axes for the distance function based on an identification and examination of the patterns in the road network, does not appear to be a consideration in the empirical studies described in the literature [782, 784, 785, 787, 1171, 1173]. In other words, the axes are chosen arbitrarily without examining the physical nature of the system. Before showing the advantages and usefulness of our approach for the weighted ℓ_p norm, some definitions are in order.

Definition 3 *A set of axes having orientation γ means that these axes are rotated counterclockwise by an angle γ from true east and north.*

Definition 4 *Let $R(\theta; \gamma_0)$ denote a function of θ, where the angle θ is measured relative to a set of axes having orientation γ_0. Then $R(\theta; \gamma_0)$ is said to have a rectangular bias if, and only if, the following conditions are satisfied:*

1. *$R(\theta + \frac{\pi}{2}; \gamma_0) = R(\theta; \gamma_0), \forall \theta \{$periodicity of $\frac{\pi}{2}\}$;*

2. *$R(\frac{\pi}{4} - \Omega; \gamma_0) = R(\frac{\pi}{4} + \Omega; \gamma_0), 0 \leq \Omega \leq \frac{\pi}{4} \{$symmetry property$\}$;*

3. *R is nondecreasing for $\theta \in [0, \frac{\pi}{4}]$ and nonincreasing for $\theta \in [\frac{\pi}{4}, \frac{\pi}{2}]$ $\{$unimodal cycle with maximum at $\theta = \frac{\pi}{4}\}$.*

Definition 5 *A transportation network has a predominant rectangular pattern relative to a set of axes with orientation γ_0 if, and only if, the following relation is satisfied: $\eta(\theta; \gamma_0) = \beta_0 + \beta_1 R(\theta; \gamma_0) + \epsilon(\theta; \gamma_0)$, where β_0, β_1 are parameters with $\beta_1 \geq 0$, η is the normalized travel distance from any point q to any point s, where vector $(s - q)$ has direction θ, R is a function with rectangular bias, ϵ is an independent error term with mean zero, and $(\theta; \gamma_0)$ denotes that the angle θ is measured relative to the axes with orientation γ_0.*

The parameters β_0 and β_1 are estimated in practice by doing a regression analysis on a data set obtained for the given network. Statistical tests can be performed on the parameters and the error terms to provide quantitative information on the nature of the transportation system [138].

An obvious example of rectangular bias occurs when

$$R(\theta; \gamma_0) = r_p(\theta; \gamma_0), \quad 0 < p \leq 2. \tag{1.15}$$

Typically we expect the road network to have an underlying rectangular grid (identified as the predominant pattern), offset by some angle γ_0 to the true east and north directions. We shall observe this condition in our case study to follow. The normalized travel distance would be modelled in this case as

$$\eta(\theta; \gamma_0) = \beta_0 + \beta_1 r_1(\theta; \gamma_0) + \epsilon(\theta; \gamma_0). \tag{1.16}$$

In a transportation network with a predominant rectangular pattern, the direction of greatest difficulty is at $\gamma_0 + m\frac{\pi}{4}, m = 1, 3, 5, 7$, while the direction of least difficulty is at $\gamma_0 + m\frac{\pi}{2}, m = 0, 1, 2, 3$. This signifies that for pairs of points separated by the same straight line distance, the actual travel distances are generally greatest at 45^0 to the set of axes with orientation γ_0 and least parallel to these axes. For the special case where $p = 2$ in (1.15), we have

$$\eta(\theta; \gamma_0) = (\beta_0 + \beta_1) + \epsilon(\theta; \gamma_0), \tag{1.17}$$

i.e., the normalized travel distance is a constant plus an error term. This signifies a highly developed network, with travel in any direction having the same degree of difficulty on average. Also note that $R(\theta; \gamma_0)$ does not necessarily represent the directional bias of a norm. For example, if the road system has one way streets or obstacles resulting in a lot of backtracking, then R might belong to a hyper-rectilinear distance function (i.e., $p < 1$ in (1.15)).

The advantage of orienting the reference axes of the distance function to comply with the physical nature of the transportation network being modelled is shown in the following important results. (For proofs, see [141]).

Property 9 *Suppose we have a transportation network with a predominant rectangular pattern as in Definition 5, and the weighted ℓ_p distance function is used to model travel distances in this network. If the reference axes of the distance function coincide with the orientation (γ_0) of the network axes, and the sample size of destinations in the data set is sufficiently large (i.e., the asymptotic limiting case $n \to \infty$), then $0 < p_j^{**} \leq 2$, where p_j^{**} is the value of p at a global minimum of criterion j, $j = 1, 2, 3$.*

Corollary 1 *If in addition $\beta_0 \geq 0$ and $R(\theta; \gamma_0) \equiv r_1(\theta; \gamma_0)$, then $1 \leq p_j^{**} \leq 2$.*

Corollary 2 *If $\beta_0 \leq 0$ and $R(\theta; \gamma_0) \equiv r_1(\theta; \gamma_0)$, then $0 < p_j^{**} \leq 1$.*

A physical interpretation applies to the parameters β_0 and β_1 in relation (1.16). The shortest route between any two points q and r on a transportation network can be approximated by a composition of straight line segments parallel to the vector $(q - r)$ and rectilinear segments. Loosely speaking, when $\beta_0 \geq 0$, the fraction $\frac{\beta_0}{(\beta_0 + \beta_1)}$ gives the proportion of the route which is Euclidean, while the remaining proportion $\frac{\beta_1}{(\beta_0+\beta_1)}$ is rectilinear. This interpretation becomes intuitively appealing when the transportation system has an underlying rectangular grid, since a typical trip moves along the grid part of the way and along diagonal roads the remainder. A value of $\beta_0 < 0$ implies that back tracking occurs in the network, so that actual distances are in general larger than rectangular distances. Back tracking is typically caused by large physical obstructions or by a predominance of one way streets.

The preceding results lead to the following general procedure for modelling travel distances in a transportation network with the weighted ℓ_p distance function.

Step 1: Verify that the transportation network has a predominant rectangular pattern with respect to a set of axes having some orientation (γ_0). If this is not the case, a different distance function should be considered.

Step 2: Rotate the reference axes of the weighted ℓ_p distance function while simultaneously determining the best fitting values of the parameters τ and p for one or more of the criteria given above, where p is restricted to the interval $(0, 2]$, $[1, 2]$ or $(0, 1]$ in accordance with Property 9 and Corollaries 1 and 2. The best fit values of τ and p will be obtained when the reference axes coincide with the orientation (γ_0) of the transportation network axes.

In brief, the procedure first involves a verification that the weighted ℓ_p function is appropriate for the network being modelled. A visual inspection of the map often suffices here to determine if the transportation network has an underlying rectangular pattern (e.g., see the following case study). Whether or not a rectangular pattern is identified visually, the angle of rotation of the reference axes is treated as an additional unknown parameter. The axes are rotated while simultaneously the best fitting values for the parameters τ and p are obtained. In this way the axes are rotated so that the distance function is in phase with the network. The advantages of such an approach can be summarized as follows:

1. the directional bias inherent in the road network is reproduced by the distance function;

2. because the reference axes coincide with those of the network, we can expect the best overall fit of the weighted ℓ_p function to actual distances; and

3. the search for the minimizing values of τ and p can be done efficiently because the nonlinear parameter p is effectively restricted by Property 9 and its corollaries to a small interval.

1.3.3   Fitting Algorithms

The purpose of this section is to derive an efficient procedure to find the best fitting values of τ and p for the criteria specified above. We first state some useful properties developed by Brimberg and Love [137].

Property 10 AD_ℓ *is a convex function of* τ, *while* SD_ℓ *and* SND_ℓ *are strictly convex functions of* τ.

Property 11 *Denote the terms in the summation defining* AD_ℓ *by*

$$g_{ij}(\tau, p) \;=\; |d_\ell(a_i, a_j) - B_{ij}| \;=\; |\tau\ell_p(a_i - a_j) - B_{ij}|, \quad i, j = 1, \ldots, n, \; i < j.$$

Consider any $g_{ij}(\tau, p)$ *as a function of* p *in the open interval* $(0, +\infty)$. *If the vector* $(a_i - a_j)$ *is parallel to an axis, then* g_{ij} *is constant. Otherwise, there are two possibilities:*

1. *if* $B_{ij} > \tau \max\{|a_{i1} - a_{j1}|, |a_{i2} - a_{j2}|\} = \tau\ell_\infty(a_i - a_j)$, *then* g_{ij} *is a unimodal function of* p, *strictly convex over the interval* $0 < p \leq p_{ij}$ *and strictly concave for* $p \geq p_{ij}$, *where* p_{ij} *is the unique value of* p *such that* $\min_p g_{ij}(\tau, p) = g_{ij}(\tau, p_{ij}) = 0$;

2. *if* $B_{ij} \leq \tau\ell_\infty(a_i - a_j)$, *then* g_{ij} *is a decreasing strictly convex function of* p *with a minimum approached asymptotically as* $p \to +\infty$.

Property 12 *Denote the terms in the summation defining* SD_ℓ *by*

$$h_{ij}(\tau, p) = \frac{[d_\ell(a_i, a_j) - B_{ij}]^2}{B_{ij}} = \frac{[\tau\ell_p(a_i - a_j) - B_{ij}]^2}{B_{ij}}, \quad i, j = 1, \ldots, n, i < j.$$

The previous result applies to any term $h_{ij}(\tau, p)$ *except when* $(a_i - a_j)$ *is not parallel to an axis and* $B_{ij} > \tau\ell_\infty(a_i - a_j)$. *For this case,* h_{ij} *is also a unimodal function of* p *with minimum at* p_{ij}. *However,* h_{ij} *is strictly convex over the interval* $0 < p \leq \rho_{ij}$ *and strictly concave for* $p \geq \rho_{ij}$, *where* ρ_{ij} *is the unique inflection point such that* $\frac{\partial^2}{\partial p^2} h_{ij}(\tau, \rho_{ij}) = 0$. *Furthermore,* $\rho_{ij} > p_{ij}$.

The shapes of g_{ij} and h_{ij} for varying p are illustrated in Figure 4 for the case where the vector $(a_i - a_j)$ is not parallel to an axis. Since AD_ℓ and SD_ℓ are the sums of terms g_{ij} and h_{ij} respectively, each term being in general neither convex nor concave in p, we obtain the following important result. (Note that SND_ℓ can be treated analogously to SD_ℓ.)

Property 13 *Consider the sums* AD_ℓ *and* SD_ℓ *as functions of* p *in the open interval* $(0, +\infty)$; *i.e., the inflation factor* τ *is fixed. In general* AD_ℓ *and* SD_ℓ *are neither convex nor concave in* p, *and may have more than one local minimum or maximum.*

FIGURE 4. General Shape of g_{ij} and h_{ij}.

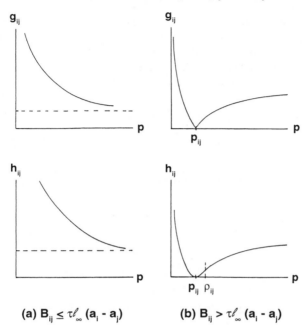

(a) $B_{ij} \leq \tau\ell_\infty (a_i - a_j)$ (b) $B_{ij} > \tau\ell_\infty (a_i - a_j)$

As a consequence of the above property, there is no easy way to find a value of p which minimizes AD_ℓ or SD_ℓ for a given τ. One is forced essentially to do a thorough numerical search over a safe range. (See section 1.3.2 and the case study in 1.3.4 for further discussion on this matter.) The following result may be useful in reducing the search interval of p.

Property 14 *Let* $K = \{(i,j) | (a_i - a_j)$ *is not parallel to an axis,* $i < j\}$, *and assume that* K *is nonempty. Also assume that* $B_{ij} > \tau\ell_\infty(a_i - a_j), \forall (i,j) \in K$, *where* τ *has some specified value. Then* $p_m \leq p_i^* \leq p_M$ *where* $p_m = \min_{(i,j) \in K} p_{ij}$, $p_M = \max_{(i,j) \in K} p_{ij}$, *and* $p_i^*, i = 1, 2$, *are the values of* p *which minimize* AD_ℓ *and* SD_ℓ, *respectively, for the given* τ.

Note that the p_{ij} can be obtained using standard techniques such as interval bisection or the Newton Raphson method [257, Chapter 6] since g_{ij} (or h_{ij}) is a unimodal function of p. Thus the values of p_m and p_M are readily determined. The range $[p_m, p_M]$ also characterizes the road network in a new way. If the width of this interval, measured by $(p_M - p_m)$, is small, we can regard the road network as being consistent with the distance function d_ℓ, in that d_ℓ approximates individual travel distances consistently with a high degree of accuracy. Thus, the width of the interval $[p_m, p_M]$ can be considered as a measure of the consistency of the road network with the distance function d_ℓ.

Let us now examine another aspect of the overall problem, namely that of finding τ_1^* and τ_2^*, the values of τ which minimize AD_ℓ and SD_ℓ respectively for a given p. The convexity results in Property 10 can be put to good use here. First we consider the sum SD_ℓ which is strictly convex in τ. Thus, a necessary and sufficient condition for minimizing SD_ℓ with p fixed is given by

$$\frac{\partial SD_\ell}{\partial \tau} = \sum_{i=1}^{n-1} \sum_{j=i+1}^{n} \frac{2}{B_{ij}} [\tau \ell_p(a_i - a_j) - B_{ij}] \ell_p(a_i - a_j) = 0,$$

which provides the closed form solution,

$$\tau_2^* = \frac{\displaystyle\sum_{i=1}^{n-1} \sum_{j=i+1}^{n} \ell_p(a_i - a_j)}{\displaystyle\sum_{i=1}^{n-1} \sum_{j=i+1}^{n} \frac{[\ell_p(a_i - a_j)]^2}{B_{ij}}} \tag{1.18}$$

(Note that a similar formula can be derived for SND_ℓ).

Next consider the sum AD_ℓ which is convex in τ. We have

$$\frac{\partial AD_\ell}{\partial \tau} = \sum_{i=1}^{n-1} \sum_{j=i+1}^{n} sign[\tau \ell_p(a_i - a_j) - B_{ij}] \ell_p(a_i - a_j),$$

wherever this derivative exists, so that AD_ℓ is piecewise linear in τ with discontinuities in the slope at

$$\tau_{ij} = \frac{B_{ij}}{\ell_p(a_i - a_j)}, \quad i, j = 1, \ldots, n, \ i < j. \tag{1.19}$$

The shape of AD_ℓ for varying τ does not permit a closed-form solution as in (1.18), since the minimizing value (τ_1^*) occurs at a nondifferentiable point. However, the following solution method can be used. Note that the complexity is $O(n^2 \log n)$ when an efficient sorting subroutine such as Quicksort is called in step 2.

Algorithm 1 {Finding τ_1^* for given p}

Step 1: Calculate each τ_{ij} using (1.19).

Step 2: Sequence and re-label the pairs $(\tau_{ij}, \ell_p(a_i - a_j))$ as $(\tau_r, \ell_r), r = 1, \ldots,$
$\frac{n(n-1)}{2}$, such that $\tau_1 \leq \tau_2 \leq \ldots \leq \tau_{\frac{n(n-1)}{2}}$; i.e., the τ_{ij} are arranged in nondecreasing order.

Step 3: (Finding τ_1^*)

Set $i = 0$, $S_i = -\sum\limits_{r=i+1}^{\frac{n(n-1)}{2}} \ell_r$.

Repeat $i \leftarrow i + 1$, $S_i \leftarrow S_{i-1} + 2\ell_i$ until $S_i \geq 0$.

If $S_i = 0$ then $\tau_1^* \in [\tau_i, \tau_{i+1}]$, else $(S_i > 0)$ $\tau_1^* = \tau_i$.

It should be clear that τ_1^* and τ_2^* are functions of p; furthermore, the curves $\tau_1^*(p)$ and $\tau_2^*(p)$ are readily obtained using Algorithm 1 and (1.18), respectively. With this information we can immediately calculate AD_ℓ^* and SD_ℓ^*, the minimum values of AD_ℓ and SD_ℓ as functions of p; i.e.,

$$AD_\ell^*(p) = AD_\ell(\tau_1^*(p), p); SD_\ell^*(p) = SD_\ell(\tau_2^*(p), p). \qquad (1.20)$$

Thus, the criteria 1 and 2 (or 3) are now reduced to minimization problems in one variable. We now outline a strategy for finding the global minimizers (τ_i^{**}, p_i^{**}), $i = 1, 2, 3$, for criteria 1, 2 and 3, respectively.

Algorithm 2 {Solving for the Criteria}:

Step 1: Determine the curve of $AD_\ell^*(p)$, $SD_\ell^*(p)$ or $SND_\ell^*(p)$, with a small enough increment Δp to identify all sub-intervals containing a local minimum. Delete those sub-intervals which obviously do not possess a global solution, and label the remaining ones $1, \ldots, M$.

Step 2: Do the following for $i = 1, \ldots, M$.
Divide sub-interval i using a smaller increment of p. With the additional points, reduce the width of the sub-interval containing the local minimum (denoted by $p^{(i)}$). Repeat this process until $p^{(i)}$ is calculated to the desired accuracy.

Step 3: Set $p_1^{**} = p^{(k)}$ and $\tau_1^{**} = \tau_1^*(p^{(k)})$, where $AD_\ell^*(p^{(k)}) = \min\limits_i \{AD_\ell^*(p^{(i)})\}$.
(For SD_ℓ, set $p_2^{**} = p^{(k)}$ and $\tau_2^{**} = \tau_2^*(p^{(k)})$, where $SD_\ell^*(p^{(k)}) = \min\limits_i \{SD_\ell^*(p^{(i)})\}$. Similarly set p_3^{**} and τ_3^{**} for SND_ℓ.)

Since the execution time required to compute AD_ℓ^*, SD_ℓ^* or SND_ℓ^* is insignificant, the increment Δp in step 1 should be chosen to ensure practically linear behaviour of the curve between adjacent calculated points. The nonlinearity between adjacent points can be evaluated by examining the second-order terms in a Taylor series expansion. However, based on our computational experience, this is unnecessary since AD_ℓ^*, SD_ℓ^* or SND_ℓ^* are observed to be gradually varying on p (e.g., see Figure 5). Thus a value of $\Delta p = 0.05$ is more than adequate for determining the curve. It then becomes a simple matter to fathom sub-intervals which do not possess a local minimum. Given three consecutive points, $p, p + \Delta p$, and $p + 2\Delta p$ over which AD_ℓ^* (SD_ℓ^* or SND_ℓ^*) is decreasing, the interval $[p, p + \Delta p]$ can be eliminated automatically. Similarly, if AD_ℓ^* (SD_ℓ^* or SND_ℓ^*) is increasing over these three points, eliminate the interval $[p + \Delta p, p + 2\Delta p]$.

TABLE 1.1. Cities Forming the Data Set

City No.	City Name	Coordinates ($\frac{1}{4}$" Unit)†	
		x_1	x_2
1	Windsor	2.0	-5.3
2	Sarnia	12.7	9.3
3	Chatham	15.35	-3.2
4	London	31.0	8.8
5	Kitchener/Waterloo	42.9	18.7
6	Brantford	46.6	11.6
7	Hamilton	53.2	13.9
8	Toronto	61.0	23.2
9	Fort Erie	68.0	6.7
10	St. Catharines	63.0	12.0
11	Stratford	35.3	17.0
12	Goderich	24.2	25.3
13	Barrie	56.0	38.7
14	Owen Sound	36.7	43.0
15	Peterborough	77.3	36.7
16	Belleville	92.1	34.3
17	Ottawa	117.7	61.6
18	Cornwall	132.5	53.8

† SCALE: 9.5 units = 50 km; Coordinates measured in base axes (east, north).

1.3.4 Case Study

In this section, we present a case study to illustrate the general concepts and algorithms presented above for fitting the weighted ℓ_p norm to a geographic region. Our example considers the road network covering the central and eastern parts of southern Ontario. Eighteen representative cities were chosen from this region to form the data set. These cities are listed in Table 1.1, with their coordinates measured in the base axes pointing true east and north. From an inspection of the official road map we can discern an approximate rectangular pattern underlying the network. The base of this rectangular grid is formed in large part by the number 401 Highway, which is a major route in the network following the north shore of Lake Ontario and the St. Lawrence River. Thus, we conclude that the weighted ℓ_p norm is a suitable distance function for this region. Using our general procedure, the reference axes of the distance function should be rotated to parallel the 401 Highway. In this way, the directional bias of the ℓ_p norm will be in phase with the directional bias inherent in the network.

The effect of rotating the reference axes on the minimum value of SND$_\ell$

FIGURE 5. Finding the Best Fit using Algorithm 2.

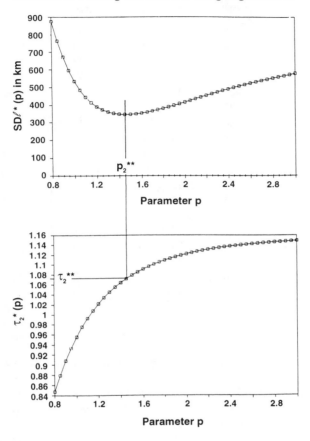

(Criterion 3), and on the corresponding parameter values τ_3^{**} and p_3^{**}, is shown in Table 1.2. The main point of interest here is the sensitivity of p_3^{**} to the axis orientation γ, and the fact that p_3^{**} takes on a rather low value (1.55) at $\gamma = 20^0$. This suggests a substantial rectangular bias in the road system, at an orientation in line with the mean direction of the 401 Highway, thus confirming our earlier conclusion based on an inspection of the map. Also note that the fit, measured by the minimum value of SND_ℓ, is optimized for practical purposes at $\gamma \simeq 20^0$. As shown in [140], an ℓ_p norm with $p > 2$ can be closely approximated by another ℓ_p norm multiplied by a scaling factor, where p now has a value in the interval $(1, 2)$ and the reference axes have been rotated 45^0 from their current position. This explains the bimodal behaviour of min SND_ℓ in Table 1.2, and the

TABLE 1.2. Effect of Orientation (γ) on Best Fit

$\gamma(^0)$	τ^{**}	p^{**}	min SND$_\ell$
0	1.122	1.80	3.252
5	1.104	1.67	3.185
10	1.098	1.61	3.122
15	1.096	1.57	3.059
20	1.096	1.55	3.031
25	1.099	1.56	3.148
30	1.108	1.63	3.248
35	1.128	1.80	3.248
40	1.153	2.05	3.277
45	1.177	2.32	3.225
50	1.197	2.57	3.135
55	1.213	2.75	3.058
60	1.222	2.82	3.027
65	1.223	2.80	3.047
70	1.216	2.67	3.109
75	1.198	2.46	3.187
80	1.175	2.23	3.252
85	1.150	2.01	3.278
90	1.122	1.80	3.252

fact that a similar degree of accuracy is obtained between pairs of results 45^0 apart.

Referring to Table 1.2, we see that if the axes are arbitrarily set parallel to the east and north directions (i.e., $\gamma = 0^0$), a value of p_3^{**} relatively close to 2.0 is obtained. This is due to the distance function being severely out of phase with the underlying pattern or structure of the network. The arbitrary specification of the reference axes in this fashion is common practice in the literature. For example, Love, Truscott and Walker [787] use the east and north directions as the reference axes in their empirical study, which covers basically the same geographic region as ours. Not surprisingly, they obtain a value of p_2^{**} (for Criterion 2) close to 2.0. This leads to the erroneous conclusion that travel distances are essentially Euclidean multiplied by an inflation factor; i.e., travel is uniformly difficult in all directions and there is virtually no rectangular bias in the road network.

Table 1.2 illustrates that the orientation of the reference axes is a critical factor in detecting the rectangular bias in a transportation network. Furthermore, the value of the parameter p (1.55) obtained after properly orienting the axes (e.g., 20^0 rotation in the case problem) provides a quantitative measure of the degree of rectangular bias in the network.

The case study confirms quite dramatically the usefulness of the general procedure discussed in section 1.3.2 for tailoring the weighted ℓ_p distance function to a geographic region. The practical implications of the general procedure are summarized as follows:

1. An efficient search for the best fitting values (τ^{**}, p^{**}), can be implemented using the algorithms derived in section 1.3.3, since the range of the nonlinear parameter p is effectively restricted to a narrow interval by Property 9 and its corollaries in section 1.3.2.

2. Because the directional bias of the network is reproduced by the distance function, a condition not guaranteed previously by the arbitrary specification of the reference axes, actual travel distances are estimated more accurately on average. Referring to the minimum values for Criterion 3 (SND_ℓ), in Table 1.2, we see that the case study supports this conclusion.

1.4   Conclusions

As a closing summary, we note that the concepts presented above for fitting the weighted ℓ_p norm are readily extended to other empirical distance functions, such as the block norms of [1171, 1173]. Furthermore, the directional bias in the transportation network does not have to be rectangular in nature. The idea is to find a distance function whose directional bias function is suitable for the network in question, to orient the reference axes so that the distance function is in phase with the network, and to adjust the parameters of the distance function to optimize a chosen criterion of fit applied to a data set obtained from the given region.

Proper orientation of the reference axes may be accomplished by printing out a table of values of the fitting criteria in relation to the angle of axes rotation similar to Table 1.2. The angle of rotation which provides the best fit can then be selected from the table. The rotation angle thus has become another unknown parameter in the fitting algorithm. A software package is now available to compute this angle to any degree of accuracy given a sample of distances from the geographical region [777]. An approximate orientation of the reference axes can sometimes be done by a visual inspection of the map. An advantage of linear models such as block norms and the weighted one-two norm developed by Brimberg and Love [138] is that a linear regression analysis can be used to find the best fitting parameter values for a specified γ. These linear distance models may also be employed as approximations for nonlinear models such as the weighted ℓ_p norm, and to provide good estimates or starting solutions for a numerical search on the nonlinear parameters [144]. Analytical results such as the convexity of the fitting criteria with respect to the inflation factor (Property 10 in section 1.3.3) should be exploited to streamline the fitting algorithms wherever possible.

We conclude with a statement of the obvious. Although multi-parameter distance functions can provide better estimates of travel distances when they are properly tailored to a given region, and although standard regression packages or specialized fitting algorithms may be used for this purpose, the use of these more complex distance functions should be tempered by the level of accuracy required in the problem.

2
REPLACING DISCRETE DEMAND WITH CONTINUOUS DEMAND

Zvi Drezner
California State University-Fullerton

2.1 Introduction

Facility location problems involve the location of one or more facilities in an area with existing demand while optimizing a certain objective function. A common simplification of this formulation is the substitution of the continuous demand in the area with a discrete set of demand points each representing a small sub-area. The main reason for such a substitution is that it is easier to construct the objective function based on a finite set of demand points rather than a continuous functional demand. Commonly, the objective function is composed as a sum of terms, one for each demand point. In the continuous formulation this sum is replaced by a double integral over the area.

In a continuous demand formulation, demand is generated over an area A with a demand density function $f(x, y)$ defined for all (x, y) in A. The demand is discretisized by dividing the area into small sub-areas each represented by a *demand point* [786]. The optimization problem is then solved assuming that all the demand in a sub-area originates at the demand point. This simplification introduces some error into the objective function and, in turn, to the location solution. The source of error in representing demand as discrete stems from the fact that the distance to a facility from a sub-area is assumed the same for all customers, whereas, in fact, this is not the case. The distance varies among customers. It is generally believed by researchers that this error does not significantly affect the calculated optimal location. This may not be generally true. In some location problems the error is quite pronounced and leads to an inferior solution. One such problem is used in this chapter as a test problem to illustrate the analysis.

This chapter offers the exact calculation of the optimal solution by double integration. The results of such an exact calculation may be used to achieve utmost accuracy. A method that offers a partial remedy to the inaccuracy problem when the calculations of the double integrals are unmanageable is offered. The distance between the demand point and the facility is modified to resemble the average distance between the facility and *all* points in the small sub-area which this demand point represents. This modification is termed *the distance correction* approach. By applying the distance correction approach the user can apply the simpler discrete models while obtaining optimal locations which are close to the ones obtained using the continuous model. This distance modification is similar to the hyperbolic approximation of the distance suggested by [1196] and [393]. It was originally suggested for preventing discontinuities in the derivative of the distance function. The exact results obtained by double integration are used as a reference for the evaluation of the effectiveness of the distance correction approach. This distance correction approach was found to be simple and effective for the test problem. It is therefore recommended that such a correction is implemented in all location problems that assume a finite set of demand points.

The particular problem investigated in this chapter is the competitive facility location problem based on Huff's formulation for capturing market share [289, 603, 604, 884]. Chapter 13 in this book provides the background on competitive facility location using Huff's estimation of market share.

2.2 Formulation and Analysis

The test problem selected for this chapter involves the selection of the best site for a new facility in the presence of existing competing facilities (see Chapter 13). The objective to be maximized is the market share captured by the new facility. The market share captured by a facility is calculated using the gravity model assumptions formulated by Huff [603, 604]. Huff postulated that the probability that a customer patronizes a facility is proportional to the attractiveness of the facility and inversely proportional to a power λ of the distance to it.

The best location for the new facility is the one at which the market share captured is maximized. Let the new facility have an attractiveness parameter S and its distance from a customer at (u, v) be $d(u, v)$. k other competing facilities numbered $j = 1, \ldots, k$ exist in the area, each with an attractiveness parameter S_j and distance $d_j(u, v)$ from (u, v). The proportion of the buying power that is attracted to the particular facility is according to gravity models:

$$p(u, v) = \frac{\dfrac{S}{d^\lambda(u, v)}}{\dfrac{S}{d^\lambda(u, v)} + \sum_{j=1}^{k} \dfrac{S_j}{d_j^\lambda(u, v)}} \tag{2.1}$$

The total buying power attracted by the new facility (also called the market share attracted by the new facility) is:

$$\text{TMS} = \iint\limits_{A} p(u,v)f(u,v)dudv \tag{2.2}$$

Standard analyses [603, 604, 289] apply the *discrete approach*. The area is represented by demand points and the integral is replaced by a sum of terms, each representing the contribution of a demand point to the market share attracted by the facility. Modeling by (2.2) is the *continuous approach*. In the following we investigate the difference between the market share captured by the new facility calculated by the continuous approach and that calculated by the discrete one. The difference in the optimal location according to these two approaches is also investigated.

The best location is defined as the site at which the market share captured is maximized. Let (x,y) be the location of the new facility. It is possible, though not likely, that the best location for the new facility is outside the area. Therefore, it is assumed that this location (x,y) can be anywhere in the plane. Let the k existing facilities be located at (u_j, v_j) for $j = 1, \ldots, k$, and let the distance between points (x,y) and (u,v) be: $d(x,y,u,v) = \sqrt{(x-u)^2 + (y-v)^2}$. The total market share captured by the new facility, to be maximized, is by (2.1) and (2.2):

$$\begin{aligned}
\text{TMS}(x,y) &= \iint\limits_{A} \frac{\dfrac{S}{d^\lambda(x,y,u,v)}\, dudv}{\dfrac{S}{d^\lambda(x,y,u,v)} + \sum_{j=1}^{k} \dfrac{S_j}{d_j^\lambda(u,v,u_j,v_j)}} \\[2mm]
&= \iint\limits_{A} \frac{S\, dudv}{S + d^\lambda(x,y,u,v) \sum_{j=1}^{k} \dfrac{S_j}{d^\lambda(u,v,u_j,v_j)}} \tag{2.3}
\end{aligned}$$

Finding the best location that maximizes $\text{TMS}(x,y)$ by (2.3) is described in [292].

2.3   Evaluating a Double Integral

The most efficient way to evaluate integrals is by using Gaussian Quadrature Formulas (GQFs) [7]. A GQF is defined for a certain weight function. Such a weight function defines a set of orthogonal polynomials. There are many known GQFs [268, 943]. For our particular application, since no special assumptions on the weight function $f(u,v)$ or the area A are assumed, the most suitable GQF is based on a weight of "1" which leads to the use of Legendre polynomials [7]. GQFs are constructed for one-dimensional integration. Since we need double integration, we use the formula twice for integration in each dimension obtaining a double sum rather than a single sum. The one dimensional GQF based on Legendre polynomials is:

$$\int_{-1}^{1} f(x)dx \approx \sum_{i=1}^{K} w_i f(x_i) \tag{2.4}$$

where K is the number of integration points used in the formula. It is simpler to define $c_i = \frac{w_i}{2}$, $z_i = \frac{1+x_i}{2}$ and obtain

$$\int_{0}^{1} f(x)dx \approx \sum_{i=1}^{K} c_i f(z_i) \tag{2.5}$$

Weights and abscissas for various values of K (K between two and 96, but not every K) are given in [7].

For general integration limits $[a, b]$ we substitute $x = a + (b - a)z$ and get by (2.5):

$$\int_{a}^{b} f(x)dx \approx (b - a) \sum_{i=1}^{K} c_i f[a + (b - a)z_i] \tag{2.6}$$

Let the area A be defined for x between a and b. For each x the area is bounded by $g_1(x) \leq y \leq g_2(x)$. This means that for a general function $F(u, v)$ (u is a variable on the x-axis and v is a variable on the y-axis):

$$\iint_{A} F(u, v)dvdu = \int_{a}^{b} \int_{g_1(u)}^{g_2(u)} F(u, v)dvdu \tag{2.7}$$

The explicit integration formulas are easily constructed. See [292] for details.

2.3.1   Error Analysis

The discussion of this section is based on [313]. The error in any Gaussian quadrature formula based on K points is proportional to the $2K^{\text{th}}$ derivative of the integrand [7]:

$$e_x = \gamma_K \frac{\partial^{2K}[G(x, y)f(x, y)]}{\partial x^{2K}} \tag{2.8}$$

where $\gamma_K = \dfrac{2^{2K+1}(K!)^4}{(2K + 1)[(2K)!]^3}$. e_x is the error in the calculation of the integral on x for a constant y and the derivative is calculated at some point

TABLE 2.1. Values for the Error Term Coefficients

K	γ_K
1	0.3333
2	$0.7407 \cdot 10^{-2}$
3	$0.6349 \cdot 10^{-4}$
4	$0.2879 \cdot 10^{-6}$
5	$0.8079 \cdot 10^{-9}$
6	$0.1541 \cdot 10^{-11}$
7	$0.2127 \cdot 10^{-14}$
8	$0.2224 \cdot 10^{-17}$
9	$0.1823 \cdot 10^{-20}$
10	$0.1203 \cdot 10^{-23}$

at the integration interval. The integral of e_x over y plus the error due to the integration over y is the total error in the integration. The coefficient γ_K decreases very rapidly with K. See Table 2.1 for values up to $K = 10$.

If the integrand $f(x, y)G(x, y)$ is a polynomial of degree not greater than $2K - 1$ in x and y in a square, then the approximation is exact. However, if the integrand is not a polynomial, then some error exists in the integration formula.

In order to select the appropriate K one needs to be able to estimate a bound on the value of the $2K^{\text{th}}$ derivative of the integrand. If this estimate is available, then the appropriate K can be selected by evaluating possible K's by (2.8). Alternatively, $G(x, y)f(x, y)$ can be fitted to a polynomial and K should be selected such that the integral of the error in the curve fitting (or more conservatively the maximum error multiplied by the integration area) is less than a required accuracy. If such estimates are difficult to obtain, it is recommended that the maximum possible K that requires reasonable effort, computer memory, and run times will be selected. The error definitely decreases with K.

2.4   Analysis of the Example Problem

To apply the double integral to the location model formulation an example problem is used. The example problem is taken from [289] and Chapter 13. The data consist of:

1. A square area of 10 by 10 miles

2. Seven existing competing facilities which are arbitrarily located in the area.

TABLE 2.2. Market Share Calculated by Various Integration Formulas

K	Market Share Captured by a Facility						
	1	2	3	4	5	6	7
2	10.03	4.19	20.71	11.69	12.25	14.69	26.44
3	10.28	28.17	11.90	11.98	11.98	12.19	13.50
4	13.10	12.37	20.78	13.09	13.02	11.22	16.42
5	11.77	18.85	16.44	12.33	12.33	11.62	16.67
6	12.12	15.80	17.76	12.94	12.97	12.13	16.30
7	12.31	16.84	17.43	12.72	12.72	11.60	16.36
8	12.31	16.51	17.44	12.78	12.79	11.73	16.43
9	12.20	16.59	17.53	12.76	12.77	11.78	16.38
10	12.26	16.56	17.45	12.78	12.79	11.75	16.40
12,16,20	12.25	16.57	17.48	12.77	12.78	11.75	16.40
The Grid	12.24	16.54	17.49	12.77	12.78	11.75	16.42

3. Demand is assumed evenly distributed in the square and all the S's are equal to one.

4. A value of $\lambda = 2$ was used.

First, we perform the integration for the distribution of the market share between the facilities for various values of K. The results are compared with the distribution of market share obtained by assuming a symmetric grid of 100 demand points arranged in a 10 by 10 pattern) as reported in [289] and chapter 13. This set of demand points is referred to as *"the grid"*.

The market share calculated by the grid is very close to the exact market share calculated by integration. However, when an optimal location for an eighth facility is sought, the result obtained by integration and that obtained by the grid differ substantially. The reason for this surprising result is explained later. The best location is found by applying the Weiszfeld procedure (see [289] and Chapter 13 for the discrete Weiszfeld procedure). In Table 2.3 we report all the 11 local optima obtained by the grid (as reported in [289]) and by integration based on 48 points. The locations of the local optima, the market share captured at each, and the number of times, n, out of 100 starting solutions that the algorithm terminated in the particular local optimum are reported for each method.

Three dimensional graphs of the surface of the market share function are shown in Figure 1 for both the discrete function and the continuous function. Apparently, there are two local optima: one around (6.5, 5.5) and one around (3,3). The grid indicates a "wrong" optimal location at (5.59, 5.54). The best location (by integration) is in fact the second best location obtained by the grid. Also, the market share is a much smoother function

FIGURE 1. Market Share Calculated by the Grid (top) and by the 48 Points Integration Formula (bottom)

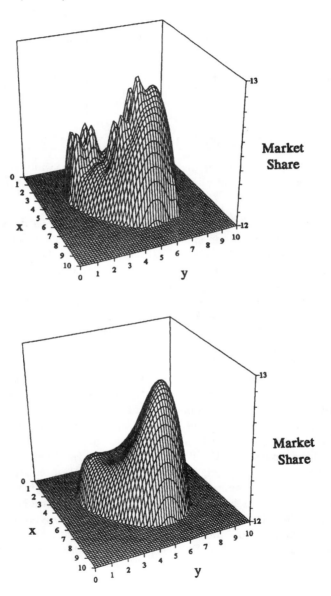

TABLE 2.3. Local Maxima Based on 100 Trials

	By The Grid			By Integration $K = 48$			
x	y	$F^\dagger$	$n^\ddagger$	x	y	$F^\dagger$	$n^\ddagger$
5.59	5.54	12.93	12	6.53	5.53	12.86	88
6.61	5.53	12.85	37	3.30	2.97	12.32	3
5.59	4.54	12.81	9	2.97	3.30	12.32	6
4.56	5.54	12.69	11	3.29	3.26	12.32	3
4.54	4.51	12.60	3				
3.50	3.50	12.53	2				
3.49	2.55	12.50	7				
2.55	3.49	12.50	3				
2.56	2.56	12.46	11				
4.75	3.59	12.35	3				
3.57	4.54	12.31	2				

$\dagger$ F is the calculated market share captured by a facility at the point (x, y).
$\ddagger$ n is the number of times the algorithm terminated at (x, y).

of the location of the new facility when continuous demand is assumed. In the grid surface ten spikes are apparent and the optimal location has a smooth shape while by the continuous function the surface of the market share is smooth showing only two clear local optima.

Since the objective function of the grid has minor spikes at demand points, almost all of the 11 local optima of the grid are very close to demand points. The Gaussian integration is also based on a grid of K by K points and the local maxima tend to be close to points in that grid. However, the distortion generated by this type of a grid is reduced. The minor spikes also explain why the results in Table 2.2 are quite accurate for the grid but are not sufficiently accurate for evaluating the optimal location. In our particular problem, the existing facilities are located away from demand points. Demand points are located at the "half mile" mark $(0.5, 1.5, \ldots, 9.5)$ while the existing facilities are located at "integer" miles. Each existing facility is located at a center of a square formed by four demand points (see the figure in Chapter 13). When the facilities are not close to demand points, the discrete formulation is quite accurate. However, when the market share is calculated for a facility located close to a demand point it is not accurate and, in fact, is biased upward forming a spike. Therefore, when the optimizing procedure is evaluating a site close to a demand point the market share is not accurately calculated (assumes an inflated value) and the algorithm may settle at the wrong location.

2.5   The Distance Correction Approach

In this section, an alternative method for by-passing the integration while approximating the continuous distribution of demand and the continuous optimal site is introduced. Each demand point represents a sub-area. The distance between a facility and the demand point is approximated by the average distance between the facility and all the points in the sub-area. Several papers deal with the locational analysis of a facility relative to area demand [77, 303, 327, 774, 1195]. The "equivalent" distance to a generally shaped area is difficult to calculate. For rectangular areas the formula is highly complicated and depends on the direction from the point to the facility [327, 774]. The simplest formula is obtained for a circular area [303, 327]. Consider a facility whose distance from a demand point (at a center of a circle) is d and the radius of the circular area is r. The average squared Euclidean distance between the facility and the circular sub-area is $d^2 + \frac{r^2}{2}$ [303]. It is shown in [292] that this result is also valid for our test problem. Therefore, as an approximation, if a demand point covers a sub-area of area s, the equivalent squared Euclidean distance is approximated by $d^2 + \frac{r^2}{2}$ where $s = \pi r^2$. This is approximately $d^2 + 0.16s$. Therefore, d^λ is replaced in all the discrete equations with $[d^2 + 0.16s]^{\lambda/2}$. This is the *distance correction* approach. Note that the area s may be different for each demand point.

The same problem was run using the discrete approach with the distance correction. One hundred randomly generated starting points were used. The sensitivity to the correction factor "0.16" was also examined. The value of 0.16 is replaced by the value α defining the distance approximation $[d^2 + \alpha s]^{\lambda/2}$. The values α =0.12, 0.16, 0.20, and 0.24 were tested. Note that if instead of equating the areas of the circle and the square we replace the square with the circle enclosed in the square or the circle enclosing the square, then the values of α are 0.125 and 0.25, respectively. This is roughly the range tested. The results for $\alpha = 0.16, 0.24$ are reported in Table 2.4. The results in Table 2.4 indicate that a larger α is more suitable for the distance correction factor. The value of $\alpha = 0.24$ yields only two local maxima. It is clear that the eleven local optima that exist in the grid dissipate as α increases. The grid solution (with its 11 local maxima) is the solution for $\alpha = 0$. In [292] the three dimensional graph of the market share surface for $\alpha = 0.2$ and $\alpha = 0.24$ are depicted. The $\alpha = 0.24$ graph still exhibits some irregularities but is much closer to the continuous integrated surface. This illustrates the effectiveness of the distance correction approach.

The distance correction approach should be applied to other location problems as well. As is discussed by Drezner [303], the squared Euclidean Weber problem is unaffected by the replacement of circular areas with their centers. As discussed there, the Weber problem with Euclidean distances

TABLE 2.4. Local Optima Obtained for Various Distance Corrections

No Distance Correction				Using Distance Correction							
				$\alpha = 0.16$				$\alpha = 0.24$			
x	y	$F^\dagger$	$n^\ddagger$	x	y	$F^\dagger$	$n^\ddagger$	x	y	$F^\dagger$	$n^\ddagger$
5.69	5.56	12.91	22	6.50	5.54	12.93	44	6.43	5.55	12.96	85
6.53	5.54	12.91	43	5.74	5.57	12.92	33	3.55	3.53	12.55	15
5.70	4.66	12.80	11	4.70	4.57	12.62	7				
4.62	4.54	12.60	8	3.52	3.51	12.52	3				
3.51	3.50	12.51	3	3.49	2.72	12.45	1				
3.48	2.65	12.45	1	2.71	3.48	12.45	3				
2.64	3.48	12.45	3	2.76	2.77	12.40	9				
2.67	2.67	12.40	9								

$\dagger$ F is the calculated market share captured by a facility at the point (x, y).
$\ddagger$ n is the number of times the algorithm terminated at (x, y).

and circular demand points resembles the hyperbolic approximation for distances in facility location [1196, 393]. This can be interpreted as a distance correction approach. As suggested there, the distance d should be replaced by $\sqrt{d^2 + \alpha s}$ with α equal to approximately 0.14.

2.6   Conclusions

The impact of modelling continuous demand as discrete demand points in all planar location problems is investigated. For a particular competitive facility location problem, the effect of discretization on the objective function seems insignificant. However, the effect on the best location for a new facility is significant. It appears that many local optima are artificially created because the objective function tends to have "spikes" in the vicinity of demand points.

An easy and efficient alternative approach is also suggested. Instead of using the distance in the formulation, a constant (depending on the sub-area covered by the demand point) is added to the square of the distance as a "correction additive". This adjustment replaces the distance to the demand point with an approximation to the average distance to all the points in the sub-area. Consequently, the objective has no longer spikes at the demand points and the number of local optima is significantly reduced. For the particular problem tested in this chapter, the optimal location obtained by this alternative approach is very close to the optimal location found by optimizing the continuous demand problem.

This correction approach is recommended for implementation in all formulations of location models which are based on discretization of demand into demand points.

3
GLOBAL OPTIMIZATION IN LOCATION

Pierre Hansen
GERAD & Ecole des Hautes Études Commerciales, Montréal

Brigitte Jaumard
GERAD & Ecole Polytechnique de Montréal

Hoang Tuy
Institute of Mathematics, Hanoi

3.1   Introduction

Global optimization aims at solving the following problem:

$$\min f(x) \tag{3.1}$$

subject to:

$$g_i(x) \leq 0 \ i = 1, 2, \ldots, m \tag{3.2}$$

where f and g_i for $i \in \{1, 2, \ldots, m\}$ are nonlinear and nonconvex functions of n real variables $x_1, x_2, \ldots, x_n$. Usually the functions f and g_i for $i \in \{1, 2, \ldots, m\}$ are assumed to be continuous or to satisfy a Lipschitz condition but are not necessarily given in explicit form (in other words, their values may be given by oracles). One seeks a globally optimum solution to problem (3.1)-(3.2) together with a proof of its optimality. In practice, it is often enough to find a globally ε-optimal value, i.e., a value within ε of the optimal one. One can also aim at localizing a vector within a distance ε of an optimal solution. Due to the nonconvexity of f, problem (3.1)-(3.2) may have many local optima. This contrasts with the usual situation in nonlinear programming where the functions are assumed to be convex and hence any local minimum is a global one. Moreover, the feasible domain, i.e., the set of values of x satisfying the constraints (3.2), need not be convex or even connected. For these reasons, problem (3.1)-(3.2) is very difficult to solve. It is, in fact, the most general problem of finite deterministic mathematical programming. Despite much recent progress in global optimization theory and algorithms, only instances with a few variables or with some particular structure are currently amenable to solution.

Not surprisingly, many problems of continuous location theory are naturally expressed as global optimization ones. This is true in particular for many extensions, studied below, of the classical Weber problem.

The present chapter has a twofold purpose:

(i) present briefly the main approaches of exact deterministic global optimization (more detailed presentations are given in [597, 595]). A few heuristic methods of global optimization, which lead to local minima or to an unproved global minimum will also be mentioned.

(ii) illustrate and discuss the use of these approaches applied to a series of problems of continuous location theory (which are, most of the time, extensions of Weber's problem). While an attempt is made to provide a fairly complete bibliography on global optimization in location, not all applications are discussed in detail. Further information on some of these problems is given in Chapter 11 in this volume.

The chapter is organized as follows: the branch-and-bound approach is reviewed in the next section. The three following sections are devoted to approximation methods: Lipschitz underestimation, primal and dual outer approximation. Decomposition and linearization are studied in the next two sections. Algorithms for problems with special structure are considered in the penultimate section. Then, conclusions as to the power and versatility of global optimization applied to continuous location theory are drawn.

The assumptions of Weber's problem are:

(H1) A single facility is to be set up at a point s of $\mathbb{R}^2$.

(H2) m origin-destination points are to be connected to the facility; the jth such point is denoted by $a^j \in \mathbb{R}^2$ for $j = 1, 2, \ldots, m$.

(H3) The distance considered between the facility at s and point a^j is the Euclidean one, noted $\|s - a^j\|$.

(H4) The transportation cost between the facility at s and point a^j is an increasing linear function of the distance covered; it is denoted by $w_j\|s - a^j\|$.

(H5) The facility is to be established at a point $s^\star$ of $\mathbb{R}$ that minimizes the total transportation cost:

$$f(s^\star) = \min_{s \in \mathbb{R}^2} f(s) = \min_{s \in \mathbb{R}^2} \sum_{j=1}^{m} w_j\|s - a^j\|.$$

All of these assumptions will be relaxed in various ways below. Weber's problem may by solved by a well-known iterative method due to Weiszfeld [1186], which exploits first order conditions. With slight modifications for iteration points where $f(s)$ is nondifferentiable (i.e., the points a^j) it converges to the global optimum [910, 142]. Alternately, an accelerated method [1015] may be used.

3.2   Branch-and-bound Methods

Branch-and-bound methods solve difficult problems by partitioning their feasible domain (or some domain containing the feasible domain, which may be unknown or difficult to characterize) into subdomains, i.e., *branching* and applying tests on the resulting subproblems which make use of evaluations of the objective function value or of the values of the constraints' left hand sides, i.e., *bounding*. For a classification of tests, see [542]. Tests may give all the information required about the current subproblem, i.e., that a particular solution is the optimal one of the subproblem (solution test) or that no solution of the subproblem has a value better than the incumbent one (optimality test) or that the subproblem has no feasible solution (feasibility test). Weaker tests provide similar information about a part of the domain of the subproblem, which can be reduced without branching.

While branch-and-bound methods have been extensively used in integer programming since its inception, their introduction in global optimization is more recent. An early instance is Horst's [594] algorithm for concave programming. In location theory, the first use of branch-and-bound to solve a continuous location problem appears to be a study of the location of an obnoxious facility [549, 556]. The proposed algorithm was expanded into the Big Square Small Square method (BSSS) for the generalized Weber problem, which is discussed next.

3.2.1   Generalized Weber Problem

The *Generalized Weber Problem* (GWP), or generalized minisum problem, considered in [547] is less restrictive and more realistic than the Weber problem stated in the introduction in several respects.

First, locational constraints forbid to establish the facility in regions already occupied or unfit for such a location. Such constraints are ubiquitous and important in practice ([1042] cites a study recommending location of a warehouse in the Alpes Glaronnaises at an altitude of 3000 meters) but rarely made explicit in location theory. Assumption (H1) is replaced by:

(H1') A single facility is to be established at a point $s \in G \subset \mathbb{R}^2$, where G is the set of possible locations. The set G is formed by the union of a finite number of convex polygons P_k, for $k = 1, 2, \ldots, n$, i.e.,
$$G = \bigcup_{k=1}^{n} P_k.$$

Note that the assumption made on the shape of G is not very restrictive as any closed domain in the plane can be approximated with sufficient precision by the union of a large number of convex polygons. More restrictive is the implicit assumption that while some regions are forbidden for setting-up the facility, travel through these regions is still allowed. Ways to take into account restrictions both on location and on travel are discussed below, in the section on algorithms for problems with a special structure.

Second, instead of using only the Euclidean norm to estimate distance, other norms are also considered. Assumption (H3) becomes:

(H3′) The distance between the facility at s and point a^j is expressed by $\|s - a^j\|_j$, where $\|.\|_j$ is an ℓ_p-norm that depends on the origin-destination point.

Third, the transportation cost functions need no more be linear in distance. Assumption (H4) becomes:

(H4′) The transportation cost between the facility established at s and point a^j is an increasing and continuous function of the distance covered; it is denoted $C_j(\|s - a^j\|_j)$.

Therefore assumption (H5) becomes:

(H5′) The facility is to be established at a point $s^\star$ of $\mathbb{R}$ that minimizes the total transportation cost

$$F(s^\star) = \min_{s \in G} F(s) = \sum_{j=1}^{m} C_j(\|s - a^j\|_j).$$

Note that while the cost functions considered belong to a very large class, they do not cover all single facility continuous location problems. With slight changes in the algorithm, several other cases can be tackled also. Obnoxious facilities give rise to utility functions decreasing in distance and the adaptation is easy. Location of semi-obnoxious facilities [1126, 335] which are desirable to some (i.e., the users) and obnoxious to others (i.e., the close residents) leads to a variant with both increasing and decreasing utility functions, discussed below. A case which does not however fit easily into this frame is that of fixed prices between zones, or in other words tariffs for transportation of goods which are constant per unit from any point in a given zone to any other point in another zone. Then a way to compute bounds based on finding the least favorable zones intersecting the region considered must be invoked.

The Big Square Small Square (BSSS) algorithm for GWP [547] proceeds by: *(i)* partitioning the smallest square containing the set G of feasible locations into subsquares; *(ii)* computing a lower bound on the values of F for those subsquares that intersect G; *(iii)* deleting the subsquares for which the lower bound exceeds the value of the best known solution, which is regularly updated; *(iv)* iterating until the length of a side of a square is smaller than a given tolerance. The rules for BSSS are as follows:

a) *Initialization.* Let Q_1 be a smallest square containing G, with sides parallel to the axes. Let ℓ be the length of the sides of Q_1 and ε the maximum side-length allowed at termination. Compute the value of F at a few feasible points, let F_{opt} be the smallest value obtained and s_{opt} the corresponding point.

b) *Branching and feasibility test.* Consider in turn each square Q_i in the current list of squares. Partition Q_i into four equal subsquares Q_k, $Q_{k+1}, Q_{k+2}, Q_{k+3}$. For each of these subsquares Q_j, check if $Q_j \cap G \neq \emptyset$. If it is the case, introduce Q_j in the new list of squares. When all squares Q_i have been considered replace the current list of squares by the new one.

c) *Improved solution.* Consider in turn each square Q_i in the current list. Let s_i be the intersection point of the diagonals of Q_i. Check if s_i is feasible and if so, compute $F(s_i)$. If $F(s_i) < F_{opt}$ set $s_{opt} \leftarrow s_i$ and $F_{opt} \leftarrow F(s_i)$.

d) *Bounding and optimality test.* For each square Q_i in the current list, determine for each point a^j the closest point $\underline{a}^j$ in the square Q_i and compute the lower bound

$$\underline{F}_i = \sum_{j=1}^{m} C_j(\|\underline{a}^j - a^j\|_j).$$

If $\underline{F}_i \geq F_{opt}$ delete square Q_i from the current list.

e) *Termination test.* If $\ell < \varepsilon$, stop: s_{opt} is the best point found and F_{opt} its value. Otherwise, set $\ell \leftarrow \dfrac{\ell}{2}$ and return to b).

Algorithm BSSS converges to an ε'-optimal solution (with $\varepsilon' = \varepsilon L$ and L a Lipschitz constant) provided that the objective function $F(s)$ has a bounded slope, i.e., is Lipschitz, and that central points s_i of squares Q_i are feasible, at least in the list obtained at termination ([957] suggests a modified procedure to find a feasible point in Q_i if it is not the case). This follows from continuity of the norms $\|.\|_j$ and functions $C_j(\|s - d_j\|)$. If the value of ε' (or of L) is given beforehand one can add the following test:

d') *Solution Test.* For each square Q_i in the current list, check if s_i is feasible and $F(s_i) - \underline{F}_i < \varepsilon'$ and delete square Q_i if it is the case.

Ways to speed up feasibility computations by keeping track of subsquares which are entirely feasible at a given iteration are discussed in [547].

Algorithm BSSS uses a breadth-first search strategy. This has the advantages of simplicity and ease of output of the regions remaining under study at each iteration, i.e., lists of squares with bounds lower than the incumbent value. Alternate search strategies would be depth-first or best-first search.

Plastria [957] recommends the former and proposes to store the current list of problems in an AVL-tree. This is a good data structure if problems for which the bound $\underline{F}_i$ is larger than or equal to the incumbent value f_{opt} are not deleted (such deletion is not mandatory, but saves space and time). Otherwise a double-ended priority queue [37] is more adequate.

Plastria [957] also proposes to determine in a second phase those squares (or, in his proposal, rectangles) for which the lower bound is within some tolerance of the best known solution. In this way the region of near optimal points can be described, which is often more useful to the decision maker than a single best point.

The lower bounds $\underline{F}_i$ used in BSSS are in many cases not very sharp, in particular when demand points are on all sides of the current region Q_i. Under assumption (H4′) it seems difficult to improve them, as they may be best possible. Nevertheless, several improvements have been proposed under more restrictive assumptions, and are discussed below.

3.2.2 Weber's problem on the sphere

Hansen, Jaumard and Krau [540] apply branch-and-bound to Weber's problem on the sphere (WS). This generalization of Weber's problem is obtained by replacing assumption (H1) by:

(H1″) A single facility is to be established at a point $s \in S$, where S is a sphere;

and assumption (H3) by:

(H3″) The distance between the facility s and point a^j is expressed by $\|s-a^j\|_g$ where $\|.\|_g$ is the geodesic distance, i.e., the shortest distance from s to a^j along a great circle of S going through these points.

Moreover, assumption (H5) is modified in an obvious way.

Weber's problem on the sphere appears to have been first studied in [678, 324]. A first exact algorithm for its solution is given in [301]. It is discussed in the next section. As the distance function on the sphere is not convex, problem WS is a global optimization one. A straightforward extension of algorithm BSSS is obtained by considering regions Q_i limited by two latitudes and two longitudes (the lateral and corner regions are more difficult to specify than in the case of the plane, but still easy to use). Improved lower bounds can be obtained by grouping points from different regions in such a way that the optimal solution for the group considered lies on a side of Q_i. However, Weber's problem restricted to a great circle is still nonconvex. Therefore the side of Q_i considered must be partitioned into a few arc segments and bounds computed for each of them. The smallest bound so obtained is valid for Q_i and the group of demand points. Summing for all groups of demand points gives a bound more precise than $\underline{F}_i$.

Another way to get better bounds is to exploit the fact that problem ws restricted to Q_i is convex for a subset of the demand points, namely those for which the maximum distance to a point of Q_i is not larger than $\dfrac{\pi r}{4}$, where r denotes the radius of the sphere. Partitioning the demand points into those which satisfy this condition, and for which the ws problem can be solved by a variant of Weiszfeld's [1186] iterative algorithm [324, 678], and the other ones for which a bound is computed as discussed above, leads to a better bound for the whole set. Using such bounds [553] problems with 500 points uniformly distributed on the sphere are solved as well as constrained problems with 1,000 points uniformly distributed on the continents.

3.2.3 Weber's Problem with Attraction and Repulsion

[798] solves Weber's problem with attraction and repulsion (WAR) by branch-and-bound, also considering a partition of the points. In this variant of Weber's problem ([1125, 1126] in the case of 3 demand points, [335, 198] in the general case, see below), assumption (H2) must be completed by:

(H2′) Points a^j correspond to either users (with indices $j = 1, 2, \ldots, m_1$) or residents (with indices $j = m_1 + 1, m_1 + 2, \ldots, m$).

Assumptions (H4) and (H5) become:

(H4″) Users want to minimize their transportation cost, i.e., $w_j \|s - a^j\|$, and residents want to minimize the disutility due to the obnoxiousness of the facility, i.e., $c_j - w_j \|s - a^j\|$, where c_j is a constant.

(H5″ The facility is to be established at a point $s^\star$ of $\mathbb{R}$ that minimizes the total transportation cost and disutility due to obnoxiousness:

$$f(s^\star) = \min_{s \in \mathbb{R}^2} \sum_{j=1}^{m_1} w_j \|s - a^j\| - \sum_{j=m_1+1}^{m} w_j \|s - a^j\| + \sum_{j=m_1+1}^{m} c_j. \quad (3.3)$$

The bound used by [798] is the sum of the optimal value of Weber's problem restricted to Q_i for the users and a bound obtained by linear overestimation of the functions $w_j \|s - a_j\|$ over Q_i for the residents. Problems with up to 10,000 users or residents are solved in moderate time.

As in combinatorial optimization, branch-and-bound methods are very versatile in continuous location theory. Several further variants and extensions of Weber's problem have been solved in that way, and many further potential applications are at hand. Among problems solved one should mention: *(i)* Weber's problem with alternate supply sources [681]; *(ii)* the Weber-Rawls problem in which both a criterion of efficiency (the transportation costs) and one of equity (the cost borne by the worst-off user) are simultaneously taken into account [548, 556], and *(iii)* the profit maximizing Weber problem [553] in which both location and price(s) must be

chosen; *(iv)* location of a polluting unit to minimize a function of obnoxiousness to inhabitants of residential zones, taking dominant winds into
account [664, 676]. Further potential applications, mentioned in [547] are:
(i) to establish a public facility in order to minimize $\sum_{j=1}^{m} [c_j \|s - a_j\|_j]^{\alpha}$ with
$\alpha > 1$ [867]: a location problem taking equity into account as large costs
are weighted more heavily than small ones; *(ii)* to find the location of a
public facility whose appeal decreases with distance, in order to maximize
the sum of the individual utilities; a location problem in public economics;
(iii) to locate a department store in order to maximize its expected number of customers [1208]: a location problem in which individual behavior
is ruled by a gravity like model; *(iv)* to locate a firm in order to maximize
the amount of amenities it enjoys [1129]: a location problem in economic
geography.

3.3 Lipschitz Optimization

Lipschitz optimization addresses global optimization problems in which it
is only assumed, in addition to definiteness, that the functions f and g_i
$(i = 1, 2, \ldots, m)$ have a bounded slope. In other words, f must satisfy a
Lipschitz condition

$$\forall s^1 \in G \quad \forall s^2 \in G \qquad |f(s^1) - f(s^2)| \le L\|s^1 - s^2\| \qquad (3.4)$$

where G is the feasible domain and L is a constant (called Lipschitz constant); if there are some constraints, the g_i must satisfy similar conditions.

The classical algorithm for Lipschitz optimization of an unconstrained
function is due to Piyavskii [946]. It proceeds to a sequential search based
on successive refinements of a lower bounding function defined as the upper
envelope of a set of circular n-dimensional cones of slope L. It starts at an
arbitrary point s^0 and defines a first lower bounding function F^0 as

$$\forall s \in G \qquad F^0(s) = f(s^0) - L\|s - s^0\|. \qquad (3.5)$$

It then builds a sequence of evaluation points and refines the function F
each time. At the kth iteration, the next evaluation point s^k is such that:

$$s^k = \arg\min_{s \in G} F^k(s) \qquad (3.6)$$

where

$$F^k(s) = \max_{i=0,1,\ldots,k-1} \{\|f(s^i)\| - L\|s - s^i\|\} \qquad (3.7)$$

This amounts to finding the lowest point of the surface defined by (3.7). The
function f is then evaluated at the corresponding point in G and the lower

bounding function updated by addition of a new lower bounding cone. At this stage, the globally optimal value $f^\star$ is known to be within the interval

$$[F_{opt}, f_{opt}] = [F^k(s^k), \max_{i=0,1,\ldots,k-1} f(s^i)] \tag{3.8}$$

and the algorithm stops when the range of this interval is smaller than ε.

Determining the lowest point of the lower bounding function is difficult. [946] observes that the globally minimum points belong to a finite set, i.e., the local minima of the lower bounding function, which are easily characterized and can be computed by solving systems of quadratic equations. [860] proposes a more practical way to find the local minima, by solving systems of n linear and one quadratic equations, each of them corresponding to the intersection point of n intersecting cones (or hyperplanes bounding the feasible domain). While the algorithm of [860] requires, at a given iteration, computing all such intersections between the new cones and the previous cones or hyperplanes, [638] shows, using geometrical arguments, that the number of systems to be considered can be drastically reduced. [638] also demonstrate that dominated cones can be identified and deleted. More details on these algorithms are given in the chapter [553] in [595].

Lipschitz optimization problems can also be solved by branch-and-bound methods, and the many algorithms proposed for that purpose are reviewed in [553]. Moreover, a computational comparison shows that the algorithms of [860, 638] take the least number of function evaluations but that branch-and-bound algorithms may be less time consuming if the function to evaluate is sufficiently simple.

3.3.1 Weber's Problem on the Sphere (again)

While Lipschitz optimization does not appear to have been invoked in location, it is worth noting that Drezner's [301] algorithm for Weber's problem on the sphere, is equivalent to Piyavskii's algorithm adapted to that context. On the sphere, the subproblem of finding the minimum point of the lower-bounding function, i.e., the minimax problem of finding a point whose smallest weighted distance to the points of a given set is maximum, is equivalent to a center, or maximin, problem: find a point whose greatest weighted distance to these points is minimum. Indeed, the optimum point of the former problem is the antipode of the optimum point of the latter one. [301] proposes to solve this subproblem by an algorithm of [297]. Alternately, another algorithm [330] could be used for that purpose. These provide two ways to solve optimally the minimax problem, which is in general nonconvex on the sphere. First a descent method [278] for minimax problems gives a local minimum and the corresponding value of the objective function. This last value determines the radius of spherical circles centered at each demand point. A specific procedure then determines if the intersection between all spherical circles reduces to one (or several) isolated point(s) or to some area(s). In the former case the algorithm stops as the global optimum of the subproblem is found. In the latter, the intersection is composed of one or several disjoint areas. A descent method is again applied from a point lying in one of these areas and the procedure is iterated until the lowest point of the underestimating function is found. The function is evaluated at this point and a new bounding function obtained. The minimax problem is then solved again and the whole procedure iterated. No computational experience is reported.

3.4   Outer Approximation

A d.-c. function is a function which may be expressed as a difference of two convex functions. D.-c. functions enjoy many properties: sums and differences, maxima, minima and absolute values of d.-c. functions remain d.-c. functions. Moreover, every function which is twice differentiable is d.-c. (although finding a d.-c. representation of it may be difficult). Therefore, d.-c. programs, i.e., programs of the form (3.1)-(3.2) in which f and the g_i for $i = 1, 2, \ldots, m$ are d.-c., have numerous applications [597, 1152]. Moreover, if the set of feasible solutions is convex, i.e., the g_i are convex for $i = 1, 2, \ldots, m$, the problem can be converted into a concave minimization problem over a convex set, i.e., a problem of the form

$$\min f(x) \quad \text{subject to:} \quad x \in D \qquad (3.9)$$

where D is a closed convex subset of R^n. In most cases of interest we can assume D compact.

Concave minimization problems with a compact convex feasible set have been the subject of numerous studies in the past decades [597]. Several methods are now available for solving these problems, based on branch-and-bound, outer approximation, polyhedral annexation or combinations of these techniques. Each method has its own advantages and disadvantages. For problems with a relatively small number of variables, as many of those encountered in continuous location theory, the outer approximation (OA) method is both simple and competitive with other approaches.

To solve problem (3.9) the basic idea of OA is to build a nested sequence of polytopes outer approximating D:

$$P_1 \supset P_2 \supset \ldots \supset D$$

such that each relaxed problem

$$\min f(x) \quad \text{subject to} \quad x \in P_k \qquad (3.10)$$

can be efficiently solved and

$$\min\{f(x) : x \in P_k\} \nearrow \min\{f(x) : x \in D\}. \qquad (3.11)$$

As initial polytope P_1 we can choose any polytope containing the feasible set D, (or at least a region known to contain an optimal solution of problem (3.9)). At iteration k, having $P_k \supset D$ in hand, we solve the relaxed problem (3.10). Let v^k be an optimal solution of the latter. If $v^k \in D$ then v^k solves (3.9) since $f(v^k) \leq f(x)\ \forall x \in D \subset P_k$. Otherwise, $v^k \notin D$, and using the convexity of D we can construct a hyperplane (cutting plane) $H_k = \{x : l_k(x) = 0\}$ cutting off x^k, i.e., an affine function $\ell_k(x)$ such that

$$\ell_k(v^k) > 0, \text{ while } \ell_k(x) \leq 0\ \forall x \in D. \qquad (3.12)$$

By adding the constraint $\ell_k(x) \leq 0$ to P_k, we define the next polytope

$$P_{k+1} = \{P_k \cap \{x : \ell_k(x) \leq 0\}. \qquad (3.13)$$

By virtue of (3.12) this polytope no longer contains v^k but still contains D and so can be used to continue the outer approximation procedure.

If things are arranged so that (3.11) holds (in which case the procedure is said to be convergent) then the procedure terminates when $v^k \in D$ or v^k has become sufficiently close to D (i.e., $f(v^k)$ has become sufficiently close to the optimal value).

Two points in this scheme must be clarified: *(i)* Solution of the relaxed problems; *(ii)* Construction of the affine function $\ell_k(x)$ satisfying (3.12). We examine them in turn.

It is well known that a concave function achieves its minimum over a polytope in at least one vertex (extreme point) of this polytope. Therefore, if V_k is the vertex set of P_k then an optimal solution of (3.10) is

$$v^k \in \operatorname{argmin}\{f(v) : v \in V_k\}.$$

At the beginning one can choose P_1 with a vertex set V_1 which is known or readily computable (e.g., P_1 may be a simplex). At iteration $k+1$, since by (3.13) P_{k+1} is obtained from P_k by adding just one linear constraint, the vertex set V_{k+1} of P_{k+1} can be derived from V_k by efficient methods, e.g. by the on-line vertex enumeration algorithm of [197] (see also [351, 598]).

For a given $v^k \notin D$ there are different ways to construct an affine function $l_k(x)$ satisfying (3.12). To ensure the convergence condition (3.11), $l_k(x)$ has to be chosen appropriately.

Let ω be an arbitrary interior point of D and let u^k be the point where the line segment $[\omega; v^k]$ cuts the boundary of D. It is then easy to see that:

Proposition 1 *If $z^k \in [u^k; v^k]$ and $p^k \in \partial g(z^k)$ (subdifferential of $g(.)$ at z^k), the affine function $\ell_k(x) = \langle p^k, x - z^k \rangle + g(z^k)$ satisfies (3.12).*

We can thus state the following OA algorithm ($\varepsilon \geq 0$ being the tolerance).

a) *Initialization.* Find a point $\omega \in \operatorname{int}D$ (i.e. such that $g(\omega) < 0$), and select a polytope $P_1 \supset D$, such that the vertex set V_1 of P_1 is known or readily computable. Set $k = 1$.

b) *Solving the current relaxed problem.* Compute

$$v^k \in \text{argmin}\{f(v): \; v \in V_k\}.$$

c) *Termination criterion.* If $v^k \in D$ then terminate: v^k is an optimal solution. Otherwise, compute the point $u^k \in [\omega; v^k]$ such that $g(u^k) = 0$. Let $\beta_k = \min\{|f(v^k)|, |f(u^k)|\}$. If

$$\frac{f(u^k) - f(v^k)}{\beta_k} \le \varepsilon,$$

then terminate: u^k is an ε-optimal solution.

d) *Constructing the next polytope.* Select a point $z^k \in [u^k; v^k]$, compute $p^k \in \partial g(z^k)$ and let

$$l_k(x) = \langle p^k, x - z^k \rangle + g(z^k), \quad P_{k+1} = P_k \cap \{x : l_k(x) \le 0\}.$$

Compute the vertex set V_{k+1} of P_{k+1}. Set $k \leftarrow k+1$ and go to step b).

One can then prove that:

Proposition 2 *If the OA algorithm generates an infinite sequence $\{v^k\}$, every accumulation point of the latter is in D and solves problem (3.9).*

Thus, if $\varepsilon > 0$, the OA algorithm must terminate after finitely many iterations.

3.4.1   Weber's Problem with Attraction and Repulsion (again)

In the Weber problem with attraction and repulsion WAR, both users and residents are considered. The objective function is given by (3.3), which is d.-c.. Introducing one new variable t, the WAR problem may be written

$$\min \quad t - \sum_{j=m_1+1}^{m} w_j \|s - a^j\| + \sum_{j=m_1+1}^{m} c_j$$

subject to:

$$\sum_{j=1}^{m_1} w_j \|s - a^j\| - t \le 0$$

$$s \in \mathbb{R}^2$$

which is a concave program. Using outer-approximation, [198] solves instances of WAR problems with up to 1000 users and residents in no more than a few tens of seconds on a SUN SPARC 2. Such computing times are of the same order of magnitude as those of [798].

The model (3.3) for the WAR problem may be criticized on the ground that while linear functions for users model reasonably well transportation costs, linear functions for residents are less adequate. Indeed, perceived obnoxiousness tends to decrease rapidly with distance. A model with exponential decay of obnoxiousness is thus more realistic. Assumption (H5) then becomes:

(H5''') The facility is to be established at a point $s^\star \in \mathbb{R}^2$ that minimizes

$$\sum_{j=1}^{m_1} w_j \|s - a^j\| + \sum_{j=m_1+1}^{m} w_j e^{-\|s-a^j\|}.$$

This function is no more in d.-c. form; however rewriting it as

$$\sum_{j=1}^{m_1} w_j \|s - a^j\| + \sum_{j=m_1+1}^{m} w_j(\|s - a^j\| + e^{\|s-a^j\|}) - \sum_{j=m_1+1}^{m} w_j \|s - a^j\|$$

yields such a form and again, adding one more variable, an equivalent concave minimization problem is obtained. Problems with up to 1000 users and residents of this last form are solved also in [198]. The increase in computing time over the linear case is moderate.

3.4.2   The Multisource Weber Problem

Two important generalizations of Weber's problem involve the simultaneous location of several facilities. In the *multifacility Weber problem*, each of these facilities produces a distinct product, each user patronizes all facilities and there are flows of given magnitude between the facilities. This problem is one of convex optimization. In the *multisource Weber problem* MW all facilities produce the same product, each user patronizes the closest facility and there are no flows between facilities. The problem is then one of global optimization, i.e., it has several local optima, as already observed by [239]. Moreover, [819] proves by reduction to the satisfiability problem, that MW is NP-complete. The problem is sometimes referred to as a *location-allocation* problem in that location of the facilities and allocation of users to these facilities are done at the same time.

The assumptions (H1), (H2) and (H5) of Weber's problem are modified as follows in the MW problem:

(H1''') There are p facilities to be located at points s^1, s^2, ..., s^p of $\mathbb{R}^2$;

(H2'') There are m origin-destination points, denoted a^j, which must each be connected to one of the facilities ;

(H5^{iv}) The facilities are to be established at points $s^{1\star}$, $s^{2\star}$, ..., $s^{p\star}$ that minimize the function

$$f(s^1, s^2, \ldots, s^p) = \sum_{j=1}^{n} w_j \min_{i=1,2,\ldots,p} \|s^i - a^j\| \qquad (3.14)$$

i.e., the sum for all users of their weighted distance to the closest facility.

Many heuristics have been proposed for the MW problem; for instance [237, 238, 239] propose to choose randomly p points of $\mathbb{R}^2$, assign each customer to the closest of these points, solve the p Weber problems for the subsets of customers assigned to each point and iterate until a stable solution is reached. References to several variants of this procedure and other heuristics for the MW problem are given in [199]. Moreover, a new heuristic and a detailed computational comparison of several such methods have recently been given by [119]. Early branch-and-bound algorithms are [722, 905, 906, 907, 908]. Ostresh observes that for $p = 2$ customers patronizing each facility are separated by a straight line. Therefore the MW problem with $p = 2$ can be reduced to $O(n^2)$ pairs of Weber problems, yielding a fairly efficient algorithm. [298] proposes another algorithm for the $p = 2$ case along similar lines. [1019, 1020] extends this line of research in an algorithm which constructs all convex sets of customer locations (or in other words sets of locations such that no location of a user outside the set is within the convex hull of the locations of users' in the set) obtained by $p - 1$ successive separations. MW then reduces to a very large set partitioning problem.

The MW problem can be expressed in d.-c. form by using the useful property that the minimum of a set of p functions is equal to their sum minus the maximum of $p - 1$ of them taken in all possible ways.

Hence, (3.14) becomes

$$\sum_{j=1}^{n}\sum_{i=1}^{p} w_j \|s^i - a^j\| - \sum_{k=1}^{n} w_j \max_{k=1,2,\ldots,p} \sum_{i=1,i\neq k}^{p} \|s^i - a^j\|.$$

Again, adding one variable reduces MW to a concave minimization problem. [199] report computational results on problems with $p = 2$ and up to $10,000$ users. They improve upon previous results for $m \geq 100$. For $p \geq 3$ the concave programming approach is not as efficient as those of [906, 908, 1019, 1020]. However, this approach can be combined with column generation, as discussed below in the section on decomposition, and is then competitive.

3.4.3   The Conditional Weber Problem

It is often the case in practice that a distribution system is not built from scratch, but that one or several facilities are already established. The resulting *conditional Weber problem* is obtained (in the case of several new facilities) by further modifying the objective function of the MW problem. Thus (H5) is replaced by

(H5^v) The facilities are to be located at points $s_1^\star, s_2^\star, \ldots, s_p^\star$ in $\mathbb{R}^2$ which minimize

$$\sum_{j=1}^{m} w_j \min\{\min_{i=1,2,\ldots,p} \|s_i - a^j\|, d_j\}$$

where d_j denotes the smallest distance from a^j to an existing facility, i.e., to minimize the sum for all users of this weighted distance to the closest existing or new facility.

[203, 205] provide heuristics for the CW problem, which generalize Chen's heuristic for the MW problem. [199] solves the CW problem by concave programming: problems with one new facility up to 20 existing facilities and up to 1000 users, or with two new facilities up to 20 existing ones, and 200 users are solved in a few minutes on a SUN SPARC2.

3.4.4   Weber's Problem with Limited Distances

A variant of the (CW) problem is the Weber problem with limited distances studied by [316], in which d_j is replaced by a given value for each $j = 1, 2, \ldots, m$. In other words, the cost function for each user considered is the same as in the Weber problem up to a given distance and then remains constant. Again this problem can be solved by concave minimization and outer approximation [199]. This problem turns out to be crucial in the solution of the MW problem by column generation discussed below.

3.5   Polyhedral Annexation

Concave minimization problems with a convex feasible set can also be solved by the polyhedral annexation (PA) method [597, 1149].

Consider again problem (3.9), where $f : \mathbb{R}^n \to \mathbb{R}$ is a concave function and $D \subset \mathbb{R}^n$ is a closed convex set. Let $x^0 \in D$ be a feasible solution which is not optimal (i.e. such that $f(x) < f(x^0)$ for some feasible solution x). By translating if necessary, we may assume $x^0 = 0$. For any real number γ denote $C_\gamma = \{x \in \mathbb{R}^n : f(x) \geq \gamma\}$. By concavity of $f(x)$ this is a closed convex set, and for any $\gamma < f(0)$, we have

$$0 \in D \cap \mathrm{int} C_\gamma. \tag{3.15}$$

Let $D^\star$ and $C^\star$ be the polars of D and C_γ, respectively, where by polar of a set $E \subset \mathbb{R}^n$ is meant the set $E^\star = \{y \in \mathbb{R}^n : \langle y, x \rangle \leq 1 \ \forall x \in E\}$. From this definition, $D^\star$ and $C_\gamma^\star$ are closed convex sets and $0 \in D^\star \cap C_\gamma^\star$. Furthermore, since $0 \in \mathrm{int} C_\gamma$, the set $C_\gamma^\star$ is compact [1012]. Clearly, as γ

decreases, the set C_γ becomes larger and the set $C_\gamma^\star$ smaller. The optimal value $\bar{\gamma}$ of (3.9) is the smallest value of γ such that $D \setminus C_\gamma = \emptyset$, or, which amounts to the same,

$$\bar{\gamma} = \max\{\gamma : \ D \subset C_\gamma\}.$$

Since D is closed and convex, we have $D \subset C_\gamma$ if and only if $C_\gamma^\star \subset D^\star$ [1012]. Therefore, problem (3.9) reduces to finding

$$\gamma = \max\{\gamma : \ C_\gamma^\star \subset D^\star\}. \tag{3.16}$$

If we define

$$\mu(y) = \max\{\langle y, x \rangle : \ x \in D\} \tag{3.17}$$

then clearly $\mu(y)$ is a convex function and $C_\gamma^\star \subset D^\star$ if and only if $\mu(y) \leq 1 \ \forall y \in C_\gamma^\star$. So problem (3.16) can in turn be rewritten as

$$\max\{\gamma : \ \max_{y \in C_\gamma^\star} \mu(y) \leq 1\}. \tag{3.18}$$

One can then show that:

Proposition 3 *Let $\gamma_1 \geq \gamma_2 \geq \ldots$ be a nonincreasing sequence of real numbers and $S_1 \supset S_2 \supset \ldots$ be a decreasing sequence of polytopes in R^n such that*

$$S_k \supset C_{\gamma_k}^\star, \quad \gamma_k \in f(D) \quad \forall k$$

$$1 < \max_{y \in S_k} \mu(y) \searrow 1 \quad (k \to \infty)$$

Then the optimal value of (3.18) is $\lim_{k \to \infty} \gamma_k$.

To solve (3.18) it suffices to construct a sequence γ_k and S_k satisfying the conditions of the above Proposition. This is done in the following PA algorithm.

a) Initialization. Let $\bar{x}^1$ be the best feasible solution available, $\gamma_1 = f(\bar{x}^1)$. By translating if necessary, make sure that $0 \in D$ and $f(0) > \gamma_1$. Find a simplex $S_1 \supset C_{\gamma_1}^\star$. (For example, take an n-simplex Q_1 containing 0 in its interior and having its vertex set in the boundary of C_{γ_1}, i.e. the surface $f(x) = \gamma_1$; then $Q_1 \subset C_{\gamma_1}$, hence, $S_1 := Q_1^\star \supset C_{\gamma_1}^\star$). Set $V_1 =$ vertex set of S_1, $k = 1$.

Iteration k One has a feasible solution $\bar{x}^k$ (current best feasible solution) with $f(\bar{x}^k) = \gamma_k$, and a polytope $S_k \supset C_{\gamma_k}^\star$ with vertex set V_k.

b) Checking $S_k \subset D^\star$. Compute $y^k \in \mathrm{argmax}\{\mu(y) : \ y \in V_k\}$. If $\mu(y^k) \leq 1$, i.e. $\max_{y \in S_k} \mu(y) \leq 1$, then terminate: by Proposition 3 γ_k is the optimal value of (3.16) and $\bar{x}^k$ is an optimal solution.

c) *Updating* γ_k. Let $x^k \in \text{argmax}\{\langle y^k, x \rangle : x \in D\}$.

If $f(x^k) \geq \gamma_k$, then let $\gamma_{k+1} = \gamma_k$, $\bar{x}^{k+1} = \bar{x}^k$.

If $f(x^k) < \gamma_k$, then let $\gamma_{k+1} = f(x^k)$, $\bar{x}^{k+1} = x^k$.

d) *Constructing* S_{k+1}. Compute $\lambda_k = \max\{\lambda \geq 1 : f(\lambda x^k) \geq \gamma_{k+1}\}$ and define

$$S_{k+1} = S_k \cap \{y : \langle x^k, y \rangle \leq \frac{1}{\lambda_k}\}.$$

Compute the vertex set V_{k+1} of S_{k+1}. Set $k \leftarrow k + 1$ and go to step b).

Using Proposition 3, one can prove

Proposition 4 *The* PA *Algorithm is convergent, i.e., if it generates an infinite sequence* $\{\bar{x}^k\}$ *then every accumulation point of this sequence solves problem (3.9).*

The algorithm can be interpreted as an outer approximation procedure for solving problem (3.16) which, in a sense, is dual to problem (3.9).

If $Q_k = S_k^*$ then $Q_1 \subset Q_2 \subset \ldots \subset C_{\bar{\gamma}}$, and it can easily be proved that

$$Q_{k+1} = S_{k+1}^* = \text{conv}(Q_k \cup \{\lambda_k x^k\}),$$

i.e. Q_{k+1} is obtained by "annexing" $\lambda_k x^k$ to Q_k. Thus the algorithm generates a sequence of expanding polytopes $Q_1 \subset Q_2 \subset \ldots$ such that

$$\gamma_k = \min\{f(x) : x \in Q_k \cap D\} \searrow \min\{f(x) : x \in D\}.$$

This explains the name polyhedral annexation given to the method.

In contrast to the OA method, the PA algorithm does not require the computation of subgradients. In several cases the dimension of C_γ^* is much smaller than n, so the PA method may work in a space of smaller dimension than the original one. However, it involves solving a sequence of subproblems of the form (3.17) which are convex programs if D is convex and nonpolyhedral.

3.5.1   Weber's Problem with Attraction and Block Norms

In some location problems [616] the objective is to maximize the total attraction of the facility to the users rather than minimizing the total cost for the users. Most often the attraction of the facility at $s \in \mathbb{R}^2$ to an user at $a^j \in \mathbb{R}^2$ can be expressed by a function of the form

$$q_j[h_j(s - a^j)],$$

where $h_j : \mathbb{R}^2 \rightarrow \mathbb{R}_+$ is a convex function (e.g. a distance function, or more generally, a gauge, which may be symmetric polyhedral, i.e., a block norm) such that $\lim_{|s| \to \infty} h_j(s) = +\infty$, and $q_j : R_+ \to R_+$ is a convex strictly decreasing function such that $\lim_{\theta \to \infty} q_j(\theta) = 0$ (the attraction decreases and tends to zero as the distance increases to infinity). The problem is then to find the location s that maximizes the total attraction, i.e.,

$$\text{maximize } \varphi(s) := \sum_{j=1}^m q_j[h_j(s - a^j)] \text{ s.t. } s \in \mathbb{R}^2. \qquad (3.19)$$

From the assumptions it easily follows that a global optimal solution always exists.

In [616] a finite descent method is proposed to find a locally optimal solution. However, since the function $\varphi(s)$ is not concave, such a solution need not be a globally optimal one. By rewriting the problem as

$$\max\{\sum_{j=1}^m q_j(t_j) : \ h_j(s - a^j) \le t_j, \ j = 1, \ldots, m\},$$

and setting

$$f(t) = -\sum_{j=1}^m q_j(t_j),$$

$$D = \{t = (t_1, \ldots, t_m) \in \mathbb{R}_+^m : \ (\exists s \in \mathbb{R}^2) \ h_j(s - a^j) \le t_j, \ j = 1, \ldots, m\}$$

we see that this is a problem of the form

$$\min\{f(t) : \ t \in D\}, \qquad (3.20)$$

where $f(t)$ is a concave function and $D \subset \mathbb{R}^m$ is a closed convex set. In [1150] problem (3.19) is solved by applying the PA method to problem (3.20).

Let $\bar{s}^1$ be the best feasible solution initially available of problem (3.19), $\bar{t}^1 = (h_1(\bar{s}^1 - a^1), \ldots, h_m(\bar{s}^1 - a^m))$ and let $t^0 \in D$ be a feasible point of (3.20) such that $f(t_0) > f(\bar{t}^1)$. Denote

$$\tilde{f}(t) := f(t + t^0), \quad \tilde{D} = D - t^0.$$

To apply the PA algorithm we first translate the origin of $\mathbb{R}^m$ to t^0, which transforms the problem into

$$\min\{\tilde{f}(t) : \ t \in \tilde{D}\}.$$

With $\gamma_1 = \tilde{f}(\bar{t}^1)$, $C_{\gamma_1} = \{t : \tilde{f}(t) \ge \gamma_1\}$, the condition $0 \in \tilde{D} \cap \text{int} C_{\gamma_1}$ needed for initialization of the PA algorithm is fulfilled.

Note that, since $q_j(.)$ is a decreasing function, $f(t) \ge f(t')$ whenever $t \ge t'$. Therefore, $\tilde{f}(t) \ge \tilde{f}(0) = \gamma_1$ for any $t \ge 0$, i.e., $\mathbb{R}_+^m \subset C_{\gamma_1}$, and hence $C_{\gamma_1}^* \subset \mathbb{R}_-^m$, so that the PA procedure can be started with a simplex

S_1 satisfying $C^\star_{\gamma_1} \subset S_1 \subset \mathbb{R}^m_-$. Such a simplex S_1 can be constructed, e.g., as follows. Denote $e = (1, \ldots, 1) \in \mathbb{R}^m$. Since $0 \in \text{int} C_{\gamma_1}$, for sufficiently small $\alpha > 0$ we have $-\alpha e \in C_{\gamma_1}$, so if we set

$$S_1 = \{t \in R^m_- : \ -\sum_{j=1}^m t_j \leq \frac{1}{\alpha}\}$$

then for every $t \in C^\star_{\gamma_1} \subset R^m_-$ the relation $-\alpha e \in C_{\gamma_1}$ implies that $-\langle \alpha e, t \rangle = -\alpha \sum_{j=1}^m t_j \leq 1$, hence $C^\star_{\gamma_1} \subset S_1$.

With this choice of S_1, every simplex S_k will be contained in $\mathbb{R}^m_-$. At iteration k one checks the inequality

$$\max\{\mu(y) : \ y \in V_k\} \leq 1,$$

where $\mu(y) = \max\{\langle y, t \rangle : \ t \in \tilde{D}\}$ and $V_k \subset \mathbb{R}^m_+$ is the vertex set of S_k. But, since $\tilde{D} = \{t \in \mathbb{R}^m : \ (\exists s \in R^2) \ h_j(s - a^j) \leq t_j + t^0_j, \ j = 1, \ldots, m\}$, it is easily seen that

$$\mu(y) = \langle y, t^0 \rangle + \nu(y),$$

where

$$\nu(y) = \max\{\sum_{j=1}^m y_j h_j(s - a^j) : \ s \in \mathbb{R}^2\}.$$

If $\langle y, t^0 \rangle + \nu(y) \leq 1$ for all $y \in V_k$ then γ_k is the optimal value. Otherwise, a cutting plane is generated which defines the next polytope S_{k+1} and the procedure continues after updating the incumbent.

Thus, the PA algorithm reduces the problem to solving a sequence of subproblems:

$$\max\{\sum_{j=1}^m y_j h_j(s - a^j) : \ s \in \mathbb{R}^2\},$$

where $y \in \mathbb{R}^m_-$. Setting $z = -y$ this subproblem is equivalent to

$$(R(z)) \qquad \min\{\sum_{j=1}^m z_j h_j(s - a^j) : \ s \in \mathbb{R}^2\}$$

which is a Weber problem depending on the parameter $z \in \mathbb{R}^m+$. Viewing $z_j \geq 0$ as a weight assigned to user j, the method can be interpreted as solving a sequence of Weber problems $R(z)$, where the weights $z_1, \ldots, z_m$ assigned to users are gradually adjusted till global optimality.

When every $h_j(.), j = 1, \ldots, m$ is a polyhedral norm, then $h_j(s - a^j)$ is affine over each of the cones vertexed at a^j and generated by a facet of the unit ball associated with this norm. Therefore, the function $\sum_{j=1}^m z_j h_j(s - a^j)$, for given $z \in \mathbb{R}^m_+$ is affine over each polyhedron of the form $\bigcap_{j=1}^m M_j$, where

M_j is a cone of the type just described in which $h_j(s - a^j)$ is affine. That is, the objective function of the subproblem $R(z)$ is a convex piecewise affine function, with the affine pieces having domains give by the elementary polyhedra as discussed in [616]. It follows that an optimal solution of a subproblem $R(z)$, and hence a global optimal solution of the location problem under consideration, is achieved at a vertex of an elementary polyhedron. This is the *intersection point* property established in [616] for this problem.

3.6 Decomposition Methods

Many global optimization problems have some structure which may be exploited in their solution, thus making it possible to solve large instances. This happens in particular when the set of variables can be partitioned into two sets of variables, i.e., x and y and the structure becomes apparent when variables of one of them are fixed. Let x correspond to the *complicating variables*, i.e., the variables such that, when they are fixed, the problem

$$\min f(x, y) \tag{3.21}$$

subject to:

$$g_i(x, y) \leq 0 \qquad i = 1, 2, \ldots, m \tag{3.22}$$

can be efficiently solved in the remaining y variables. Then, the problem (3.21)-(3.22) can be viewed as a global optimization one in the variables x only where the values of $f(x, y)$ for fixed x are given by an oracle. In other words, the problem may be projected on the space of the x variables.

When we say that the problem (3.21)-(3.22) is easy to solve, we mean that it may be reduced, e.g., to a linear or convex minimization problem or to a simple plant location problem, which is NP-hard but easily solvable in practice. The projected problem may be solved by branch-and-bound or by outer approximation [410, 449].

3.6.1 Profit Maximizing Weber Problems and Price Policies

Decomposition has been applied to the study of plant location for profit maximization under various price policies. Assuming a finite set of potential locations for facilities, discriminatory pricing [555] leads to a reduction to the simple plant location problem; uniform delivered pricing [558] and uniform mill pricing [536] lead to models with a single complicating variable (the price) and zone pricing [552] to a model with several (the prices in each zone). Similar problems can also be studied in the case of location of a single facility in the plane [553]. A more detailed presentation of the models and the corresponding algorithms is given in Chapter 15 in this volume.

The technique of column generation has been introduced in [476] to solve large-scale linear programs, leading to approximate solution of difficult combinatorial problems such as the cutting-stock problem. This technique applies to location-allocation problems, where the columns correspond to feasible sets of users patronizing the same facility. Combining column generation with branch-and-bound leads to optimal algorithms. This was first done in location in [886], in a study of the capacitated plant location problem with single source constraints. [541] applies this approach to the multisource Weber problem, discussed above. The column generation subproblem to be solved at each iteration turns out to be a Weber problem with limited distances, [316] which, as mentioned above, can be expressed as a concave minimization problem and solved by outer approximation and vertex enumeration. Problems with 100 points and 10 facilities are solved exactly for the first time.

3.7 Linearization Methods

As linear, convex and mixed integer programming are powerful tools, it is not surprising that linearization techniques have been long used in global optimization. They apply most naturally to the case of separable programs, in which both the objective function and the constraints can be written as sums of univariate functions [395, 1091]. Then such functions can be partitioned into convex and concave parts and a convex relaxation obtained by linear outer approximation of concave parts. It can be refined by branching on such parts, i.e., choosing an interior point and using two more precise linear outer approximations. Indeed, such an approach was one of the first ones used to solve the uncapacitated simple plant location problem UFLP [365] but other techniques such as dual ascent methods [390] proved to be more efficient in practice.

The linearization method is however basic to the algorithm of [1092] for facility location with concave costs, i.e., a generalization of the UFLP in which set-up and operating costs for facilities are concave.

In some cases where the program considered is not separable, it can be reduced to that form by a suitable transformation. This holds for quadratic programs with indefinite objective functions and linear constraints. Finding the eigenvalues of the objective function matrix and performing an adequate rotation gives a separable program to which the above mentioned technique applies [916].

In the general case, nonlinear terms are under or over estimated by linear functions. For instance, it can be shown [16, 17] that the convex envelope of $x_i x_j$ with $\ell_i \leq x_i \leq u_i$ and $\ell_j \leq x_j \leq u_j$ is $\max\{\ell_i x_j + \ell_j x_i - \ell_i \ell_j, u_i x_j + u_j x_i - u_i u_j\}$ and the corresponding concave envelope $\min\{\ell_i x_j + u_j x_i - \ell_i u_j, u_i x_j + \ell_j x_i - \ell_j u_i\}$. Thus $x_i x_j$ can be approximated by adding a new variable $t = x_i x_j$ and using the above relations to bound it.

Moreover, the powerful *linearization-relaxation technique* [1066] can be brought to bear. Assuming variables x_j to be bounded by $\ell_j \leq x_j \leq u_j$ one can consider products of factors $(x_j - \ell_j) \geq 0$ and $(u_j - x_j) \geq 0$ or of such factors with (linear) constraints. This gives new implied constraints, which can then be linearized by introducing new variables and combined to obtain tight constraints in low-dimensional space. This technique is used by [1070, 1071] for polynomial programming problems and for the multisource Weber problem with minimization of the sum of squared distances to the closest facility as objective.

3.7.1 The Weber Multisource Problem with Rectilinear Distances and Capacities

[1068] considers a version of the multisource Weber problem where production (or service) at the facilities is bounded. Moreover, it assumes the transportation to take place parallel to the axis, i.e., following the Manhattan norm. Hypothesis (H3) and (H5) becomes

(H1^{iv}) The distance considered between the facility at $s = (s_1, s_2)$ and the point $a^j = (a_1^j, a_2^j)$ is the rectilinear one, noted

$$|s - a^j| = |s_1^j - a_1^j| + |s_2^j - a_2^j|.$$

(H5^{vi}) The facilities are to be established at points $s_1^\star, s_2^\star, \ldots, s_p^\star$ of $\mathbb{R}$ that minimize the total transportation cost

$$f(s_1^\star, s_2^\star, \ldots, s_p^\star) = \min_{\substack{s_1, s_2, \ldots, s_p \\ w_{ij}, \, i = 1, 2, \ldots, p \\ j = 1, 2, \ldots, m}} \sum_{i=1}^{m} \sum_{j=1}^{m} w_{ij} |s_i - a^j|$$

where

$$\sum_{j=1}^{m} w_{ij} = c_i \qquad i = 1, 2, \ldots, p$$

$$\sum_{i=1}^{p} w_{ij} = w_j \qquad j = 1, 2, \ldots, m$$

$$w_{ij} \geq 0 \quad i = 1, 2, \ldots, p; j = 1, 2, \ldots, n$$

the c_i being the capacities of the facilities.

One way to transform the problem so-defined to a more tractable form [1154] is to replace each quantity within an absolute value sign by the difference of two non-negative variables. The absolute value itself is then the sum of these variables and the problem one of *bilinear programming*. [1069] uses this formulation to develop a polar cutting plane algorithm and [1073] further strengthens this procedure through deeper negative edge extension polar cuts.

[1068] considers a different reduction to a bilinear program, based upon [1187]: optimal locations can always be found at intersection points of vertical and horizontal lines going through the points a^j. Then introducing variables z_{ik} equal to 1 if facility i is located at intersection point k and 0 otherwise yield a bilinear program where one set of variables are 0-1 and the other continuous. A branch-and-bound algorithm using the linearization-relaxation technique leads to solve problem with up to 5 facilities and 20 points.

3.8   Specialized Methods

As mentioned above, it is frequent in practice that location or travel is forbidden in some regions. Constraints on location are taken into account in branch-and-bound methods such as BSSS, discussed above. However, this method does not exploit possible convexity of the objective function.

3.8.1   Weber's Problem with Constraints on Location

Hansen, Peeters and Thisse [550] consider a constrained Weber problem (CWP) obtained by replacing assumption (H1) by (H1′), i.e., by assuming G to be equal to the union of a finite set of convex polygons.

This algorithm is based on [607]. It consists in a search for the unconstrained solution followed by an exploration of some of the boundary parts of some of the polygons defining the feasible region. Let s_1 be an infeasible point; then a feasible point s_2 is *visible* from s_1 if the line segment $]s_1, s_2[$ contains no feasible point. (A slightly different concept will be used in the algorithm: the point s_2 of P_j is j-visible from the point s_1 if the line segment $]s_1, s_2[$ contains no point of P_j). Then it is easy to show that:

Proposition 5 *The optimal solution $s^\star$ to the* CWP *either coincides with the optimal solution $\tilde{s}^\star$ of the corresponding Weber problem or is visible from $\tilde{s}^\star$.*

Moreover, another property will be useful in seeking the best solution on the boundary of a polygon:

Proposition 6 *The objective function of* CWP *restricted to the visible (or j-visible) boundary of a polygon is quasi-convex.*

The rules of the algorithm are the following:

a) Unconstrained solution. Determine the solution $\tilde{s}^\star$ of the unconstrained Weber problem associated with the (CWP). Check if this solution is feasible. If so, stop, as it is optimal. Otherwise, choose an arbitrary feasible point as incumbent.

b) *Ranking polygons.* Rank the polygons by increasing distance to $\tilde{s}^\star$.

c) *Exploration of j-visible boundary.* Select the unexplored polygon closest to $\tilde{s}^\star$. If none remain, stop, the incumbent solution being optimal. Find by Fibonacci search the best extreme point e_j of this polygon. Check if e_j is identical to the best point $s_j^\star$ on the visible boundary using directional derivatives, and if not determine that best point by Newton-Raphson search. Update the incumbent if necessary.

d) *Elimination of dominated polygons.* If $s_j^\star$ is at a corner point of P_j eliminate all polygons in the cone pointed at $s_j^\star$ and containing P_j; if $s_j^\star$ is on a side of P_j eliminate all polygons in the half plane defined by extending that side which does not contain $\overline{s}^\star$. Return to Step c).

Using this algorithm, problems with up to 200 users and 100 feasible polygons were solved in a few seconds on a IBM 370/158.

[29] address a similar constrained Weber problem, in which the infeasible region is the union of the interiors of a finite set of (possibly nonconvex) polygons. Then a rectangle containing an optimal solution is obtained by taking the minimum and the maximum coordinates of all demand points and corners of polygons. The feasible part of this rectangle has a piecewise-linear (but not connected) boundary. It is therefore easy to decompose this region into a set of convex polygons. The problem thus reduces to the previous one.

[528] consider similar constrained Weber problems for which the objective function is such that the unconstrained problem may be solved in polynomial time: minimize the sum of rectilinear distances, squared Euclidean distances or Chebychev distances. After finding the unconstrained solution one also checks for its feasibility and if it is not feasible explores parts of the visible boundary of feasible polygons.

3.8.2 Weber's Problem with Barriers to Travel

[29] consider also the case where both travel and location are forbidden in some regions. [679] previously proposed a method based on the calculus of variations for the case of one forbidden disk. Of course, travel being prohibited implies location is too but the converse is not necessarily true. In Weber's problem in the presence of forbidden regions and/or barriers to travel the assumptions (H1) and (H3) are replaced by

(H1^v) A single facility is to be set up at a point p of $\mathbb{R}^2 \setminus S'$, where S' is a set of open polygons.

(H3''') The distance considered between the facility at p and point a^j is
the Euclidean length of the shortest path joining these points and
crossing none of the polygons of S'.

Due to this last assumption an auxiliary problem must be solved: find the
shortest path between two points of $\mathbb{R}^2$ with some regions forbidden for
travel (or, in other words with barriers). The shortest path problem with
barriers [547] reduces to a shortest path problem in a graph, with origin,
destination and corner points of polygons as vertices and edges joining pairs
of points among these which are mutually visible, i.e., which can be joined
by a straight line crossing no forbidden region. The edges are weighted by
the Euclidean distance between their endpoints.

[29] use a subroutine for shortest paths with barriers and a variable origin
(or facility location) within a simulated annealing approach to Weber's
problem with forbidden regions. Recall that in simulated annealing [697,
285] an initial (feasible) solution is drawn at random and a move to a
point in its neighborhood is done also at random; if the resulting point
improves the value of the objective function it is kept unconditionally, i.e.,
it becomes the new current solution; if it deteriorates this value it is kept
with a probability decreasing in the amount of deterioration involved and
in the number of iterations already performed. This procedure is iterated
until convergence occurs. Under adequate conditions the algorithm may be
shown to converge to an optimal solution with probability one. However,
the time before this happens may be enormous, and on many types of
problems, simulated annealing tends to be outperformed in practice by
other heuristics such as tabu search. It is therefore useful to keep track of
the best solution currently found, as proposed by [29]. The results they give
for a few test problems appear to be good, although the optimal solutions
are unknown (or unproved). Such optimal solutions could be obtained by
an adaptation of the BSSS algorithm taking into account the various paths
from a destination point to a square containing the location of the facility.

3.9   Conclusions

Most models of continuous location theory are expressed as convex mathematical programs. Their numerical solution is then fairly easy and large
instances can be tackled. However, the necessary assumptions are quite
restrictive. There are several reasons why nonconvex mathematical programs are needed for realistic modelling. First, locational constraints are
ubiquitous in practice: large areas are usually already occupied or unfit for
location. Such restrictions are most often ignored, possibly under the implicit assumption that they can be taken into account *ex post* in an *ad hoc*
way. Second, restrictions to travel, due to natural obstacles and again to
occupancy of the soil are also frequent. While network location models ad-

dress such constraints, much of continuous location theory does not. Third, concave objective functions are needed in many situations e.g., to express economies of scale in transportation or exponential decrease in obnoxiousness of polluting facilities. Fourth, when several facilities are simultaneously located and users patronize the closest one, market areas must be considered. Discrete location theory accommodates for this by restricting possible locations for facilities to a finite set of points. Such restrictions may lead to neglect many possibilities.

Techniques from global optimization are numerous and diverse. They have been extensively studied in the last two decades and applied to location since about ten years. This has led to substantial progress in the solution of long standing problems (e.g., Weber's problem on the sphere, the multisource Weber problem) and, to the formulation and efficient solution of many new ones as (e.g., Weber's problem with attraction and gauges, Weber's problem with attraction and repulsion, Weber's problem with limited distances).

Branch-and-bound methods, such as BSSS and its extensions, appear to be well-suited to the solution of single facility location problems. Indeed, this approach is highly versatile as illustrated above. Its efficiency depends largely on obtaining precise bounds, for which specific procedures must be devised. When such procedures are used branch-and-bound appears to be as least as efficient as other global optimization approaches to single facility location. This does no more seem to be true for simultaneous location of several facilities in continuous space.

Due to the richness in properties of d.-c. functions many location problems can be expressed as d.-c. programs and a notable proportion of them as concave programming problems. The latter ones can be solved by outer approximation and vertex enumeration, provided the number of variables, equal to twice the number of facilities plus one, is fairly small, say not larger than 7. Other techniques to solve concave programs [597] do not appear to have been yet applied to location and a comparison between solution methods seems worthy of further study.

Very large problems can be solved through decomposition and global optimization if there are a small number of complicating variables (the fixation of which leads to a problem which can be easily solved in practice, possibly even if it is NP-hard). Moreover, the combination of column generation, branch-and-bound and global optimization, which has recently been applied to the notoriously difficult (and NP-hard) multisource Weber problem, appears to be very promising.

Thus, despite the limitations on the size of problems amenable to exact solution, global optimization methods play an important role in continuous location theory, both to formulate more realistic models and to solve them.

4
INFERRED IDEAL WEIGHTS FOR MULTIPLE FACILITIES

Morton E. O'Kelly
The Ohio State University

4.1 Introduction

In the location theory literature, it is common to seek an optimal location for one or several new facilities, given a set of assumptions about the distribution of demand, and the nature of costs of overcoming spatial separation. There exists a large set of results concerning the optimal location of facilities, and there are systematic reviews of this literature [127, 470, 786].

An interesting variant on the traditional approach was proposed by Eastin [356]. He took the location of a single facility as given, and the locations of demand points served by the facility are also fixed. However, the *weights* served at these fixed demand points are unknown. Then, under the assumption that the facility represents a Weberian point of least cost location, [356] devised methods to *deduce* the weights of the fixed demand points. The weights must be chosen so that the fixed facility is an "optimal" location. In the specific case of Eastin's analysis, the facility satisfies the optimality conditions for the euclidean two dimensional Weber least cost location problem. In this case there are two such conditions, which together with a normalization constraint, result in a system of three equations. If there are more than three unknown weights, the equation system is underdetermined, and a formalism such as "entropy maximization" is needed to fill in the missing information. Eastin's [356] idea is an excellent candidate for the application of information theoretic and entropy methods, in that there may be a genuine uncertainty about the sizes of the demands that support a fixed facility, and so it is necessary to deduce the missing information in a manner that is consistent with the facts. In deducing the weights however, the modeler need not make any biased assumptions about the missing data. As [1180] has shown, Jaynes' entropy maximizing procedure creates estimated probability distributions under exactly the kinds of conditions outlined here.

In this chapter, Eastin's model is repeated in the more general form of a Kullback information minimizing problem [1180]. This allows prior additional data about the size of the fixed demand to be used, in conjunction with the first order conditions of the Weber problem, to arrive at the least biased estimate of the unknown demand weights. The bulk of the chapter then deals with various extensions and applications of Eastin's concept in the context of (a) multifacility, (b) location-allocation, (c) spatial interaction, and (d) hub and spoke variants of the basic case. The overall goal of the chapter is to use the Eastin analysis as a vehicle to re-visit and review many aspects of the conventional location analysts suite of tools, in a way that emphasizes the demand distribution needed to support an optimal facility. The chapter thus complements others in this volume by focussing attention on the nature of the demand distribution which supports the optimality of the location of a facility.

4.1.1   Notation

Single Facility Case

$Q = (X, Y)$ is the location of a facility which serves the n demand points [it is an optimal location by hypothesis]

$p_i = (x_i, y_i)$ is the location of the fixed demand point i.

w_i = the unknown demand for "service" at fixed location i.

q_i = a prior estimate of w_i

Θ, ϕ are parameters chosen to satisfy non-linear equations

Multiple Facility Case

w_{ij} = the demand for facility j from i [W_{ij} if demand is exogenous]

$d_{ij} = \sqrt{(X_j - x_i)^2 + e + (Y_j - x_i)^2 + e}$ the distance from p_i to Q_j

$c_{jk} = \sqrt{(X_j - X_k)^2 + e + (Y_j - X_k)^2 + e}$ the distance from Q_j to Q_k

v_{jk} = the demand for interaction between facilities j and k [V_{jk} if demand is exogenous]

4.2   Information Minimizing Model

To motivate and provide a background for the main idea of this chapter, Eastin's [356] technique is restated, in a generalized fashion, to include prior information about the size of the unknown demands. [In the original paper, there was no allowance made for the prior estimate of the demand for the service existing at node i.] Let $Q = (X, Y)$ be the location of a fixed single "optimal" facility. Let $p_i = (x_i, y_i), i = 1, \ldots, n$ be the sites of demand for the services of the facility. The demand at p_i is defined as

w_i and is assumed to be unknown, or unobservable. There exists a prior estimate of the demand at i, and this is defined as q_i. Let d_i be the straight line distance between p_i and Q. $d_i = \sqrt{((X - x_i)^2 + e) + ((Y - y_i)^2 + e)}$ which is the euclidean distance from the single facility location Q to the fixed demand location p_i, and the small constant e is introduced if necessary to ensure continuity in the $\partial(d_i)/\partial X$ derivatives. [The small constant e prevents the denominator of the following expressions from going to zero even if Q coincides with any p_i.] Then, $\partial(d_i)/\partial X = (X - x_i)/d_i$ and $\partial(d_i)/\partial Y = (Y - y_i)/d_i$. Define $x'_i = (X - x_i)/d_i$ and $y'_i = (Y - y_i)/d_i$. By hypothesis, Q is the solution to the problem MIN $Z = \sum_i w_i d_i$. Therefore, Q satisfies the following conditions: $\partial Z/\partial X = 0, \partial Z/\partial Y = 0$. With the addition of the normalizing condition that the weights sum to one, the system of equations to be satisfied by the unknown demand weights is:

$$\sum_i w_i = 1 \tag{4.1}$$

$$\sum_i w_i \frac{X - x_i}{d_i} = 0 \tag{4.2}$$

$$\sum_i w_i \frac{Y - y_i}{d_i} = 0. \tag{4.3}$$

In the usual location model, Q is unknown, and therefore equations (4.2) and (4.3) have to be solved by an iterative scheme, as discussed in Love, Morris and Wesolowsky [786, pp.14-15]. In essence (X, Y) cannot be isolated in equations (4.2) and (4.3) because of the dependence of d_i on the location of the facility. Now, however, when the facility is fixed, and hypothesized to be optimally located, the need for iteration disappears, and the major remaining degrees of freedom lie in the unknown weights. [Incidentally, the same kind of question could be posed as one of determining the positions of the fixed demand points, given their weight.] To illustrate and to fix ideas, consider the example in Figure 1 which will be used repeatedly in the sections to follow. There are 4 demand points, at $a = (0,0); b = (0,10); c = (10,0);$ and $d = (10,10)$, and these points have "unknown" weights, and there is a supposedly optimal facility at coordinates (8,3). Now, since there are four unknowns $[w_a, w_b, w_c, w_d]$ and just three equations (4.1)-(4.3) above, the system is underdetermined. Thus, one simple strategy would be to add a constraint (or equation) stating that $w_a = K$, some other constant, and then to explore the solutions to the equations as K is varied parametrically. The set of four equations to be satisfied in the case of this numerical example is then included in the following linear program.

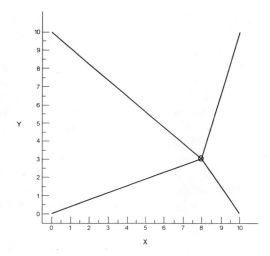

FIGURE 1. Optimal Facility at (8,3), with Four Unknown Demands

	x	y	d	x'	y'
a	0	0	8.5440	0.9363	0.3511
b	0	10	10.6301	0.7526	−0.6585
c	10	0	3.6056	−0.5547	0.8321
d	10	10	7.2801	−0.2747	−0.9615

The information in Figure 1, recalling that $(X, Y) = (8, 3)$, gives a formulation of the problem as:

$$Max \ \{ \ w_b + w_c + w_d \ \}$$

Subject to

$$w_a + w_b + w_c + w_d = 1$$
$$w_a = K$$
$$0.9363w_a + 0.7526w_b - 0.5547w_c - 0.2747w_d = 0$$
$$0.3511w_a - 0.6585w_b + 0.8321w_c - 0.9615w_d = 0$$

Solutions to this problem for various K are given in Table 4.1: the problem is infeasible for $K = 0.4$. The technique used in the preceding example, resolves the indeterminacy in the equations by making the unjustified assumption that one of the weights is fixed to a prespecified value. Apart from the obvious bias inherent in such an approach, it is flawed for two other reasons. One is that for a general problem of n weights, $n - 3$ new equations would have to be introduced. Second, even with an equal number of equations and unknowns, there is the possibility that the system of

TABLE 4.1. Solutions for Various Values of K

	$K = 0.1$	$K = 0.2$	$K = 0.3$
w_a	0.1	0.2	0.3
w_b	0.26355	0.13179	0.0
w_c	0.41837	0.36744	0.31652
w_d	0.21808	0.30077	0.38345

equations has no solution, as in the case of the infeasible $w_a = 0.4$ above. The general issue then, is to determine in an unbiased way, the most likely pattern of the unknown weights.

While there exist many possible strategies for inferring the missing weights, Eastin [356] uses an entropy maximizing model which is closely related to the following information theoretic approach. The essence of this approach is that the unknown weights are derived in a manner which is consistent with the known information or data (e.g. equations (4.1-4.3) must be satisfied), but any other influence on the data distribution must be encoded in the form of prior expectations, or in the form of constraints which must be met by the inferred weights. In this scheme, the weights are to be chosen to satisfy the convex information minimizing problem:

$$\text{MIN } I = \sum_i w_i \ln(\frac{w_i}{q_i}) \text{ subject to (4.1), (4.2) and (4.3).}$$

Forming the Lagrangean

$$\mathcal{L} = I + \Theta(\sum_i w_i \frac{X - x_i}{d_i} - 0) + \phi(\sum_i w_i \frac{Y - y_i}{d_i} - 0) + \mu(\sum_i w_i - 1) \quad (4.4)$$

and setting $\partial\mathcal{L}/\partial w_i = 0$ we see the solution to this problem occurs when:

$$\frac{\partial\mathcal{L}}{\partial w_i} = 1 + \ln\frac{w_i}{q_i} + \Theta x_i' + \phi y_i' + \mu = 0 \quad (4.5)$$

then

$$w_i = q_i e^{-1-\mu-\Theta x_i'-\phi y_i'}, \text{ and since } \sum_i w_i = 1, \quad (4.6)$$

$$\sum_i q_i e^{-1-\mu-\Theta x_i'-\phi y_i'} = 1, \quad (4.7)$$

$$\text{implying } e^{-1-\mu} = \frac{1}{\sum_i q_i e^{-\Theta x_i'-\phi y_i'}}, \quad (4.8)$$

substituting, we eliminate the multiplier μ to give:

$$w_i = \frac{q_i e^{-\Theta x_i'-\phi y_i'}}{\sum_k q_k e^{-\Theta x_k'-\phi y_k'}} \quad (4.9)$$

Note that Θ and ϕ are chosen to satisfy (4.2) and (4.3) respectively. In general, this pair of unknown parameters has to be found by solving a system of two nonlinear equations. (If an arbitrary pair of parameters is inserted in the w_i equation it would be fortuitous if the equations (4.2) and (4.3) are met exactly, so a numerical adjustment such as Newton- Raphson search is used to refine Θ and ϕ until (4.2) and (4.3) are met within some prespecified tolerance.)

This is the basic result in Eastin's [356] analysis, and there are many useful consequences of his idea, even in the context of a single facility location. Some of these consequences are spelled out here, before moving to the main contribution of this chapter, which is to extend the method to various multiple facility problems. A property of this kind of model which is exploited in sensitivity analysis, is that the objective function is related to the value of the multipliers that solve (4.1), (4.2), and (4.3), in the following way: substitute from (4.6) into the objective function, and for clarity denote the solution values of the parameters as μ^*, Θ^*, and ϕ^* respectively:

$$I = \sum_i w_i \ln \frac{w_i}{q_i} = \sum_i w_i(-1 - \mu^* - \Theta^* x_i' - \phi^* y_i')$$

Result: $I = -1 - \mu^*$.

To see this, note that $I = (-1 - \mu^*) \sum_i w_i - \Theta^* \sum_i w_i x_i' - \phi^* \sum_i w_i y_i'$ and the result follows because $\sum_i w_i x_i' = 0, \sum_i w_i y_i' = 0$, and $\sum_i w_i = 1$ at an optimally chosen set of parameters. Note that the objective function is equal to the log of the "balancing factor" (i.e. that term which guarantees that the weights sum to one), because from (4.8)

$$-1 - \mu^* = \ln \frac{1}{\sum_i q_i e^{-\Theta^* x_i' - \phi* y_i'}},$$

which provides a useful cross-check on the calculations of the parameters. Under what circumstances will w_i reveal the "correct" weights, for a particular instance of an optimally located Q?

Result: If the location is fixed at $Q = (X^*, Y^*)$, then the information minimizing choice of $w_i = q_i$ if and only if q_i solves the set of equations (4.10)-(4.11):

$$\sum_i q_i \frac{X^* - x_i}{d_i} = 0 \tag{4.10}$$

$$\sum_i q_i \frac{Y^* - y_i}{d_i} = 0. \tag{4.11}$$

Proof: If q_i solves (4.1)-(4.3), $w_i = q_i$ is a feasible solution also, because the parameters Θ and ϕ may be set arbitrarily (e.g. to zero). The objective function I achieves its minimum value when $w_i = q_i$, so no better solution can be found. The "only if" part comes from the observation that if q_i does not solve the equations (4.1) − (4.3), the parameters Θ and ϕ are not both zero, and so $w_i \neq q_i$.□

An alternative way to view this result, is to ask under what circumstances we would expect the prior weights q_i to be an accurate predictor of the final inferred weights w_i. The answer is that w_i is the same as q_i if (X^*, Y^*) is a minimizer of the Weberian problem MIN $Z = \sum_i q_i d_i$. In other words, weights are set at q_i such that the coordinates (X^*, Y^*) yield the minimum cost facility. If the inferred weight model is run with a pair of data vectors q_i and Q which do not satisfy the equations (4.2) and (4.3), then w_i cannot equal q_i.

As a corollary, at the solution when $w_i = q_i$, the parameters of the problem are especially easy to calculate: $\mu^* = -1, \Theta^* = 0$, and $\phi^* = 0$.

What are the implications for the inferred ideal weights of making various erroneous assumptions about the optimality of the location, or about the nature of the prior weights of the facilities? If Q is a true optimum location for a given set of weights F_i but is also optimal under another regime $q_i \neq F_i$ we can make incorrect inferences about the ideal weights - for instance if the entropy model is run with q_i as the priors and q_i solves the equations, then $w_i = q_i$ and will of course differ from F_i. This aspect of the Weberian location model needs to be better understood - while the convexity of the objective function ensures a unique optimum in the location domain (i.e. $Q = (X^*, Y^*)$), this solution is consistent with a variety of alternative weights. If we propose a location, and set some prior estimate on the weights, it should be no surprise that the solution values for the w_i are different from q_i (unless we get very lucky and suggest weights that support the optimality of that location), and indeed the weights derived are likely to be a compromise between the distribution of x_i', y_i' and q_i.

A brief numerical example can be considered. First set $(X^*, Y^*) = (6.0171, 2.8378)$, and note that $q^1 = [.2\ .3\ .4\ .1]$ is a solution to the set of equations (4.1)-(4.3). Then the inference model, yields $w^1 = [.2\ .3\ .4\ .1]$. On the other hand with the same location but $q^2 = [.25\ .25\ .25\ .25]$, the ideal weights are $w^2 = [.2780\ .1865\ .3384\ .1970]$, with $w_i^2 > w_i^1$ whenever $q_i^2 > q_i^1$ and $w_i^2 < w_i^1$ whenever $q_i^2 < q_i^1$.

In the single facility case, the inferred ideal weights may differ from their prior estimates. This deviation is a measure of the extent to which the facility ought to adjust its in- and out-flows in order to come into compliance with the Weberian notion of optimality. Any surprise that the analyst feels concerning the chosen set of "missing" values, might prompt a reappraisal of the assumptions that go into the choice of constraints in the model. The strength of the information theoretic and entropy method is that it forces the analyst to adopt a least biased approach. The use of the information minimizing formalism also introduces many opportunities for sensitivity analysis, as is discussed in Webber [1180, Chapter 5]. For example, it is reasonable to ask about the impact of a small increment in the distance of a certain demand point from a fixed central facility: if that facility is to remain "optimal" the ideal weight of the demand must decrease. Any further "facts" about the probable distribution of the unknown weights can also be incorporated in the form of constraints on the original model. For example if there is a revenue associated with a demand for the facility of size w_i in zone i, of r_i, and the analyst wishes to ensure that the expected value of revenue matches some preset level R, a constraint of the form $\sum_i w_i r_i = R$ is appended to the model, and this will simply introduce another parameter to be solved in the final model (4.9): $w_i = \dfrac{q_i e^{-\Theta x_i' - \phi y_i' - \alpha r_i}}{\sum_k q_k e^{-\Theta x_k' - \phi y_k' - \alpha r_k}}$ where the numerical search would now include three unknowns, and three equations.

4.2.1 Examples

Before going on to the discussion of multiple facilities, we consider some further numerical examples of the single facility model. With reference to the same layout as in Figure 1, consider several alternative "optimal" facilities.

TABLE 4.2. Results of Numerical Example: Three alternative sites for a single "optimal" facility.

Facility at (8,3)	Facility at (1,4)	Facility at (5,8)
Uniform prior weights [$q_j = 1/n$]		
$w_a = 0.1711$	$w_a = 0.4183$	$w_a = 0.1522$
$w_b = 0.1697$	$w_b = 0.3911$	$w_b = 0.3477$
$w_c = 0.3821$	$w_c = 0.0891$	$w_c = 0.1522$
$w_d = 0.2769$	$w_d = 0.1013$	$w_d = 0.3477$
$\Theta = 0.5061$	$\Theta = -1.3468$	$\Theta = 0.0000$
$\phi = -0.1005$	$\phi = 0.0194$	$\phi = 0.6771$
Non-uniform prior weights [$q_a = 0.2, q_b = 0.3, q_c = 0.4, q_d = 0.1$]		
$w_a = 0.0952$	$w_a = 0.3916$	$w_a = 0.0722$
$w_b = 0.2699$	$w_b = 0.4247$	$w_b = 0.3934$
$w_c = 0.4208$	$w_c = 0.1467$	$w_c = 0.2324$
$w_d = 0.2141$	$w_d = 0.0370$	$w_d = 0.3020$
$\Theta = 0.6949$	$\Theta = -1.3950$	$\Theta = 0.4492$
$\phi = 0.5045$	$\phi = -0.1100$	$\phi = 1.2050$

Table 4.2 reports the results of placing a single optimal facility at each of the following alternatives in turn: (8,3) or (1,4) or (5,8). These are not multiple facility problems, rather three separate applications of the single facility case. That is, suppose the facility is at (8,3) and is optimally located. What is the ideal distribution of demand to support this location? Similar analyses are then performed if the location is at (1,4) or (5,8).

Facility 1 at (8,3) is closest to demand point c, Facility 2 at (1,4) is closest to demand point a, and Facility 3 is equidistant from demand points b and d. When the prior weights are uniform, the major influence on the inferred ideal weights is the spatial distribution of the demand points around the facility. Thus, with reference to Table 4.2, for facility 1, $w_c = 0.3821$ is the largest value; for facility 2, $w_a = 0.4183$ is the largest value; and for facility 3, the values of w_a and w_b are symmetric to w_c and w_d. The Θ and ϕ parameters in the uniform prior case show that for facility 3, there is a zero value for the "x" dimension, which is expected as the data are symmetric in that dimension. When non-uniform priors are introduced, the results are quite interesting. Suppose that the prior estimates of the weights are $q_i = [.2\ .3\ .4\ .1]$. In the case of facility 1, the ideal weight of w_c is reinforced by the relatively large prior probability ($q_3 = 0.4$), which increases w_c from 0.3821 to 0.4208. In the case of facility 2, the slightly larger prior probability for demand "b" (i.e. 0.3 compared to the uniform value of 0.25) pushes the ideal weight of w_b above that of demand point "a". Finally, for the third facility, which enjoyed symmetry of w_a and w_c to w_b and w_d in the uniform prior case, now has demands which are more weighted than before (w_b and w_c) and two others that are less weighted than before (w_a and w_d). The details of the outcome of the changes in the case of this analysis are shown in Figure 2. A closed circle is positioned along the line joining each demand point to the facility, in proportion to the weight of the demand (the circle is placed closer to the facility if the weight is large). The open circles show the impact of the prior probabilities. The results are no longer symmetric in the same way. Note also that the non-uniform priors redirect the demand weights to be larger in the cases of demand points b and c, while demand points a and d are adjusted downward.

4.3   Extensions to Multiple Facilities

The first section of this chapter laid out the basic technique for inferring ideal weights, and suggested uses of this method in the case of a single "optimally" located facility. What follows is a natural (but previously untested) extension of Eastin's [356] idea to multiple facilities. Four separate kinds of multiple facility location models are considered: (a) multifacility; (b) location-allocation; (c) spatial interaction; and (d) multiple hub.

4.3.1   Multifacility Problem

There is a standard model for the multifacility location problem which essentially says that there are fixed flows from a set of existing demand points to facilities, whose locations are to be determined. [See Love, Morris and Wesolowsky, [786, Chapter 4].] The objective in the standard literature

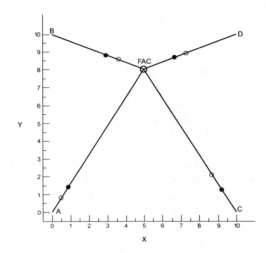

FIGURE 2. Role of Priors in Reweighting the Demands (Re Fac 3)

	weight uniform prior	non uniform prior	weight with non-uniform prior
a	0.1522	0.2	0.0722
b	0.3477	0.3	0.3934
c	0.1522	0.4	0.2324
d	0.3477	0.1	0.3020

is to choose the locations for p facilities, so as to minimize

$$T = \sum_i \sum_j w_{ij} d_{ij} + \sum_j \sum_k v_{jk} c_{jk} \qquad (4.12)$$

where w_{ij} are the flows between the new facilities and the existing demand points, and v_{jk} are the flows between the new facilities. d_{ij} is the distance between fixed demand point i, and fixed facility j; and c_{jk} is the distance between Q_j and Q_k. The v_{jk} values indicate that there are expected to be flows between the newly located central facilities, and these play an important role in determining the optimality conditions for the locations. Typically these interfacility flows are exogenously given, and do not depend at all on the location of the facilities. (See the hub case below for a variant on this restriction.) An intuitive example might be a set of manufacturing plants, which are required to exchange certain quantities of material, regardless of the actual site chosen for the plants.

The analysis of ideal weights for these problems is conveniently broken into a number of variants:

1. If v_{jk} is <u>known</u> (possibly zero), we could be asked to find w_{ij}. In this case the influence of the known interfacility flows will act as

further information which determines the ideal inferred weights for the origin-to-facility flows.

2. Alternatively, a given set of known w_{ij} values could serve as a backdrop for the determination of unknown v_{jk} flows.

3. Finally, a situation of complete uncertainty could exist with respect to both the flows from the fixed demand points, and between the fixed facilities: in this case we are asked to find both w_{ij} and v_{jk}, given only the locations of the facilities and the demand points.

Many of the most important aspects of the problem can be illustrated by focussing attention on the first of these cases: i.e. v_{jk} is known and w_{ij} is required to be found, and this is the first case presented below.

To distinguish exogenous "known" and endogenous "unknown" values, denote <u>known</u> values as V_{jk} and W_{ij} respectively, while <u>unknown</u> values are v_{jk} and w_{ij}. Table 4.3 provides a quick overview of the analyses to follow.

TABLE 4.3. The Three Cases

CASE 1	CASE 2	CASE 3
Find $\{w_{ij}\}$	Fix $\{W_{ij}\}$	Find $\{w_{ij}\}$
Fix $\{V_{jk}\}$	Find $\{v_{jk}\}$	Find $\{v_{jk}\}$

Case 1: V_{jk} known

In the case of known interfacility flows V_{jk} (possibly $V_{jk} = 0$), it can easily be shown that the analysis decomposes into a set of p separate applications of the single facility inference problem.

Let $Q_j = (X_j, Y_j)$ be the locations of a set of fixed facilities. Let $p_i = (x_i, y_i), i = 1, \ldots, n$ be the sites of demand for the services of these facilities. The demand at p_i for the jth facility is defined as w_{ij} and is assumed to be unknown, or unobservable. There exists a prior estimate of the demand at i, however, and this is defined as q_{ij}. Let d_{ij} be the straight line distance between p_i and Q_j. By hypothesis, the set of facilities Q is the solution to the problem:

$$\text{MIN } Z = \sum_i \sum_j w_{ij} d_{ij} + \sum_j \sum_{k \neq j} V_{jk} c_{jk}$$

With the addition of the normalizing condition that the weights sum to a fixed value, W_j for each facility, the system of equations to be satisfied by the unknown demand weights is simply repeated for each j and so Q

satisfies the following conditions:

$$\sum_i w_{ij} = W_j = 1 \tag{4.13}$$

$$\sum_i w_{ij} \frac{X_j - x_i}{d_{ij}} + \sum_{k \neq j} [V_{jk} + V_{kj}] \frac{X_j - X_k}{c_{jk}} = 0 \tag{4.14}$$

$$\sum_i w_{ij} \frac{Y_j - y_i}{d_{ij}} + \sum_{k \neq j} [V_{jk} + V_{kj}] \frac{Y_j - Y_k}{c_{jk}} = 0. \tag{4.15}$$

In order for a given set of facilities to be optimally located with respect to a set of demands w_{ij}, the above system of equations must be satisfied. If each equation (4.14) is rewritten $a_j + b_j = 0$, and each equation (4.15) is rewritten $r_j + s_j = 0$, then there are limitations on the magnitude of b_j and s_j terms, which depend on the magnitudes of the feasible interfacility flows. It can be demonstrated that there is a reduced set of optimality conditions for this problem, when V_{jk} is known: i.e. $\sum_j a_j = 0$, and $\sum_j b_j = 0$. In other words $\sum_j \sum_i w_{ij} \frac{X_j - x_i}{d_{ij}} = 0$, and $\sum_j \sum_i w_{ij} \frac{Y_j - y_i}{d_{ij}} = 0$, at optimality. [This result should be compared with the condition for the original single facility case.]

A very interesting aspect of this problem is the possibility that a given set of interfacility flows is <u>inconsistent</u> with the optimality of the locations, assuming that the unknown weights are scaled to sum to one. In simple terms, the values of the interfacility flows may make it impossible for a set of fixed facility locations to be optimal with respect to the other data in the problem. In this case, it will be impossible to find a solution to the equations above. Assume for now that suitable values of V_{jk} have been chosen. The issue of determining feasible values will be discussed further in case (3) below. Once a certain set of feasible interfacility flows is provided, the information minimizing model can be solved to find a set of w_{ij} terms. That is, the weights are to be chosen to satisfy the information minimizing objective: MIN $\sum_i w_{ij} \ln \frac{w_{ij}}{q_{ij}}$ subject to (4.13), (4.14) and (4.15). The solution to this problem is:

$$w_{ij} = \frac{q_{ij} e^{-\Theta_j x'_{ij} - \phi_j y'_{ij}}}{\sum_k q_{kj} e^{-\Theta_j x'_{kj} - \phi_j y'_{kj}}}$$

but now Θ_j and ϕ_j are chosen to satisfy (4.14) and (4.15), and where $x'_{ij} = \frac{X_j - x_i}{d_{ij}}$ and $y'_{ij} = \frac{Y_j - x_i}{d_{ij}}$. To the extent that the new facilities are badly located vis-a-vis the fixed interfacility flows V_{jk}, the inferred weights from the demand locations to the facilities are adjusted to compensate. The extent of this distortion can be measured by comparing the results for the w_{ij} values with $V_{jk} = 0$, and with non-zero V_{jk}.

Example (continued)

TABLE 4.4. Multifacility Numerical Example; case 1: fix interfacility flows to be $V_{12} = 0.11, V_{13} = 0.12$, and $V_{23} = 0.12$, with three "optimally" sited facilities.

Facility at (8,3)	Facility at (1,4)	Facility at (5,8)
Uniform prior weights $[q_j = 1/n]$		
$w_a = 0.1098$	$w_a = 0.5346$ ^	$w_a = 0.0741$
$w_b = 0.1091$	$w_b = 0.4534$ ^	$w_b = 0.4129$ ^
$w_c = 0.5034$ ^	$w_c = 0.0048$	$w_c = 0.0765$
$w_d = 0.2778$ ^	$w_d = 0.0073$	$w_d = 0.4365$ ^
$\Theta = 0.9630$	$\Theta = -4.1139$	$\Theta = 0.0298$
$\phi = -0.1812$	$\phi = 0.0801$	$\phi = 1.4185$
Non-uniform prior weights $[q_a = 0.2, q_b = 0.3, q_c = 0.4, q_d = 0.1]$		
$w_a = 0.0574$	$w_a = 0.5327$ ^	$w_a = 0.0326$
$w_b = 0.1781$	$w_b = 0.4557$ ^	$w_b = 0.4367$ ^
$w_c = 0.5300$ ^	$w_c = 0.0088$	$w_c = 0.1180$
$w_d = 0.2344$ ^	$w_d = 0.0027$	$w_d = 0.4127$ ^
$\Theta = 1.1884$	$\Theta = -4.0827$	$\Theta = 0.5613$
$\phi = 0.5036$	$\phi = -0.1239$	$\phi = 1.9800$

Table 4.4 presents data concerning the ideal weights for the demand for three simultaneously located facilities, with a certain fixed level of interfacility flow. The elements of this table marked with the symbol "^" indicate an increase over the corresponding single facility case in Table 4.2. Because of the weighted connection between facility 1 and 2, 1 and 3 and, 2 and 3: $V_{12} = 0.11, V_{13} = 0.12$, and $V_{23} = 0.12$. Since these weights would tend to pull the optimal facilities closer together, relatively larger w_{ij} values in the vicinity of the fixed location are required to maintain their optimality. Thus, in the case of Facility 1 at (8,3), the demands from w_c and w_d increase. The second facility at location 2, experiences increased weight from w_a and w_b, and facility 3 experiences greater values of w_b and w_d. The introduction of priors leaves the substance of the analysis largely unchanged, because of the strong role of the data concerning the interfacility flows. The facility weighted flows are shown in Figure 3 for case 1.

Case 2: v_{jk} unknown, but W_{ij} known

Given locations for the new fixed facilities, and the demand to facility flows, the first terms in (4.14) and (4.15) a_j and r_j are constants, and the only adjustment that is possible to bring the fixed locations into optimality, is to configure the interfacility flows. The maximum set of such values may

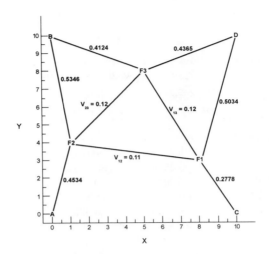

FIGURE 3. Case 1:
  4 demand points
    $a = (0,0)$; $b = (0,10)$; $c = (10,0)$; and $d = (10,10)$.
  3 facilities
    $1 = (8,3)$; $2 = (1,4)$; and $3 = (5,8)$.

be obtained from the following linear program, where the fixed demand-to-facility flows are clearly denoted W_{ij}:

$$Max \sum_j \sum_k v_{jk}$$

subject to

$$\sum_i W_{ij} \frac{X_j - x_i}{d_{ij}} + \sum_{k \neq j} [v_{jk} + v_{kj}] \frac{X_j - X_k}{c_{jk}} = 0 \qquad (4.16)$$

$$\sum_i W_{ij} \frac{Y_j - y_i}{d_{ij}} + \sum_{k \neq j} [v_{jk} + v_{kj}] \frac{Y_j - Y_k}{c_{jk}} = 0. \qquad (4.17)$$

where we have simply introduced into the linear program, specific coefficients for the W_{ij} values.

Case 3: w_{ij} and v_{jk} unknown

Consider the issue of "suitable" values of v_{jk}: clearly, if arbitrarily large values of interfacility flows are introduced into (4.14) and (4.15) there will be no feasible values of w_{ij} to satisfy (4.13)-(4.15). A simple feasibility test is therefore recommended before attempting to solve for the information minimizing flows. This is the topic of Case 3. To determine the permissible range of values of the exogenous interfacility weights, set up the following linear program: Max $\sum_j \sum_k v_{jk}$ subject to (4.13) - (4.15). This will provide the values of the maximum interfacility flows and a consistent set of demand weights. Further analysis can be performed to set the weight of one or two of the interfacility flows arbitrarily, and then solve for the remaining values.

Example (continued)

Before going on to the discussion of location-allocation, let us consider briefly, some numerical examples of the multifacility model. With reference to the four fixed demand points, and the three facilities analyzed in case 1, the following table (see Table 4.5) reports the results of case 2 and 3 analyses.

TABLE 4.5. Results of Numerical Example

	CASE 2			CASE 3		
	Fix $\{W_{ij}\}$			Find $\{w_{ij}\}$		
W_a	0.1	0.52	0	0.04	0.49	0
W_b	0	0.48	0.57	0	0.51	0.62
W_c	0.65	0	0	0.73	0	0
W_d	0.25	0	0.43	0.23	0	0.38
	Find $\{v_{jk}\}$			Find $\{v_{jk}\}$		
v_{1j}	—	0.1488	0.3656	—	0.2000	0.4300
v_{2j}	—	—	0.0819	—	—	0.0000
v_{3j}	—	—	—	—	—	—

4.3.2   Location-Allocation Problem

The location-allocation model differs from the multifacility one, in that there is no predetermined set of allocation variables between the origins and the facilities. That is, the aim is to simultaneously locate facilities, and to assign the demand to these facilities on the basis of cost minimization.

One of the very useful properties of the location allocation model is that when a set of facilities is given, the appropriate assignment of demand to each facility is easily determined. That is, since nearest center assignment is an optimal strategy, the "trade area" of each facility is found by constructing a proximal area, of all those demand points which are closer to the facility than to any other. Thus, in inferring the ideal weights for a set of supposedly optimal facilities, we perform a spatial decomposition of the problem so that for each trade area, the weights of the fixed demands are determined. This method reveals that the weights of each of the service regions are determined only up to a multiplicative constant, since the given facility locations induce a set of spatially independent trade regions. It is interesting to note that if a set of facilities are optimally located in the location-allocation model, then they remain optimal under an arbitrary rescaling of the demand weights in each service region. Thus, we may arbitrarily set the weight of each sub region to a fixed level.

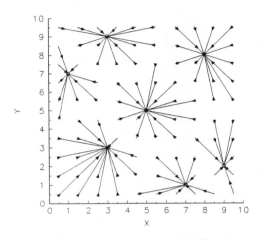

FIGURE 4. Location Allocation Results

An example of this kind of analysis is shown Figure 4. There, the set of 7 facilities is fixed, and the proximal region surrounding each is determined for j as the set of all points i, for which $d_{ij} < d_{ik}$ where $k \neq j$. Given each of these sets of points, we simply apply the single facility inference model developed above. Figure 4 illustrates the solution values of the weights as arrow heads which are placed closer to the facility when they are large, and further away when they are small.

4.3.3   Extension to Spatial Interaction Based Flow

As an application of the facility inference problem, to a case where the allocations are not "all-or-nothing" as in the above section, consider the recent model of Drezner [289]. She devised a set of conditions to make a single new facility provide an optimal addition to a set of existing facilities. These existing facilities belong to two sets - one a set of "competitor" locations, and the other, a set of "our chain" locations. The new facility, and the existing facilities, share in a demand distribution, according to a gravitational interaction model. This model states that the interaction between the demand point i, and the jth facility is

$$P_{ij} = \frac{W_i \frac{S_j}{d_{ij}^\lambda}}{\frac{S}{d_i^\lambda} + \sum_j \frac{S_j}{d_{ij}^\lambda}} \quad \text{for } i = 1, \ldots, n; \; j = 1, \ldots, k.$$

$$P_{i,new} = \frac{W_i \frac{S}{d_i^\lambda}}{\frac{S}{d_i^\lambda} + \sum_j \frac{S_j}{d_{ij}^\lambda}} \quad \text{for } i = 1, \ldots, n. \tag{4.18}$$

where w_i is the demand for the service at i, S_j is the share of attraction located at j, and d_{ij} is the distance between the demand origin and the fixed facility j. In turn the newly located facility has attraction S (no subscript) and the distance from the demand point to the new facility is d_i. The parameter λ is the exponent on the distance deterrence term, and the larger the value of λ the more sharply interaction falls off with distance. The idea in Drezner's model is to position the new facility at (X, Y) so as to minimize the share of total demand going to the competitor's chain. If the first c of the k facilities are ours, the competition's total demand is $\sum_i \sum_{j=c+1}^k P_{ij}$, and our chains share of demand is $\sum_i (\sum_{j=1}^c P_{ij} + P_{i,new})$.

After a lot of analysis, Drezner [289] shows that the new facility must satisfy the following condition, rearranged slightly to suit the terms here:

$$\sum_i \frac{X - x_i}{d_i} w_i \frac{a_i d_i^{\lambda-1}}{(1 + b_i d_i^\lambda)^2} = 0 \tag{4.19}$$

$$\sum_i \frac{Y - y_i}{d_i} w_i \frac{a_i d_i^{\lambda-1}}{(1 + b_i d_i^\lambda)^2} = 0 \tag{4.20}$$

where $a_i = \frac{1}{S} \sum_{j=c+1}^k [S_j / d_{ij}^\lambda]$, and $b_i = \frac{1}{S} \sum_{j=1}^k [S_j / d_{ij}^\lambda]$.

In Drezner's paper [289] the (X, Y) coordinates of the location are unknown, so there has to be an iterative solution of the set of two equations. In our case, we hypothesize that the facility is at least a local optimum, and so the above equations hold. Then, the target of inference becomes in our case the w_i values for the unknown weights of demand. The equations above are quite complex, but careful scrutiny shows that they are

$$\sum_i \frac{X - x_i}{d_i} w_i Z_i = 0$$

$$\sum_i \frac{Y - y_i}{d_i} w_i Z_i = 0 \tag{4.21}$$

where Z_i are constant for a given facility location:

$$Z_i = \frac{d_i^{\lambda-1} a_i}{(1 + b_i d_i^\lambda)^2} \tag{4.22}$$

In these equations d_i is the distance to the new facility, from zone i, and d_{ij} is the distance from the demand zones to each of the existing k facilities. In Drezner's notation the first c of the k centers belongs to our own chain. The solution for the entropy maximizing values of the unknown weights is:

$$w_i = \frac{e^{-\Theta x'_i - \phi y'_i}}{\sum_k e^{-\Theta x'_k - \phi y'_k}} \qquad (4.23)$$

where Θ and ϕ are chosen to satisfy (4.19) and (4.20) respectively, and $x'_i = Z_i(X - x_i)/d_i$ and $y'_i = Z_i(Y - y_i)/d_i$.

As an example of the application of this model, the sample problem discussed in Drezner [289, 288] and in Chapter 13 of this book is re-visited. That problem has a 10 by 10 grid of demand nodes, and in the original research they are all weighted with equal demand of 1 unit. With a distance decay parameter of $\lambda = 2$, and a weight of 1 for every facility ($S = S_j = 1$) the analysis shows the optimal site for a new facility, when one of the existing centers belongs to our own chain. The impact of allowing the facility to join an existing one is examined (Table 4.6). A total of seven different analyses are performed: in each case a fixed location is chosen for our existing store, and an optimal addition is made to that location by picking a second site to maximize our chain's market share captured. Following Drezner [289] we can then tabulate how much the existing store would sell [without any addition]; how much is sells with the addition of the new branch [a lower amount because of "cannibalism"]; and the increment to our chain sales as the difference between the combined sales of the existing store and the new location, and the previous sales by the existing store.

The numerical results are organized into two main sections. First to reproduce the Drezner result, the incremental share of demand going to our chain is computed. Then the same analysis is performed with the endogenously determined weights. The top half of Table 4.6 exactly corresponds to Drezner's tables and the bottom half matches it well in most respects. However, in two cases, the introduction of ideal weights greatly changes the solution. These cases are when the existing store is located in "2" and in "7". When this is the case, the optimally located additional store is in the lower left hand corner of the study region, and given this opportunity, the ideal weight model places a large share of the demand in the immediate vicinity of that location, and since this is relatively isolated from the rest of the competitions locations, the new facility gets a very large market share. Figure 5 shows the demand surface (w_i) for one example application. In this case the existing facility is number 4, (at $X = 5.61, Y = 4.46$). The demand surface increases in the vicinity of the new facility, decreases around existing competitive stores, and stays about the same in the vicinity of fixed store 4.

TABLE 4.6. Drezner's Model Compared to Inferred Ideal Weights

Selected two store locational plans

$w_i = 1/n$ for all n points						
			$j =$			
Existing Store	1	2	3	4	5	6	7
(A) Existing without new	12.24	16.54	17.49	12.77	12.78	11.75	16.42
Additional X, Y if our store is j	5.54 5.62	2.51 2.51	6.90 6.03	5.61 4.46	6.86 5.36	4.49 5.56	2.54 2.54
(B) Existing with new store	11.14	14.75	16.16	11.78	11.92	10.89	15.68
(C) New Store	12.92	12.46	12.76	12.80	12.83	12.68	12.46
(D)Competition Share with new store	75.94	72.79	71.08	75.42	75.25	76.43	71.86
(E) Additional Market Share = (B)+(C)-(A)	11.82	10.67	11.43	11.81	11.97	11.82	11.72
$w_i =$ ideal weight determined to make the new location optimal						
			$j =$			
Existing Store	1	2	3	4	5	6	7
(A) Existing without new	12.37	15.96	17.63	12.81	12.85	11.87	16.51
Additional X, Y if our store is j	5.54 5.62	2.51 2.51	6.90 6.03	5.61 4.46	6.86 5.36	4.49 5.56	2.54 2.54
(B) Existing with new store	11.58	7.73	16.14	11.63	11.82	10.68	19.69
(C) New Store	12.72	40.73	12.45	12.80	13.09	12.53	26.13
(D)Competition Share with new store	75.70	51.54	71.41	75.57	75.09	76.79	54.18
(E) Additional Market Share = (B)+(C)-(A)	11.93	32.50	10.96	11.62	12.06	11.34	29.31

4.3.4 Hub Location Problem

The hub location problem is a hybrid of the location-allocation and the multifacility problems [901, 45]. That is, the assignment of the fixed demand to the facilities is determined by the model, but there are expected to be interfacility flows. These flows are not pre-determined (as in the case of the multifacility flow problem) as the amount of inter facility interaction is endogenously determined by the allocation variables. As a further complicating factor, it is known that the optimal assignment variables do not necessarily obey a "nearest is best" policy, and so the device of constructing the spatial decomposition of the problem (as in (b) above) is not guaranteed to be optimal.

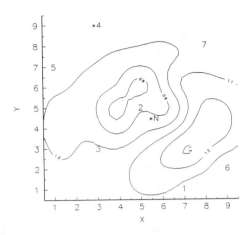

FIGURE 5. New inferred demand surface, when weights are inferred.
Facilities are located at (3,9) and (5.6, 4.6). Original weights were $w_i = 1$ in
Drezner's analysis. Here the weights are recalculated to ensure that the two lo-
cations satisfy the optimality conditions for the maximization of share capture.

An application of the inferred ideal weight model to a hub analysis problem is now offered. The idea of this application is that a hub location has
been chosen to service a set of fixed cities. The question then is for what
pattern of interaction between these fixed cities is the hub an optimal location? A linked set of information minimizing models can be used to answer
this puzzle. Assume the inferred ideal weights determine the total volume
shipped in and out of each node; then, from among the wide variety of
detailed interaction patterns which are *consistent* with these marginal to-
tals, one could be deduced using a doubly constrained interaction model as
in [415]. The method can be used to diagnose problems in a hub system,
because the empirical/actual flows between the cities may turn out to be
quite different from the inferred ideal flows. In that case the hub either
ought to be relocated, or the hub operator should solicit business so as to
bring the actual interactions served more into line with the ideal pattern.

4.4   Conclusions

This chapter has used the Eastin inferred ideal weight model as a vehicle
for the review of several types of demand estimation. Underlying many
of these models is the goal of either making least biased estimates of unknown information, or the derivation of feasible solutions to problems with
multiple linear constraints.

5

CONJUGATE DUALITY IN FACILITY LOCATION

Carlton H. Scott
University of California, Irvine

Thomas R. Jefferson
University of California, Los Angeles

Soheila Jorjani
California State University, San Marcos

5.1 Introduction

The concept of duality has proven to be a valuable notion in the analysis of linear and (convex) nonlinear programs. The dual program often has an interpretation different to the primal program from which it is defined that adds further insight. In some cases the dual program may be easier to solve than the primal and may be useful as a stopping criterion for an algorithmic procedure. The best known approach to duality is undoubtedly the Lagrangian method whereby one appends the constraints of a mathematical program to the objective by the use of multipliers which then become the variables of the corresponding dual program. There are however alternative approaches to duality which may also be used in the analysis of location problems. In this paper we pursue the conjugate function approach to the generation of dual programs and illustrate the methodology by its application to various problems in location. Specifically, we will overview an approach to duality using conjugate function that parallels the approach of geometric programming. This provides a complete duality theory for convex mathematical programs. Since both the well known minisums and minimax location problems are convex programs, the duality theory is applicable and dual programs may be derived. This will be the subject of the remainder of the paper. Explicit duals will be given for a variety of minisum models and a minimax model in order to generate a set of representative primal and dual programs. Other combinations are possible and may be found in the literature or derived using the principles and ideas presented here.

Duality has long been of interest to researchers in location theory. The single facility minisum model was first analyzed by [74] who used quasi-linearization to find a dual. Other early papers were [723, 1214]. Subsequently, the single facility model was extended to the multifacility case by [420, 781] and by [775] who used the ℓ_p metric in place of the more usual Euclidean metric. Juel and Love [651], using the conjugate duality theory, dualized the linearly constrained model with arbitrary norms. Minimax location models have been developed by [381] using Lagrangian methods. Subsequently, Scott et. al [1058] used conjugate duality to obtain the dual of a linearly constrained minimax model while [653] found the dual of a generalized minimax model.

5.2   Conjugate Duality Theory

The approach to duality presented in this paper uses the concepts of conjugate functions [1012]. Due to its great flexibility, we use the generalized geometric programming version of conjugate duality theory, which was developed in [939].

The theory of generalized geometric programming pairs the following two programs:

(P):
$$\text{Minimize } \{\, g_0(x^0) \,\} \tag{5.1}$$

Subject to:

$$\begin{array}{lll} \text{explicit constraints} & g_i(x^i) \leq 0, \; i \in I, & (5.2) \\ \text{implicit constraints} & x^0 \in C_0, \; x^i \in C_i, \; i \in I, & (5.3) \\ \text{and cone condition} & x \in \chi \subset R^n, & (5.4) \end{array}$$

and the geometric programming dual of P,

(D):
$$\text{Minimize } \left\{\, g_0^*(x^{0*}) \,+\, \sum_I g_i^{*+}(x^{i*}, \lambda_i^*) \,\right\} \tag{5.5}$$

Subject to:

$$\begin{array}{lll} \text{implicit constraints} & x^{0*} \in C_0^*, \; (x^{i*}, \lambda_i^*) \in C_i^{*+}, \; i \in I, & (5.6) \\ \text{and cone condition} & x^* \in \chi^* \subset R^n. & (5.7) \end{array}$$

The relations between programs (P) and (D) are as follows: χ is a closed convex cone; χ^* is the dual cone of χ, i.e., $\chi^* = \{x^* | \langle x^*, x \rangle \geq 0, \forall x \in \chi\}$; I is the index set of explicit constraints in program (P); $x = (x^0, x^I)$ is the cartesian product of vectors x^0 and x^I; $x* = (x^{0*}, x^{I*})$ is similarly defined. Here x^I is the cartesian product of vectors $x^i, i \in I$, and x^{I*} is similarly defined. $\langle \cdot, \cdot \rangle$ denotes a finite dimensional inner product. $[g_i(x_i), C_i]$, $i \in \{0\} \cup I$, is a pair of closed convex functions g_i defined over a convex set $C_i \subset R^{n_I}$. $[g_i^*(x^{i*}), C_i^*]$, $i \in \{0\} \cup I$ is a pair of closed convex functions g_i^* defined over the convex set C_i^* and is the conjugate transform of $[g_i(x^i), C_i]$, i.e., g_i defined over the convex set of C_i.

$$g_i^*(x^{i*}) \triangleq \sup_{x^i \in C_i} \left(\langle x^{i*}, x^i \rangle - g_i(x^i) \right) \tag{5.8}$$

and

$$C_i^* = \left\{ x^{i*} \;\middle|\; \sup_{x^i \in C_i} \left(\langle x^{i*}, x^i \rangle - g_i(x^i) \right) < \infty \right\} \tag{5.9}$$

$\left[g_i^{*+}(x^{i*}, \lambda_i^*), C_i^{*+} \right]$, $i \in I$, is the positive homogeneous extension of $\left[g_i^*(x^{i*}), C_i^* \right]$ with

$$g_i^{*+}(x^{i*}, \lambda_i^*) \triangleq \begin{cases} \sup_{x^i \in C_i} \langle x^{i*}, x^i \rangle \text{ if } \lambda_i^* = 0 \text{ and } \sup_{x^i \in C_i} \langle x^{i*}, x^i \rangle < \infty \\[2mm] \lambda_i^* g_i^*(x^{i*}/\lambda_i^*) \text{ if } \lambda_i^* > 0 \text{ and } x^{i*}/\lambda_i^* \in C_i^* \end{cases} \tag{5.10}$$

$$C_i^{*+} = \left\{ (x^{i*}, \lambda_i^*) \mid \lambda_i^* = 0 \text{ and } \sup_{x^i \in C_i} \langle x^{i*}, x^i \rangle < \infty \right\}$$
$$\cup \left\{ (x^{i*}, \lambda_i^*) \mid \lambda_i^* > 0 \text{ and } \sup_{x^i \in C_i} \langle x^{i*}, x^i \rangle < \infty \right\} \tag{5.11}$$

It is interesting to note that all the explicit constraints in the primal (program P) are transferred to the objective function in the dual (program D). Under mild conditions concerning feasibility and relative interiors [939], the primal and dual programs are related at optimality in the following manner:

$$g_0(x^0) + g_0^*(x^{0*}) + \sum g_i^*(x^{i*}, \lambda_i^*) = 0, \tag{5.12}$$

$$x^{0*} \in \partial g_0(x^0), \tag{5.13}$$

$$x^{i*}/\lambda_i^* \in \partial g_i(x^i), \lambda_i^* > 0, \; i \in I. \tag{5.14}$$

These optimality conditions allow an optimal point for one program to be calculated from an optimal point of the other. Here $\partial g_i(x^i)$ denotes the subgradient set of g_i at the point x^i, i.e.

$$\partial g_i(x^i) \triangleq \left\{ x^{i*} | g_i(x^i) + \langle x^{i*}, z^i - x^i \rangle \leq g_i(z^i), \; \forall z^i \in C_i \right\}. \tag{5.15}$$

Note that this theory gives rise to a Min-min duality as opposed to the usual Min-max duality.

In order to illustrate how the duality theory is applied we consider the following two dimensional program.

$$\text{Minimize} \quad (x_1 - 2)^2 + 2(x_2 - 1)^2$$
$$\text{subject to:} \quad x_1^2 + x_2^2 \le 4$$
$$x_1 + x_2 \ge 2$$
$$x_1 \ge 0, \ x_2 \ge 0.$$

Note that the primal program (P) is in a separable form in the sense that the variables in the objective variables x^0 are different to the variables in the explicit nonlinear constraints x^i, $i \in I$. Their relation occurs in the cone χ. hence the first step in constructing the dual is to separate the variables. Linear constraints are generally captured in the cone. This results in a program of the form:

$$\text{Minimize} \quad (x_1^0 - 2)^2 + 2(x_2^0 - 1)^2$$

subject to:

explicit constraints: $x_1^{1^2} + x_2^{1^2} - 4 \le 0$

implicit constraints: $\alpha \in \{2\}$

and cone condition: $x_1^0 + x_2^0 - \alpha \ge 0$

$$x_1^0 - x_1^1 = 0$$
$$x_2^0 - x_2^1 = 0$$
$$x_1^0 \ge 0, \ x_2^0 \ge 0.$$

Note that the new variable α with one point domain has been introduced to transform the linear constraint into a cone. Our program is now in the form of the primal program (P).

We now compute the dual using the prescription of program (D). The conjugate transform of the objective function is readily calculated as

$$\frac{x_1^{0*^2}}{4} + 2x_1^{0*} + \frac{x_2^{0*^2}}{8} + x_2^{0*} + 2\alpha^*$$

with $x_1^0 = \frac{x_1^{0*}}{2} + 2$ and $x_2^0 = \frac{x_2^{0*}}{4} + 1$ at optimality. Similarly, the conjugate transform of the explicit convex inequality is $\frac{x_1^{1*^2}}{4} + \frac{x_2^{1*^2}}{4} + 4$ with positive homogeneous extension $\frac{x_1^{1*^2}}{4\lambda} + \frac{x_2^{1*^2}}{4\lambda} + 4\lambda$ and $x_1^1 = \frac{x_1^{1*}}{2\lambda}$, $x_2^1 = \frac{x_2^{1*}}{2\lambda}$, $\lambda \ge 0$ at optimality. The case $\lambda = 0$ implies that the constraint is inactive. Further, the dual core is readily calculated as

$$x_1^{0*} = u + v_1 + w_1$$
$$x_2^{0*} = u + v_2 + w_2$$

$$x_1^{1*} = -v_1$$
$$x_2^{1*} = -v_2$$
$$\alpha^* = -u$$
$$u \geq 0, \ w_1 \geq 0, \ w_2 \geq 0.$$

Collecting these results into prescription (D) we obtain the following dual program.

Minimize $\quad \dfrac{x_1^{0*^2}}{4} + 2x_1^{0*} + \dfrac{x_2^{0*^2}}{8} + x_2^{0*} - 2u + \dfrac{x_1^{1*^2}}{4\lambda} + \dfrac{x_2^{1*^2}}{4\lambda} + 4\lambda$

subject to: $\quad x_1^{0*} + x_1^{1*} - u \geq 0$

$\qquad\qquad\quad x_2^{0*} + x_2^{1*} - u \geq 0$

$\qquad\qquad\quad u \geq 0, \ \lambda \geq 0.$

at optimality the primal solution is constructed from the dual solution from the relation $x_1 = \frac{x_1^{0*}}{2} + 2, \ x_2 = \frac{x_2^{0*}}{4} + 1$.

5.3   The Minisum Model

5.3.1   The Minisum Problem: Unconstrained

Given n fixed locations $a_j = (a_{j1}, a_{j2})^T, j = 1, \ldots, n$ and weights w_j associated with location j, the minimum problem is to find the location $x = (x_1, x_2)^T$ so as to minimize the weighted sum of distance between the unknown location and each fixed location. Specifically, we have the following optimization problem.

$$\text{Minimize} \ \left\{ \ \rho(x) \ = \ \textstyle\sum_{j=1}^{n} w_j \parallel x - a_j \parallel \ \right\}$$
$$x$$

where

$$\parallel x - a_j \parallel \ = \ \sqrt{(x_1 - a_{j1})^2 \ + \ (y_1 - a_{j2})^2}$$

For simplicity, we use the ℓ_2 Euclidean norm; it is straightforward to use the more general ℓ_p norm. The procedure for this is given in the sequel.

   In order to invoke the duality theory of generalized geometric programming, we need to separate the variables in order to facilitate computation of the conjugate transform. Hence we have the following program

(P1):

$$\text{Minimize} \ \sum_{j=1}^{n} w_j \parallel x^j - a_j \parallel \qquad\qquad (5.16)$$

with $x^1 = x^j$ for $j = 1, \ldots, n$.

The conjugate transform of $w_j \parallel x^j - a_j \parallel$ is:

$$\sup_{x^j} \left(\langle x^j, x^{j*} \rangle \right) - w_j \parallel x^j - a_j \parallel$$

$$= \langle a_j, x^{j*} \rangle + \sup_{x^j} \left(\langle x^j, x^{j*} \rangle - w_j \parallel x^j \parallel \right)$$

$$\leq \langle a_j, x^{j*} \rangle + \sup_{x^j} \left(\parallel x^j \parallel \left(\parallel x^{j*} \parallel -w_j \right) \right)$$

$$(\text{since } \langle x^j, x^{j*} \rangle \leq \parallel x^j \parallel \cdot \parallel x^{j*} \parallel)$$

$$= \langle a_j, x^{j*} \rangle \quad \text{with } \parallel x^{j*} \parallel \leq w_j.$$

Hence the dual objective is the sum of the conjugate transforms

$$\sum_j \langle a_j, x^{j*} \rangle \quad \text{with} \quad \parallel x^{j*} \parallel \leq w_j \quad \text{for } j = 1, \ldots, n. \tag{5.17}$$

Further, we need the dual cone of the cone defined by (5.16). This is defined by,

$$\sum_j \langle x^j, x^{j*} \rangle \leq 0; \quad \text{i.e.,} \quad \sum_j \langle x, x^{j*} \rangle \leq 0$$

which implies that

$$\sum_j x^{j*} = 0. \tag{5.18}$$

Combining (5.17) and (5.18) and expanding, we obtain the dual program to (P1) as

(D1)

$$\text{Minimize} \sum_j \langle a_j, x^{j*} \rangle \tag{5.19}$$

subject to: $\sum_j x^{j*} = 0$

and $\sqrt{(x^{j*_1})^2 + (x^{j*_2})^2} \leq w_j \quad \text{for } j = 1, \ldots, n.$

The primal (P1) and dual (D1) programs are related at optimality in the following way

$$\sum_j w_j \parallel x - a_j \parallel + \sum_j \langle a_j, x^{j*} \rangle = 0$$

i.e., the primal and dual objectives sum to zero. Further, the complementary slackness and sub-gradient conditions imply that,

if $x \neq a_j, j = 1, \ldots, n,$ then $x = a_j + \dfrac{x^{j*}}{w_j} \parallel x - a_j \parallel, j = 1, \ldots, n,$

$$\text{and } \| x^{j*} \| = w_j, \ j = 1, \dots, n$$

and if $x = a_r$ for some r, then $\| x^{r*} \| < w_r$

These conditions allow the primal solution to be obtained from the solution of the dual program. It is seen that the dual variables x^{j*} are direction vectors such that lines through a_j in the direction u^{j*} go through the optimal facility location. We will see that these basic results carry over to the extensions that follow.

5.3.2   The Minisum Problem: Linear Constraints

In this case, we assume that the unknown location is limited to a convex polyhedron defined by a finite set of linear inequalities.

$$Cx \ge b \tag{5.20}$$

The addition of these constraints affects both the dual objective and the dual cone. The first step is to transform, the polyhedral set to a cone by introducing a new variable β with one point domain $\{b\}$. Hence, the primal program may be written as

(P2):

$$\text{Minimize } \rho(x) = \sum_{j=1}^{n} w_j \| x^j - a_j \|$$
$$\text{with } x^1 = x^j \text{ for } j = 1, \dots, n.$$

subject to the implicit constraint

$$\beta \in \{b\}$$

and the cone condition

$$x^1 = x^j, \text{ for } j = 2, \dots, n; \ Cx^1 - \beta \ge 0.$$

Calculation now shows that dual program to be

(D2)

$$\text{Minimize } \left\{ b^T \beta^* + \sum_j \langle a_j, x^{j*} \rangle \right\}$$

$$\text{subject to:} \qquad \sum_j x^{j*} + C^T \beta^* = 0$$
$$\beta^* \le 0.$$

Here β^* is the Lagrange multiplier associated with the primal constraint (5.20).

Consequently, programs (P2) and (D2) are related as previously with the additional complementary slackness condition $(Cx - b)^T \beta^* = 0$.

5.3.3   The Minisum Problem: Nonlinear Metric Constraints

We consider the following model.

(P3):

$$\text{Minimize} \left\{ \sum_{j=1}^{n} w_j \parallel x^j - a_j \parallel \right\}$$

subject to: $\parallel x - a_j \parallel \le d_j$ for $j = 1, \ldots, n$.

In this case, we are requiring that the distance from each fixed location j to the unknown location not exceed d_j, for $j = 1, \ldots, n$. Separating variables, program (P3) may be written as

$$\text{Minimize} \sum_j w_j \parallel x^j - a_j \parallel$$

subject to: $\parallel y^j - a_j \parallel \le d_j$ for $j = 1, \ldots, n$

$x^1 = x^j = y^j$ for $j = 1, \ldots, n$

The conjugate transform of $\parallel y^j - a_j \parallel -d_j$ is $d_j + \langle a_j, y^{j*} \rangle$ subject to $\parallel y^{j*} \parallel \le 1$ with positive homogenous extension $d_j \lambda_j + \langle a_j, y^{j*} \rangle$ subject to $\parallel y^{j*} \parallel \le \lambda_j$, $\lambda_j \ge 0$. Hence, following the prescription in Section 5.2, the dual to program (P3) is

(D3)

$$\text{Minimize} \sum_j \left(\langle a_j, (x^{j*} + y^{j*}) \rangle + d_j \lambda_j \right)$$

subject to: $\sum_j x^{j*} + y^{j*} = 0$

$\parallel x^{j*} \parallel \le w_j$ for $j = 1, \ldots, n$

$\parallel y^{j*} \parallel \le \lambda_j$ for $j = 1, \ldots, n$

$\lambda_j \ge 0$ for $j = 1, \ldots, n$

An Alternative dual may be constructed by treating the nonlinear metric constraint as an implicit primal constraint. In this case we need the conjugate transform of $[f(x), F] = [w \parallel -a \parallel, \{x | \parallel x - a \parallel \le d\}]$. This is given by

$$
\begin{aligned}
f^*(x^*) &= \sup_{x, \parallel x - a \parallel \le d} \left(\langle x, x^* \rangle - w \parallel x - a \parallel \right) \\
&= \langle a, x^* \rangle + \sup x, \parallel x \parallel \le d \left(\langle x, x^* \rangle - w \parallel x \parallel \right) \\
&= \langle a, x* \rangle + \sup x, \parallel x \parallel \le d \left(\parallel x \parallel \cdot \parallel x^* \parallel -w \right) \\
&= \langle a, x^* \rangle + \max \left\{ 0, d(\parallel x^* \parallel -w) \right\}
\end{aligned}
$$

Consequently, the dual program is

$$\text{Minimize } \sum_{j=1}^{n} \left(\langle a_j, x^{j*} \rangle + \max \left\{ 0, d_j(\| x^{j*} \| - w) \right\} \right)$$

subject to:
$$\sum_{j=1}^{n} x^{j*} = 0$$

5.3.4 The Minisum Problem: Multiple Facility Location With Interactions

We consider the following model in which m facilities are to be located.
(P4)

$$\text{Minimize } \sum_{i=1}^{m} \sum_{j=1}^{m} w_{ij} \| x^i - a^j \| + \sum_{i=1}^{m} \sum_{j=1}^{m} \frac{v_{ij}}{2} \| x^i - x^j \|$$

Separating variables, we obtain the equivalent form

$$\text{Minimize } \sum_{i=1}^{m} \sum_{j=1}^{m} w_{ij} \| x^{ij} - a^j \| + \sum_{i=1}^{m} \sum_{j=1}^{m} \frac{v_{ij}}{2} \| y^{ij} \|$$

subject to:

$$x^{i1} = x^{ij} \text{ for } i = 1, \ldots, m; \ j = 1, \ldots, n \qquad (5.21)$$
$$y^{ij} = x^{i1} - x_{ij} \text{ for } i = 1, \ldots, m; \ j = 1, \ldots, n; \ i \neq j. \qquad (5.22)$$

By analogy with the previous sections, it is straightforward to see that the dual cone to the cone defined by (5.21) and (5.22) is

$$\sum_{j=1}^{n} x^{j*} + \sum_{j=1 \neq i}^{m} y^{ij*} - \sum_{k=1 \neq i}^{m} y^{ki*} = 0 \text{ for } i = 1, \ldots, m$$

Hence, using the prescription in section 5.3.1, the dual program is
(D4)

$$\text{Minimize } \sum_{i=1}^{m} \sum_{j=1}^{n} \langle x^{ij*}, a_j \rangle$$

subject to:

$$\| x^{ij*} \| \leq w_{ij} \text{ for } i = 1, \ldots, m; \ j = 1, \ldots, n$$
$$\| y^{ij*} \| \leq \frac{v_{ij}}{2} \text{ for } i = 1, \ldots, m; \ j = 1, \ldots, n; \ i \neq j$$

$$\sum_{j=1}^{n} x^{j*} + \sum_{j=1 \neq i}^{m} y^{ij*} - \sum_{k=1 \neq i}^{m} y^{ki*} = 0 \text{ for } i = 1, \ldots, m$$

Further, at optimality, the primal and dual programs are related by

$$\sum_{i=1}^{m}\sum_{j=1}^{n}\langle x^{ij*}, a_j \rangle + \sum_i \sum_j w_{ij} \| x^i - a_j \| + \sum_i \sum_k \frac{v_{ik}}{2} \| x^i - x^k \| = 0$$

$$\sum_{ij}\langle x^j, x^{ij*} \rangle + \sum_i \sum_j \langle x^i - x^j, y^{ij*} \rangle = 0 \tag{5.23}$$

$$x^{ij*} = 0 \quad \text{or} \quad w_{ij}\frac{x^i - a_j}{\| x^i - a_j \|} \quad \text{for } i = 1, \ldots, m; \; j = 1, \ldots, n$$

$$y^{ij*} = 0 \quad \text{or} \quad \frac{v_{ij}}{2}\frac{x^i - x^j}{\| x^i - x^j \|} \quad \text{for } i = 1, \ldots, m; \; j = 1, \ldots, m; \; i \neq j$$

5.3.5   The Minisum Problem: ℓ_p norm

All the previous results can be readily generalized to the cone of the ℓ_p norm

$$\| x - a_j \|_p = \sqrt[p]{(x_1 - a_{j1})^p + (x_2 - a_{j2})^p}$$

It is readily shown, that this has conjugate transform $\langle a_j, x^{j*} \rangle$ with $\| x^{j*} \|_q \leq w_j$ where $\frac{1}{p} + \frac{1}{q} = 1$ and this forms the basis for the generalization. Merely replace $\| x - a_j \|$ or $\| x - a_j \|_2$ with $\| x - a_j \|_p$ and the conjugate transform as appropriate. As an example, consider the following model involving the sum of two norms.

(P6)

$$\text{Minimize} \sum_j w_j \| x - a_j \|_{p_1} + \sum_j v_j \| x - a_j \|_{p_2}$$

Separating the variables as usual, we have

$$\text{Minimize} \sum_j w_j \| x - a_j \|_{p_1} + \sum_j v_j \| y^j - a_j \|_{p_2}$$

subject to:

$$x^1 = x^j = y^j \quad \text{for } j = 1, \ldots, n.$$

The dual program is

(D6)

$$\text{Minimize} \sum_{j=1}^{n} \langle a_j, x^{ij*} + y^{j*} \rangle$$

subject to:

$$\| x^{j*} \|_{q_1} \leq w_j \quad \text{for } ij = 1, \ldots, n$$
$$\| y^{ij*} \|_{q_2} \leq v_{ij} \quad \text{for } j = 1, \ldots, n$$
$$\sum_{j=1}^{n} x^{j*} + y^{j*} = 0$$

with $\quad \frac{1}{p_1} + \frac{1}{q_1} = 1, \; \frac{1}{p_2} + \frac{1}{q_2} + 1.$

The primal and dual programs are related by

$$\sum_j w_j \parallel x - a_j \parallel_{p_1} + \sum_j v_j \parallel x - a_j \parallel_{p_2} + \sum_j \langle a_j, x^{j*} + y^{j*} \rangle = 0$$

$$x^{j*} = 0 \quad \text{or} \quad w_j \frac{x - a_j}{\parallel x - a_j \parallel_{p_1}} \quad \text{for } j = 1, \ldots, n$$

$$y^{j*} = 0 \quad \text{or} \quad v_j \frac{x - a_j}{\parallel x - a_j \parallel_{p_2}} \quad \text{for } j = 1, \ldots, m; \ i \neq j$$

5.4   The Minimax Model

In the location of an emergency service, it is often preferable to locate the service facility so that the maximum distance to any user is minimized. Consequently, the server is more "equitably" located. Once again, we consider the Euclidean distance metric noting that the generalization to the ℓ_p metric is straightforward.

Specifically we consider the following model.

(P7)
$$\text{Minimize} \quad \text{Maximum} \quad \{w_i \parallel x - a_i \parallel^2\}$$
$$x \in R^2 \quad i = 1, \ldots, n$$

subject to the constraint set defined by

$$Qx \geq b$$

Here a_i, $i = 1, \ldots, n$ denote a fixed set of localities in the plane and x is the position of the facility to be located. w is the weight vector.

Since the maximum of a finite number of convex functions is convex, program (P7) is convex and nondifferentiable. Consequently we rewrite it in the following differentiable form.

(P8)
$$\text{Minimize} \ \{\alpha\}$$
$$x, \alpha$$

subject to

$$w_i \parallel x - a_i \parallel^2 \leq \alpha \quad \text{for } i = 1, \ldots, n \tag{5.24}$$

$$Qx \geq b$$

Consider the nonlinear constraints (5.24). These may equivalently be written as

$$x^2 + \beta \leq 0$$
$$\beta + \frac{\alpha}{w_i} + 2a_i^t x - a_i^t a_i \geq 0 \quad \text{for } i = 1, \ldots, m$$

with only one nonlinear constraint. Hence we write program (P8) in the form required for use of the duality theory.

(P9)

$$\text{Minimize } \{\alpha\}$$
$$x, \alpha, \beta, \gamma$$

subject to implicit constraints $\gamma \in \{b\}$, $\delta_i \in \{a_i^t a_i\}$
and explicit constraints $x^2 + \beta \leq 0$
and cone condition

$$\beta + \frac{\alpha}{w_i} + 2a_i^t x - \delta_i \geq 0 \quad \text{for } i = 1, \ldots, m \tag{5.25}$$

$$Qx - \gamma \geq 0 \tag{5.26}$$

Hence, using the prescription in section 5.2, the conjugate transform of the primal objective is

$$\sup(\alpha\alpha^* + \langle \delta, \delta^* \rangle + \langle \gamma, \gamma^* \rangle - \alpha)$$
$$\gamma \in \{b\}$$
$$\delta_i \in \{a_i^t a_i\} = b^t \gamma^* + \sum_i a_i^t a_i \delta_i^* \quad \text{with } \alpha^* = 1$$

Further, we need the conjugate transform of $x^2 + \beta$ which is $\frac{1}{4}x^{*2}$, $\beta^* = 1$. The positive homogeneous extension, $\frac{1}{4}\frac{x^{*2}}{\beta^*}$ with $\beta^* \geq 0$, is the conjugate transform of $x^2 + \beta \leq 0$.

Finally, we require the dual cone of the primal cone defined by inequalities (5.25) and (5.26). Calculation shows this to be defined by

$$\beta^* = \sum_i u_i$$
$$\alpha^* = \sum_i \frac{u_i}{w_i}$$
$$x^* = 2Au + Q^t v$$
$$\delta^* = -u$$
$$\gamma^* = -v$$

where $u \geq 0$, $v \geq 0$ are dual variables corresponding to the primal constraints (5.25) and (5.26). Combining, we have the following dual program,

(D7)

$$\text{Minimize } \left\{ \frac{x^{*2}}{4\sum_i u_i} - \sum_i a_i^t a_i u_i - b^t v \right\}$$
$$x^*, u, v$$

subject to

$$x^* = 2AU + Q^t v$$
$$\sum_i \frac{u_i}{w_i} = 1$$
$$v \geq 0, \ u \geq 0$$

where A is the matrix of fixed locations $(a_1, \ldots, a_n)$. This is, of course, a convex program, which is readily solvable by conventional methods. Elzinga et al.[381] make the point that convex programs of this type are more readily solved than the primal formulation.

At optimality, the primal and dual programs are related by

$$\frac{Au + \frac{1}{2}Q^t v}{\sum_i u_i} = x$$

and

$$u_i(\beta + \tfrac{\alpha}{w_i} + 2a_i^t x - a_i^t a_i) = 0 \quad \text{for } i = 1, \ldots, n$$
$$v^t(Qx - b) = 0$$

Special cases of this model are of interest.

Case 1. Equal weights (say, $w_i = 1, i = 1, \ldots, n$)

It follows that $\sum_{i=1}^{n} u_i = 1$ and hence program (D7) is the quadratic program

$$\text{Minimize} \left\{ \frac{1}{4}x^{*2} - \sum_i a_i^t a_i u_i - b^t v \right\}$$

subject to   •

$$x^* = 2Au + Q^t v$$
$$\sum_i u_i = 1$$
$$v \geq 0, \ u \geq 0$$

Case 2. No linear constraints ($Q = 0$, $b = 0$)

In this case, we have the following simple dual program.

$$\text{Minimize} \left\{ \frac{u^t A^t Au}{\sum_i u_i} - \sum_i a_i^t a_i u_i \right\}$$

subject to:

$$\sum_i \frac{u_i}{w_i} = 1, \ u \geq 0$$

6
USING VORONOI DIAGRAMS

Atsuo Suzuki
Nanzan University

Atsuyuki Okabe
The University of Tokyo

6.1 Introduction

In the OR and spatial economics literature, we find many kinds of location problems. Although these problems differ from one another, we may classify them by space into two groups: the *discrete location problem* [853] (or the *network location problem*) and the *continuous location problem* [311]. The former problem assumes that facilities can be located at a node or a link of a network. The latter problem assumes that facilities can be located at any point in the continuum of a plane. The discrete location problem has been extensively studied in OR since Hakimi [509] using efficient computational methods developed in graph and network theories. The study of the continuous location problem also dates back many years (e.g., Weber's problem [1181]), but its development was relatively slow. This slow development partly resulted from analytical difficulties in handling geometrical computation in a continuous plane. Recently this difficulty is overcome to a great extent by the progress in computational geometry [972]. In particular, efficient computational methods for constructing the Voronoi diagram have enabled us to design efficient algorithms for the solution of continuous location problems. In this chapter, we survey a class of continuous location problems solved by using Voronoi diagrams.

The chapter consists of eight sections. In the next section (Section 6.2), we briefly introduce the concept of the Voronoi diagram together with notation and terminology to be used in this chapter. In Section 6.3, we consider the *continuous p-median problem* that corresponds to the p-median problem studied in the discrete location problem. We show in detail a method for solving this problem because this method is fundamental to many continuous location problems. In fact, problems in Sections 6.5 and 6.6 are variations of the continuous p-median problem.

In Section 6.4, we discuss the *continuous p-center problem* that corresponds to the p-center problem studied as a discrete location problem. Although the objectives for the continuous p-center and p-median problems are quite different, both problems can be solved using the Voronoi diagram. The numerical method for the solution of the p-center problem is slightly different from that for the p-median problem, because a near optimal solution of the former method is worse than that of the latter method. To improve it, we adopt a finishing-up process.

In Section 6.5, we consider a continuous location problem, termed the *time-space p-median problem*, in which facilities provide periodical service. Although time and space have different dimensions, we can transform this problem into the continuous p-median problem by introducing a transformation parameter. By this transformation, we can solve the time-space p-median problem using a Voronoi diagram.

In Section 6.6, we analyze the *mobile facility location problem*. A mobile facility travels in a region and stops at several service points. Since the travel distance per day is limited, we introduce a distance constraint. This problem becomes a constrained continuous p-median problem.

In Section 6.7, we briefly review several continuous location problems and in Section 6.8 available free computer programs are listed.

6.2   The Voronoi Diagram

A popular tool for solving continuous location problems is the Voronoi diagram. The Voronoi diagram is a tessellation of the plane. Given a set of p points, $P = \{p_1, \ldots, p_p\}$ in the plane, we associate locations in the plane with the closest member of the point set P. The result is a tessellation of the plane into a set of polygons, $V_1, \ldots, V_p$, associated with $p_1, \ldots, p_p$, respectively. An example is shown in Figure 1. We call this tessellation the *Voronoi diagram*, the polygons $V_1, \ldots, V_p$ constituting the Voronoi diagram *Voronoi polygons*, the set P a *generator set*, and the members in P *generators* or *generator points*. By definition of a Voronoi polygon, if a point p is in V_i, then p_i is the nearest generator point to it . Because of this property, we sometimes say that region V_i is the territory of p_i, or p_i dominates region V_i. This simple diagram has many applications in various areas, such as biology, ecology, geography, physics, archaeology, crystallography [897].

To state the above definition mathematically, let $\boldsymbol{x}_i = (x_i, y_i)$ be the location vector on the two-dimensional Euclidean space, $\boldsymbol{R}^2$, representing the location of p_i, and $\|\boldsymbol{x} - \boldsymbol{x}_i\|$ be the Euclidean distance between $\boldsymbol{x}$ and $\boldsymbol{x}_i$. The Voronoi polygon associated with p_i is written as

$$V_i = \bigcap_{j:j \neq i} \{\boldsymbol{x} \in \boldsymbol{R}^2 \mid \|\boldsymbol{x} - \boldsymbol{x}_i\| < \|\boldsymbol{x} - \boldsymbol{x}_j\|\}. \qquad (6.1)$$

Voronoi polygon V_i is an open set. We sometimes use a closed Voronoi polygon, denoted by $\bar{V}_i$, i.e. the closure of V_i. Let ∂V_i be the boundary of $\bar{V}_i$, i.e. $\partial V_i = \bar{V}_i \backslash V_i$, and

$$W_{ij} = \partial V_i \cap \partial V_j, \qquad (6.2)$$

$$U_{ijk} = W_{ij} \cap W_{jk} \cap W_{ki}. \qquad (6.3)$$

W_{ij} (the boundary shared with two Voronoi polygons $\bar{V}_i$ and $\bar{V}_j$) is a *Voronoi edge*, and U_{ijk} (the point where three or more Voronoi edges meet) is a *Voronoi point* (the case in which exactly three Voronoi edges meet is called a non-degenerate case; in the following analysis, only the non-degenerate case is treated to avoid lengthy treatments). Voronoi edges and Voronoi points have a number of interesting properties [897].

To construct the Voronoi diagram efficiently, many algorithms have been developed in computational geometry [412, 895, 1063]. Two major methods are the divide-and-conquer method and the incremental method [972]. The computational time of the former method is $O(p \log p)$ in the worst case and on the average. The later method, more specifically, the quaternary incremental algorithm by [895] runs in time $O(p^2)$ in the worst case, but runs in $O(p)$ on the average. Actually it constructs the Voronoi diagram in about 0.3 ms per one generator point with Sun SPARC2 computer. This efficiency is indispensable in solving continuous location problems because we often use the Voronoi diagram iteratively in such problems.

6.3   The Continuous p-median Problem

One of the most basic problems in the discrete location problem is the p-median problem. This problem is to find a set of the locations of p facilities that minimizes the total weighted distance from users to their nearest facility. The p-median problem is easily defined in a continuous space. However, methods for solving these two problems are quite different. We first show the mathematical formulation of the p-median problem on a continuous plane, and then develop a computational method for solving it.

6.3.1   Mathematical Formulation

We consider a set of p facilities that supply the same service in a region S (say a square region represented by a unit square on a plane). Demand for the service by users is distributed continuously over S. The density of this demand at x is $\phi(x)$. Users use their nearest facility among the p facilities, and the access cost of a user is the function, $f(t^2)$, of the distance t, between the user and his(her) nearest facility. Under these assumptions, the problem is to determine the location of the p facilities so that the total cost of the users is minimized. The objective function is given by the total access cost of the users:

$$F(\boldsymbol{X}) = F(\boldsymbol{x}_1, ..., \boldsymbol{x}_p) = \int_S f(\min \|\boldsymbol{x} - \boldsymbol{x}_i\|^2)\phi(\boldsymbol{x})\mathrm{d}\boldsymbol{x}. \qquad (6.4)$$

Minimizing $F(\boldsymbol{X})$ is a non-linear unconstrained minimization problem with $2p$ variables.

We now reformulate equation (6.4) in terms of the Voronoi diagram. Since users are assumed to use their nearest facility, the region in which users use facility i is given by V_i. Thus we can decompose the integral of the objective function (6.4) into

$$F(\boldsymbol{X}) = \sum_{i=1}^{p} \int_{V_i} f(\| \boldsymbol{x} - \boldsymbol{x}_i \|^2)\phi(\boldsymbol{x})\mathrm{d}\boldsymbol{x} = \sum_{i=1}^{p} F_i, \qquad (6.5)$$

which is more tractable than (6.4) (note that F_i represents the total access cost to the nearest facility from all users in V_i). Once the Voronoi diagram is constructed, we can numerically compute the integral of (6.5).

6.3.2 Solving the continuous p-median problem

By equation (6.5) the objective function is non-convex, and has non-differentiable points. We therefore adopt a numerical iterative method that gives an approximate solution rather than employ analytical methods. Following [624], we employ a simple descent method. Starting from an initial configuration of facilities, we first compute the objective function (6.5), and determine a descent direction as a function of the partial derivatives of equation (6.5) and the other auxiliary quantities. A line search along that direction is performed, and the solution updated. This process is continued until the change in the solution is less than a given tolerance.

In the above process, it is crucial to obtain the partial derivatives of the objective function and the Hessian. The details of the derivation can be found in [624, 897]. In the following the results are summarized.

$$\frac{\partial F}{\partial x_i{}^\lambda} = \int_{V_i} (x_i{}^\lambda - x^\lambda)f'(\| x - \boldsymbol{x}_i \|^2)\phi(\boldsymbol{x})\mathrm{d}\boldsymbol{x} = \frac{\partial F_i}{\partial x_i^\lambda} \quad (\lambda = 1, 2) \quad (6.6)$$

$$\frac{\partial^2 F}{\partial x_i{}^\lambda \partial x_j{}^\kappa} = \begin{cases} H_{\lambda\kappa}^i + G_{\lambda\kappa}^i & (j = i), \\ G_{\lambda\kappa}^{ij} & (j \neq i, W_{ij} \neq \emptyset), \\ 0 & (j \neq i, W_{ij} = \emptyset), \end{cases} \qquad (6.7)$$

where

$$H_{\lambda\kappa}^i = \int_{V_i} [f'(\| x - \boldsymbol{x}_i \|^2) + 2(x_i{}^\kappa - x^\kappa)(x_i{}^\lambda - x^\lambda)f''(\| x - \boldsymbol{x}_i \|^2)]\phi(\boldsymbol{x})\mathrm{d}\boldsymbol{x},$$

$$G_{\lambda\kappa}^i = - \sum_{\substack{j:j \neq i \\ W_{ij} \neq \emptyset}} K_{\lambda\kappa}{}^{iji},$$

$$G^{ij}_{\lambda\kappa} = K_{\lambda\kappa}{}^{jji},$$

$$K_{\lambda\kappa}{}^{kji} = \int_{W_{ij}} \frac{1}{\alpha_{ij}}(x_i^{\kappa} - x^{\kappa})(x_k^{\lambda} - x^{\lambda})f'(\parallel x - x_i \parallel^2)dx, \quad \lambda, \kappa = 1, 2.$$

Calculating this Hessian requires extensive computational effort. To reduce the computational effort, [624] use only the diagonal elements of $H^i_{\lambda\kappa}$ of (6.7) as an approximate Hessian. The descent direction is given by

$$d_i^{\lambda} = (H^i_{\lambda\lambda})^{-1}\frac{\partial F}{\partial x_i^{\lambda}} \quad (i = 1, \ldots, n; \lambda = 1, 2). \tag{6.8}$$

When $f(t^2)$ is given by $f(t^2) = t$, the direction is determined by $d_i^{\lambda} = -(x_i^{\lambda} - \overline{x}_i^{\lambda})$ using the following quantities:

$$\frac{\partial F}{\partial x_i^{\lambda}} = \int_{V_i}(x_i^{\lambda} - x^{\lambda})\phi(x)dx = \mu(V_i)(x_i^{\lambda} - \overline{x}_i^{\lambda}), \tag{6.9}$$

$$\mu(V_i) = \int_{V_i}\phi(x)dx, \tag{6.10}$$

$$\overline{x}_i^{\kappa} = \int_{V_i}x^{\kappa}d\mu(x)/\mu(V_i), \tag{6.11}$$

$$H^i_{\lambda\kappa} = \mu(V_i)\delta_{\lambda\kappa}. \tag{6.12}$$

Note that this direction is toward the centroid of the Voronoi region. Thus, in each iteration, each facility moves toward the centroid of its Voronoi region. Consequently, when the procedure ends (i.e. the partial derivatives (6.6) are close to zero), each facility is near the centroid of its Voronoi region. Mathematically, this means that each facility is near the point at which F_i (with $f(t^2) = t$ in equation (6.5)) is minimized.

When $f(t^2)$ is given by $f(t^2) = \sqrt{t}$, the partial derivatives are given by

$$\frac{\partial F}{\partial x_i^{\lambda}} = \int_{V_i}\frac{(x_i^{\lambda} - x^{\lambda})}{\parallel x - x_i \parallel}\phi(x)dx. \tag{6.13}$$

In the final state of the iteration, these values are very close to zero. This means that each facility is near the point at which F_i (with $f(t^2) = \sqrt{t}$ in equation (6.5)) is minimized. Note that for V_i, the point at which F_i is minimized is called the *continuous Weber point*. Thus, each facility moves to the Weber point of its region. We use this property to develop a heuristic method. We now state the above procedure as an algorithm.

1. Initialization: Set an initial value $x_i^{(0)}$, $i = 1, \ldots, p$ and $\nu = 0$.

2. Construct the Voronoi diagram generated by a set of points $x_i^{(\nu)}$'s

3. Compute the value of F, the partial derivatives of F and the auxiliary quantities by numerical means.

4. Determine the descent direction $\boldsymbol{d}^{(\nu)}$ by (6.8).

5. Replace $\boldsymbol{x}_i^{(\nu)}$ with $\boldsymbol{x}_i^{(\nu+1)}$ using the formula $\boldsymbol{x}_i^{(\nu+1)} = \boldsymbol{x}_i^{(\nu)} + \hat{\alpha}\boldsymbol{d}^{(\nu)}$, where $\hat{\alpha}$ is a parameter determined in the line search.

6. If the difference between $\boldsymbol{x}^{(\nu)}$ and $\boldsymbol{x}^{(\nu+1)}$ is small enough, stop. Otherwise, increase ν by 1 and go to step 2.

An example is shown in Figure 1 which is drawn by the program coded by [895] under the assumption that the population density is uniform over the unit square S, and $f(t^2) = \sqrt{t}$. Figure 1 (a) shows the initial configuration of $p = 100$ facilities; panel (b) shows a near optimal configuration of the 100 facilities. As is seen in this panel, the resulting configuration is a bee-hive pattern. An actual application is studied by [899] where the configuration of public mail boxes in Koganei, Tokyo is optimized taking the actual population distribution into account.

Recently, Iri et al. applied their automatic differentiation program, called PADRE2, to this problem and obtained good results [623, 718]. The program generates the FORTRAN subroutine of the partial derivatives (6.6) and the Hessian (6.7) of the objective function (6.5) from the subroutine of the objective function (6.5). Their method reduces the difficulty in computing the partial derivatives of the objective function (6.5) analytically. In addition, their method gives a good estimate of the rounding errors. This estimate is useful in determining a stopping criterion.

6.4   Continuous p-center Problems

In the p-center problem we need to find a set of facilities which minimize the maximum distance from a user to its nearest facility. Examples are emergency facilities, such as fire stations, hospitals, and patrol car centers for a security guard company. In the discrete location problem, these location problems are called the *p-center problem*. In this section we discuss the continuous version of the p-center problem.

6.4.1   Mathematical formulation

We assume, as in the p-median problem, that users use the nearest facility among p facilities, and that their demand is continuously distributed over a region S (say a square region). Let R be the maximum distance among the distances between users and their nearest facility. We wish to find a configuration of p facilities, $\boldsymbol{x}_1, \ldots, \boldsymbol{x}_p$, that gives the minimum value of R. Stated differently, we wish to find a configuration of the centers of p disks that cover S whose radius is the smallest. Mathematically, we minimize the following objective function.

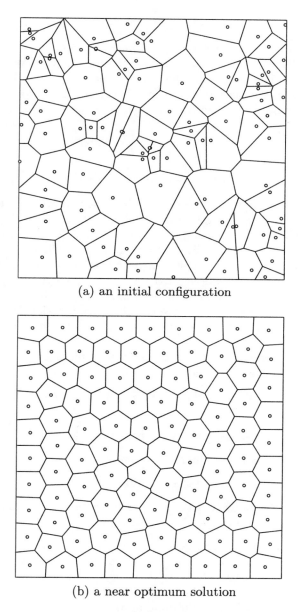

(a) an initial configuration

(b) a near optimum solution

FIGURE 1. An approximate solution of the continuous p-median problem for $p = 100$

$$F(\boldsymbol{x}_1, \ldots, \boldsymbol{x}_p) = \max_{\boldsymbol{x} \in S} \min_{i=1,\ldots,p} ||\boldsymbol{x} - \boldsymbol{x}_i||. \tag{6.14}$$

Since a user in V_i uses the facility at $\boldsymbol{x}_i$, and a Voronoi polygon is convex, a user whose distance is the longest among the users in V_i is located at the boundary of V_i. The maximum of the objective function (6.14) is hence attained at one of the Voronoi points or a point on the boundary of S. Let U_{ijk} be a Voronoi point. Then the objective function (6.14) is written as

$$F(\boldsymbol{x}_1, \ldots, \boldsymbol{x}_p) = \max_{i=1,\ldots,p} \max_{\boldsymbol{x} \in (U_{ijk} \cup (\overline{S} \cap V_i))} ||\boldsymbol{x} - \boldsymbol{x}_i||. \tag{6.15}$$

6.4.2 Solving the Continuous p-center Problem

As mentioned in the preceding section, in the p-median problem, each facility in a near-optimal configuration is placed near the centroid (if $f(t^2) = t$) or the continuous Weber point (if $f(t^2) = \sqrt{t}$) of its Voronoi polygon. In the p-center problem, each facility in a near-optimal configuration is placed near the center of its Voronoi polygon (note that the center of a polygon is the point from which the longest distance to a point in the polygon is minimized). Computational methods for solving these problems are different, because the partial derivative of the objective function (6.15) cannot be used. Thus we employ a heuristic method for the continuous p-center problem.

Our method consists of two phases: an iterative phase and a finishing-up process. The iterative process is:

1. Generate p centers randomly in S (an initial configuration of p-centers).

2. Construct the Voronoi diagram generated by a set of these p centers.

3. For each Voronoi polygon, find its center.

4. Relocate the p-centers to the resulting centers.

5. If the change produced by the relocation is greater than a pre-specified tolerance level, go to step 1; otherwise, stop.

Since a near optimal configuration depends on an initial configuration, we repeat the above process m times and choose the best one as a near optimal solution. Experiments show that the best configuration may not always be close to the true optimal configuration even for a fairly large m. Thus we employ the following finishing-up process.

Let $\boldsymbol{x}_{il} = (x_{il}, y_{il})$ be the lth Voronoi point of the ith Voronoi polygon V_i (or $V_i \cap S$) where $l = 1, \ldots, l_i$; l_i is the number of points on V_i (or $V_i \cap S$). Then we solve the following non-linear programming problem.

$$\min_{\boldsymbol{x}_i, \boldsymbol{x}_{il}, l=1,\ldots,l_i, i=1,\ldots,p} R^2,$$

subject to: $(x_i - x_{il})^2 + (y_i - y_{il})^2 \leq R^2$; $\forall l = 1, \ldots, l_i$; $i = 1, \ldots, p$

Note that a feasible value for (x_{il}, y_{il}) is restricted when (x_{il}, y_{il}) is on the boundary of S. To solve this non-linear programming problem, we employ AMPL [418], an optimization package for non-linear programming, (a sample program an data is given in [1109]). Figure 2 depicts two examples ($p = 30$ and $p = 49$) obtained by this program [1109]. The resulting configurations are similar to those of the continuous p-median problem.

6.5   The Time-Space p-Median Problem

In the above sections, we implicitly assume that facilities provide uninterrupted service. In this section, we deal with a location problem of facilities which provide their service only on a specific day in a week (month, or year), or facilities which open periodically. A typical example is a market place which opens with regular time intervals. Since this location problem involves not only space but also time, the problem looks quite different from the continuous p-median problem. We show, however, that this location problem may be regarded as an extension of the continuous p-median problem, called the *time-space p-median problem*.

6.5.1   *Mathematical formulation*

We consider a region S in which there are m facilities and facility i provides its service at time $t_i + ks$, $k = \ldots, -1, 0, 1, \ldots$ where s is the length of an interval time period and $0 \leq t_i \leq s$. The period in which a facility provides its service is so short in comparison with s that the service period may be regarded as a point in time. To illustrate this type of choice behavior, suppose that a consumer at x has a habit of eating fish at dinner every Friday. If market i at x_i opens every Monday, he(she) has to wait for three days and travels $||x - x_i||$. If market j opens every Thursday, he has to eat fish a day earlier and travels $||x - x_j||$. In both cases, we may measure consumer's dissatisfaction by the number of days between the day when he wants to eat fish and the day when the market opens. Suppose that market j is farther than market i from the consumer. Then which store is chosen? If the customer does not mind travel that much but is impatient with the day discrepancy, he will choose market j. On the other hand, if he does not like travel but is patient with the day discrepancy, he will choose market i. These choices are two extremes. Usually a consumer chooses a store considering the travel distance and the day discrepancy, that can be stated as choosing according to a general distance consisting of a spatial distance and a time distance. The general distance d, called the *time-space distance*, between a user at location x at time t and facility i at location x_i that opens at t_i is given by

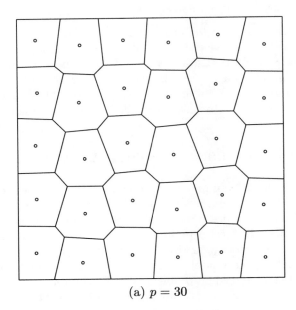

(a) $p = 30$

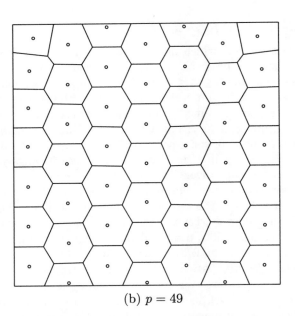

(b) $p = 49$

FIGURE 2. Example solutions of the continuous p-center problem

$$d^2 = ||\boldsymbol{x} - \boldsymbol{x}_i||^2 + \alpha^2(t - t_i)^2, \tag{6.16}$$

where α is a balancing factor. In the p-median problem, we assume that every user chooses the nearest facility with respect to the Euclidean distance. In the time-space p-median problem, we assume that every user chooses the nearest facility with respect to the time-space distance.

For ease of exposition, we assume that region S is represented by a unit line segment. Facility i at location x_i that opens at time t_i is represented by the coordinates (x_i, t_i) on the plane. We denote the demand density at location x at time t by $\phi(x, t)$.

From the choice behavior assumed above the total time-space distance of users in region S over a unit interval s, is given by

$$F(x) = \int_0^1 \int_0^s f(\min_{1 \leq i \leq m} [||x - x_i||^2 + \alpha^2(t - t_i)^2])\phi(x, t)\mathrm{d}x\mathrm{d}t. \tag{6.17}$$

In this equation, we may set $\alpha = 1$ by changing the unit of time. Once this transformation is made, we can construct the Voronoi diagram generated by points $(x_i, t_i), i = 1, \ldots, m$ on the x-t plane (Figure 3). In terms of a Voronoi polygon of this Voronoi diagram, the above choice behavior is stated as: a user in V_i chooses facility i. Thus, the objective function (6.17) is written as

$$F(x_1, \ldots, x_m) = \sum_{i=1}^m \int_{V_i} f(||x - x_i||^2 + \alpha^2(t - t_i)^2)\phi(x, t)\mathrm{d}x\mathrm{d}t. \tag{6.18}$$

When region S is given by an area on the plane, we should modify equation (6.18) by replacing x_i with $\boldsymbol{x}_i$. In this case, the Voronoi diagram becomes a 3-dimensional Voronoi diagram (which is difficult to show on the plane).

6.5.2 Solving the Time-Space p-Median Problem

A method for solving the time-space p-median problem is similar to the method for the continuous p-median problem except in two respects. First, we use a $(k + 1)$-dimensional Voronoi diagram when the market region is k-dimensional space. Second, in the iterative process, since t_i is assumed to be fixed, a variable $(\boldsymbol{x}_i, t_i)$ is variable only with respect to $\boldsymbol{x}_i$.

Figure 3 shows some examples obtained by the above method in [1118]. In these examples, we treat only linear markets and two market places are opened each time. We use 2-dimensional Voronoi diagram whose horizontal and vertical axes represent the linear market and time, respectively. These examples show how a near optimal configuration varies with respect to the

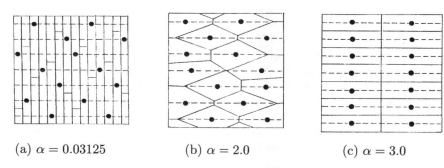

(a) $\alpha = 0.03125$ (b) $\alpha = 2.0$ (c) $\alpha = 3.0$

FIGURE 3. Examples of the solution of the time space p-median problem : horizontal and vertical axes represent linear market and time, respectively

value of α. In Figure 3, the length of a linear region is 1 km, $m = 14$ and $s = 7$ (a week). A value, $\alpha = 2$, implies that the disutility of traveling 2 km is equivalent to the disutility of waiting one day.

As shown by [1113], the above time-space p-median problem can be modified in a few ways. First, the time-space distance may be modified by considering that time flows only in one direction (we cannot go back time). The time-space distance given by equation (6.16) allows $t > t_i$ as well as $t < t_i$, but it may be defined only for $t < t_i$ (or $d = \infty$ if $t > t_i$). Second, the service period may be extended from a very short time (a point) to a certain length of time. In this case, a facility in the time-space dimension is represented by a line segment and a Voronoi diagram for line segments (instead of points) is to be used (see [897]).

6.6   Mobile Facility Location Problem (Constrained p-median)

In the preceding sections, we dealt with fixed facilities, but generally facilities may include mobile facilities. An example is a mobile library. Usually a mobile facility travels in a region and stops at several service points where users can receive service. The problem is to find a configuration of the service points to minimize the total distance of users to their nearest service points. Thus the objective function is the same as that of the p-median problem. The formulation differ because a mobile facility can travel a limited distance in one day. We call the p-median problem with this constraint the *constrained p-median problem*.

6.6.1   Mathematical Formulation

Let x_i be the location of the ith service point and V_i be the Voronoi polygon associated with the ith service point. Then the objective function is

$$F(x_1, \ldots, x_n) = \int f(\min_i \| x - x_i \|^2)\phi(x)dx = \sum_{i=1}^{n} \int_{V_i} f(\| x - x_i \|^2)\phi(x)dx.$$

$$(6.19)$$

Since the total travel distance is limited, a constraint $\sum_{i=1}^{n-1} \|x_i - x_{i+1}\| \leq D$ is added, or,

$$\|x_i - x_{i+1}\| \leq d, \quad i = 1, \ldots, n-1. \qquad (6.20)$$

Obviously, if the constraint is loose enough (i.e. d is very large relative to S), an optimal configuration becomes the same as that of the continuous p-median problem.

6.6.2   Solving the mobile facility location problem

A method for solving the mobile facility problem is similar to that for the p-median problem if we can convert the above constrained programming problem into a non-constrained one. Several methods are available for this conversion [401]. Following [1118], we employ the multiplier method [572, 971] which is a penalty function method. We incorporate the constraint (6.20) into the objective function with Lagrange multipliers and penalty parameters. The extended Lagrangean function is

$$L_t(X, \lambda) = F + \sum_{i=1}^{n-1} \frac{1}{2t_i}[\max\{0, \lambda_i + t_i(\|x_i - x_{i+1}\| - d)\}^2 - \lambda_i^2], \quad (6.21)$$

where λ_i's are Lagrangean multiplier and t_i's are penalty parameters. With this extended Lagrangean function, we can numerically solve the constrained p-median problem in the following manner.

1. Set $t_i^{(0)} > 0, \lambda_i^{(0)} = 0, i = 1, \ldots, n; \epsilon > 0, c^{(0)} = \infty, \nu = 0$.

2. Solve the unconstrained minimization problem $\min_X L_{t^{(\nu)}}(X, \lambda^{(\nu)})$. Let $X^{(\nu+1)}$ be the resulting solution.

3. Replace $c^{(\nu+1)}$ with

$$c^{(\nu+1)} = \min[\max_i | \max\{\|x_i^{(\nu+1)} - x_{i+1}^{(\nu+1)}\| - d, -\frac{\lambda_i^{(\nu)}}{t_i^{(\nu)}}\}|, c^{(\nu)}].$$

If $c^{(\nu+1)} < \epsilon$, stop. Otherwise, go to step 4.

4. Replace λ_i and t_i with

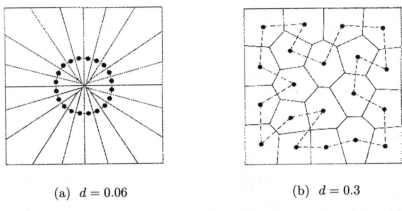

(a) $d = 0.06$ (b) $d = 0.3$

FIGURE 4. An example of the solution of the mobile facility location problem

$$\lambda_i^{(\nu+1)} = \begin{cases} \max\{0, \lambda_i^{(\nu)} + t_i^{(\nu)}(\|x_i^{(\nu+1)} - x_{i+1}^{(\nu+1)}\| - d)\} & (c^{(\nu+1)} < c^{(\nu)}), \\ \lambda_i^{(\nu)} & (\text{otherwise}), \end{cases}$$

$$t_i^{(\nu+1)} = \begin{cases} 10t_i^{(\nu)} & (|\max(\|x_i^{(\nu+1)} - x_{i+1}^{(\nu+1)}\| - d, -\frac{\lambda_i^{(\nu)}}{t_i^{(\nu)}}| > \beta c^{(\nu)}), \\ t_i^{(\nu)} & (\text{otherwise}) \end{cases} \quad (\beta = 0.25)$$

and go to step 2.

Figure 4 shows near optimal configurations of service points with respect to d. These figures show that service points becomes more dispersed as d becomes longer.

6.7   Other Continuous Location Problems

In addition to the continuous location problems formulated in the preceding sections, many other continuous location problems are found in the literature. These problems are briefly reviewed in this section.

In the continuous p-median problem, we assume that users behave individually: when a user chooses a facility, he does not consider the convenience of other users. Users, however, do not always behave individually, they sometimes behave as a group. A group of users chooses a facility considering the convenience of all the users in the group. An example is a choice of a tennis court or a baseball field. Baseball requires two teams of nine members each. To choose a baseball field, we consider the convenience of 18 members. This type of location problems is considered by [894], who formulate it as a variation of the p-median problem. They show that as the number of members in a group increases, facilities tend to be more centralized.

Facilities considered in the above continuous location problems are final destinations, but some kinds of facilities may not be final destinations. Consider, for example, bus stops on a bus route connected to a railway station. In this case, a final destination is the railway station and bus stops are intermediate destinations. A commuter walks to a bus stop and takes a bus to a railway station. Obviously the commuter chooses a bus stop from which he(she) can reach the station with the shortest commuting time. Thus, the commuter may not choose the nearest bus stop. As a result, the area in which commuters use the ith bus stop is not given by Voronoi polygon V_i of the ordinary Voronoi diagram generated by the bus stops. Instead, the area is given by a Voronoi region of the weighted Voronoi diagram [897]. Using this Voronoi diagram, [1112] obtained an optimal configuration of bus stops along a bus route. Although the weighted Voronoi diagram is different from the ordinary Voronoi diagram, the computational method adopted by [1112] is almost the same as that for the continuous p-median problem. An empirical study by [1112] shows that an existing configuration of bus stops does not achieve the minimum total travel time of commuters; rather it maximizes the revenue of the bus company.

A method for solving a continuous location problem may be utilized to understand spatial competition of retail stores. [898] formulates a two-dimensional spatial competition model as an extension of a one-dimensional Hotelling model [599]. Consumers are assumed to choose their nearest stores, and retailers are assumed to maximize their profits independently. Under the assumption of a uniform demand over a region, maximizing the profit of a store is equivalent to maximizing the market area of the store which is given by the Voronoi polygon associated with the store. The problem is to find a stable configuration, if it exists, of stores competing for market share. The numerical simulation of [898] shows that the configuration becomes fairly stable and it is similar to a bee-hive pattern. This dispersed pattern is different from the concentrated pattern obtained in Hotelling's one-dimensional spatial competition model, suggesting that the dimension of space is crucial to understand location behavior [31, 896].

Mobile facilities discussed in Section 6.6 provide their service at several points, but some kinds of mobile facilities may provide their service at any point on a traveling route. Consider, for example, a peddler wagon selling ice cream ringing a bell (or sounding a music). When a consumer wants to have a cone of ice cream at his(her) home, he goes to the nearest point of the service route along which that wagon travels. The peddler wants to determine the service route that achieves the maximum profit with the constraint that the wagon can travels a limited distance in a day. Thus the location problem is to find an optimal route that maximizes the profit with the constraint. This problem is studied by [1118]. He treats the service area of the wagon in terms of a Voronoi diagram generated by a chain of line segments. The computational method becomes very similar to the constrained continuous p-median problem by approximating a line segment constituting the service route by two end points and a middle point.

In the real world we find many phenomena that show a Voronoi-like pattern. For example, the territories of the fish *tillapia mossambica* look like a Voronoi diagram. In this case, we wish to know to what extent an observed tessellation is close to a Voronoi diagram. The problem is thus to fit a Voronoi diagram to a given tessellation. This problem is investigated by [1110]. The objective function is given by the difference in the area between a given tessellation and a Voronoi diagram, and this objective function is to be minimized. The computational method for solving this problem is similar to that for the p-median problem.

This Voronoi fitting method can be applied to a school location problem. Students are supposed to attend the schools in their school district. Since school districts are pre-determined, some students do not attend their nearest school. The problem is to find a configuration of schools that minimizes the number of students who do not attend their nearest school. This problem can be solved by the Voronoi fitting method.

A dual diagram of the Voronoi diagram is called the *Delaunay triangulation*. We can use this diagram to solve some network problems. One example is the *Euclidean Steiner problem*. Given a set of points, the problem is to obtain the shortest network which connects all the given points adding, if necessary, relay points. This problem is proved to be an NP hard problem. We should hence employ a heuristic method when we solve a large number of points. [1111] develops such a method which is similar to the computational method adopted in the continuous location problem mentioned above.

6.8   Conclusions

Many continuous location problems can be solved using Voronoi diagrams: the ordinary 2-dimensional Voronoi diagram (Sections 6.3, 6.5, 6.6), the ordinary 3-dimensional Voronoi diagram (Section 6.5), the line Voronoi diagram (Section 6.7), the weighted Voronoi diagram (Section 6.7). When continuous location problems are solved, efficient computer programs for constructing these Voronoi diagrams are essential. Some programs are open to the public. VORONOI2 developed by [1107, 1108] constructs the 2-dimensional Voronoi diagram. This program can be modified to construct the weighted Voronoi diagram [1106]. SEGVOR developed by [619, 620] construct the line Voronoi diagram. L1VOR developed by Suzuki constructs the Voronoi diagram of ℓ_1 metric. These programs are available upon request.

Part II

VARIOUS OBJECTIVES IN FACILITY LOCATION

7
LOCATION WITH MARKET EXTERNALITIES

Margaret L. Brandeau
Stanford University

Samuel S. Chiu
Stanford University

Shiv Kumar
Applied Insurance Research

Thomas A. Grossman, Jr.
University of Calgary

7.1 Introduction

Research on location decisions has generally assumed away market externalities, overlooked the latitude available to customers in choosing a facility to use, and ignored the effects of transportation and congestion costs on consumer behavior. However, congestion externalities, user choice, and demand elasticities do exist in practice and, therefore, it is quite appropriate that they be incorporated rigorously into locational analysis.

A congestion externality exists when the utility of one consumer is affected negatively or positively by the actions of other consumers. Examples of negative congestion externalities include waiting time in service facilities, traffic delay in transportation systems, the blocking probability in telecommunication systems, the student-to-teacher ratio in schools, the bed occupancy level in hospitals, and noise pollution in residential areas. Congestion may also act as a positive externality when consumers are crowd lovers (e.g. nightclubs).

Consumers often select a facility to minimize their own total cost of receiving service. This cost is likely to include both travel cost and a cost associated with the congestion externality. We refer to this situation as a "user-choice" environment. Since the utilization of a facility depends on the decisions of all the users in aggregate, the user-choice environment has at its core an equilibrium problem. Facility utilization decisions are made in this manner for many types of facilities, including post offices, grocery stores, bank branches, automated teller machines, and retail outlets. In other cases, consumers are assigned to facilities by a central authority. We call this a "dictatorial" environment. Customer assignments are made in this manner for voters' polling places and jobs run on networked computer systems.

If facilities provide essential services (such as selling groceries, issuing driver's licenses, or providing polling places on election day), consumer demand may be fixed. However, for facilities that provide non-essential services (for example, fast food restaurants or retail stores), consumer demand may be a function of the total cost of receiving service.

This chapter describes models for the efficient location of facilities and the allocation of a limited resource (e.g. service capacity) among facilities in an environment with negative congestion externalities. We consider models that incorporate user choice and models that assume a dictatorial environment, as well as models with elastic and inelastic demand. We describe several published models, present some new models and results, and highlight areas where further research is needed.

In Section 7.2, we introduce our notation and assumptions. In Section 7.3, we study the user-choice equilibrium and provide some existence and uniqueness results. In Section 7.4, we consider the location of facilities with market externalities under fixed and elastic demand environments. For the case of fixed demand, we present two location models – one for public facilities that locate to minimize total consumer cost, and one for competing private facilities that locate to maximize their own market share. Both models assume a user-choice environment. For the case of elastic demand, we present two location models – one for a single private facility in a user-choice environment, and one for multiple public facilities in a dictatorial environment. In Section 7.5, we consider the related problem of optimal resource allocation in the presence of market externalities. A limited resource (e.g. service capacity) that can be used to reduce congestion at facilities is to be allocated among facilities. We present two models – one for the optimal allocation of resources in a user-choice environment and one for a dictatorial environment. Both models assume inelastic demand. We conclude in Section 7.6 with suggestions for further research.

As a preliminary, we briefly review the existing location literature, with particular emphasis on models that incorporate externalities and models that allow for elastic demand. For our purposes, we classify the location literature into four groups depending upon the presence or absence of externalities and whether demand is elastic or inelastic.

7.1.1   Models with Inelastic Demand and No Externalities

The considerable literature on location models with inelastic demand and no externalities follows the four classical prototype models of the p-median, p-center, simple plant location, and quadratic assignment problems (and extensions and refinements thereof). This literature has been thoroughly reviewed over the years by various authors (e.g. [127]).

7.1.2   Models with Elastic Demand and No Externalities

Research on location decisions in the presence of elastic demand has been more limited. [990] presented gravity models for shopping center locations when consumers will buy from the nearer of the two equally attractive centers. In an early paper that explored the optimal location of public facilities with price-sensitive demands, [1168] defined demand as a function of delivered price at various locations and then maximized the net social benefit. They considered a simple step function for demand which is constant as long as the price is below a certain willingness to pay, and zero otherwise. [555] solved a similar problem for private facilities with the objective of maximizing profit. [389] generalized these models and included the case of quasi-public enterprises that maximize net social benefit subject to a constraint ensuring sufficient revenue to cover costs.

[833, 593] observed empirically that the market share of a company branch is a decreasing convex function of the travel distance. [99] used this observation to analyze the location of a single private facility on a network with distance-dependent demand. They considered the demand at each node to be a Poisson distributed random variable whose mean is a negative exponential function of the shortest distance between the node and the facility.

Some research has also been done on competitive location under elastic demand with no externalities. [1090] generalized Hotelling's [599] theory of spatial competition to include distance-dependent demand. [511, 513] formulated and analyzed several competitive location models for non-essential services where customers patronize a facility in inverse proportion to its distance. [731, 746] have developed competitive models where demand is elastic with respect to both distance and price.

7.1.3   Models with Inelastic Demand in the Presence of Externalities

Location models incorporating externalities have mostly been developed in the context of emergency service facility location and have almost universally assumed the externality to be queuing delay. [737] initiated the study of queuing in facility location with the hypercube model. [93] later generalized the p-median problem to include queuing-like congestion of the facilities.

[95] examined a mobile facility location problem on a general network with the objective of minimizing average response time to customers. [211] formulated the p-server single-facility loss median (p-SFLM) and stochastic queue median (SQM) versions of this problem depending on whether waiting customers are lost or put in an M/G/1 queue, respectively. More recently, [129] formulated and analyzed the stochastic queue center (SQC) location problem whose objective is to minimize expected response time to

the farthest customer.

[94] gave efficient heuristics for service territory allocation between two servers operating as $M/G/1$ queuing systems without cooperation. [98] later generalized these heuristics to solve for optimal location-allocation decisions for p mobile servers acting separately as $M/G/1$ systems. [97] used the hypercube model to solve the stochastic queue p-median problem with server cooperation.

There has been some recent interest in analyzing competitive locations for congested facilities in a user-choice environment. [708] showed that, when customers are uniformly distributed on a line, minimum differentiation occurs for two facilities but no locational equilibrium exists for three or more facilities. [125] analyzed the locational equilibrium when two congested public facilities are located by competing governmental jurisdictions. [131, 132] studied the equilibrium user choice pattern and location decisions for public and private facilities in the presence of negative externalities. Their location models and results will be described in Section 7.4.

7.1.4   Models with Externality-Elastic Demand

The literature on location models with elastic demand and externalities is sparse. The fact that demand can be a function of an externality such as waiting time was observed and modeled by [626] in planning for the capacity of service operations. [156, 917] studied the effects of congestion externalities on the attractiveness of supermarkets and hospitals, respectively. [750] characterized the equilibrium in disaggregate facility choice systems subject to congestion-elastic demand. [462] considered a model in which customers select a facility based on both distance and facility attributes such as size, service quality, and cleanliness. [87] presented the first facility location model on a network that assumed mean demand to be a function of service delay. [726, 727, 728, 729] have recently studied location decisions for private facilities under externality-elastic demand. Some of this research will be described in Section 7.4 of this chapter.

We now present the notation and assumptions used in the remainder of this chapter.

7.2   Notation and Assumptions

Let $G(N, A)$ be an undirected network with node set $N(|N| = n)$ and arc set A. Let M denote the set of the m facilities and let $F_j, j \in M$, be the location of facility j.

We assume that demand is nodal. Demand per unit time at each node $i \in N$ is an independently distributed Poisson random variable D_i with mean λ_i. Let the part of demand at node i that visits facility j be D_{ij} with mean λ_{ij}. The D_{ij}'s are assumed to be independent Poisson variables.

Then the total mean demand originating from node i, for all $i \in N$, is

$$\lambda_i = \sum_{j=1}^{m} \lambda_{ij}.$$

The facilities are immobile: users travel to the facilities for service. Travel between any two points is along the shortest path. Consumers visiting a particular facility constitute only a small fraction of the network users, so congestion effects on travel times can be ignored. Let $d(a,b)$ denote the shortest path distance between points a and b. Let the function $T_i(d(i, F_j))$ denote the corresponding round trip travel cost incurred by the consumer at node i when obtaining service at facility j. We assume that $T_i(\bullet)$ for all $i \in N$ is a strictly increasing, twice differentiable, nonnegative concave function. The travel cost function is indexed by i to model the fact that the means of transportation at various nodes may not be the same.

The total demand visiting facility j is a random variable $D_j = \sum_{j=1}^{n} D_{ij}$ For a queuing-based facility, since the D_{ij}'s are independently Poisson distributed, the arrivals at facility j are the output of an $M/G/\infty$ process. In this case, D_j is also a Poisson distributed random variable with mean $\Lambda_j = \sum_{i=1}^{n} \lambda_{ij}$. The quantity Λ_j is our measure of the congestion at facility j.

Let $P_j(\Lambda_j)$ denote the negative externality at facility j due to congestion Λ_j. We assume that $P_j(\bullet)$, $j \in M$, is a strictly increasing, twice differentiable, nonnegative convex function. A common example of an externality function satisfying these assumptions is the queuing delay at a facility. If the jth facility operates as an M/G/1 system, then $P_j(\Lambda_j) = \dfrac{\Lambda_j \overline{S_j^2}}{2(1 - \Lambda_j \overline{S_j})}$ where $\overline{S_j}$ is the mean service time and $\overline{S_j^2}$ is the second moment of service time at facility j. Note that each facility may have a different externality function. This reflects the fact that facilities may not be identical even though they provide similar services.

We assume that consumers are homogeneous in that they all assign an identical cost to a negative externality. Let $C(P_j(\Lambda_j))$ denote the externality cost incurred by a user visiting facility j where $C(\bullet)$ is a strictly increasing, twice differentiable, nonnegative convex function. To simplify notation, we henceforth denote $C(P_j(\Lambda_j))$ by $E_j(\Lambda_j)$. It is readily seen that the externality function $E_j(\bullet)$ is a strictly increasing, twice differentiable, nonnegative convex function for all $j \in M$.

Users patronize the facility (or facilities) for which the sum of their travel cost plus externality cost is minimized. Define the variable u_i, $i \in N$, as follows: $u_i = \min_{j \in M} \{T_i(d(i, F_j)) + E_j(\Lambda_j)\}$. Consumers at node i will visit a facility for which the sum of their travel cost and externality cost is u_i, the minimum cost over all facilities.

If demand is elastic, the dependence of the mean demand rate λ_i at node

i on the total cost u_i is given by

$$\lambda_i = \overline{\lambda}_i h_i(u_i) \tag{7.1}$$

where $\overline{\lambda}_i$ is the maximum possible demand rate at node i (this may be thought of as a surrogate for the total population at node i), and $h_i(\bullet)$: $\mathcal{R}_+ \to [0,1]$ is a strictly decreasing, twice differentiable, convex function for each $i \in N$, where $\mathcal{R}_+ = \{z \in \mathcal{R} | z \geq 0\}$. The demand function in (7.1) satisfies the three requisite conditions suggested by [626]: negative slope, finite maximum value, and long right tail. Two examples of such a function are $h(x) = e^{-ax}$ and $h(x) = \frac{a}{a+bx}$. Clearly, $\lambda_i \in [0, \overline{\lambda}_i]$ for all $i \in N$. For inelastic demand, $\lambda_i = \overline{\lambda}_i$, so $h_i(u_i) = 1$.

7.3   Analysis of User-Choice Equilibrium

In this section we define and characterize the user-choice equilibrium given fixed locations and attributes of the m facilities. The exposition here is based on [728].

7.3.1   Equilibrium Specification

Let $\lambda \in \mathcal{R}^{nm}$ be the vector of $\{\lambda_{ij}\}$ and $\mathbf{u} \in \mathcal{R}^n$ be the vector of $\{u_i\}$.

Definition 1 *A demand vector λ is a user-choice equilibrium if there exists a vector $\mathbf{u}$ such that $\lambda, \mathbf{u}$ solve the following system of equations:*

$$
\begin{align}
\text{(DE):} \quad & \lambda_{ij}[T_i(d(i, F_j)) + E_j(\lambda_{1j} + \ldots + \lambda_{nj}) - u_i] = 0 \ \forall i \in N, j \in M \tag{7.2} \\
& T_i(d(i, F_j)) + E_j(\lambda_{1j} + \ldots + \lambda_{nj}) - u_i \geq 0 \ \forall i \in N, j \in M \tag{7.3} \\
& (\lambda_{i1} + \ldots + \lambda_{im}) - \overline{\lambda}_i h_i(u_i) = 0 \ \forall i \in N \tag{7.4} \\
& \lambda_{ij} \geq 0 \ \forall i \in N, \ j \in M \tag{7.5} \\
& u_i \geq 0 \ \forall i \in N. \tag{7.6}
\end{align}
$$

Equation (7.3) implies that u_i is the minimum cost incurred when user i patronizes any facility $j \in M$. Equation (7.2) represents the equivalent Wardrop condition that if a positive amount of demand from node i is serviced at facility j, then the cost incurred must be u_i; also, if the cost of servicing user i's demand by the jth facility is greater than u_i, then no demand from node i will go to facility j. Equation (7.4) states that user i's demand is a function of the minimum cost u_i, which in turn depends on the congestion at the facilities. Equations (7.5) and (7.6) give the nonnegativity conditions.

(DE) can be interpreted as a traffic equilibrium problem with congestion-elastic demand [2]. The equivalent traffic network is shown in Figure 1. Some properties of this network deserve mention.

FIGURE 1. Equivalent traffic network for demand equilibrium

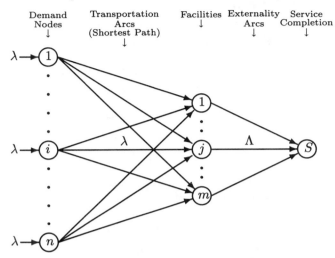

1. The cost of traversing an arc is either a function of the corresponding arc flow only (as in the case of externality arcs) or is constant (as in the case of travel arcs).

2. The network follows the additive model: total cost on any path of the network is the sum of the costs on all the arcs of that path.

3. The demand between any origin-destination (i.e. demand-facility) pair depends only on the smallest path cost u between that origin-destination pair.

7.3.2   Equivalent Nonlinear Complementarity Problem

[2] have shown that, under certain conditions, a general traffic equilibrium problem can be converted to a Nonlinear Complementarity Problem (NCP). We employ their ideas to formulate an NCP that is equivalent to our demand equilibrium problem (DE). Consider the system of equations (DE') which are the same as (DE) but with equation (7.4) replaced by:

$$u_i((\lambda_{i1} + \ldots + \lambda_{im}) - \overline{\lambda_i} h_i(u_i)) = 0 \; \forall i \in N$$
$$(\lambda_{i1} + \ldots + \lambda_{im}) - \overline{\lambda_i} h_i(u_i) \geq 0 \; \forall i \in N \qquad (7.7)$$

Theorem 1 below is based on Proposition 4.1 of [2].

Theorem 1 *The systems of equations (DE) and (DE') have the same solution set.*

We now define the vector $\mathbf{z} = (\lambda, \mathbf{u}) \in \mathcal{R}^{nm+n}$ and further let

$$f_{ij}(\mathbf{z}) = T_i\left(d(i, F_j)\right) + E_j(\lambda_{1j} + \ldots + \lambda_{nj}) - u_i \; \forall i \in N, \; j \in M \quad (7.8)$$

$$g_i(\mathbf{z}) = (\lambda_{i1} + \ldots + \lambda_{im}) - \overline{\lambda_i} h_i(u_i) \; \forall i \in N \quad (7.9)$$

$$\mathbf{R}(\mathbf{z}) = (f_{ij}(\mathbf{z}) \; \forall i \in N \text{ and } j \in M, \; g_i(\mathbf{z}) \; \forall i \in N). \quad (7.10)$$

Clearly, $\mathbf{R}$ is a vector valued function from $\mathcal{R}^{nm+n}$ into itself. It is easy to see that (DE') can be rewritten as

(DE") :
$$\mathbf{z}^T \mathbf{R}(\mathbf{z}) = 0 \quad (7.11)$$
$$\mathbf{R}(\mathbf{z}) \geq 0 \quad (7.12)$$
$$\mathbf{z} \geq 0 \quad (7.13)$$

(DE") is a nonlinear complementarity problem. Solving for the user-choice equilibrium is equivalent to finding a vector $\mathbf{z}$ that satisfies this system of equations.

7.3.3   Existence and Uniqueness of Equilibrium

We now establish conditions under which a user-choice equilibrium exists and is unique.

Theorem 2 *If the locations and attributes of all facilities are fixed, a user-choice equilibrium exists.*

Proof Consider the system of equations (DE'). $T_i(\bullet)$ and $E_j(\bullet)$ are non-negative continuous functions for all $i \in N$, $j \in M$. Also, $\lambda_i = \overline{\lambda_i} h_i(u_i)$ is bounded from above by $\overline{\lambda_i}$ for all $i \in N$. Therefore, Theorem 5.3 of [2] applies and (DE') has a solution. By Theorem 1, the systems (DE) and (DE') have the same solution set. Therefore, a demand equilibrium exists. □

The following theorem states that the level of demand λ_i at node i and the level of demand Λ_j visiting facility j, for all $i \in N$ and $j \in M$, are uniquely determined when the locations of all the facilities are fixed. [131] have similarly proved the uniqueness of Λ_j for the inelastic demand case (i.e. when $h_i(u_i) = 1$).

Theorem 3 *If the locations and attributes of all the facilities are fixed and demand is elastic, then $\{\Lambda_j\}$, $\{\lambda_i\}$ and $\{u_i\}$ are unique.*

Proof Consider the formulation (DE"). Suppose two different vectors $\mathbf{z}^1$ and $\mathbf{z}^2$ exist which solve this system of equations. Then

$$(\mathbf{z}^1 - \mathbf{z}^2)^T (R(\mathbf{z}^1) - R(\mathbf{z}^2)) \leq 0. \quad (7.14)$$

Substituting $\mathbf{z} = (\lambda, \mathbf{u})$, $\lambda_i = \sum_j \lambda_{ij}$, $\Lambda_j = \sum_i \lambda_{ij}$ and $\mathbf{R}$ as in (7.10) and carrying out the vector multiplication, we obtain

$$\sum_i \sum_j \left(\lambda_{ij}^1 - \lambda_{ij}^2\right) \left(E_j(\Lambda_j^1) - E_j(\Lambda_j^2) - u_i^1 + u_i^2\right) +$$

$$\sum_i \left(u_i^1 - u_i^2\right) \left(\lambda_i^1 - \lambda_i^2 - \overline{\lambda}_i h_i(u_i^1) + \overline{\lambda}_i h_i(u_i^2)\right) \le 0. \qquad (7.15)$$

Further simplification yields

$$\sum_j \left(\Lambda_j^1 - \Lambda_j^2\right) \left(E_j(\Lambda_j^1) - E_j(\Lambda_j^2)\right) +$$

$$\sum_i \left(u_i^1 - u_i^2\right) \left(-\overline{\lambda}_i h_i(u_i^1) + \overline{\lambda}_i h_i(u_i^2)\right) \le 0. \qquad (7.16)$$

$E_j(\bullet)$ is a strictly increasing function for all $j \in M$ and $h_i(\bullet)$ is a strictly decreasing function (for the elastic demand case) for all $i \in N$. Therefore, (7.16) is satisfied if and only if $\Lambda_j^1 = \Lambda_j^2$ for all $j \in M$ and $u_i^1 = u_i^2$ for all $i \in N$. Since (7.1) defines λ_i as a function of u_i, we have $\lambda_i^1 = \lambda_i^2$. Therefore, the vectors $\{\Lambda_j\}$, $\{\lambda_i\}$ and $\{u_i\}$ are unique. $\square$

We emphasize that the user-choice equilibrium $\{\lambda_{ij}\}$, and hence the solution vector $\mathbf{z}$ of (DE") may not be unique. However, when multiple demand equilibria exist, solving for any one of them will give us the unique vectors $\{\Lambda_j\}$ and $\{\lambda_i\}$. We now turn to the problem of computing a demand equilibrium.

7.3.4   Computation of Equilibrium

As noted, our demand equilibrium can be interpreted as traffic equilibrium on a congested network. This allows us to directly apply existing methods for computing traffic equilibria under congestion-elastic demand. [1] has solved for traffic equilibria by using Newton's method on the nonlinear complementarity system of the form (DE"). Given a vector $\mathbf{z}^k$, we can find a vector $\mathbf{z}^{k+1}$ by solving the following linear complementarity subproblem (LCP) using the pivoting scheme of [753]:

(LCP) $\mathbf{z} \ge 0$, $w(\mathbf{z}) = R(\mathbf{z}^k) + \nabla R(\mathbf{z}^k)(\mathbf{z} - \mathbf{z}^k) \ge 0$, $\mathbf{z}^T w(\mathbf{z}) = 0$.

The sequence $\{\mathbf{z}^k\}$ converges locally and quadratically to the solution of (DE").

[71] pointed out that sometimes the equilibrium equations can be considered to be the Kuhn-Tucker conditions of an optimization problem that can be solved using mathematical programming tools. This equivalence is not possible in general but holds when certain special conditions, e.g. the properties (a) - (c) listed in Section 7.3.1, are satisfied. Since the network

of Figure 1 satisfies these properties, the system (DE) corresponds exactly to the following optimization problem:

$$\text{(DE''')} \ \text{Min} \sum_i \sum_j T_i[d(i, F_j)]\lambda_{ij} + \sum_j \int_0^{\Lambda_j} E_j(y)dy - \sum_i \int_0^{\lambda_i} h_i^{-1}\left(\frac{v}{\lambda_i}\right) dv$$

s.t.

$$\lambda_i = \sum_j \lambda_{ij} \ \ \forall \, i \in N$$

$$\Lambda_j = \sum_i \lambda_{ij} \ \ \forall \, j \in M$$

$$\lambda_{ij} \geq 0 \ \ \forall \, i \in N, \ j \in M.$$

DE''' has a convex objective function [409] and linear constraints, and can be solved by the Frank-Wolfe method, for example.

7.4   Facility Location with Market Externalities

We now focus on location models for private and public facilities with market externalities in user-choice and dictatorial environments.

7.4.1   Inelastic Demand Models

We first discuss two models that assume inelastic demand and a user-choice environment.

Location of Competing Private Facilities in a User-Choice Environment

Consider two competing grocery stores with similar products and prices. Each store wishes to locate to maximize its market share. When deciding which store to patronize, customers consider the distance they must travel as well as the time they must wait for service. Each facility locates to maximize its own equilibrium market share. This can be modeled as a competitive location problem in a user-choice environment with congestion externalities.

We assume that the competitive location problem is a zero-sum Stackelberg game with two agents. [511] considered a similar problem with a leader and a follower who can locate p and r facilities, respectively, and [296] considered a special case in which $p = r = 1$. However, in those analyses, customers are assumed to patronize the closest facility. We assume that customers also consider congestion externalities when selecting the facility they will patronize.

[708] analyzed a competitive location problem with congestion externalities. He considered identical facilities on a line with uniformly distributed

customers who select a facility based not only on travel time but also on waiting time for service. For the case of two facilities, the optimal locations occur at the midpoint of the line. When there are more than two competing facilities, [708] showed that a locational equilibrium does not exist. The model we consider here differs in that we allow for nonidentical facilities (with differing externality functions and different capacities) and discrete demands.

The competitive location problem with user choice can be thought of as a three-stage sequential game: the leader locates first; the follower locates second; and then, given those facility locations, a simultaneous choice game is played by the users, determining the user-choice equilibrium. This model, introduced in [131], can be stated as:

Problem 1 $$\min_{F_1} \max_{F_2} M(F_2|F_1)$$

where $M(F_2|F_1)$ represents the equilibrium market share for the follower, given location F_1 for the leader and F_2 for the follower. The market share of each facility is the fraction of the total (fixed) customer demand Λ that patronizes that facility. The follower wants to find an optimal location, given F_1, which we denote by $F_2^*|F_1$. The leader wants to find an optimal location F_1^*, given that the follower will select $F_2^*|F_1^*$. We let $p* = M(F_2^*|F_1^*)$ represent the user-choice equilibrium market division (market share of the follower) associated with the optimal locations.

[131] characterized the user-choice equilibrium given inelastic demand and two fixed facilities. Because of the special structure with inelastic demand and only two facilities, the result is much simpler than that obtained in [728]. [131] showed that for any fixed facility locations, at most one customer will be marginal. They defined, for each user i,

$$\Delta_i = T_i(d(i, F_1)) - T_i(d(i, F_2)), \qquad \forall i \in N. \tag{7.17}$$

Assume that users with identical values of Δ_i have been aggregated, and the Δ_i's have been indexed so that $\Delta_1 > \Delta_2 > \Delta_3 > \ldots > \Delta_n$. Define

$$f_k = \sum_{i=1}^{k} \lambda_i, \qquad k = 1, \ldots, n \tag{7.18}$$

as the aggregated demand up to and including user k, and

$$D(p) = E_2(p) - E_1(f_n - p) = E_2(p) - E_1(\Lambda - p) \tag{7.19}$$

as the difference in externality cost when p of the total demand chooses facility 2, where $f_n = \Lambda$ is the total customer demand. The customer demand division corresponding to a user-choice equilibrium satisfies one of the following two conditions [131]:

$$\Delta_1 > \ldots > \Delta_{i-1} > D(p) > \Delta_i > \Delta_{i+1} > \ldots > \Delta_n, \quad p = f_{i-1} \tag{7.20}$$

$$\Delta_1 > \ldots > \Delta_{i-1} > D(p) = \Delta_i > \Delta_{i+1} > \ldots > \Delta_n, \; f_{i-1} \leq p < f_i \quad (7.21)$$

In the first case, the equilibrium market division is $p = f_{i-1}$: customers $1, 2, \ldots, i-1$ patronize facility 2, and customers $i, i+1, \ldots, n$ use facility 1. In the second case, customers in group i are split between the facilities (user i is a "marginal user"), while customers with a lower index patronize facility 2, and customers with a higher index patronize facility 1; thus, the equilibrium market division p falls between f_i and f_{i-1}. [131] presented an $O(n \log n)$ algorithm for computing the equilibrium, given fixed facility locations; this is the effort required to sort the customers by Δ_i.

[132] analyzed the competitive location problem for the special case of a tree network. They characterized the optimal location for the follower, given a location for the leader, and then analyzed the leader's location decision, given knowledge of the follower's behavior. The following theorem [132] reduces the search for the follower's optimal location.

Theorem 4 *For Problem 1, an optimal location $F_2^*|F_1$ on the tree network can always be found at a node or at the location of the leader.*

This result holds because of the tree network structure: If the follower can gain market share by moving away from the leader into an adjoining subtree, this means that all customers served by the follower occur in that subtree. Thus, movement to the next adjacent node brings the follower closer to all the customers he serves, and cannot decrease the follower's market share.

The optimal location for the leader is not necessarily nodal. However, [132] showed that leader's optimal location satisfies the intuitively reasonable condition that the follower's best possible market share must be exactly balanced in two subtrees about the leader's location. Given a location F_1, we define a partition of the tree at F_1 into two subtrees, $T_{F_1}^+$ and $T_{F_1}^-$, both rooted at F_1, such that $T_{F_1}^+ \cap T_{F_1}^- = F_1$, and $T_{F_1}^+ \cup T_{F_1}^- = T$. Also define $F_2^+|F_1^*$ as the best location for facility 2 in subtree $T_{F_1}^+$, and $F_2^-|F_1^*$ as the best location in $T_{F_1}^-$ given F_1^*. Then:

Theorem 5 *For Problem 1, a location F_1^* is a global optimum solution for the leader on a tree network if and only if $M(F_2^+|F_1^*) = M(F_2^-|F_1^*)$ for some choice of $T_{F_1}^+$ and $T_{F_1}^-$.*

This condition is necessary and sufficient for optimality of F_1^*.

To characterize the optimal locations, we introduce the notion of (distance-blind) attractiveness: we say that one facility is more attractive than the other if the facility attracts more market share when both facilities are located at the same place (i.e., when there is no distance differential between them). [132] showed that if both facilities are equally attractive, then both are optimally located at the 1-median; this is the same as the optimal solution when there are no externalities. However, if the leader is less attractive than the follower, then the leader's optimal location is always at the median, while the follower's optimal location is not necessarily at the median (but is nodal). When the leader is more attractive than the follower, we cannot determine in advance where the leader should locate. For this case, an $O(n^2)$ algorithm for determining the optimal locations is presented [132]. This is comparable to algorithms for the 2-median problem on a tree network, which require $O(n^2)$ computations [660, 854].

Define p_0 as the market division such that $D(p_0) = 0$ (see equation (7.19)). The value p_0 is a measure of the relative attractiveness of the two facilities in the absence of a distance differential: when there is no difference in cost of travel to the two facilities (i.e., $\Delta_i = 0 \ \forall i$), the leader captures market share $(\Lambda - p_0)$, while the follower captures p_0. [131] showed that the follower can increase his market share above p_0 by moving away from F_1 only when some subtree about F_1 has greater demand than p_0. If $p_0 \geq \frac{1}{2}$ (the follower is the more attractive facility in the absence of locational differentiation) then the leader can locate at the median, where (because of the median weight condition) no adjacent subtree generates more than half the demands, thus holding the follower to p_0. On the other hand, if $p_0 < \frac{1}{2}$ (the leader is more attractive), wherever the leader locates (including at the median), there may be an adjacent subtree containing more than p_0 of the total demand. If so, the follower is guaranteed a greater market share than p_0 by moving toward that subtree. Thus, there is an incentive for the less efficient facility to be the follower.

A similar "first entry paradox" has been identified by other authors who analyzed competitive location problems on tree networks. For example, [461] showed that, for a sequential competitive location game on a line with equally attractive facilities, those facilities that locate later may be able to gain more market share than those that locate earlier. [372] also identified a follower's advantage in their analysis of a two-facility competitive location problem on a tree network where facilities may not be equally attractive and where customer choice levels are proportional to the inverse of distances to the facilities. It appears that the first entry phenomenon occurs because of the tree network structure: since the network contains no loops, the leader cannot always guard against the follower carving out his own market niche, and thus capturing a disproportionately large market share.

[132] considered the case of two competitors, each of whom locates a single facility. The locational analysis is simplified by the fact that there is at most one marginal customer when there are two facilities, and this marginal customer can be readily identified using (7.20,7.21). Furthermore, they showed that, on a tree network, the identity of the marginal customer changes in a predictable way as the relative locations of the two facilities change. If more than two facilities are located, a unique user-choice equilibrium still exists [132]. However, on a network with m facilities, as many as $m - 1$ customers may be marginal. In order to analyze the location problem with more than two competitors or facilities, further analysis of corresponding changes in the user-choice equilibrium is needed.

[132] restricted their analysis to a tree network. For a general network, the existence and uniqueness properties of the user-choice equilibrium, given fixed facility locations, still hold [131]. However, the locational results for a tree network do not carry over directly. On a general network, the follower may no longer have an optimal nodal location. In addition, it is possible that the follower may no longer have a strategic advantage. Following [95] we can identify primary regions and breakpoints of a network; these are, respectively, portions of links over which shortest distance travel paths do not change, and points (at the boundaries of primary regions) at which shortest distance paths change. To analyze the competitive location problem within primary regions, the existing tree network analysis could be applied. Further analysis is needed to characterize properties of the solution across breakpoints and on the overall network.

Location of Public Facilities in a User-Choice Environment

We now consider a public facility location problem in the user-choice environment, again with inelastic demand. A natural location objective for public facilities is to minimize total customer cost ("social cost"). In our framework, this consists of travel cost plus externality cost. This is similar to the social welfare type of objective considered by [389, 1168]. The difference is that we also incorporate congestion externalities and the resulting user-choice equilibrium. Examples of public facilities where customers can select the facility they patronize include post offices, Social Security Administration offices, and Department of Motor Vehicles offices.

The public facility location problem, introduced by [131], is:

Problem 2
$$\min_{\{F_j\}} \sum_{i=1}^{n} \lambda_{ij}[T_i(d(i, F_j)) + Q_j(\Lambda_j)]$$

s.t.
$$\Lambda_j = \sum_{i=1}^{n} \lambda_{ij} \quad \forall j \in M$$

$\{\Lambda_j\}$ is a user-choice equilibrium.

This problem is a mixed integer bilevel program. Given locations $\{F_j\}$, the user-choice equilibrium is unique. When there are two facilities, the market division satisfies the equilibrium condition (7.20,7.21).

[131] analyzed this problem for the special case of two facilities on a tree network. They first analyzed the optimal location for facility 2, given a fixed location for facility 1 (denoted by $F_2^*|F_1$) and then analyzed the optimal location for facility 1 (denoted by F_1^*) with knowledge of facility 2's best corresponding location.

Assume that F_1 is fixed and, if F_1 is not at a node, assume that we create an artificial node there. [131] analyzed facility 2's optimal location using the directional derivative of the objective function with respect to F_2. They showed that the directional derivative of the objective function with respect to either of the two facility locations (given that the other facility is in a fixed location) has "breakpoints" (points at which it is continuous but not differentiable) across nodes and where a customer becomes marginal or non-marginal. Furthermore, they showed that, in regions with no marginal customer, the objective function is linear, and in regions with a marginal customer, the objective function is convex. They also showed that the optimal location for at least one of the facilities is nodal. This implies the following [131, Theorems 2 & 3]:

Theorem 6 *For Problem 2:*

(i) An optimal location for one of the two facilities can always be found at a derivative breakpoint or at a point that sets the derivative to zero.

(ii) An optimal location for the other facility can always be found at a node.

Theorem 6 (i) restricts the location of one of the facilities to one of (at most) $4n$ possible locations: these are the n nodes, the derivative breakpoints (there are at most $2n$ of these given a fixed F_1), and possible intermediate points in marginal regions where the derivative is zero (there are at most n such points). Theorem 6 (ii) restricts the other location to at most one of n possible locations. However, when the facilities are not identical, it is not possible to determine in advance which of the two facilities should be located at a node.

Using these localization results, [131] developed an algorithm for determining the optimal facility locations that considers $O(n^2)$ possible locations. The algorithm essentially involves complete enumeration, but examines a finite set of potential locations rather than an infinite set because of the localization results. The complexity is comparable to algorithms for the 2-median problem on a tree network, which require $O(n^2)$ computations; see [660, 854]. For each candidate pair of locations, an additional $O(n)$ computations are required to determine customer allocations, leading to $O(n^3)$ steps.

As for the competitive location problem in the user-choice environment, it is likely that one could extend the locational analysis for the public facility problem to the case of a general network by analyzing the objective function over primary regions and across breakpoints. Furthermore, for the case of more than two facilities, knowledge of how the user-choice equilibrium changes with changes in facility locations could be used to analyze the public facility location problem.

7.4.2   Elastic Demand Models

We now consider the more general case of location models where demand is a function of the cost. This situation arises naturally in many applications where facilities provide non-essential services. Examples include fast food restaurants, gas stations, banks, and retail stores. The development here is abstracted from [727, 728, 729].

Location of a Single Private Facility in a User-Choice Environment

[727] considered the decision of a profit maximizing firm (m^{th} facility) that wants to optimally position itself on the transportation network $G(N, A)$. Let a be the profit per unit of demand served. We assume that $(m - 1)$ other competitors are already located on the network. We assume a user-choice environment, and model consumer cost as the sum of travel cost plus the cost due to some negative congestion externality such as waiting time at a facility. Consumer demand is elastic with respect to this cost. This situation results in a demand equilibrium that is parameterized by the location of the market entrant.

We define the *node i breakpoint* on any link (a, b) of length l to be a point at a distance s_i, where $a < s_i < b$, from node a such that the travel distance from it to node i through node a is the same as the travel distance from it to node i through node b. A *primary region* (s_r, s_t) is defined to be the portion of the link (a, b) between two adjacent breakpoints s_r and s_t, and N_1 and N_2 are defined to be the sets of nodes from which consumers travel to the primary region through nodes a and b, respectively. That is, $d(i, a) + s_i = (\ell - s_i) + d(b, i)$, $N_1 = \{i \in N : x + d(a, i) \le (\ell - x) + d(b, i)\}$, and $N_2 = N - N_1$.

We would like to know how the equilibrium market share of the mth facility varies with its location. Suppose the location F_m of the mth facility is at a distance x from the breakpoint s_r on some primary region (s_r, s_t) of length L. The demand equilibrium (DE") can be rewritten as

$$(\text{DE"}(x)) \qquad \mathbf{z}^T R(\mathbf{z}, x) = 0, R(\mathbf{z}, x) \ge 0, \mathbf{z} \ge 0 \qquad (7.22)$$

where
$R(\mathbf{z}, x) = (f_{ij}(\mathbf{z}, x) \; \forall i \in N \text{ and } j \in M, \; g_i(\mathbf{z}, x) \; \forall i \in N) \in \mathcal{R}^{nm+n}$,
$f_{ij}(\mathbf{z}, x)$: $\forall i \in N, j = 1, 2, \ldots, m - 1$ as defined in (7.8),
$g_i(\mathbf{z}, x)$: $\forall i \in N$ as defined in (7.9),
$f_{im}(\mathbf{z}, x) = T_i(d(i, s_r) + x) + E_m(\Lambda_m) - u_i \; \forall i \in N_1$, and
$f_{im}(\mathbf{z}, x) = T_i(D(i, s_t) + L - x) + E_m(\Lambda_m) - u_i \; \forall i \in N_2$.

If we consider a primary region different from (s_r, s_t), then the sets N_1 and N_2 and hence the definition of $R(\mathbf{z}, x)$ will change correspondingly. Therefore, we carry out our analysis separately over each primary region. The optimal location for the mth facility over the entire network is then found by comparing the optimal locations over all primary regions.

(7.22) is a nonlinear complementarity system parameterized by x. The continuity and differentiability of the parametric solution $\mathbf{z}(x)$ of such systems has been studied by [730, 816], and others. Unfortunately, their results cannot be applied here directly since they assume uniqueness of the solution vector $\mathbf{z}$ which may not be true for our problem, as noted before. We introduce some more notation in order to get around the non-uniqueness of $\mathbf{z}$.

A closer look at the vector-valued function $R(\mathbf{z}, x)$ reveals that even though its argument is $\mathbf{z} = (\{\lambda_{ij}\}, \{u_i\})$, it in fact only depends upon $\{\Lambda_j\}, \{\lambda_i\}, \{u_i\}$ and not on the individual elements $\{\lambda_{ij}\}$. We define a new vector $\mathbf{y} = (\{\Lambda_j\}, \{\lambda_i\}, \{u_i\}) \in \mathcal{R}^{m+2n}$. Then we can express $R(\mathbf{z}, x)$ as $R(\mathbf{y}, x)$. We are interested in the variation in $\mathbf{y}$ as x changes over the primary region between 0 and L.

Consider a specific value of x, say x^*, such that $0 \leq x^* \leq L$. By Theorem 2, the corresponding solution vector $\mathbf{z}^*$ to (7.22) exists. Let the vector $\mathbf{y}$ associated with $\mathbf{z}^*$ be $\mathbf{y}^*$. Assuming strict complementarity holds, we define the index sets I and J at $\mathbf{y}^*$ as

$$I = \{k|R_k(\mathbf{y}^*, x^*) > 0, z_k^* = 0\} \text{ and } J = \{k|R_k(\mathbf{y}^*, x^*) = 0, z_k^* > 0\}.$$

The equations $\mathbf{z}_I^* = 0$, where I is the index set defined above, may directly cause certain elements of the vector $\mathbf{y}^*$ to be zero. We denote the index set of these elements of $\mathbf{y}^*$ by K. Now consider the following set of equations:

$$\sum_{j \in M} \Lambda_j - \sum_{i \in N} \lambda_i = 0 \qquad \text{(Demand Conservation)} \qquad (7.23)$$

$$y_k^* = 0 \qquad \forall k \in K \qquad (7.24)$$

$$R_k(\mathbf{y}^*, x^*) = 0 \qquad \forall k \in J. \qquad (7.25)$$

From Theorem 3, $\mathbf{y}^*$ is unique. This implies that we can isolate a set of $(m + 2n)$ independent equations from the system (7.23) - (7.25) that can be solved for $\mathbf{y}^*$. We denote this set of equations as

$$Q(\mathbf{y}^*, x^*) = 0. \qquad (7.26)$$

Let the Jacobian matrix of Q with respect to $\mathbf{y}$ evaluated at $(\mathbf{y}^*, x^*)$ be $\nabla_\mathbf{y} Q(\mathbf{y}^*, x^*)$ and with respect to x evaluated at $(\mathbf{y}^*, x^*)$ be $\nabla_x Q(\mathbf{y}^*, x^*)$.

Theorem 7 *Suppose $\nabla_\mathbf{y} Q(\mathbf{y}^*, x^*)$ is nonsingular. Then, (i) $\mathbf{y}$ can be expressed as a continuous and differentiable function $\mathbf{y}(x)$ in the neighborhood of x^*, and (ii) the derivative of $\mathbf{y}(x)$ with respect to x in the neighborhood of x^* is given by $\nabla_x \mathbf{y}(x^*) = -[\nabla_\mathbf{y} Q(\mathbf{y}^*, x^*)]^{-1} \nabla_x Q(\mathbf{y}^*, x^*)$.*

Proof *(i)* Follows directly from the application of the Implicit Function Theorem to (7.26), and *(ii)* follows from the total differentiation of (7.26) with respect to x. $\square$

In view of Theorem 7, we can compute the local derivative of $\mathbf{y}(x)$ with respect to x at any point in the primary region. Since $\Lambda_m(x)$ is just an element of $\mathbf{y}(x)$, the gradient information about $\Lambda_m(x)$ is directly obtained.

The optimization problem of finding the profit maximizing location for the mth facility over the primary region (s_r, s_t) can now be expressed as

Problem 3   Max $J_1(x) = \alpha \Lambda_m(x)$
s.t. $0 \leq x \leq L$,
          $\Lambda_m(x)$ is obtained from the solution of (7.22) for any value of x.

Mathematical programs of this type with nonlinear complimentarity systems in their constraints are special cases of bilevel programming problems [710]. Good solution methods for bilevel problems generally do not exist. We pursue a gradient projection ascent approach here similar to the Normalized Sensitivity Descent Algorithm - Predetermined (NSDAP) of [436]. However, this method has some drawbacks. Since we have no information regarding the second derivative of $\Lambda_m(x)$, we cannot say anything about the convexity of Problem 3. A gradient projection method may therefore lead to a local optimum and will only be a heuristic. Such a method will also tend to be computationally expensive since it involves repeatedly solving the embedded NCP and repeated gradient evaluations. Moreover, knowing the gradient vector at any point does not help in finding the appropriate step length, so predetermined, monotonically decreasing step sizes must be employed.

We discussed above the computation of the local gradient of $\Lambda_m(x)$ at x^*. This directly gives the gradient with respect to x of the objective function of Problem 3 at x^* as $\nabla_x J_1(x^*) = \alpha \nabla_x \Lambda_m(x^*)$, and hence the normalized gradient $\delta(x^*)$ as

$$\delta(x^*) = \frac{\nabla_x J_1(x^*)}{|\nabla_x J_1(x^*)|}.$$

We use the predetermined sequence $\{\theta_k\}$ of monotonically decreasing step sizes where

$$\theta_k = \frac{0.25L}{2^k}, \quad k = 0, 1, 2, \ldots \qquad (7.27)$$

This step size sequence was found to be the most efficient one by [436] among the many they investigated. It should be noted that $\sum_{k=0}^{\infty} \theta_k = 0.5L$, so the algorithm is capable of finding the most distant feasible point in the limit if the search is started at the midpoint of the primary region.

[727] suggested a heuristic for locating the mth facility on the transportation network in the presence of $m - 1$ existing facilities.

Location of Multiple Facilities in a Dictatorial Environment

We now consider the location-assignment decisions of m facilities on a transportation network with negative congestion externalities and elastic demand. Assume that the decision maker has the authority to make system-wide consumer assignment decisions. The objective is to locate facilities and assign customers to facilities so as to maximize the total utilization of the facilities. Such a system-optimizing framework would apply to preventive child care (e.g. inoculation) units, polling places, or franchising decisions.

An interesting aspect of this framework is that it also gives rise to an equilibrium in demand that is parameterized by facility locations. The overall location-allocation problem, therefore, becomes a generalized p-median model with an embedded demand equilibrium. [729] proved a nodal optimality theorem and showed that the profit maximization problem can be formulated as a nonlinear integer convex programming problem and can be solved using branch-and-bound.

Let M denote the set of m identical facilities and let $F_j, j \in M$, be the as yet undetermined location of the jth facility on the network. Define the following decision variables:

$$y_{ij} = \begin{cases} 1 & \text{if demand point } i \text{ is allocated to facility } j, \\ 0 & \text{otherwise.} \end{cases}$$

Assume that each demand node is assigned to exactly one facility, i.e.

$$\sum_{j=1}^{m} y_{ij} = 1 \quad \forall i \in N. \tag{7.28}$$

The demand arrival rate at facility $j, j \in M$, is a Poisson distributed random variable with mean

$$\Lambda_j = \sum_{i=1}^{n} \lambda_i y_{ij}. \tag{7.29}$$

We assume that the function $h_i(\bullet)$ is exponential. If the ith node is assigned to the jth facility, the mean demand rate λ_i generated at node i is given by

$$\lambda_i = \overline{\lambda_i} e^{-\beta(T_i(d(i,F_j)) + E_j(\Lambda_j))}, \tag{7.30}$$

where β is a positive constant. Exponentially decreasing demand has been observed empirically by [833, 626] and has been employed previously in location models by [87]. Since the demand nodes are assumed to be homogeneous, the negative exponential function above is identical for each of them. However, we do allow the nodes to differ in their total population (as reflected in $\overline{\lambda_i}$) and in their means of transportation (as reflected in $T_i(\bullet)$). Clearly, $\lambda_i \in [0, \overline{\lambda_i}]$ for all $i \in N$. We also assume that two or more facilities cannot be located at the same node.

Suppose that the locations of the m facilities are fixed and their respective nodal allocations are specified. Since demand is elastic, we would like to characterize the resulting equilibrium and level of demand that visits each facility.

Definition A *facility demand equilibrium* Λ^* exists when, for a given location-allocation, the following condition is satisfied for each $j \in M$:

$$\Lambda_j^* = \sum_{i=1}^{n} y_{ij} \overline{\lambda_i} e^{-\beta(T_i(d(i,F_j)) + E_j(\Lambda_j^*))}. \tag{7.31}$$

Theorem 8 *If facility locations and consumer assignments are specified, a facility demand equilibrium exists.*

Proof Consider the level of demand Λ_j visiting the jth facility. Clearly, $0 \leq \Lambda_j \leq \sum_{i=1}^{n} y_{ij} \overline{\lambda_i}$. Define a new function

$$H(\Lambda_j) = \sum_{i=1}^{n} y_{ij} \overline{\lambda_i} e^{-\beta(T_i(d(i,F_j)) + E_j(\Lambda_j))}. \tag{7.32}$$

For a given location-allocation, y_{ij} and $T_i(d(i, F_j))$ are constants. Obviously, $0 \leq H(\Lambda_j) \leq \sum_{i=1}^{n} y_{ij} \overline{\lambda_i}$ so $H(\Lambda_j)$ is a continuous mapping from a closed convex set onto itself. Therefore, by Brouwer's Fixed Point Theorem [1040] there exists a value Λ_j^* such that $0 \leq \Lambda_j^* = H(\Lambda_j^*) \leq \sum_{i=1}^{n} y_{ij} \overline{\lambda_i}$. This implies that (7.31) holds for the jth facility. The same argument can be repeated for all j, implying that a facility demand equilibrium exists.□

We define a new function $Q_j(\Lambda_j)$ such that

$$Q_j(\Lambda_j) = \Lambda_j e^{\beta E_j(\Lambda_j)}. \tag{7.33}$$

The following lemma follows from direct differentiation of (7.33).

Lemma 1 $Q_j(\Lambda_j)$ is a strictly increasing, convex function for $\Lambda_j \geq 0$.

Theorem 9 If facility locations and consumer assignments are specified, the facility demand equilibrium is unique.

Proof Consider the jth facility. At equilibrium, (7.31) is satisfied and can be rewritten as

$$Q_j(\Lambda_j^*) = \sum_{i=1}^{n} y_{ij} \overline{\lambda_i} e^{-\beta T_i(d(i,F_j))}, \tag{7.34}$$

where $Q_j(\bullet)$ is as defined in (7.33). Note that for a given location-assignment, the right hand side of (7.34) is a constant. Also, by Lemma 1, the left hand side is a strictly increasing function of Λ_j. Therefore, the equality in (7.34) must be satisfied for a unique Λ_j^*. The same argument can be repeated for all $j \in M$, implying that the facility demand equilibrium is unique.□

The equilibrium facility demand levels Λ_j^*, $j \in M$, can be computed in a straightforward fashion. We define

$$\phi_j(\Lambda_j) = \left| Q_j(\Lambda_j) - \sum_{i=1}^{n} y_{ij} \overline{\lambda_i} e^{-\beta T_i(d(i,F_j))} \right|. \tag{7.35}$$

It is easily seen [727] that $\phi_j(\Lambda_j)$ is a non-differentiable, strongly quasi-convex function over $[0, \sum_{i=1}^{n} y_{ij} \overline{\lambda_i}]$. Also, over this interval, $\phi_j(\Lambda_j)$ achieves its minimum at Λ_j^*. Thus, computing the equilibrium demand level Λ_j^*

is equivalent to finding the minimum of $\phi_j(\Lambda_j)$. An efficient way to find the minimum of a non-differentiable, strongly quasiconvex function over a closed interval is the Golden Section Search method [64].

We henceforth drop the superscript * and refer to the equilibrium levels simply as demand. We are now ready to analyze the location-assignment problem. This can be stated as:

Problem 4a $$\text{Max } \sum_{j=1}^{m} \Lambda_j$$

s.t.

$$Q_j(\Lambda_j) = \sum_{i=1}^{n} y_{ij} \overline{\lambda}_i e^{-\beta T_i(d(i,F_j))} \quad \forall\ j \in M \qquad (7.36)$$

$$\sum_{j=1}^{m} y_{ij} = 1 \qquad\qquad\quad \forall\ i \in N \qquad (7.37)$$

$$F_j \in G \qquad\qquad\qquad \forall\ j \in M \qquad (7.38)$$

$$y_{ij} \in \{0, 1\} \qquad\qquad\quad \forall\ i \text{ and } j. \qquad (7.39)$$

The function $Q_j(\Lambda_j)$ is as defined in (7.33), so (7.36) gives the facility demand equilibrium condition. Constraint (7.37) requires each demand node to be serviced by exactly one facility, (7.38) constrains the locations of all the facilities to be on the network, and (7.39) gives the integrality condition on the assignment variables.

The next theorem is analogous to the classical Nodal Optimality Theorem for the p-median problem [509, 853]. We first state the following two lemmas without proof.

Lemma 2 *If $f(g)$ is a nondecreasing concave function of g, and $g(z)$ is a concave function of z, then $f(g(z))$ is concave in z.*

Lemma 3 *If $f(g)$ is a nonincreasing convex function of g, and $g(z)$ is a concave function of z, then $f(g(z))$ is convex in z.*

Theorem 10 *At least one set of optimal locations for Problem 4a falls on the nodes of the network.*

Proof Suppose the optimal locations and allocations have been determined. Consider a facility j located on link (a, b) of length L at a distance x from node a. Suppose V_j is the set of all demand nodes allocated to this facility. Clearly, the shortest distance from node i, $i \in V_j$, to facility j is

$$d(i, F_j) = \min(d(i, a) + x, d(i, b) + L - x). \qquad (7.40)$$

Since $(d(i, a) + x)$ and $(d(i, b) + L - x)$ are linear in x, they are both concave in x. Since the minimum of two concave functions is concave, (7.40) implies that $d(i, F_j)$ is a concave function of x. Therefore, by Lemma 2, $T_i(d(i, F_j))$ is a concave function of x. Hence, by Lemma 3, $\overline{\lambda}_i e^{-\beta T_i(d(i,F_j))}$ is convex in x for all $i \in V_j$. Since the sum of convex functions is convex, (7.34) implies that at equilibrium $Q_j(\Lambda_j(x))$ is convex in x over the link (a, b).

Since a convex function achieves its maximum at an extreme point of the feasible region, $Q_j(\Lambda_j)$ is maximized when the jth facility is located at either node a or node b. Since $Q_j(\Lambda_j)$ is a strictly increasing function of Λ_j by Lemma 1, Λ_j is also maximized at either of the two nodes. Therefore, the jth facility can be moved to a node without decreasing the objective function value in Problem 4a.

This argument can be repeated for each of the facilities, implying that at least one set of optimal locations must fall on the nodes of the network. □

Theorem 10 restricts the search for the optimal locations to just the nodes of the network, thus converting the problem to a combinatorial one. Define the following decision variable:

$$z_j = \begin{cases} 1 & \text{if a facility is established at the node } j \text{ of the network,} \\ 0 & \text{otherwise.} \end{cases}$$

If the demand from node i is serviced from the facility at node j, the distance traveled is simply $d(i,j)$. We can define a constant c_{ij} for every pair of the nodes i and j of the network as

$$c_{ij} = \overline{\lambda_i} e^{-\beta T_i(d_i(i,j))}. \tag{7.41}$$

Using (7.41), the first set of constraints in Problem 4a becomes

$$Q_j(\Lambda_j) = \sum_{i=1}^{n} c_{ij} y_{ij} \quad \forall j \in M. \tag{7.42}$$

After incorporating (7.42) into the objective function, the profit maximization Problem 4a can be rewritten as the following nonlinear integer programming problem:

Problem 4b
$$\text{Max} \sum_{j=1}^{n} Q_j^{-1} \left(\sum_{i=1}^{n} c_{ij} y_{ij} \right)$$

s.t.

$$\sum_{j=1}^{n} y_{ij} = 1 \quad \forall i \in N \tag{7.43}$$

$$y_{ij} \leq z_j \quad \forall i,j \in N \tag{7.44}$$

$$\sum_{j=1}^{n} z_j = m \tag{7.45}$$

$$y_{ij}, z_j \in \{0,1\} \quad \forall i,j \in N. \tag{7.46}$$

Equation (7.44) requires that the demand be serviced from a node where a facility is established, (7.45) constrains the total number of established facilities to m, and (7.46) implies that all the location-assignment decisions are binary. Note that the subscript j now varies over the node set N rather than over the facility set M. If no facility is located at a node, the constraints ensure that the total demand visiting that facility is zero.

We now establish concavity of the objective function of Problem 4b in the allocation variables, leading to an integer convex programming problem. Let $\mathbf{y}$ denote $\{y_{ij}\}$.

Lemma 4 $Q_j^{-1}\left(\sum_{i=1}^{n} c_{ij} y_{ij}\right)$ is concave in $\mathbf{y}$.

Proof Let $\sum_{i=1}^{n} c_{ij} y_{ij} = t_j$. We first prove that $Q_j^{-1}(t_j)$ is a concave function of t_j. From (7.42), $Q_j(\Lambda_j) = t_j$. Differentiating both sides with respect to t_j gives $\dfrac{d^2 Q_j}{d\Lambda_j^2}\left(\dfrac{d\Lambda_j}{dt_j}\right)^2 + \dfrac{dQ_j}{d\Lambda_j}\dfrac{d^2\Lambda_j}{dt_j^2} = 0$. Rearranging terms and substituting $\dfrac{d\Lambda_j}{dt_j} = 1/\dfrac{dQ_j}{d\Lambda_j}$ yields

$$\frac{d^2 Q_j^{-1}(t_j)}{dt_j^2} = \frac{d^2\Lambda_j}{dt_j^2} = -\frac{\frac{d^2 Q_j}{d\Lambda_j^2}}{\left(\frac{dQ_j}{d\Lambda_j}\right)^3} \leq 0. \qquad (7.47)$$

The last inequality in (7.47) follows from Lemma 1. Hence $Q_j^{-1}(t_j)$ is a concave function of t_j. Since the term t_j is a linear function of $\mathbf{y}$, the lemma follows. $\square$

Since the sum of concave functions is concave, Lemma 4 implies that Problem 4b involves maximization of a concave function subject to linear and integrality constraints. Problem 4b is therefore an integer convex programming problem that can be solved using an enumerative approach. To guide the search process, [729] established upper and lower bounds on the solution value.

If the integrality conditions are relaxed, the solution of the resulting problem will yield an upper bound for the objective function value of Problem 4b. The relaxed problem is defined as **Problem 4c.**

Since Problem 4c is a convex nonlinear programming problem with linear constraints, the most appropriate method for solving it is the Frank-Wolfe algorithm, which requires the repeated linearization of Problem 4c with a line search procedure. The specialized Frank-Wolfe procedure is described in [729].

If the optimal solution to Problem 4 is required, the bounds computed above can be used in any general purpose branch-and-bound solution procedure for integer programming problems. [456] provides a good survey of the available enumerative algorithms. We recommend the one due to [258].

7.5   Resource Allocation with Market Externalities

Changing facility locations may be an expensive strategic option if facilities are already located. Another alternative is to change the characteristics of a facility – e.g., to increase the number of servers, speed up the service of individual servers (e.g., through training or technological improvements), decrease the variance of service time, etc. – in order to reduce congestion externalities. The cost of such changes will depend on the extent to which the facility's characteristics are altered. The question is how to balance the cost of change with the associated benefits (i.e., increased market share in the private facility case, or lowered social cost in the public facility case)

In this section we examine two resource allocation models for the case of public facilities (in fixed locations) with inelastic customer demands. Both models assume that a fixed facility-improvement budget is available to invest among the facilities. The objective is to determine the allocation of the service resource improvement budget across facilities that minimizes total social cost, where social cost is defined as in the public facility location model in a user-choice environment (Problem 2) to be the sum of travel cost plus externality cost for all customers. An example of such a budget allocation decision is an organization with separate mainframe computer systems where users decide which system to use. Service improvement resources could be expended on incremental processors, improved software, or improved operator training. Another example is offices of public agencies (e.g., the Social Security Administration): service could be improved by increasing automation at service offices, adding more service representatives, providing advanced training of personnel, or increasing office hours.

7.5.1   Resource Allocation for Public Facilities in a User-Choice Environment with Inelastic Demands

Several researchers have considered facility investment decisions for an environment with facility congestion costs or other negative externality costs, but the analyses have not considered a user-choice environment. For example, [11] examined the allocation of a fixed number of servers to a given set of facilities to minimize the overall cost of service delay; however, in that model each user can receive service only at a particular facility, so no user choice problem arises. [279] examined service price and capacity decisions for a single facility with negative externalities where demand is a function of mean queue delay. [1155] analyzed capacity decisions in a system in which homogenous users (or heterogeneous users whose utility has a particular distribution) choose between a public facility with a congestion cost and a private facility with a congestion cost and a toll. Our analysis differs in that we consider facility investment in a user-choice environment.

Assume that a fixed budget B is to be invested among m facilities to improve their ability to provide service. Investment in facility j reduces the mean time to receive service for any given level of user demand (Λ_j) at the facility. Denote the investment in facility j by α_j, where $\sum_{j=1}^{m} \alpha_j = B$, $\alpha_j \geq 0$, and $\underline{\alpha} = (\alpha_1, \ldots \alpha_m)$. We assume that $\underline{\alpha}$ can be varied continuously. The objective is to allocate the budget among the facilities (i.e., determine $\underline{\alpha}$) so as to minimize the total cost of all users receiving service. Since the locations of facilities are fixed, we write the travel cost term $T_i(d(i, F_j))$ as a constant c_{ij}.

The externality function at facility j, $E_j(\bullet)$, is now written as $E_j(\Lambda_j, \alpha_j)$. As before, we assume that $E_j(\bullet)$, $j = 1, \ldots, m$, is a continuous, differentiable, non-negative function, increasing and convex in Λ_j, and as $\Lambda_j \to K_j$ (the capacity of facility j), $E_j(\bullet)$ approaches ∞. We also assume that $E_j(\bullet)$ is a continuous, strictly decreasing function of α_j (i.e., increasing the investment in facility j continuously decreases the mean waiting time, given any fixed Λ_j.)

Since a change in the investment distribution among the m facilities will change the relative congestion levels at the facilities, the user-choice equilibrium is a function of $\underline{\alpha}$, so we write $\{\Lambda_j(\underline{\alpha})\}$, and denote the amount of demand from user i that is served by facility j as $\lambda_{ij}(\underline{\alpha})$. We write the expected waiting time for service at facility j as $E_j(\Lambda_j(\underline{\alpha}), \alpha_j)$.

Using this notation, the investment problem, introduced in [504], is:

Problem 5 $\min_{\underline{\alpha}} Z(\underline{\alpha}) = \sum_{i=1}^{n} \sum_{j=1}^{m} \{\Lambda_{ij}(\underline{\alpha})[c_{ij} + Q_j(\Lambda_j(\underline{\alpha}), \alpha_j)]\}$

$$\text{s.t.} \quad \sum_{j=1}^{m} \alpha_j = B$$

$$\Lambda_j(\underline{\alpha}) = \sum_{i=1}^{n} \lambda_{ij}(\underline{\alpha}) \quad \forall j \in M$$

$$0 \leq \alpha_j \leq B \quad \forall j \in M$$

$$\{\Lambda_j(\underline{\alpha})\} \text{ is a user-choice equilibrium}$$

For any fixed investment vector $\underline{\alpha}$, the user-choice equilibrium can be readily calculated using the methods described in Section 7.3. However, to analyze the investment decision, we need to need to know which customers frequent which facilities for a given budget allocation, and we need to know how the user choice pattern changes as the investment distribution among the facilities changes.

Let $\Delta_i^{pr} = c_{ip} - c_{ir}$ represent the relative attractiveness of facilities p and r to user i in the absence of an externality cost difference, and $q^{pr} = Q_r(\Lambda_r) - Q_p(\Lambda_p)$ represent the relative attractiveness of the two facilities (to any user) in the absence of a travel cost difference. When $\Delta_i^{pr} = q^{pr}$, user i is indifferent between facilities p and r; otherwise, the user prefers one facility to the other. Define the *marginality equation* for a user-choice equilibrium $\{\Lambda_j\}$ for user i and facilities p and r as the comparison of the magnitude of q^{pr} and Δ_i^{pr}. There are $nm(m-1)$ marginality equations. For any user i, either

$$\Delta_i^{pj} < q^{pj} \quad \forall j \in \{1, \ldots, m\} \backslash \{p\} \text{ for some } p \in \{1, \ldots, m\} \qquad (7.48)$$

or

$$\Delta_i^{pr} = q^{pr} \quad \forall p, r \in T_i^+ \text{ and } \Delta_i^{pr} < q^{pr} \quad \forall p \in N_1, r \in N_2,$$
$$\text{where } N_1 \cap N_2 = \emptyset \text{ and } N_1 \cup N_2 = \{1, \ldots, m\}. \qquad (7.49)$$

We henceforth simplify notation by writing $\{\Lambda_j\}$ and $\{\lambda_{ij}\}$ instead of $\{\Lambda_j(\underline{\alpha})\}$ and $\{\lambda_{ij}(\underline{\alpha})\}$. Any marginality equation that holds with equality in (7.49) is said to be an *active marginality equation*, and an active marginality equation is said to be an *effective marginality equation* if $\lambda_{ip} > 0$ and $\lambda_{ir} > 0$.

A user-choice equilibrium $\{\Lambda_j\}$ has between 0 and $nm(m-1)$ active marginality equations. If there is more than one active marginality equation, there can be more than one realization $\{\lambda_{ij}\}$. It is easy to create examples where a user-choice equilibrium $\{\Lambda_j\}$ contains an arbitrary number of marginal users (in fact, it is easy to construct examples where all users are marginal users). [504] established the following:

Theorem 11 *For any user-choice equilibrium* $\{\Lambda_j\}$, *at most* $m-1$ *of the effective marginality equations are independent.*

The feasible region of the solution space for Problem 5 is the positive portion ($\alpha_j \geq 0 \;\; \forall j$) of the hyperplane $\sum_j a_j = B$. Define the *set of independent effective marginality equations* (SIEME) to be the set that spans the set of active marginality equations. [504] showed how to partition the feasible region into at most $m-1$ primary regions, each of which is characterized by a unique SIEME (i.e., a unique user-choice equilibrium facility utilization pattern), and break curves that define the boundaries of primary regions.

They showed that objective function is continuous across the feasible region, and that objective function derivatives exist within but not across the primary regions. They also established that the objective function is non-convex within primary regions and break curves. Thus, to find the optimal budget allocation, they find local optima within all primary regions and break curves.

A local minimum of the objective function $Z(\underline{\alpha})$ occurs at a value of $\underline{\alpha}$ such that $dZ(\underline{\alpha})/d\alpha_p = 0$ for $p = 1, \ldots, m-1$. Finding such a local minimum is complicated by the underlying equilibrium problem, since the Λ_j's that satisfy $dZ(\underline{\alpha})/d\alpha_p = 0$ must be consistent with the user-choice equilibrium for $\underline{\alpha}$. Therefore, one must insure that the E independent marginality equations are satisfied. Hence, to find the local minima, one must simultaneously solve a system of $(m-1+E)$ equations with $(m-1+E)$ unknowns.

There are $m-1$ equations of the form $dZ/d\alpha_p = 0$ and E independent effective marginality equations. These equations contain $m-1$ unknown α_p's, and E unknown Λ_j's.

For each local minimum, the objective function value can be found by computing the user-choice equilibrium and corresponding total cost. The optimal investment distribution is then given by the best of these local optima. [504] presented an algorithm for determining the optimal investment pattern, and showed that the complexity of the algorithm is at least $O(n^m m^{n+8})$. Much of this complexity is due to the effort to partition the feasible region into primary regions and break curves.

Since the exact solution method may be computationally unwieldy for large problems, they also proposed several heuristic solution methods. For example, one could limit the primary regions that are examined (thereby avoiding the task of completely partitioning the feasible region) and use a descent method within each of these regions. One could consider only primary regions without a marginal user; however, it may be difficult to locate such regions (some problems may not have such regions). One could also limit analysis to the primary regions corresponding to a limited set of $\underline{\alpha}$ chosen in consultation with a decision maker.

One could use numerical descent methods directly on the overall problem, starting from an initial $\underline{\alpha}$ (thereby avoiding the partitioning task completely). Bi-level programming methods could be used [710], as could a projected gradient method [790]. Although such an approach may not find the global optimum, it would provide at least a local minimum even for very large problems.

Another heuristic approach is to reduce the size of the problem before attempting to solve it, using the analyst's intuition about likely "good" solutions. One could reduce the number of customers by aggregating users with similar c_{ij}'s (e.g., households in a neighborhood). One could reduce the number of facilities by aggregating similar facilities (e.g., networked workstations); this would involve developing appropriate values for $E_j(\Lambda_j, \alpha_j)$ and c_{ij}'s for the aggregate facility, as well as an appropriate restatement of the budget constraint. One could limit the facilities available to each user by assigning users to facilities a priori or providing users with a small choice of facilities (similar to what many health maintenance organizations do currently). Finally, one could bound the budget allocation at each facility to reduce the size of the feasible region, based on perceived fairness, expected "poor" solutions, or even political considerations.

7.5.2   Resource Allocation for Public Facilities in a Dictatorial Environment with Inelastic Demands

In this section we examine the service resource budget allocation problem in a dictatorial environment. This problem, introduced by [505], can be expressed as:

Problem 6

$$\min_{\{\lambda_{ij}\},\underline{\alpha}} Y(\{\lambda_{ij}\},\underline{\alpha}) = \sum_{i=1}^{n} \sum_{j=1}^{m} \lambda_{ij}[c_{ij} + E_j(\Lambda_j,\alpha_j)]$$

s.t.

$$\Lambda_j = \sum_{i=1}^{n} \lambda_{ij} \ \forall j \in M$$

$$\sum_{j=1}^{m} \lambda_{ij} = \lambda_i \ \forall j \in M$$

$$\lambda_{ij} \geq 0 \ \forall i \in N, \ j \in M$$

$$\sum_{j=1}^{m} \alpha_j = B$$

$$0 \leq \alpha_j \leq B \ \forall j \in M$$

The decision variables are an m-dimensional budget allocation vector, denoted by $\underline{\alpha}$, and the set of nm user facility assignments, denoted by $\{\lambda_{ij}\}$. As before, since the facilities are fixed, we have written the travel cost term $T_i(d(i, F_j))$ as a constant c_{ij}.

For any particular value of $\underline{\alpha}$, the problem reduces to one of assigning customers to facilities so as to minimize total social cost. The resulting "dictator's assignment problem" has a convex objective function and linear constraints. Hence the Frank-Wolfe method is a ready solution candidate. One can also use standard non-linear programming techniques [790]. [505] showed that the optimal solution is unique in $\{\Lambda_j\}$ but not in $\{\lambda_{ij}\}$ (nonuniqueness can occur when two different customers have the same travel costs).

The objective of the dictator's assignment problem is to minimize the total cost to all users of receiving service. [503] has pointed out that one can reinterpret this problem as a description of the actions of a set of sovereign users. However, the users are a special breed. Unlike the user-choice case, where users act to myopically self-optimize, users in this case are altruistic and seek to do the greatest good. A user at i will select the most "attractive" facility, where attractiveness is measured by $c_{ij} + E_j(\Lambda_j) + \Lambda_j \frac{\partial Q_j(\Lambda_j)}{\partial \Lambda_j}$. The notion of attractiveness represents the direct cost to user i of frequenting facility j (i.e., $c_{ij} + E_j(\Lambda_j)$), with the addition of the $\Lambda_j \frac{\partial Q_j(\Lambda_j)}{\partial \Lambda_j}$ term. This latter term represents the impact of user i's actions on all users frequenting facility j; it is the extra cost that accrues to all users if user i increases his usage of facility j. Hence, on the margin, user i will act to minimize the total cost of all users. This is indeed a special breed of user!

To optimally solve Problem 6, [505] presented an approach similar to that for Problem 5. As before, the feasible region is the portion of the hyperplane $\sum_j \alpha_j = B$ contained in the first quadrant of the m-dimensional Cartesian Space. Analogous to the user-choice case, they partition the feasible region into primary regions and break curves that separate the primary regions. Each primary region is characterized by a unique facility utilization pattern. Analogous to the user-choice case, they showed that the facility utilizations $\{\Lambda_j(\underline{\alpha})\}$ change continuously with $\underline{\alpha}$; that objective function derivatives exist within primary regions and break curves, but not across primary regions; and that the objective function is not necessarily convex within primary regions or break curves. Thus, their solution approach examines objective function derivatives within each primary region and break curve to identify all local minima. The best of these solutions is the optimal solution. [505] presented an algorithm for finding the optimal solution, and showed that its complexity is at least $O(n^m m^{n+6})$. Because of this difficulty, they suggest that heuristics similar to those proposed for the user-choice case could be applied to solve the problem.

We comment briefly on the relationship between the user-choice environment and the dictator's environment. The dictator's assignment will always yield the minimum total social cost. [506] showed that by imposing a toll on the use of each of the facilities, the self-optimizing users would achieve the same facility usage pattern as the lowest-cost assignment that the dictator would choose. They showed how to compute the toll without knowledge of the user-choice equilibrium.

7.5.3 Areas for Further Research on Resource Allocation Problems

The resource allocation models presented here have assumed a fixed service improvement budget. One could extend the analysis by making the budget variable rather than fixed. To formulate this problem we can add the budget B (weighted by an appropriate conversion factor) to the objective function, add a constraint $0 \leq B$, and then minimize over $\underline{\alpha}$ and B. This problem has a clear bi-level structure (by fixing B the problem reduces to those considered in this section) that might be exploited in its solution.

One could consider a budget allocation problem in the user-choice environment when the objective is to maximize revenue or profit rather than minimize total cost. This could be done for a single entity that controls all the facilities (which corresponds to a problem faced by developing African nations that finance expansion of their health care systems using clinic fees; see [30]) or for the owner of a single competitive facility who must make an investment in the presence of a competing facility. The competitive case could be extended to see if there exists a Stackelberg equilibrium for the investment decision for both facilities.

Another extension would be to consider users with elastic demand. Such research would have application to health care in developing countries (see [30]) where transport costs to clinics (and even bribes) are important in the individual's decision to seek health care. Issues of existence of equilibrium would be key.

7.6   Future Research and Conclusions

Location analysis in the presence of market externalities provides a fertile area for research and there is much scope for further work. The models discussed in this chapter can be extended and applied to more complex situations. We briefly discuss several generalizations.

Load Balancing Problem For the location-assignment problem 3a, the objective could be to maximize the minimum facility utilization. [726] has discussed how this problem can be formulated and solved as a linear integer program.

Other Customer Choice Models The user-choice model considered in this chapter could be extended to proportionate choice systems where a consumer's probability of choosing a facility is inversely related to the corresponding cost. [750] have studied the equilibrium and [728] have analyzed location decisions under such an assumption.

Another possible extension is to consider an externality function that depends on the entire user-choice pattern $\{\lambda_{ij}\}$. This type of model is relevant when users from node i are not the same as users from node k in terms of their service time requirement. One example is a grocery store serving different areas of a community: one might expect condominium and apartment dwellers to require less service time in the grocery store than people from suburban neighborhoods. Before addressing the question of appropriate facility locations in this environment, one must first analyze the user-choice equilibrium for this type of model.

Facility Location and Resource Allocation Finally, another line of research would be to consider the question of facility location in the context of the budget allocation problem. If a solution to the optimal budget problem were available, it would probably be possible to optimize facility size given facility location by developing the special case where the budget is to be provided only to a single location. From this point, research to consider simultaneous optimization of facility location and facility size might be fruitful.

8
OBJECTIVES IN LOCATION PROBLEMS

H.A. Eiselt
University of New Brunswick
Gilbert Laporte
University of Montreal

8.1 Introduction

Location models have been around for a long time. Modern location theory is usually said to have begun with Weber's [1181] treatise on the locations of industries. For many years hence, his treatment of location problems in general and objective functions in particular has dominated the literature. It was Hakimi [509], whose landmark contribution first distinguished between cost-minimizing minisum objectives, which had dominated the literature thus far, and minimax objectives. This also marks the first major attempt by an operations researcher to solve location models. Throughout the 1960s and the early to mid-1970s, minisum objectives were applied to private sector applications, whereas minimax objectives were suggested for public sector applications. These two types of problems reigned unchallenged for about a decade.

This changed dramatically in the mid-1970s. Goldman [492] and Church and Garfinkel [221] seem to have been the first to suggest what later became known as "obnoxious location models", i.e., models in which customers no longer consider the facility desirable and try to have it close to their own location, but instead avoid the facility and stay away from it. Typical applications are optimal locations of nuclear reactors, garbage dumps, or water purification plants.

Another line of investigations commenced with [811] and later [524]. Their research is based on the work of researchers such as [867, 1038] and deals with "fairness" or "equity". The philosophical background was provided by [985] who claimed that "there is an analogy between the two principles [of justice] and the maximin rule for choice under uncertainty." Making the least well-off element as well off as possible would provide maximum fairness. A large number of "equity" objectives have been developed recently and attempt to address these and similar questions and offer solutions based on the value system of the decision maker.

In this chapter, we not only survey what we consider the most important contributions to location models with different objectives, but we also propose a framework for them. For that purpose, it is best to recall the forces that move facilities into certain directions. As an example, consider the simple Weber problem of finding a single point in the two-dimensional plane that minimizes the sum of Euclidean distances to a number of given points. A mechanical analog of this model is the Varignon apparatus [1194]. The resulting concept of "pull" is central to our discussion. More generally, we will cast our discussion in the framework of attracting and repelling forces, similar to those of a magnet. Customers or users may either try to attract (or pull) desirable facilities closer to them, or repel (or push) undesirable facilities from them. These concepts facilitate an explanation of the locations resulting from the use of different objectives.

The remainder of this chapter is organized as follows: in the second section, we "set the stage" for arbitrary location models by discussing their major components and describing a framework of decision making. Section three elaborates on the location of desirable facilities, while section four discusses undesirable facilities. Section five reviews objectives that try to keep a balance while locating facilities. We conclude our investigation in section six by summarizing our results and pointing out a number of potential future research directions.

8.2   Elements of Location Models

In this section we first introduce and discuss what we consider the most important components of location models. We then provide a decision-making framework whose output is any of the known location models. We restrict ourselves to single-objective, single-decision maker location problems in order to facilitate the study of behavior of different objectives by avoiding overlapping and interfering pushes and pulls of the objectives.

A number of attempts have been made to summarize the major components of location models in taxonomies, not unlike those well known in queuing or scheduling. The first such attempt was made by [533] for center problems, followed by [862, 385] who suggested a very elaborate scheme, and most recently, by [525]. In this paper, it is sufficient to characterize the most important features of location models without a formal scheme.

Space. It is customary to distinguish between *continuous* location models, i.e., models in which the location of facilities in some d-dimensional space $\mathcal{R}^d$ is to be planned, *discrete* location problems, in which facilities may only be located at some prespecified points, and *network* location models. In the last case, it may be useful to further distinguish between models in general networks as opposed to specialized networks, like tree networks, as a good number of location problems are NP-hard in general, but solvable in polynomial time on trees. Furthermore, the space available to facilities to locate may be restricted by the introduction of "forbidden zones", i.e., areas in which facilities may not locate. Such areas are, indeed, meaningful as they can model zoning laws or other restrictions governing the problem.

The number of new facilities to be located by the decision maker. In the simplest cases, only one facility is to be located. In general, the decision maker has $p \in N$ facilities to locate. It is also possible that the number of facilities is endogenous, in which case the establishment of each facility carries a given fixed cost, such as the well-known simple facility location problem. In case of $p > 1$ facilities, it is no longer sufficient to consider only the location aspect of the problem: now one must also decide how to *allocate* customers to facilities. In some cases, the facility planners decide which facility customers patronize, e.g., when school boards allocate students to schools. However, most allocation decisions are made by customers who decide where to shop or obtain a service. When locating multiple new facilities, interactions between them may have to be considered. Some models allow interactions between facilities, i.e., terms in the objective include (weighted) distances between new facilities. Examples of reasons for doing so include military installations that are spread out so as to limit the damage by a direct hit, or franchises of gas stations or fast food restaurants that the decision maker tries to disperse so as to avoid direct competition with each other. Another application concerns the positioning of brands in some feature space where the decision maker attempts to position brands so as to minimize competition among his own brands.

The number of existing facilities. Two different types of models exist. In the first model the number of facilities to be located by the decision maker is fixed, whereas it is variable in the second. While both types certainly have their applications, we concentrate on the former. The reason is that adding another variable to the problem, *viz.* the number of facilities, complicates the problem and adds very little to the understanding of how optimal locations are found. It does, however, provide insight into the inherent tradeoffs between increased benefits due to shorter distances in case of a larger number of facilities and higher costs in such a case. Another aspect is that in many location models it is assumed that the decision maker who plans the location of his facilities faces an empty space without any similar or competing facilities. There may be cases in which this is a reasonable assumption, but in general similar facilities already exist and the task is to add new ones in an optimal fashion. The existing facilities may be the decision maker's own or those of competitors. Such conditional location problems were introduced by [852]. Also in this context belong the concepts of *medianoid* and *centroid* as defined by [511]. In particular, an $(r|X_p)$-medianoid is defined as a set of r locations, given that p facilities at specific locations in the set X_p already exist. On the other hand, an $(r|p)$-centroid is defined as a set of r optimal locations for the facilities of competitor A, given that subsequently competitor B will locate p facilities optimally, i.e., assuming that B will solve a conditional location problem.

Two new references in the field of conditional location models are [104, 513].

The decision maker's objective. That is the focus of this paper. Here, we distinguish between three different types of objectives: (1) minimization of a function of distance, (2) maximization of a function of distance, and (3) balancing of functions of distance. We will discuss this issue in much more depth below; for now, it should suffice to say that there is more to objectives than the simple minisum and minimax objectives.

Customers. Although there are a number of location models without customers, these are present in most models. If so, we need to know their distribution, their actual demand, as well as customer behavior. For instance, it may be assumed that customers are either distributed uniformly over a given set, or that they are located at specific points in space or vertices in a network. Secondly, even though demands in deterministic models are known with certainty, they may not necessarily be fixed or inelastic. There are important models in which demands are functions of distance or other externalities [131, 132]. Finally, it is important to know about customer behavior. As already mentioned above, customers may be assigned to a facility or may be free to choose, in which case the question is whether they always patronize the closest facility or use another utility function. As an example, if we have to locate a facility that sells a required good for which there is no substitute, then customers have to buy the good, regardless of how far or undesirable the facility is, and thus their demand is fixed. Most goods or services are, however, not of that type. A case in point is restaurants. An individual who wants to patronize a restaurant may refrain from doing so if the nearest such facility is too far away - there is usually the choice of eating at home. Locating facilities in such a model means that decision makers face variable demands, which has important ramifications regarding the decision maker's objective. In such a case it may, for instance, be useful to maximize demand capture or coverage.

Given our above discussion of the major elements of location models, we are now able to derive the main categories of objectives. Many such classes have been devised. One is the distinction between private and public facilities. In the literature, "private" is often associated with "minisum" and "public" with "minimax" [1141]. Reasons for doing so seem compelling: given linear transport cost functions, the minisum objective will locate a facility so as to minimize the sum of transport costs. On the other hand, when locating a public facility such as a swimming pool or a library, one would like to determine a location that is close to even the most distant potential user. Things are, however, not quite as simple. Consider the location of a public library on a network. Let the number of users be represented as weights at the nodes of the network. Furthermore, assume that all users make the same number of trips to the library. (This assumption is not restrictive - if one user patronizes the library twice as often as another, he could simple be represented by twice the weight of the other). The minisum objective will then find a solution that minimizes the total distance traveled by all library users. This can be considered as an expression of "collective welfare" in which a so-called "private" objective optimizes the location of public facilities. For a further discussion of this argument, see [1004].

One possibility to choose an objective is to derive it from a set of axioms or value statements [686, 687]. This route has been taken by [590] who specifies three "reasonable" axioms which are then applied to a simple model with equally-weighted vertices on a tree embedded in the plane. The result is that the minimization of the sum of squared Euclidean distances is the unique objective function that satisfies all three axioms. [152] also proceeds from value statements to a specific mix of objectives.

In contrast to the above, we consider the choice of objective as a two-stage game involving two types of agents called planners and users. Users are normally customers or, in other location models, patrons (e.g., of public facilities). For simplicity, we assume homogeneity of customers, i.e., all users view the facilities similarly. This assumption is made for convenience and is not mandatory. Considering any existing or planned facility, a user has essentially four choices: find it desirable, undesirable, partly desirable and partly undesirable, or be indifferent. In location terms, if a user considers a facility desirable, then he would like to have it as close as possible to his own location. In such a case, the user would like to exert a certain "pull" on the facility. If the facility is judged undesirable, the user would like to "push" the facility away - the time-honored NIMBY (not in my back yard) principle. Users who are indifferent can be left out of the model as they will not have any objections either way. The case of judging a facility partly desirable and partly undesirable is rather interesting. It was first mentioned by [492] in the context of customers wanting a facility "close, but not too close". Take a supermarket, for instance. On the one hand, customers may wish to live close to the facility for matters of convenience. On the other hand, there will be traffic and noise associated with the supermarket which makes this facility undesirable. When dealing with such a situation, users have essentially three choices: either they exert a pulling force on the facility, given a constraint that restricts proximity to a minimum, exert a pushing force with an added constraint that guarantees the facility is within a certain distance, or create a mixed "push-pull" objective, i.e. a net force such as a linear convex combination of the two forces. All three approaches have been used in location modeling (see, e.g., the discussion of the cent-dian in the next section), albeit in different contexts.

Suppose now that users' wishes and behavior are known to the planners. In a sequential (Stackelberg) game, the users are the followers and the leaders, i.e., the planners, are aware of the followers' preferences. Taking those preferences into account, planners can now decide what is best. In the case of private enterprise, this would mean what is best for the planner. In the case of public facilities, it means what the planners judge to be best for the users, undoubtedly a more precarious and difficult decision. Let us apply this user-planner framework to a number of well-known scenarios. Assume the case of inelastic demand and no competition. Users who consider the facilities desirable will patronize any one of the planner's facilities, most likely the closest facility. If the planner has to pay for transporting the goods to the customers, he will open as many facilities as necessary in order to minimize the sum of facility costs and transport costs; the standard simple facility location problem results. If the number of facilities to be located is given, minimization of costs results in the standard p-median problem. If the facilities are franchises, the planner (headquarters) may wish to balance business between franchises, a balancing objective. If demand is elastic, the number of facilities to be located is variable, and the planner follows a cost-minimization objective; we obtain a min-cost covering problem. If, in the same scenario, the number of facilities is fixed, the planner may follow a max covering objective. Similarly, other objectives can be derived in the context of a sequential two-stage game.

8.3   Pull Objectives

The class of objectives discussed in this section are all based on the assumption that the facility or facilities to be located are desirable. As a result, the planner wishes to locate close to the customers. As a result, these models are said to have "pull" objectives. They include the oldest known location problems. Among them are the original Weber problem [1181], as well as median, center, and covering problems. A number of original contributions are surveyed in this section along with some of the more prominent results.

Minimizing the sum of weighted distances with so-called *minisum objectives* is an obvious choice for public and private objectives as long as cost functions are linear. However, as we show below, they can also arise in profit-maximizing contexts. Our discussion follows closely that of [440]. Consider the location of a private facility under profit maximization. Assume for now that delivered pricing is applied, i.e., the decision maker pays for the transport of the goods to the customers. Define p_i as the price charged at location i (in case uniform pricing rather than spatial price discrimination is applied, $p = p_i$ for all i), let $d_i(p_i)$ denote the demand at location i given price p_i, suppose that the demand will be satisfied exactly, and let F be the fixed and c be the variable costs in a linear cost function. Furthermore, denote by t_{ij} the unit transport costs between a customer located at i and the facility located at j. The profit function is then $Profit = \sum_i p_i d_i(p_i) - c \sum_i d_i(p_i) - F - \sum_i t_{ij} d_i(p_i)$ where the first sum denotes the revenue, the second specifies the variable production costs, and the last sum defines the transport cost. For any set of fixed prices p_i, revenue and production costs are fixed, so that profit is maximized whenever transport costs are minimized.

Another possible pricing scheme is mill pricing. It requires the decision maker to set a price p which is charged at the facility; customers then pay their own transport costs. In this case, the profit function is $Profit = (p - c) \sum_i d_i(p + t_{ij}) - F$. For any fixed price p, the function simplifies to a maximization of sales. The problem thus becomes one of *maximum capture*. Maximum capture objectives also arise when a decision maker attempts to maximize sales, revenue, or the number of customers reached, e.g., by advertisements or radio signals. Wherever they originate, minisum and maximum capture objectives will locate facilities close to the user and hence belong to the class of "pull" objectives.

Consider first the case of minisum objectives. To formalize matters, suppose that a decision maker plans to locate a fixed number of facilities, say, p. The locations of the facilities are denoted by $x_1, x_2, \ldots, x_p$. Customers are located at points $v_1, v_2, \ldots, v_n$. Locations of customers and facilities can either be in some space (such as $\mathcal{R}^d$) or on a network. Assume that customers always patronize the closest facility, i.e., customers at v_i will deal with facility x_ℓ if $d(v_i, x_\ell) = \min_k \{d(v_i, x_k)\}$. (Other choice rules are possible). Associating weights w_i, $i = 1, \ldots, n$ to the customers, the minisum objective is $Min f_{sum}(x) = \sum_i w_i d(v_i, x_l) = \sum_i w_i \min_k \{d(v_i, x_k)\}$.

Single-facility location problems with minisum objective in the plane are also referred to as *Weber problems* or, sometimes, *Steiner-Weber problems*. Good historical accounts of the problem are found in [237, 419, 1020] and, more recently, [1194]. The objective of the problem is convex and differentiable almost everywhere. The problem was first solved by Weiszfeld [1186]. His iterative solution technique was later rediscovered by [872, 725]. The method is known to converge quickly with problems occuring only at the given demand points at which derivatives of the function do not exist. Simple perturbations provide an easy remedy. Applying squared Euclidean, rather than Euclidean, distances results in a closed form solution. The optimal point equals the center of gravity. As far as forces are concerned, squared Euclidean distances introduce a higher penalty for far away customers than standard Euclidean distances, thus resulting in locations that avoid distant customers, even those with smaller weights. Using rectilinear, rather than Euclidean, distances simplifies the problem significantly [786]. A number of extensions have been proposed. Cooper [240] considered the single Weber problem with Euclidean distances raised to the k-th power. For $k \geq 1$, the objective is convex, whereas for $k < 1$, it is neither convex nor concave. He suggests an iterative technique similar to Weiszfeld's for the solution of the problem. Another possible extension puts the 1-Weber problem into d-dimensional space; an adaptation of Weiszfeld's method as a solution technique was proposed by [241].

In case $p > 1$ facilities are to be located, the problem is frequently referred to as the *multi-Weber problem*. The first to use it was [872], but

[237] appears to have been the first to formally describe the problem. Two versions exist. In the first, shipments between facilities as well as individual facilities and customers are known. In the second version it is not known which customers are going to be served by an individual facility until facilities are actually located. The latter version gives rise to two major issues: locating facilities and allocating customers to them. This prompted Cooper's [238] alternate location - allocation heuristic: starting with any initial locations of p facilities, allocate each customer to its closest facility, and then optimally relocate a facility among the customers allocated to it. The procedure is repeated until it converges. One of the difficulties related to multi-Weber problems is that its objective function is neither convex nor concave. A solution method that employs a quasi-Newton method was proposed by [200] who also reviews a number of exact and heuristic techniques. For further details, see [786].

The first results regarding minisum location problems on networks are found in Hakimi's [509] seminal paper. One of the major contributions of this paper is the characterization of the set of solutions for minisum problems. In particular, Hakimi proves that for problems on general graphs with customer demands occuring at vertices, the minisum objective is concave and thus its minimum is attained at one of its boundaries (a vertex of the graph). This "Hakimi property" also holds for $p > 1$ facilities as shown by [510]. It is an extremely useful property as it enables users to restrict the search for optimal solutions to the vertices. For 1-medians, this gives rise to a simple solution technique, requiring $O(n^2)$ operations, provided that the shortest path matrix is known. For $p > 1$ facilities, i.e., a p-median problem, [660] have shown that such problems on general graphs are NP-hard. On the other hand, they show that p-medians on trees can be found in $O(n^2 p^2)$ time. Some early heuristics, one the equivalent of Cooper's alternate location - allocation, are described and tested by [1123]. Another early paper on the subject is by [1004]. Some newer references are [116, 212].

An intriguing result is due to [489] who solves the 1-median problem on a tree network. It turns out that the distances associated with the edges are immaterial. Define subtrees spanned by a vertex v as those subtrees that result if v and all its incident edges are deleted as well as the weight of a (sub-) tree as the sum of weights of its vertices. Then a vertex $v*$ is a median only if the weight of its largest subtree does not exceed half the weight of the entire tree. Interestingly enough, this result goes back to [644] and [600]. Such a vertex $v*$ can be determined in linear time.

With Hakimi's results known, a number of authors have investigated related problems. For instance, [493] show that the Hakimi property also holds for some problems with concave objective functions, i.e., models in which unit travel costs are nonincreasing with distance.

[488] offers two new formulations. In both, weights are no longer associated with vertices, but with origin - destination pairs. The idea is to first ship supplies that turn up at some origin to one of the processing cen-

ters and, after having processed them, direct them to their destination. The weights of the pre- and post-processed products may be different. The objective is to locate p processing facilities so as to minimize the sum of weighted distances. A second formulation allows additional shipments between facilities. This formulation permits shipments between facilities which can be considered as hubs. Both formulations have in common that they do possess the Hakimi property and thus allow efficient, albeit complete, enumeration on the nodes of the network. [516] generalize Goldman's results.

Another major class of problems attempts to locate a facility so as to make the longest customer-facility distance as short as possible. This *minimax objective* was also discussed by [509]. One of the foundations of this objective is the "Theory of Justice" [985] which generalizes the classical theory in the *contractarian tradition*. It essentially equates justice or fairness with the objective of making the least well off element as well off as possible. [563] points out a number of serious flaws in Rawls's arguments and suggests instead the use of expected values in the *utilitarian tradition*. Formally, a p-facility problem with minimax objective can be written as $Minf_{max} = \max_{i}\{w_i d(v_i, x_l)\} = \max_{i}\{w_i \min_{k}\{d(v_i, x_k)\}\}$. Locations satisfying the above relation are usually referred to as p-centers. In the plane, locating one center can be accomplished by geometrically inspired methods developed by [883, 379]. The latter coin the term "delivery boy problem" for the unweighted objective that minimizes the maximum distance and "messenger boy problem" for an objective that maximizes the sum of customer-facility distance plus a customer-specific constant. The cases of rectilinear and Chebyshev norms are dealt with in a similar fashion, except that the circles are replaced by squares and diamonds, respectively. The relation between center problems in the plane and so-called "covering problems" is outlined below.

In graphs, it is easy to see that even the unweighted version of the minimax problem does not have the Hakimi property. On a single edge with node weights of one, the point that minimizes the longest distance is the midpoint on the arc. This gives rise to what are known as *vertex centers*, as opposed to *absolute centers*, on graphs. A vertex center is defined as the node to which the longest weighted distance from any other node is as short as possible, whereas an absolute center is any point on the graph with this property. A 1-vertex center on a graph can easily be determined by complete enumeration in $O(n^2)$ time, again provided the matrix of shortest paths is available. [490] describes a solution technique for the messenger boy problem on trees and [532] delineates efficient methods for 1-absolute and 1-vertex centers on trees.

A variety of complexity results are surveyed by [1119]. Weighted absolute 1-center problems on networks are solved by [515] in $O(|E|n^2 \log n)$ time. This bound is improved by [659] to $O(|E|n \log n)$. The same authors solve the equivalent unweighted problem in $O(|E|n)$ time. On trees, complexities improve dramatically. The unweighted absolute 1-center on trees was first determined by [490] in $O(n^2)$ time, a bound to be improved later by [532, 517] to $O(n)$ time. [659] describe an $O(n \log n)$ algorithm for the corresponding weighted problem. The same authors also determine the p-center problem to be NP-hard on general graphs. They provide algorithms for the weighted and unweighted problems with complexities $O[|E|^p(n^{2p-1})(\log n)/(p-1)!]$ and $O[(|E|^p(n^{2p-1})/(p-1)!]$, respectively. Weighted absolute p-centers on trees are determined by the same authors in $O(n^2 \log n)$ time, whereas vertex p-centers on trees are computed in $O(n \log^2 n)$ time by [821]. [659] solve unweighted absolute and vertex p-centers on trees in $O(n \log^{p-2} n)$ and $O(n \log^{p-1} n)$ time, respectively. They also note that trees always have a unique absolute 1-center and at most two vertex 1-centers.

Some researchers have tried to combine the efficiency of the median and the "fairness" of the center objective. One way of accomplishing this is by way of a constrained p-median problem, i.e., a formulation with a minisum (= median) objective that has upper bound constraints on all facility - customer distances. Problems of this type were first introduced by [1141, 694]. Let d_{max} denote the maximal distance allowed between any customer and its closest facility. Clearly, for d_{max} sufficiently large, the medi-center equals the median. [531] describes algorithms for single-facility *medi-center* problems. More recently, [105] develop algorithms for p-medi-center problems as well as related uncapacitated medi-center location problems, in which facilities have a fixed cost associated with them that is incurred if they are established at any of the nodes. Then the number of facilities to be located is endogenized and thus becomes part of the solution. Another possibility of incorporating median and center components in a single problem is to employ a minimax objective with a bound on the sum of weighted distances. The single "budget"-type constraint is much more difficult to handle than individual upper bounds and we are not aware of any work on this problem.

[520] introduced *cent-dians* whose objective is a linear convex combination of minisum and minimax objectives, *viz.* $f = \lambda f_{sum} + (1-\lambda)f_{max}$ with $\lambda \in [0; 1]$. Halpern's results concern the last problem in a tree network. He shows that the cent-dian is located at one of the vertices on the path connecting center and median. In addition to a procedure that determines optimal cent-dian locations, it is shown that if all weights are integers, then for all $\lambda \geq \frac{1}{2}$, the cent-dian coincides with the median. The author also demonstrates how the cent-dian problem can be reduced to a regular median problem: create a new vertex for the center, split the edge with the center accordingly, and let the weight of the new vertex be $(\lambda^{-1} - 1)$. The median in the new graph equals the cent-dian in the original. [522] shows the tradeoff curve between median and center objectives. In yet another paper on the subject, Halpern [521] investigates 1-cent-dians on general undirected graphs. He proves that the objective function is continuous,

piecewise linear, and concave. Again, the cent-dian is located on a path connecting center and median. It is also shown how to solve the problems as an equivalent median problem.

[1087] formulated a problem of which 1-medians and 1-centers are special cases. A single facility is called a *k-centrum* if for any given integer k, the objective is to minimize the sum of distances to the k farthest customers. Clearly, for $k = 1$, the problem reduces to that of a 1-center; similarly, for $k = n$, the problem turns into a 1-median problem. In that sense, p-medians and p-centers are one generalization of 1-medians and 1-centers, whereas k-centra are a generalization in a different direction. Slater's model is based on a tree network with all edge lengths equal to one. He proves that k-centra either are one node or two adjacent nodes. Furthermore, k-centrum and $(k+1)$-centrum sets have a nonempty intersection. Slater's results were generalized [26, 27] for tree networks with arbitrary positive edge lengths. Most importantly, they prove that in any given tree, the k-centrum is an elementary path. Moreover, if k is odd, then the set of nodes in a k-centrum cannot exceed two. The eccentricity function, i.e. the function that defines a k-centrum, is convex on trees but not on general graphs.

A different type of objective is used in *covering problems*. The idea is to locate facilities so as to "cover" customers and thus "capture" their demand. There are two versions of covering problems, one in which the number of facilities is fixed and another in which it is variable. With a fixed number of facilities, we obtain a *max cover* problem in which we locate facilities so as to maximize the demand captured by the facilities - the maximum capture problem derived earlier from a general problem formulation. With a variable number of facilities, we may wish to cover the entire demand with the smallest number of facilities, given that no customer is farther than a prespecified distance d^* from its closest facility. This is the *min (cost) cover* problem.

Min cost cover problems in the plane were first described by [1141]. The authors used a linear programming relaxation and applied cuts whenever necessary to obtain integer solutions. A similar approach is described by [1139], who use results about set covering problems known from integer programming. [1170] describes heuristics to minimize the number of tower ladders to fire houses.

Some instances of max cover problems date back a long time. One of these problems locates a single facility so that the Euclidean distance from the facility to its farthest customer is as small as possible. Such covering for a single facility is equivalent to what is known as the *smallest enclosing circle*. This problem goes back to [1116]. It is apparent that the 1-max cover problem also solves the 1-center problem in the plane. [1063] have derived $O(n \log n)$ algorithms to solve such problems. Their method is based on Voronoi diagrams, for details see [897]. Note the relationship between p-max cover problems and p-center problems. Set all demands equal to one and let $r*$ denote the maximum radius in a given p-center problem. Then the corresponding p-max cover problem will attempt to locate p facilities so as to maximize the number of nodes covered within $r*$ distance. As $r*$ solves the p-center problem, it is possible to cover all nodes within that distance. In that sense, the p-center problem is a special case of the max cover problem with covering distance $r*$. Variations on $r*$ also allow sensitivity analyses on p-center problems.

Max covering problems on tree networks were first described by [223] who also considered the possibility of additional proximity constraints. The authors describe algorithms and report computational results on Swain's 55-node network. In a follow-up paper, [222] relax the assumption that facilities must be located at nodes of the network. For the "absolute max cover problem," they derive NIPS, or network intersection point sets, and prove that at least one solution is comprised entirely of NIP locations. [821] prove that max cover on general graphs is NP-hard, a result derived via reduction from *Minimum Dominating Set*. For tree networks, they describe an $O(n^2 p)$ algorithm. The authors also show that the min cost cover problem on trees is easy as the matrix of coefficients is balanced and hence the problem can be solved as a linear program. [514] introduce "Voronoi p-centers" on networks which are locational patterns designed to capture weight or demand on the basis of distance. The problems are difficult; even for a fixed number of facilities they are solvable only for small values of p.

[84] notes that even if facility locations at NIPS are allowed, non-NIPS solutions may exist with shorter max facility-customer distances. As a remedy, he defines the *p-partial center problem* which locates p facilities so as to minimize the maximum distance to the closest facility from demands that it covers, given a covering distance $d*$. Clearly, it is not suggested to use p-partial center problems in isolation as that may result in degenerate solutions in which only a few outlying customers are served. Instead, pareto-optimal solutions for the two objectives, max cover and p-partial center, are determined. For the case of tree networks, the author outlines an $O(n^2 \log n)$ algorithm.

A few other results regarding pull objectives are worth mentioning. [375] introduce the concept of a *minimax flow center*, defined as a point p on a network which minimizes the maximal arc flow in a flow pattern that satisfies all demands at the vertices from p. The concept combines location and routing aspects in one model. The authors show that on trees, minimax flow centers coincide with medians. The weighted version of the problem with positive arc weights is solved in linear time by [370]. Other types of location-routing problems, e.g., those originating in the context of mail delivery, are described by [120]. For a survey on the subject, see [734].

[921] take a fresh look at the location of non-emergency facilities. In essence, emergency facilities such as hospitals have an inelastic demand, i.e., the demand for service at some point will not decrease if the facility were to move farther away. This is different for non-emergency public facilities such as libraries. Demand for their services is elastic - if they were

FIGURE 1. Supply and Demand Curves

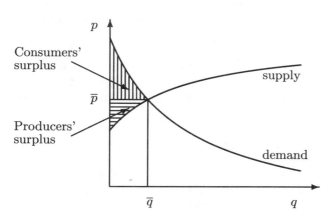

to move far away, potential patrons would seek other ways of learning and different forms of entertainment. This is the type of facility the paper focuses upon. Following standard economic theory, the appropriate objective in case of public facilities is to maximize the gain to users, i.e., the value of the service minus the cost of travel minus the cost of supplying the service. Under some mild assumptions, it is shown that this gain to users equals *consumers' surplus*, i.e., the difference between the aggregate price consumers are prepared to pay for a product and the amount they are actually paying. The concept is illustrated in the quantity-price graph of Figure 1. The resulting equilibrium price is $\bar{p}$ at which $\bar{q}$ units of the good will be sold. Suppose now that all potential customers are ordered according to their willingness to pay, starting with those willing to pay the highest price for the good. Then all customers to the left of $\bar{q}$ are prepared to pay the corresponding prices on the demand curve, but actually pay only the lower price $\bar{p}$. Their agglomerated benefit is the difference between what they are prepared to pay and what they actually pay. This is the consumers' surplus, shown as the diagonally hatched area in Figure 1. Similarly, we can order producers according to their willingness to sell for a given price starting with the one who would sell for the lowest price. Then the producers' surplus is the agglomerated difference between what producers would sell for and the price they actually receive. This is the horizontally hatched area in Figure 1.

In their analysis, Pearn and Ho use three types of demand functions: linear, modified constant elasticity, and modified exponential. They prove that consumer surplus is convex under any nondecreasing demand function. As the Hakimi property holds, only nodes have to be considered as potential solutions. The computational part of the study considers four problems: (1) Locate a prespecified number of facilities so as to maximize consumers' surplus, (2) maximize net consumer benefit with a variable number of facilities, (3) the usual p-median, and (4) minimize total cost with variable p. The main results are: comparing solutions to problems (1) and (2) with those of (3) and (4), differences in locational behavior increase as the size of the network increases. The steeper the slope of the demand function, the larger the difference between fixed and inelastic demand models. This is not surprising at all as inelastic demand has zero slope, so all this says is that the results are getting more different the more different the problems are. Also, the maximization of net benefit tends to result in fewer facilities than the minimization of total cost.

The work of [1168] follows similar lines. The authors present two models, both with an objective to maximize Marshallian welfare, i.e., the sum of consumers' surplus plus producers' surplus. The first model represents "public fiat", where consumers are assigned to facilities and service may be denied. The second model requires that all customers are served. The price-quantity relation is simple: as long as the price stays below a given constant $\bar{p}$ (a number representing the maximal willingness to pay) the demand is a constant. Once the price exceeds $\bar{p}$, the demand drops to zero. Both situations are modeled as integer programming problems, but are not solved.

In order to visualize the differences between different "pull" objectives, we consider the Weber problem, the 1-center problem, and the max cover problem. In order to simplify the discussion, we restrict ourselves to locations on a line segment of unit length. Let two customers A and B be located at the ends of the line segment at points 0 and 1, respectively. A third point C is first located at 0 as well. Moving C to the right towards "1", we examine changes of the optimal facility location with the given objective. All customer weights are assumed to be unity. Denoting C's location on the line by C and the facility location by X, the sum of customer-facility distances is $(1 + X - C)$ if $X \geq C$ and $(1 + C - X)$ if $X \leq C$. At optimum, $X = C$ in both cases. In other words, the facility location follows facility C and hence changes even for very small changes.

This is no longer true for center problems. In the same example, the center will be at $X = \frac{1}{2}$, regardless of C's location. In a more general context, as long as a customer relocates within the smallest enclosing hypershpere specified by the center location and the longest customer-facility distance, it does not exert any pull at all. Only if the customers leave the hypersphere will the facility relocate. Similarly, the facility will relocate if a customer on the hull of the hypersphere moves. In our one-dimensional example, if B were able to move to the right by some positive distance ϵ, then the facility location would move to the right as well, albeit only by $\frac{1}{2}\epsilon$. In summary, minimax objectives are less responsive than minisum objectives.

Finally, consider the max cover objective. As customers are not distributed evenly across space but are located at a number of given points, one may expect discontinuities in the capture function. To show that this is indeed true, take again the one-dimensional example above and let the customer weights be $w_A = 4$, $w_B = 3$, and $w_C = 2$, and assume that the covering distance is 0.2. For $C \in [0; 0.2]$, $X \in [0; 0.2]$ and 7 units are captured. For $C \in [0.2; 0.4]$, $X \in [C - 0.2; 0.2]$ and again, 7 units are captured. For $C \in [0.4; 0.6]$, $X = 0$ and only the 4 demand units of A are captured. For $C \in [0.6; 0.8]$, $X \in [0.8; C + 0.2]$ and the 5 demand units of B and C are captured, and finally, for $C \in [0.8; 1]$, $X \in [0.8; 1]$ and the captured demand is still 5. This example shows that not only are there discontinuities, but as one customer moves steadily in one direction, the facility jumps back and forth, making the optimal solution hard to predict. As mentioned above, the reason for this behavior is the discrete locations of customers.

8.4   Push Objectives

Location problems involving undesirable facilities have been discussed since the early 1970s. [879] outline a model that incorporates not only building costs, but also expected costs to mollify opposition to a facility. However, the paper presents no formal model and no procedural suggestions are made. First attempts to formalize were made by [492, 221]. The latter investigate the aptly named "maxian" model: an equivalent to the median, but with a maxisum objective. Other early references are [1063, 1074]. Models of this type later became known as noxious or obnoxious location models, the difference between the two being restricted to semantics rather than objective [385]. Whereas the former refers to potentially dangerous facilities, e.g., those that may be poisonous in case of a leak, the latter type of facility represents a mere nuisance. In both cases, however, customers or the public at large try to stay as far away as possible. Following Erkut and Neuman's suggestion, we use the term "undesirable" instead. As we will see below, work on undesirable facility location models represents one of the major fields of research today.

To visualize the difference between minisum and maxisum objectives, consider a simple example. Let four equally weighted customers be located at the vertices of a square in $\mathcal{R}^2$ with Euclidean distances. The unique minisum location of a single new facility is at the intersection of the diagonals. The maxisum objective, on the other hand, is optimized by locating the facility as far away as possible towards infinity, clearly showing the forces that push the facility. In order to avoid such unbounded optimal solutions, it is necessary to define a set within which the facility can be located, something unnecessary with minisum objectives for which it is an elementary exercise to prove that it is always better to locate the facility inside the convex hull spanned by the customers. If we were to restrict in our above example the location of a new facility to the convex hull of customers, the maxisum objective would be optimized by locating at any of the four customers locations.

Problems involving the location of undesirable facilities in the plane were investigated independently by researchers in computational geometry [1063]. Locating a facility so as to maximize the distance between a single new facility and any of the customers is equivalent to the geometrical problem of finding the *largest empty circle*, i.e., the largest circle that does not include any given point. The center of such a circle is the optimal location for such an undesirable facility. Such problems were later referred to as 1-maximin problems. Standard references in the field of computational geometry are the books [972, 897]. From a strictly locational point of view, the first to deal with maxisum facility location in the plane are [260] and [326]. The authors of the former reference show that the objective of the problem is not convex and may have local optima. They also determine a finite set of candidates for the optimal solutions and describe an algorithm. [326] present a binary search technique for the 1-maxisum problem subject to constraints that guarantee that all facility-customer distances are "away, but within reach". This is again a case in which a facility provides services and disservices, e.g., work and pollutants, at the same time. [862] called facilities of this type "semiobnoxious".

[827] also solves the 1-maximin problem with Euclidean distances in the presence of constraints that limit the locations to a bounded region. The author presents a geometric method based on circles with increasing radii. Based on a mathematical formulation of the problem, Kuhn-Tucker conditions are established and a Lagrangean approach is used for the nonconvex problems. An extension of this problem is presented by [829] in which not only does the facility have to be inside the bounded region, but it must also be at a certain minimum distance from the customers; constraints similar to those by [260]. The authors delineate an algorithm and give an example of the location of a toxic dump. In a further paper, [831] investigate the problem of locating a single facility in the plane so as to minimize the maximum weighted inverse square distance from the facility to the customers. This objective is justified as various types of pollution tend to decrease with the square of the distance between polluting facility and "customer". Examples of pollutants that behave in this manner are noise and radiation. Define S as a bounded set in two-dimensional Euclidean space and let P denote the convex hull of given points. Then the authors derive the following properties: (1) the local minimum is attained either in P or (2) on the boundary of S. If the local optimum is in P, then at least three weighted squared distances between facility and customers are equal and binding, and (3) if the local optimum occurs at a vertex of S, then at least one weighted squared distance is binding. For the special case of a convex polygon S, the authors describe an $O(n^4)$ algorithm. For the general case, an interactive procedure is suggested. [388] consider the same problem and show it to be equivalent to the maximization of the minimum of weighted squared distances. Their proof uses a more general formulation with arbitrary exponents.

[830] provide a survey of results concerning maximin and maxisum objectives with Euclidean and rectilinear distances. In the latter case, let S' be the convex hull of a polygon S. Then the set of vertices of S' includes the optimal solution, which gives rise to an $O(mn)$ method, with n being the number of customers and m symbolizing the number of vertices of S.

[661] considers a model without customers. He requires again that the single facility must be located in a polygonal region, but protected areas exist in which the facility may not be located. The objective is to maximize the minimum distance between the facility and any point in one of the prespecified protected regions. This means that the undesirable facility should be as far away as possible from nature preserves, parks, or any other protected areas. This problem reduces to the geometric problem of finding a circle of maximum radius with its center in the permissible region, so that the circle does not intersect any of the protected regions.

Finally, [656] reduce a number of "reasonable" objectives in the context of undesirable facilities to the maximization of the sum of the logarithms of Euclidean distances. The authors prove that extremal values occur on the boundary of the domain of definition. A physical equivalent in the form of an electrostatic problem is noted and a solution procedure is suggested.

[384] is one of the few papers dealing with the location of $p > 1$ undesirable facilities in the plane. They show that the p-maxian is NP-hard and devise an exact branch and bound method as well as a heuristic method that finds optimal solutions most of the time. Another paper is [549]. The authors define *anti-Weber* and *anti-Rawls problems* as the counterparts of p-Weber and p-Rawls problems in the plane. It is assumed that costs are decreasing and continuous in distance. For linear nuisance costs, the set of extreme points of the feasible region contains at least one solution to the anti-Weber problem. The problem is solved with a branch and bound method similar to an earlier "big square - small square" method. The anti-Rawls problem is solved with a simple graphical method. It turns out that facilities will frequently be located optimally along the boundaries of the feasible set. As an example, many of France's nuclear power plants are located along the borders with Germany and Belgium. An alternative approach is to create a compensation scheme for customers based on their exposure to pollutants and include efficiency considerations in the location model.

Consider now maxisum objectives on networks. In the simplest case, only one facility is to be located so as to maximize the weighted sum of distances (which are determined as the shortest paths) from a facility to all customers located at the vertices of the network. For the case of a single facility, this is the *1-maxian* or *antimedian*. Assuming that the facility can locate anywhere on the network, i.e., not necessarily at a vertex, [221] observe that the Hakimi property does not hold, i.e., the facility may best be located on an edge rather than at a node. The authors show that there exists an optimal solution point either on a cycle or at a pendent vertex.

A corollary of this result is that on a tree network at least one maxian is located at a pendent vertex. A procedure for the 1-maxian problem is then described. [1136] delineates a linear-time algorithm for 1-maxians on trees. [849] further investigates anticenters and antimedians. The suggested procedure for antimedians is similar to that of [221]. The case of anticenters in mixed graphs is also investigated.

[862] further discuss and classify models with push objectives. The authors identify different categories of problems, *viz.* anti-covering models, maxians, anticenters, and multiobjective problems. They pay special attention to potential constraints of the problems. Two categories are outlined: one in which distances between facilities are constrained, and one which constrains distances between facilities and customers. The former category is relevant in case a decision maker locates multiple facilities or when some facilities already exist. Some complexity results are surveyed and a number of unstudied problems are outlined. It is noted that distance constraints may result in nonconvexities and, in general, the Hakimi property does no longer hold. [195] further discuss anti-covering problems. The problem, which is equivalent to finding maximum independent sets, is then solved by four heuristic methods. Worst-case analyses on the objective values of the heuristic methods are performed and computational results are reported.

[1074] introduced a new type of problem which is quite distinct from other location models. It does not include customers and focuses exclusively on the facilities. The goal is to locate p new facilities on a network so as to maximize the minimum interfacility distance, where distances are measured as the length of the shortest path between facilities. The model is referred to as a *p-dispersion problem* or, as [386] called it, a maxminmin problem. Military and civil applications exist for this model.

The p-dispersion problem is well studied. Consider the simple case of a tree and two facilities. It is easy to envisage that the two facilities should be located at the pendent vertices which mark the end of the longest path in the tree. As the 1-center would be halfway between the two, the objective function value of this particular 2-dispersion problem is exactly twice the value of the corresponding 1-center problem. [1074] proves that this relation is true in general on trees. Denoting by $z(*)$ the objective value of problem $*$, he proves that the relation $z(p\text{-center}) = \frac{1}{2}z((p+1)\text{-dispersion})$ holds. This is an interesting "duality" result between problems belonging to different classes, one being a pull and the other a push objective. [187] point out that Shier's duality results do not extend to general networks. [719] shows that $\frac{1}{2}z((p+1)\text{-dispersion})$ is still a lower bound for $z(p\text{-center})$ in general networks. Noting this duality between center and dispersion problems, [187] develop a polynomial algorithm for the p-dispersion problem on trees with complexity $O(n^2 \log p \log n)$. [719] formulates the problem on a general network as a mixed integer program and solves some sample problems. [382] shows that the p-dispersion problem is NP-hard (via transformation from CLIQUE) and presents two exact branch-and-bound methods and one heuristic algorithm. Computational experience for p-dispersion problems on general networks is reported. The author also notes the close connection between p-dispersion problems and anti-cover problems whose objective is to find facility locations that are at least a fixed distance apart. One result in this context is a minimax theorem by [1074] who proves that the smallest number of sets of some given radius r required to cover the tree equals the largest number of points of the tree with mutual distances between them of at least 2r.

Another model was first suggested by [862]. It is called the *p-defense problem* and its objective is to maximize the sum of distances between each facility and its nearest neighbor. In Erkut and Neuman's descriptive terminology, this is a maxsummin problem. Moon and Chaudhry classify most p- defense and p-dispersion problems with lower bounds and/or upper bounds on distances as computationally "impossible". A related problem is a "maxsumsum" or p-defense-sum problem, discussed by [695]. The problem was shown to be strongly NP-hard on general networks, but can be solved in polynomial time on trees.

[386] present another problem in their survey. It is the maxminsum problem in which the decision maker maximizes the smallest sum of distances between any of the facilities and their customers. In other words, customers are first allocated to facilities and the sum of (shortest) facility-customer distances is determined for each facility. The smallest such sum is to be made as large as possible. A damper placement problem on trusses investigated by [696] appears to be the only reference. For two facilities, the problem on trees is identical to finding the diameter of the tree, a task that can be accomplished in linear time. The authors characterize solutions for $p > 2$ and devise a heuristic solution method.

The forces pushing the locations can again be examined as shown in the simple scenario at the end of the previous chapter. Anti-Weber, anti-Rawls, and anti-cover problems behave similarly to their pulling counterparts: the anti-Weber location responds immediately to customer movements, the anti-Rawls facility location does not change as long as a customer moves outside the largest empty hypersphere and responds only if relocated inside the sphere, and the anti-cover problem exhibits jumps.

8.5   Balancing Objectives

There seems to be little doubt that objectives that attempt to balance (weighted) distances between facilities and customers are the most contentious type of objective. The main goal of balancing is to achieve "equity", defined in the American Heritage Dictionary as "the state, ideal, or quality of being just, impartial, and fair." Thus commences the elusive search for fairness, following the adage "fairness is whatever you want it to be." [914] statement "That the issue of rightful discrimination or coercive redistribution undertaken for the purpose of equality is far from being philosophically resolved..." highlights the problems related to equity.

Most applications that employ equity objectives are part of public decision making where the objective is to serve the population fairly. There are, however, applications of balancing objectives in the private sector as well. Consider, for instance, the objective of balancing workload among police officers as suggested by [114]. Many other labor relations situations would fit similar objectives. Another example relates to a machine that consists of a variety of components. The quality of each component then determines the quality of the final product. Often, the final product is no better than the worst job done by any of the components. Hence, the objective would be to use components in the machine so as to balance the quality.

Balancing objectives were first introduced to location analysis by [811] and later by [867]. The first algorithmic treatment of the subject was by [524]. Today, there is a substantial number of papers on location models with equity objectives. What is disturbing is that most authors use the terms equity and equality synonymously, as pointed out by [383]. As location scientists are relative newcomers to the subject of equity, it appears prudent to search for guidance in areas which have dealt with the topic for a long time. Two such areas are sociology and economics. As an example, for centuries economists have tried to determine which income distributions are "fairer" than others, welfare functions being one approach to the problem.

At the heart of the problem lies the question of measurement. The scenarios discussed in the previous two sections dealt with decision makers who had a single, one-dimensional objective: they either wanted to minimize distance (as a proxy expression of cost or disutility) or maximize distance (as a surrogate criterion of some utility). Distances and costs are, of course, easy to measure. In other words: if we consider the distribution of all facility - customer distances for any given solution, push and pull objectives optimize some function of the mean, i.e., the first central moment, of this distribution. In contrast, most equity objectives attempt to minimize the variability of the distribution of distances. It is by no means clear why that should be the case. Surely, it seems to fit the objective of equity if all users have the same distance to walk or drive in order to reach a public facility. However, as [877] points out, equality is only one aspect of equity. One could, for example, think of minimizing skewness, i.e., the third moment, rather than variation, e.g., variance (i.e., the second moment). Such an objective can easily be justified in the same manner it is justified for income distributions: why should few users have very short walks to a public facility, whereas many users have longer-than-average distances to walk (or vice versa)? However, as [1059] points out, minimizing skewness attempts to achieve symmetry, not equality. It depends whether or not symmetry is among the decision maker's goals. Objectives of this type could possibly be used as secondary objectives.

Another point appears worth mentioning. In the "private" objectives, the differences between efficiency and effectiveness are quite clearly understood. Simply put, whereas efficiency deals with a simple measure of output such as profit or cost, effectiveness puts efficiency in relation to some expression of perceived need. For instance, emergency vehicles may be stationed optimally and do the best possible job so that they must be considered very efficient; they may just be not effective if there is not a sufficient number of them. On the other hand, even a very large number of poorly positioned ambulances may effectively serve the population; they may, however, not be very efficient in doing so. Similarly, we can distinguish between equality, an objective that, similar to efficiency, tries to achieve balance without any reference to need, and equity, which, similar to effectiveness, relates equality to need in some way. In that sense, we can write

$$\text{efficiency/effectiveness} = \text{equality/equity}.$$

In the location literature, virtually all "equity" models really have equality objectives, see [88, 836] and others. A notable exception is [1144] and his discussion of the locations of day-care centers in Toronto.

Examining balancing objectives, even such an extensive list as compiled by [806], one will notice that there are only two different categories. The first attempts to minimize the spread of the distribution, while the second minimizes the deviation from a central point. The former class includes only few objectives. One of these is the minimization of the maximum (distance) which was already discussed in a previous chapter. However, it should be mentioned here as some authors consider maximin objectives as part of the balancing class, justified, e.g., by the above product quality example. The second is the minimization of the range. [321] appear to be the only authors to discuss a weighted range objective in detail. Minimax objectives can be viewed as a special case of range minimization with a fixed lower bound. Neither objective is very satisfactory, as both concentrate exclusively on extreme values and ignore all others. The other class of objectives uses some function of the deviation of distances from a given norm. They all either employ squared deviations or absolute deviations. In our following discussion, we will concentrate on such objectives.

[1144] provides a justification for equity and distinguishes between "horizontal equity", i.e., equal treatment of individuals in similar circumstances, and "vertical equity", i.e., equal treatment of individuals in different circumstances. He then uses benefit incidence, defined as benefits in relation to needs where need is equated with demand. [1038] identifies and discusses four surrogate criteria for equity: equal payments (either in absolute dollars or equal ability to pay), equal outputs, equal inputs (per area or capita), and equal satisfaction of demand (equal input per unit of demand). In location analysis, interest would appear to focus on the last two criteria with weights suitably defined as population or potential demand. A covering objective with a properly defined covering distance d^* could solve the equal input/satisfaction of demand problem in the sense that no potential customer would be farther away than d^* from the nearest facility such as ambulance or fire station. Most importantly, the choice of the surrogate criterion is a matter of value. Without a value system balancing objectives have little, if any, meaning.

Another problem with balancing objectives was pointed out by [321]. It can be visualized as follows: consider two scenarios, one in which a facility is located at a distance of, say, 10 miles from each of the n customers, and another which is 3 miles from $(n-1)$ customers and right next to the remaining customer. Equality objectives would favor the former solution, even though everybody is better off in the second. However, favoring the second solution (which most people would) means the incorporation of some efficiency component in the objective in addition to the equalizing factor. [383] points out that for this reason equity considerations should not be the only objective in a model. Similar concerns were raised by [836, 794]. One way out of the dilemma is to use multiple objectives and consider efficiency - equity tradeoffs [524, 867, 797].

One final word of caution before we discuss details of balancing objectives in general. Even if the planners were to decide which of the objectives to use (which, Savas claims, is rather unlikely in the case of government planning as that would invite opposition), the problem may still persist: individuals will not be served well at all if for one aspect of their lives an equity objective were to be used whereas for another, a different and possibly nonbalancing objective is employed. For example, it may sound "fair" at first glance if downtown electricity users are forced to subsidize suburban electricity rates, but is it still fair when, at the same time, suburban real estate taxes are lower than their downtown counterparts? Such "departmentalization" may create more unfairness than its balancing objective purports to resolve.

Today's discussion of balancing objectives has lost much of its initial arbitrariness. Again borrowing from other disciplines such as economics, reasonable axioms have been used to evaluate balancing objectives. The first of the two most frequently used axioms is *scale invariance*. Scale invariance is satisfied if the degree of equality does not change with the type of measure applied to the problem. In other words, an income distribution does not become any more unequal because incomes are measured in pounds Sterling rather than in dollars. In terms of location problems, the distribution of facility - customer distances is no more unequal if we decide to measure distances in feet rather meters. Even though this seems like a very reasonable assumption, a number of popular inequality measures fail on this count. The second popular axiom is the *principle of transfers*, also called Pigou-Dalton condition, according to the two researchers outlining the principle in 1920. The principle of transfers was originally designed for income distributions where it states that an income distribution becomes less unequal if one dollar from an above-average income is transferred to an income that is below average. In our discussion, we simply replace income by "facility-customer distance".

[806] outline a number of additional criteria for the selection of balancing criteria. Among them are *analytic tractability*, a requirement not uniquely associated with balancing objectives, *normalization* of measures, e.g., in the [0, 1] interval, and *impartiality* which is guaranteed in location models as demand nodes are numbered completely arbitrarily. The issue of *appropriateness* can probably only be decided on a case-by-case basis and depends on the goals and values of the decision maker(s). *Sensitivity* of a coefficient deals with the coefficient's changes for a small change of any of the problem parameters. Very high or low sensitivities are generally undesirable. Another important criterion is *pareto-optimality*. As an example, consider three almost collinear points in the plane. The distances to a single facility are equal if the points are located on the circumference of a circle with its center at the facility. The closer the three points are to a line, the larger the radius of the circle (and the facility is no longer located in the convex hull of the three points). The facility-customer distances approach infinity as the customer locations approach the line. It is clear that such a location, while equally distant from the three given points, would be highly inefficient. Customers at all three locations would benefit if the facility were to move closer towards them, until it is located at the central point. While pareto-optimality is an efficiency rather than equality objective, this example shows the importance of Erkut's [383] observation that equality objectives alone may not be meaningful. For some additional comments on pareto-optimality, see [321].

In the following paragraphs, we provide a listing of balancing objectives along the lines of [806]. We first survey equality objectives whose function is to locate facilities so as to equalize all customer-facility distances. The first seven objectives are non-normalized. All objectives discussed in this section attempt to minimize measures of inequality or inequity, i.e., they are of type Min $f(x)$. Denote by d_i the distance between the facility and the i-th customer and let $\overline{d}$ denote the average customer-facility distance. These distances would have to be further refined in case more than one facility is to be located. Following tradition, we may allocate customers to their closest facility, even though that is not necessarily the case.

Objectives (1) and (2) consider only the extreme values. They are

(1) $f(x) = \max_i \{d_i\}$ (center)

(2) $f(x) = \max_i \{d_i\} - \min_i \{d_i\}$ (range)

As [383] points out, neither objective satisfies scale invariance or the principle of transfers, rendering these objectives questionable for balancing purposes in spite of their analytical tractability.

The following equality measures include terms denoting deviations from the mean distance. They are

(3) $f(x) = \sum_i |d_i - \overline{d}|$ (MAD: mean absolute deviation)

(4) $f(x) = \sum_i (d_i - \overline{d})^2$ (variance)

(5) $f(x) = \max_i |d_i - \overline{d}|$ (maximum deviation)

Maximum deviation could possibly be redefined to only consider distances longer than average (for desirable facilities) or shorter than average (for undesirable facilities).

(6) $f(x) = \frac{1}{n} \sum_i (\log d_i - \log \overline{d})^2$ (variance of logarithms).

[383] shows that (3), (4), and (5) satisfy neither scale invariance nor the principle of transfers, and [18] shows that (6) does satisfy scale invariance, but not the principle of transfers.

The last class of non-normalized equality objectives measures differences between all pairs of distances, *viz.*

(7) $f(x) = \sum_i \sum_j |d_i - d_j|$.

Again, [383] shows that neither scale invariance nor the principle of transfers is satisfied, thus severely limiting this measure's usefulness. The author suggests a number of related measures such as maxsum and summax, which have not yet been discussed in the literature.

The next class consists of normalized equality objectives. In essence, they are similar to the above objectives, but divided by a measure of central tendency, typically the mean. One such measure is

(8) $f(x) = \frac{1}{2n\overline{d}} \sum_i |d_i - \overline{d}|$ (Schutz's index)

As [18] shows, the index does satisfy scale invariance, but violates the principle of transfers. Another popular measure is

(9) $f(x) = \frac{1}{\sqrt{\overline{d}}}[\sqrt{\sum_i (d_i - \overline{d})^2}]$ (coefficient of variation)

The coefficient of variation is easy to use and compute as it is derived directly from the variance. Among the most popular measures of equality is the Gini index. It is defined as twice the area between the Lorenz curve and the 45^0 line, the latter denoting total equality. Formally, we can write

(10) $f(x) = \frac{1}{2n^2\overline{d}} \sum_i \sum_j |d_i - d_j|$ (Gini coefficient)

The Gini coefficient is described via the Lorenz curve known from economics [52]. It first orders the customers in increasing order of their distance to the facility that serves them and then plots the cumulative proportion of customers against the cumulative proportion of distance, see Figure 2. A point (a, b) on the Lorenz curve then means that the first $100a$ percent of customers are a total of $100b$ percent of the total distance away from

FIGURE 2. The Lorenz Curve and the Gini Coefficient

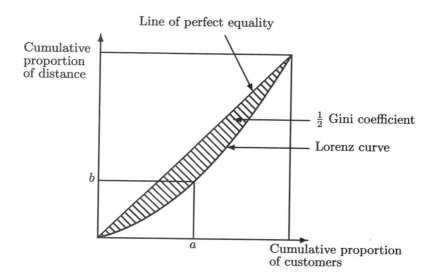

the facility. Clearly, the straight line segment between $(0,0)$ and $(1,1)$ includes points of complete equality. The Gini index then expresses the area between the Lorenz curve and the line of perfect equality in relation to the area below the line of perfect equality. This expression is proportional to the hatched area in Figure 2. Some problems associated with the Gini index are pointed out by [261].

Another measure in this set derives from information theory. It is

$$(11) \quad f(x) = \frac{1}{n\overline{d}} \sum_i |d_i \log d_i - \overline{d} \log \overline{d}| \qquad \text{(Theil's entropy coefficient)}$$

Finally, there is Atkinson's coefficient which was designed for social welfare functions. It can be written as

$$(12) \quad f(x) = 1 - n^{1/(\delta-1)} \left[\sum_i (\frac{d_i}{\overline{d}})^{1-\delta} \right]^{1/(1-\delta)} \qquad \text{(Atkinson's coefficient)}$$

Among the indices of this class, Schutz's coefficient does satisfy scale invariance but not the principle of transfers [18], whereas coefficients (9) - (12) satisfy both scale invariance as well as the principle of transfers. A good discussion of equality measures is provided by [1059].

All indices discussed above measure inequality; none of them is designed to be used as an expression for fairness. To do so, we require another parameter. This parameter is designed to put the distance into relation to need. Following [806], we will call it an attribute and denote it by a_i,

$i = 1, \ldots, n$. Examples for attributes could be population size, or demand for a specific service. The development of meaningful proxy expressions for need is nonexistent in the locational context, which is of little surprise as virtually all location models with balancing objectives use equality rather than equity objectives. The simplest equity objective is

(13) $f(x) = \sum_i |d_i - a_i|$

a simple measure comparing distances with attributes. It is reminiscent of the mean absolute deviation; in fact, it reduces to MAD if all attributes are equal to $\bar{d}$. All remaining objectives consider relative distances, i.e., distances in relation to the fixed attributes. A variance-like measure is

(14) $f(x) = \sum_i [\frac{d_i}{a_i} - \frac{\bar{d}}{\bar{a}}]^2$

with $\bar{a}$ being the mean of attributes a_i. Hoover's concentration index expresses the average sum of absolute differences of the proportions of distance and need, i.e.,

(15) $f(x) = \frac{1}{n} \sum_i |\frac{d_i}{\bar{d}} - \frac{a_i}{\bar{a}}|$ (Hoover's concentration index).

whereas Coulter's coefficient squares the differences, viz.

(16) $f(x) = \sqrt{\frac{1}{n} \sum_i (\frac{d_i}{\bar{d}} - \frac{a_i}{\bar{a}})^2}$ (Coulter's coefficient)

A pairwise comparison between the distance-to-attribute ratios similar to (7) is

(17) $f(x) = \sum_i \sum_j |\frac{d_i}{a_i} - \frac{d_j}{a_j}|.$

Adam's coefficient is written as

(18) $f(x) = \sum_i |\frac{d_i - a_i}{a_i}|$ (Adam's coefficient)

Finally, there is the sociospatial version of Schutz's coefficient [1144], viz.

(19) $f(x) = \sum_i |\frac{d_i a_i}{\sum_k d_k a_k} - \frac{a_i}{n\bar{a}}|$ (Schutz's sociospatial index)

A number of other indices can be thought of. For instance, the sum in (15) and (17) could be replaced by the "max" operator. Also, one may wish to normalize individual coefficients.

[877] compares a number of the equality objectives delineated above. In particular, he examines the Gini coefficient, the mean deviation, Hoover's index, the variance, and Theil's entropy index. His locational model is extremely simple, it consists of only a line segment with three customer points, one at each end and one in between. Travel costs are assumed to be linear, there are no economies of scale, the demand is completely inelastic, and customers have complete information. One of the results is that the

mean deviation criterion, Hoover' concentration index, and variance criterion generate very similar sets of optimal solutions. Whereas efficiency objectives tend to be shallow near optimum, equality objectives are not.

A similar study was performed by [805]. Their chosen five different equality objectives are the center, variance, sum of absolute deviations from average, sum of absolute deviations between all pairs, and range. The objectives were tested on one sample network with 54 nodes. Three facilities were to be located and, in lieu of a suitable algorithm, 500 potential solutions were chosen and the top-ranking solutions were determined for each objective. The result was quite unexpected: except for the center objective, the other four objectives ranked the top contenders very similarly. If this result were to be confirmed in large-scale studies, it would eliminate the need for elaborate searches for "the best" equality criterion.

There is little work on problems with balancing objectives in the plane. One exception is the paper by [811], who assumes that customers are uniformly distributed in the plane and facilities will be equally-sized and spaced so that their market or service areas are honeycomb-shaped. The intrinsic question is: will there be few large or many small facilities to serve the market? Employing the standard deviation of the distances between customers and their nearest facilities, facility size and location - given that facilities remain located in a regular pattern - are optimized. The trade-off between equality and efficiency (measured in service consumption) is displayed. Similar to the conclusion above [877], the equality objective is steeper in the vicinity of its optimum that the efficiency objective near its optimum. Among the very few reports concerning balancing objectives for the location of facilities in the plane is by [321] who minimize the range of facility-customer distances in the plane.

Location models with balancing objectives on networks have received a bit more attention. [524] were the first to algorithmically discuss balancing objectives. The objectives in their paper minimize variance of facility - customer distances and the Lorenz point. In another paper, [836] describes a linear-time algorithm for the minimization of the sum of weighted variances of the distances on trees. It is readily apparent that the model does not possess the Hakimi property. In order to incorporate efficiency as well as equality, the author suggests searching for points with low weighted variances only in the ϵ-neighborhood of an efficient, i.e., median, solution for some prespecified ϵ.

The concept of a Lorenz point was suggested by [524] and further discussed by [794]. A Lorenz point is a point that minimizes the proportion of travel distance to the facility in relation to the proportion of customers who patronize the facility. The author of the second study demonstrates that the Hakimi property does not hold and he then develops an $O(n^3 \log n)$ algorithm for Lorenz points on trees. Lorenz points may be quite distant from customers, again implying the necessity to incorporate some sort of efficiency objective along with the chosen balancing objective.

[83] formally incorporates mean and variance in one model. More specifically, he investigates three versions of the mean-variance problem: (1) minimize the mean distance subject to a constraint that requires the variance not to exceed a given limit, (2) minimize the variance subject to an upper bound on the mean distance, and (3) minimize a linear convex combination of mean and variance. Similar problems arise in portfolio analysis models. The author delineates $O(n^3)$ algorithms for the three problems on general networks that reduce to linear-time algorithms on tree networks.

[88] devise a tax-subsidy system in which all customers are provided with the same benefit, defined as (subsidy) - (tax) - (travel cost). Maximizing the sum of weighted customer benefits, the solution is identical to that of a p-median with a well-defined system of taxes, subsidies, and side payments. The problem becomes difficult when travel costs are nonlinear. In case such a tax-subsidy system cannot be devised for practical or political reasons, the authors derive an $O(n^4)$ algorithm which locates a single facility that minimizes the sum of weighted absolute deviations of travel distance from the average travel distance. The problem does not possess the Hakimi property.

Based on problems associated with the use of balancing objectives without considering efficiency, a number of authors have considered equality - efficiency tradeoffs. Among the general references are [811, 114]. [524] consider median, center, variance, and Lorenz objectives on tree networks. The main result is that the Lorenz solution is quite removed from the others and at the Lorenz point, the other three objectives assume values much different from their respective optima. Also, the Lorenz measure is the most sensitive of the four. Choosing optimal center and variance points, the median objective suffers a bit, whereas center and variance measures suffer moderately if the median is chosen.

Changes of facility locations based on customer movement given balancing objectives are much more tedious to analyze than in the case of pull and push objectives. Here, we have chosen the variance objective to demonstrate the changes of facility location based on customer movements. Again, let three customers A, B, and C be located on a line segment at $A = 0$, $B = 1$, and C initially at 0. We will again examine what happens if C moves towards "1". For a facility located at X, the mean customer-facility distance is $\frac{1+X-C}{3}$ if $X \geq C$ and $\frac{1+C-X}{3}$ if $X \leq C$. Computing the expressions for the variance and setting the derivative with respect to X to zero yields $X_1^* = \frac{1}{2} + \frac{C}{4}$ with variance$(X_1^*) = \frac{C^2}{2}$ for $X^* \geq C$ and $X_2^* = \frac{1}{4} + \frac{C}{4}$ with variance$(X_2^*) = \frac{1}{2} - C + \frac{C^2}{2}$ for $X^* \leq C$. It is an elementary exercise to show that for $C \leq \frac{1}{2}$, X_1^* is optimal, whereas for $C \geq \frac{1}{2}$, X_2^* minimizes the variance. This leads to the following result: for $C = 0$, $X^* = X_1^* = \frac{1}{2}$. As C moves towards $\frac{1}{2}$, the facility also moves to the right. At $C = \frac{1}{2}$, $X^* = \frac{5}{8}$. As C moves to the right of $\frac{1}{2}$, X^* jumps back to $\frac{3}{8}$ and, as C moves towards 1, the facility moves towards $\frac{1}{2}$. At $C = 1$, the facility is optimally located at $\frac{1}{2}$. In summary, as C moves from 0 to 1, the facility X moves from $\frac{1}{2}$ to $\frac{5}{8}$, jumps to $\frac{3}{8}$, and moves back to $\frac{1}{2}$. It is worth noting that the rather dramatic move of a customer through the entire range from 0 to 1 results in only moderate changes in the location of the facility from $\frac{3}{8}$ to $\frac{5}{8}$. In other words, whereas the customer relocates across 100 percent of the given space, the facility moves only across 25 percent of the space.

8.6   Conclusions

In this chapter we have examined location models with different objective functions. It was shown that objectives can be seen as the result of a Stackelberg game involving users and planners. Special attention was paid to the forces that govern the location process. On that basis three different types of objectives were identified: pull, push, and balance objectives. For different objectives within these classes we have shown how the forces move facilities to their optimal locations. For selected objectives we have shown how facility locations react to customer movements.

Applying the concept of pulling, pushing, or balancing forces to any of the objectives in the individual contributions mentioned in this discussion enables the user to predict the outcome of the location process. However, it is questionable if an appropriate objective should be obtained by means of a recursive process of the type: this is the location that we want, so that will be the objective that achieves such a location. Instead, it appears appropriate to derive objectives either from a set of reasonable axioms or by way of a set of value statements by the decision maker. Initial attempts in this direction have been made and we hope in the future to see steps towards a unified theory of defining objectives for location problems.

8.7   Glossary

Absolute center Point on network that minimizes the maximum customer-facility distance [509].

Absolute median Point on network that minimizes the sum of customer-facility distances, equals vertex median.

Anticenter Point that maximizes the minimum weighted customer-facility distance [260].

Anticenter dispersion Location of p new facilities among m old facilities to maximize the minimal distance between new and old facilities [382].

Anticenter-maxian Linear convex combination of anticenter and maxian objectives.

Anticovering Locational pattern of the maximum number of facilities such that all facility-facility distances $\geq \delta$ [260, 862].

Antimedian Point on a network that maximizes the sum of distances = maxian [221], $\rightarrow$ maxian.

Anti-Rawls → anticenter.

Anti-Weber Point in the plane that maximizes the sum of customer-facility distances.

Cent-dian Linear convex combination of center and median [520, 531].

Center Point in network that minimizes the maximum distance [509].

Constrained p-median p points on a network that minimize the sum of customer-nearest facility distances, s.t. all distances $\leq \delta$ [1141], → medi-center.

Covering → Min (cost) cover, max cover.

Defense Locational pattern of p facilities so as to maximize the sum of closest facility-facility distances (maxsummin problem) [862].

Delivery boy Point that minimizes the maximum customer-closest facility distance [379], → min covering problem, smallest enclosing circle [1116].

Dispersion Locational pattern of p facilities so as to maximize the minimal facility-facility distance (maxminmin problem) [1074].

Dispersion-Defense Linear convex combination of dispersion and defense objectives.

General center Point on network that minimizes the maximum distance between it and any point on the network [851].

General median Point on network that minimizes the sum of distances between it and any point on the network [851].

k-centrum Point on network that minimizes the sum of weighted distances from the k farthest customers [1087].

Largest empty circle Solution to 1-maximin problem [1063].

Lorenz point Point that minimizes the proportion of travel distance to the proportion of customers who patronize the facility [524, 794].

Max cover p points that maximize the sum of weights covered within a given distance δ [223].

Maxian Point on network that maximizes the sum of weighted distances [221].

Maxisum dispersion Solution to → dispersion problem, that maximizes the sum of facility-facility distances [719].

Median p points on network that minimize the sum of customer-nearest facility distances [509, 510].

Medi-center Point on a network that minimizes the average distance, s.t. all distances $\leq \delta$ [1139, 694]. → constrained p-median.

Messenger boy Point on network that minimizes the maximum {customer-facility distance + customer-specific constant} [379].

Min cost cover The minimal number of facilities that cover all customers, given a covering distance δ [850].

Min covering sphere → smallest enclosing circle, center.

Multi-Weber problem p points in the plane that minimize the sum of customer-closest facility distances [872, 237].

Partial center p points that minimize the maximum customer-facility distance, considering only distances for covered customers [84].

Smallest enclosing circle Circle that solves the → 1-center problem in the plane [1116, 883, 378, 1139].

Vertex center Vertex on network that minimizes the maximum distance.

9
DISTRIBUTION SYSTEM DESIGN

Arthur M. Geoffrion
University of California, Los Angeles

James G. Morris
University of Wisconsin-Madison

Scott T. Webster
University of Wisconsin-Madison

9.1   Introduction

Since the 1950's, formal computer models have been used for answering distribution system design questions such as:

- How many stocking points should there be, and where should they be located? Should they be owned?

- Should all stocking points carry all products or specialize by product line?

- Which customers should be served by each stocking point for each product?

- Where should plants be located?

- What should be produced at each plant and how much?

- Which suppliers should be used and at what levels?

- What should the annual transportation flows be throughout the system? Should pool points be used, and if so where should they be?

Warehouse location problems, posed as mixed-integer programming (MIP) models with binary variables for fixed charges related to site choice and continuous variables for the flow of goods, began to appear in the operations research literature with such papers as [63] (which describes a heuristic for a nonlinear warehouse location model). About two decades ago the structural design of distribution systems using an optimizing approach became technically feasible. Geoffrion and Powers [458] provide perspective on algorithmic and associated evolutionary developments, stating that

"It is now possible for companies of all sizes to find distribution system designs that are optimal for all practical purposes even, in many cases, when the scope of the design problem is extended to the complete logistics system in the broadest contemporary sense."

Yet a challenge remains. Although perils attend non-optimizing (heuristic) approaches [459], optimization has failed to become the dominant approach to distribution system design. The slow demise of non-optimizing approaches in actual practice is considered "a lesson in humility to the MS/OR profession [458]." Our intention is to establish a base for embracing powerful optimizing approaches by presenting a guide on tools conductive to fast analysis of distribution systems. Specifically, we wish to: (i) convey the power of highly simplified models for quick diagnosis and exploratory purposes, as well as for developing insights into cost relationships in distribution systems; and (ii) show how algebraic modeling language systems are convenient for quickly specifying a model and for solving aggregate models of distribution system design problems through access to general commercial solvers.

In order to provide a context for our discussion, we begin with a case study in the form of a series of memos that illustrates a systematic response to a question on the suitability of an existing system, from initial diagnosis through results of formal analysis. We then consider each tool in turn, beginning with a highly simplified model known as the general optimal market area model, then moving on to algebraic modeling language systems. Our emphasis is on tools that can be easily and inexpensively implemented in industry. Algorithms are not our focus. As a convenient reference for readers interested in more detail on methodology and applications, we conclude with a brief annotated bibliography. Attention is restricted to distribution planning models wherein location of facilities is a central question.

9.2   A Case Study

——————————[memo 1]——————————————
TO: Director of Management Science
FROM: Vice President of Logistics
DATE: June 1

As a newly appointed Vice President with responsibility for our network of distribution centers, I would like your advice concerning the suitability of this network in the current business environment.

We've had the same 13 DC locations for several years now, although the volume of business has been growing and freight rates, especially on the out-bound side, have been increasing with alarming rapidity. In the most recent 12-month period, our costs have been as follows on a total volume of 25 million CWT:

Distribution center fixed	$ 6,150,000
Distribution center variable	18,500,000
Inbound transportation	42,850,000
Outbound transportation	29,100,000
	$96,600,000

Do you think it would be worthwhile to build a model of our distribution system in order to help us reduce costs? I know that you could lead such an effort, but I would need some assurances concerning the potential for cost savings and where those savings might come from.

—————————————[memo 2]————————————————

TO: Vice President of Logistics
FROM: Director of Management Science
DATE: June 10

Thank you for your inquiry concerning the possibility of a distribution network modeling study. It is always difficult to predict in advance what the cost savings will be for a major modeling effort, but I will do the best I can.

An evaluation of the merits of our current distribution network breaks down into three fundamental questions:

Q1. How well are we using our current distribution network? That is, how well are we loading plants, allocating production to the DC's, and assigning customers to DC's?

Q2. How good are our current 13 DC locations?

Q3. Do we have too many or too few DC's?

I list the questions in this order because it is the natural one not only for diagnosis, our present concern, but also for redesign of the network. We cannot think usefully about changing DC locations (Q2) until we understand how any given set of locations should be used (Q1), and we cannot think usefully about changing the number of DC's (Q3) until we understand how any given number of DC's should be located (Q2) and used (Q1). Each question is considered in turn.

Q1. How well utilized is the current distribution network?

The growth of demand has, as you know, been putting a real strain on our capacity expansion program. For most product lines, nationwide production capacity is only about 10% greater than nationwide demand and the location of this capacity is quite poorly matched to the location of demand. This poses a fairly easy problem of plant loading, as there is little latitude to decide how much of each product to make at each plant, but it does present a rather intricate problem of allocating production to DC's.

For a given set of customer-to-DC assignments, the DC's look like demand points to the plants. It is possible to calculate the plant loadings and plant-to-DC allocations for each product so as to minimize the sum of production and inbound transportation costs while honoring plant capacities and DC demands. This could be done easily using an optimization technique you have no doubt encountered before called linear programming. But what if the customer-to-DC assignments are changed? That would require resolving the loading/allocation problems and would also change the outbound transportation costs. Simultaneous cost minimization for loading, allocation, and customer assignment presents a very difficult optimization problem owing to the gigantic number of possible ways to assign customers to DC's. The only good news is that our DC's have sufficient unused equipment and space that customers probably do not have to compete significantly for limited DC capacity, but even so the problem is beyond the reach of the general purpose optimization software presently available.

The bottom line is this: if the joint plant loading/product allocation/customer assignment problem is so difficult for present day computers and generally available software, we should not expect that our company is solving it particularly well. Indeed, we have never taken a system-wide look at this problem. We have only nibbled at it in manual studies of limited scope. A thorough analysis would not only capture the previously inaccessible savings, but it would also capture the accessible savings that might otherwise escape notice for a year or two or even more. There is always a lag between the constant creation of savings opportunities by the turbulent business environment and our discovery of these opportunities in the normal course of running our business. A thorough analysis would eliminate this lag for the time being.

I would offer a conservative guess that, if we were to acquire the special purpose computer software to solve this problem properly, transportation costs would drop by about 2%. This savings factor seems modest enough and is in keeping, so I am told by consultants who specialize in this line of work, with the experience of other firms which have solved the joint plant loading/product allocation/customer assignment problem with tightly constrained supply on an integrated, system-wide basis for the first time. Thus I estimate that our cost breakdown would have been as follows had we utilized our current 13 DC's optimally:

DC fixed costs	$ 6,150,000
DC variable costs	18,450,000
Inbound transportation	42,000,000
Outbound transportation	28,500,000
	$95,100,000

Additional savings would probably accrue for production costs, but I will omit consideration of that for the sake of simplicity.

Q2. How good are the current 13 DC locations?

A small random sample of outbound freight rates shows that delivery costs average $0.0075 per CWT-mile. Total annual delivered volume is 25 million CWT, of which about $\frac{1}{4}$ is in the very cities in which our DC's are located. Suppose we assume an average delivery distance of 10 miles for the demand that is collocated with our DC's. That comes to $\frac{1}{4} \times 25,000,000$ CWT $\times 10$ miles $\times$ $0.0075(\text{CWT-mi}) = \$468,750$. For the non-collocated demand, suppose that the average delivery distance is 156 miles [453, Figure 5]. That comes to $\frac{3}{4} \times$ $25,000,000$ CWT $\times 156$ miles $\times \$0.0075(\text{CWT-mi}) = \$21,937,500$. Total predicted outbound transportation cost is the sum of these two numbers, or $22,406,250. This prediction is $6,093,750 below the adjusted figure of $28,500,000 given at the conclusion of the discussion of Q1. This suggests that the average delivery distance must actually be considerably greater than 156 miles, perhaps as much as 200 miles, which in turn suggests that our current DC's are not well situated with respect to current concentrations of demand.

To be conservative, I will assume that improved DC locations could achieve half of the savings potential indicated above, namely $3 million. This may, however, be gained at the price of increased inbound freight cost. Inbound costs should rise much less than outbound costs will fall. One reason is that our plants are scattered all over the country, so any DC move **away** from one plant will tend to be **toward** another plant. Cost changes thus will tend to cancel out. Another reason is that empirical studies with simplified one plant models show that inbound transportation cost tends to change very little with the number and location of DC's provided there are at least 10 of them ([453, equation (14)]). Thus, I believe that no more than half of the outbound cost savings would be lost in increased inbound costs. This leads to the following revision of the cost breakdown given at the conclusion of the discussion of Q1:

DC fixed costs	$ 6,150,000
DC variable costs	18,450,000
Inbound transportation	43,500,000
Outbound transportation	25,500,000
	$93,600,000

Q3. Do we have too many or too few DC's?

I believe that we have too few DC's. My reasons are several. First of all, demand growth implies that new DC's should be added over time because there is greater volume over which to spread the DC fixed costs. Second, more DC's are needed when outbound freight rates inflate faster than inbound freight rates because more DC's reduce the amount of outbound shipping. We are behind the times on both counts, as we have had 13 DC's for 4 years. Another tipoff that we have too few DC's is the fact that total outbound freight cost is more than 4 times total DC fixed cost, even after the adjustments made in my discussion of Q1 and Q2. An idealized analysis suggests that this ratio should be more like 2 than 4, and hence that we need to increase the number of DC's (to decrease the numerator and increase the denominator of this ratio). The idealized analysis I have in mind is as follows:

If a DC has fixed cost f \$/yr, lies in a part of the country having demand evenly distributed with density ρ CWT/(mi^2-yr), can deliver for t \$/(CWT-mi), and has a circular service area of S square miles, then it is an elementary exercise to show that the total associated annual cost is $f + .377\sqrt{S}(\rho S)t$. Total annual cost per CWT throughput (divide by ρS) is minimized when $S^* = 3.05(\rho t/f)^{-2/3}$. Note that the outbound transportation cost at S^* is $.377\rho[3.05(\rho t/f)^{-2/3}]^{3/2}t = 2f$.

This analysis can be applied to our problem with these numbers: $f = \$6,150,000/13 = \$473,077$; $\rho = 25,000,000$ CWT/2,000,000mi^2 = 12.5 CWT/mi^2 (2 million square miles is actually $\frac{2}{3}$ of the land area of the continental U.S., to account for the absence of any significant population in much of the western central part of the country); $t = \$0.0075$(CWT-mi). The result is $S^* = 89,730mi^2$. To cover 2 million mi^2, this requires **22 DC's**.

This estimate of 22 for the optimal number of DC's, besides being based on extremely simplified assumptions, does not consider the influence of inbound transportation costs or the fact that DC's tend to coincide with major demand concentrations. These considerations introduce cost tradeoffs that tend both to increase and decrease the optimal number of DC's. I will not pursue such refinements here.

Suppose for the sake of argument that 22 is the best number of DC's for us. Let's examine the cost consequences. Total fixed cost should go up by about $473,077 \times (22-13) = \$4,257,700$. Total DC variable cost should not change significantly, as the variation between individual DC variable cost rates is not great. Total inbound cost should not change much, for reasons described under Q2. I'll assume a generous $1 million increase. Total outbound cost should decrease significantly. The idealized analysis discussed above predicts that total outbound cost varies as one over the square root of the number of DC's (See, for example, [451, equation (3)]). It follows that total outbound cost will go down to a fraction of $\frac{1/\sqrt{22}}{1/\sqrt{13}} = 0.7687$ of the former level. But this neglects the fact that many of the 9 new DC's will fall on major demand concentrations, thereby reducing outbound freight even more. Suppose that the 9 new DC's are collocated even half as well as the current 13. Then an additional $\frac{1}{2} \times \frac{9}{13} \times .25 = 9.375\%$ of all demand will be collocated, which will reduce the 0.7686 factor to $(1-0.09375) \times 0.7686 = 0.6965$. Applying this to the $25,500,000 figure given at the end of the discussion on Q2, we obtain a savings of $7,740,000. This is enough to overcome the increases in fixed and inbound costs, with a net annual savings of $2,480,000. With these assumptions, the cost breakdown given at the end of the discussion of Q2 becomes:

DC fixed costs	$10,400,000
DC variable costs	18,450,000
Inbound transportation	44,500,000
Outbound transportation	17,761,000
	$91,110,000

Summary

I have argued that possible improvements in our current distribution system can be viewed as occurring in three phases corresponding to the three questions stated at the outset. The cost breakdown is summarized as follows (costs in millions):

	Current	(Q1) Better Utilization	(Q2) Relocate Current DC's	(Q3) Add DC's
DC fixed costs	$ 6.150	$ 6.150	$ 6.150	$10.400
DC variable costs	18.500	18.450	18.450	18.450
Inbound transp	42.850	42.000	43.500	44.500
Outbound transp	29.100	28.500	25.500	17.760
Total	$96.600	$95.100	$93.600	$91.110
Savings over previous system		$1.500	$1.500	$2.490

Adding more DC's (about 9 of them) will make marketing happy, as each new DC would increase our "market presence" in a new area and probably lead to improved market penetration. Notice also that adding DC's should greatly reduce outbound average shipping distances, which tends to reduce our vulnerability to continued rapid increases in motor carrier rates.

The total potential savings of about 5.5%, an estimate that I view as quite conservative, falls within the $5 - 15\%$ range which distribution consultants claim is typical for a full scale modeling study.

I hope that these "back of the envelope" arguments will suffice to help you decide whether to sponsor a modeling effort. Comparable studies at other firms suggests that the costs of such a study would be under $100,000 and take about 6 months to complete. It looks to me like a high payoff venture.

—————————[memo 3]—————————

TO: Director of Management Science
FROM: Vice President of Logistics
DATE: July 1

Your response to my June 1 memo makes a good case for undertaking a thorough analysis. With it, I can justify making a substantial commitment in labor and resources. Let's go ahead with the project. The main target issues I would like to address are:

- How many DC's should we have?
- Where should the DC's be located?
- What size should each DC be?
- How should our 7 plants be loaded, and how should their output be allocated to the DC's?
- Which customers should be assigned to each DC?

Although cost minimization is the basic objective, customer service should also be considered. The physical proximity of DC's to their markets is an important factor.

Please conduct the analysis so as to continue our present policy of assigning each customer to a single private full line DC.

One more thing. My last employer had a real disaster with a computer-based model for inventory management. Finished goods inventory went up and up and up some more instead of down as promised. That experience taught me that it is important for executives to keep in close touch with their fancy models. Consequently, I will ask that you explain to me in non-mathematical terms the model as it evolves, and whatever results come out of the model after it is built. I want you to help me discover not only *what* to do, but also *why* to do it.

—————————[memo 4]—————————

TO: Vice President of Logistics
FROM: Director of Management Science
DATE: October 1
RE: Distribution Planning Model Synopsis

The data development stage of the project is virtually complete now. Soon we will be commencing verification and validation. You mentioned in your July 1 memo your desire to "keep in close touch" with the model and to understand why, as well as merely know what it is trying to tell us. I'm glad that you feel this way, as it coincides exactly with my own philosophy.

An important step is for you to have at your fingertips a clear and concise synopsis of the model. Such a synopsis is attached, together with a report on the supporting data. [The model structure is included here, but for brevity the detailed data are omitted.] After scrutinizing it for possible errors, you can keep it at hand for ready reference as we move into the formal analysis stage of the project.

THUMBNAIL MODEL SKETCH and DATA ELEMENT CHECKLIST

7 PLANTS   ⟹   35 CANDIDATE DC's ⟹   150 CUSTOMER ZONES

(1) List of product groups (10)

(2) List of plants (7)

(3) Plant capacities in CWT /yr (plant × product)

(4) Unit production cost in $/CWT (plant × product)

(5) List of candidate sites for DC's (35); all private and full line

(6) Lower limit on each DC throughput is 100,000 CWT/yr

(7) Fixed cost for each DC in $/yr

(8) Variable cost in $/CWT by DC (same for all products)

(9) List of customer zones (150 most populous S&MM metropolitan markets)

(10) Annual customer demand in CWT/yr (customer zone × product)

(11) Single sourcing rule: each customer zone assigned uniquely

(12) Net selling price (customer zone × product); omitted

(13) List of permissible inbound links and freight rates in $/CWT (same for all products, nearly all rail)

(14) All possible outbound links up to 1500 miles long included (a total of 3728 DC/customer pairs); freight charge is $0.0075/(CWT-mi) for any product (nearly all truck)

_____[memo 5]_____

TO: Vice President of Logistics
FROM: Director of Management Science
DATE: December 10
RE: Formal Analysis: Summary of Key Findings

Since completing the verification and validation exercises about a month ago, we have as you know been busy using the model as a tool to study how our distribution system can be improved. Your active participation in these studies has been invaluable. Although you have seen most of the results already in raw form, we are now in a position to summarize the key findings in a coherent way.

I want to stress that nothing in this memo is intended as a recommendation. We are, however, at the point where the findings and insights summarized here must be interpreted in light of various considerations outside the scope of the formal analysis. The result of that exercise will be specific, prioritized recommendations for action.

This memo makes three passes at summarizing the key findings. The first establishes that current annual distribution costs can (according to the model) be reduced from $96,600,000 to $88,810,000. This represents a projected savings of $7,790,000. The second pass establishes that the projected savings can be attributed to three distinct types of improvement carried out sequentially:

Types of Improvement	Savings
Improve plant loading, product allocation to DC's, and customer assignments (keeping the current DC's)	$2,269,000
Improve DC locations (keeping the number at 13)	2,412,000
Move to a network with 21 DC's	3,109,000
	$7,790,000

The third pass attempts to achieve a deeper understanding of each of the three types of improvement.

The figures mentioned above are exclusive of production costs. You will recall that we included no unit production costs at all in the model for product groups

4-10, owing to the severe limitations of our cost accounting systems. It turns out that the unit production costs we did include for products $1 - 3$ exert only a negligible influence on the target issues of interest. Hence this memo intentionally disregards the role of production costs.

FIRST PASS: The Bottom Line

The bottom line is that projected annual savings amount to $7,790,000. Here is a comparison of current annual costs (in millions) with what is projected under the least cost distribution network, which has 8 more DC's:

	Current	Least Cost Network	Savings (Loss)	% Change from Current
DC fixed costs	$ 6.150	$ 9.737	($3.587)	58
DC variable costs	18.500	18.727	(0.227)	1
Inbound transp	42.850	43.428	(0.578)	1
Outbound transp	29.100	16.918	12.182	-42
Total	$96.600	$88.810	$7.790	-8

Considering our corporate challenge to reduce logistics costs, I think you will be pleased to see that massive savings in outbound trucking materialized as predicted. These more than offset increases in other cost categories caused (mainly) by the addition of new DC's. Profits may well go up <u>more</u> than $7,790,000 because our analysis assumed demand to be fixed, whereas it should actually increase somewhat owing to the addition of 8 DC's. Marketing claims that each new DC improves local market share through improved "market presence." One measure of market presence is average delivery distance, which decreases from about 150 miles with the current network to about 90 miles with the least cost network.

The answers to all of the target issues listed in your July 1 memo are available in a detailed set of reports describing the least cost network.

SECOND PASS: Three Steps to a Least Cost Network

The cost comparison given above shows how the projected $7,790,000 savings is distributed by cost category, but does not tell us <u>why</u> the cost categories change as they do in going from the current to the least cost network. My aim now is to identify the major reasons why the costs change as they do.

The explanation has to do with the different kinds of changes which might be made in our current distribution network. There are 5 possible kinds of changes:

A. change plant loadings (how much of each product is made at each plant);

B. change product allocations (the shipping pattern from plants to DC's);

C. change the assignment of customer zones to DC's;

D. change the DC locations (keeping the number the same);

E. change the number of DC's.

If we allow no changes, the Distribution Planning Model will simply mimic (simulate) our current distribution network and will therefore yield a total cost of $96,600,000. As more and more freedom is allowed to make changes, the model will be able to drive the total cost closer and closer to the $88,810,000 floor.

It is instructive to examine what happens when the model is given only partial freedom to make changes. In this way we can better understand where the ultimate total savings come from. We have examined two such cases. These, along with the current and "full freedom to change" cases, can be portrayed as follows:

```
              A        B        C        D        E
"Current"   ===============fixed=================
"Step 1"    *****optimize*****************=======fixed=====
"Step 2"    ****************optimize****************===fixed==
"Step 3"    *********************optimize*********************
```

The model also lends itself to the case where optimization occurs only over A and B, but we did not pursue this one. The managerial interpretation of these cases is clear. Step 1 asks "How much improvement is possible by better utilization of the current 13 DC locations?" Step 2 asks "What is the least cost distribution network having 13 DC's?" Step 3 asks "What is the least cost distribution network?"

The <u>differences</u> are particularly revealing, as they enable the total $7,790,000 savings to be decomposed into components associated with utilization alone, locational choice, and a change in the number of DC's. These differences and the cases themselves are summarized below. You may wish to make a comparison with a similar summary at the end of my June 10 memo.

This summary reveals that improved utilization of the current 13 DC's could save $2,269,000, improved locations of the 13 DC's could save another $2,412,000, and going to a 21 DC network could save $3,109,000 more.

THREE STEPS TO THE LEAST COST DISTRIBUTION NETWORK

		— Step 1 —			— Step 2 —			— Step 3 —		
	Current	Optimal	Savings		Optimal	Savings		Optimal	Savings	
Cost Element	Network	Util.	(Loss)	%Δ	Loca's	(Loss)	%Δ	Number	(Loss)	%Δ
DC fixed costs	$ 6.150	$ 6.139	$0.011	-0.2	$ 6.000	$ 0.139	-2.3	$ 9.737	($3.737)	62.3
DC var. costs	18.500	18.462	0.038	-0.2	18.716	(0.254)	1.4	18.727	(0.011)	0.1
Inbound tran.	42.850	41.777	1.073	-2.5	42.693	(0.916)	2.2	43.428	(0.735)	-1.7
Outbound tran.	29.100	27.953	1.147	-3.9	24.510	3.443	-12.3	16.918	7.592	-31.0
Total	$96.600	$94.331	$2.269	-2.3	$91.919	$2.412	-2.6	$88.810	$3.109	-3.4

Legend:

1. All costs in millions.

2. Savings (loss) and % are with reference to the preceding step.

————————[end of memo 5]————————

9.3   Diagnostic Tools

Managers are unlikely to accept analysis they do not satisfactorily understand. Mathematical programming models can deliver an optimal solution for a given set of input data, but they do not explain *why* the solution is what it is. That is why our case study took pains to prepare the way for, and then to interpret, the optimal results by intuitively reasonable explanations. In this section we consider diagnostic tools that, among other functions, promote understanding of the numerical results.

Diagnostic tools are methods suitable for quickly detecting high potential opportunity areas worthy of further study. Given the complexity of the real-world, these tools tend to be based on stripped-down models that attempt to capture the essence of a problem with minimal data requirements (typically achieved through a combination of aggregation and simplifying assumptions). As such, diagnostic tools serve to build and complement intuition. Both the importance and the lack of such methods in the literature are noted in [451, 453, 500], [566, chapter 7], [1014, see section 2.6.2 in particular].

Memo 2 illustrates the use of a simple model for both diagnosis and developing insight into the nature of cost trade-offs. In the remainder of this section we briefly present a generalization of this model, known in the operations research literature as the general optimal market area (GOMA) model, and comment on its potential range of application. We begin with the assumptions.

1. Demand per year is evenly distributed over the distribution territory with density ρ per square mile.

2. Given market area S served by a facility, the annual facility volume is ρS and the facility operating cost is $f(\rho S)^\alpha + v\rho S$. Economies of scale are reflected in $f(\rho S)^\alpha$, where $0 \leq \alpha < 1$, with smaller values of α implying greater relative economies of scale.

3. Unit transportation cost as a function of miles traveled, δ, is $t\delta^\beta$. Economies of distance exist when $\beta < 1$. If $\beta = 1$, then t represents the unit transportation cost per mile.

4. Various regular two-dimensional market shapes may be specified (e.g., hexagon, circle, diamond).

5. Various distance norms may apply (e.g., Euclidean, rectilinear).

6. The objective is to minimize cost.

For many combinations of β, market shape, and distance norm, transportation cost is proportional to the market area raised to the $\frac{\beta}{2}$ power; that is, average transportation cost per unit $= t\sigma S^{\frac{\beta}{2}}$ where σ is the *configuration factor*. The value of σ depends on β, the market shape, and the distance norm. For example, with a circular market shape and the Euclidean distance norm, $\sigma = \dfrac{2}{(2+\beta)\pi^{\frac{\beta}{2}}}$. See [391] for other values of σ. When $\beta = 1$, σ is simply the average distance from the market center to customers for a one square mile market area.

Based on the above information, we may express the average facility operation and transportation cost per unit as

$$f(\rho S)^{\alpha-1} + v + t\sigma S^{\frac{\beta}{2}}, \tag{9.1}$$

which is minimized at $S^* = \left\{ \dfrac{2(1-\alpha)f}{\beta t\sigma} \rho^{\alpha-1} \right\}^{\frac{2}{\beta+2(1-\alpha)}}$. $\tag{9.2}$

The ratio of transportation cost to fixed facility cost at S^* is

$$\frac{t\sigma(S^*)^{\frac{\beta}{2}}}{f(\rho S^*)^{\alpha-1}} = 2\frac{1-\alpha}{\beta} \ . \tag{9.3}$$

Expression (9.1) gives an indication of the cost structure present in distribution system design problems. As illustrated in memo 2, expressions (9.2) and (9.3) are helpful for evaluating the sensitivity of an efficient design to changes in demand and cost rates, and for quickly assessing the potential value of redesigning an existing system.

We have limited our discussion to a basic form of the GOMA model. Other features that can be incorporated into the model include inbound transportation costs [453], demand that is dependent upon distance from the facility [391], a network comprising primary and secondary facilities [1182], and demand that changes over time [1183].

9.4   Algebraic Language Tools

Algebraic modeling languages interfaced with standard optimizers provide an intermediate tool for quickly and easily developing prototype mathematical programming representations of distribution system design models. Examples include AMPL [418], GAMS [148], and LINGO [1050]. An algebraic programming language overcomes some of the awkwardness that attends standard commercial general-purpose MIP software. Such software usually requires coding front and back end software for its use. Changes in the parameter values for sensitivity analysis are often difficult and changes to the model structure can require excessive labor. Porting to new computer environments poses additional challenges. It is important to stress, however, that specialized algorithms will still be needed to solve large-scale distribution system design problems.

We use LINGO in the following illustration of algebraic modeling languages, but there is no intention to favor any particular software product. It is applied to a model described by Pooley [966] for a study of facilities at Ault Foods Limited. Super-LINDO was used as a solver. (LINDO is the background solver for LINGO.) The model, in which both plants and warehouses are located is given by:

Minimize: $\sum_j [e_j a_j + \sum_{ik} p_{ij} x_{ijk}] + \sum_k [f_k z_k + v_k \sum_{il} D_{il} y_{kl}]$

$\qquad + \sum_{ijk} c_{ijk} x_{ijk} + \sum_{ikl} t_{ikl} D_{il} y_{kl}$

subject to:

production capacity (for all j): $\sum_{ik} x_{ijk} \leq P_j a_j$

warehouse capacity (for all k): $\sum_{il} D_{il} y_{kl} \leq W_k z_k$

production meets demand (for all i and k): $\sum_j x_{ijk} = \sum_l D_{il} y_{kl}$

single-sourcing (for all l): $\sum_k y_{kl} = 1$

with $x_{ijk} \geq 0$; a_j, z_k, and $y_{kl} = 0, or\, 1$;

where:

i	index for commodities;
j	index for plant sites;
k	index for warehouse sites;
l	index for customer zones;
e_j	fixed portion of possession and operating costs for plant j;
f_k	fixed cost for warehouse k;
v_k	average unit variable cost of throughput for warehouse k;
p_{ij}	average unit cost of producing commodity i at plant j;
c_{ijk}	average unit cost of shipping commodity i from plant j to warehouse k;
t_{ikl}	average unit cost of shipping commodity i from warehouse k to customer zone l;
D_{il}	demand for commodity i in demand zone l;
P_j	production capacity of plant j;
x_{ijk}	amount of commodity i shipped from plant j to warehouse k;
a_j	zero-one variable equal to 1 if a plant is established at site j, and equal to 0 otherwise;
z_k	zero-one variable equal to 1 if a warehouse is established at site k, and equal to 0 otherwise;
y_{kl}	zero-one variable equal to 1 if warehouse k is assigned to demand zone l, and equal to 0 otherwise;

and all parameters measured on an annual basis. Solution times were 15-90 minutes using a 386 PC. There were 6 commodity groups, 10 plant sites, 13 warehouse sites, and 48 customer zones. The distribution network had a sparse structure as viable outbound links were determined through discussions with marketing representatives to satisfy service constraints, and not all possible plant-warehouse combinations were included [967].

We developed the following LINGO expression of the *dense network* form of the model (with data based on the Huntco Foods case in [1210]):

```
MODEL:
1] SETS:
2] PLANTS/1..3/:PCAP,FCP,PSITE;
3] WARE/1..5/:WCAP,FCW,WCOST,WSITE;
4] CUST/1..6/;
5] ITEMS/1..2/;
6] INBOUND(ITEMS,PLANTS,WARE):INCOST,VOL;
7] OUTBOUND(ITEMS,WARE,CUST):OUTCOST;
8] COMBIJ(ITEMS,PLANTS):PCOST;
```

```
 9] COMBIK(ITEMS,WARE);
10] COMBKL(WARE,CUST):YWC;
11] COMBIL(ITEMS,CUST):DEM;
12] ENDSETS
13] DATA:
14] FCP = 1,1.1,1.5;
15] FCW = .899,.899,.899,.899,.899;
16] WCOST = .6,.5,.4,.7,.9;
17] PCAP = 15,12,13;
18] WCAP = 9,9,9,9,9;
19] DEM = 2,2.5,4,2,3.5,5,
20]     1,2.5,3.5,3,4.5,2;
21] INCOST = .20,.57,.58,.53,.69,
22]     1.04,.91,.93,.91,.80,
23]     .59,.34,.47,.31,.32,
24]     .85,.76,.67,.58,.49,
25]     .29,.38,.47,.56,.65,
26]     .15,.16,.17,.18,.19;
27] OUTCOST = .34,.63,.48,.82,.76,1.07,
28]     .62,.46,.33,.63,.61,.92,
29]     .64,.39,.53,.75,.56,1,
30]     .6,.42,.33,.65,.59,.95,
31]     .75,.42,.55,.49,.44,.8,
32]     .25,.35,.45,.55,.65,.75,
33]     .98,.88,.78,.68,.58,.48,
34]     .15,.25,.35,.45,.55,.65,
35]     .91,.81,.71,.61,.51,.41,
36]     1.1,1.2,.89,.65,.75,.34;
37] PCOST = 2,4,3,
38]     5,3,2;
39]ENDDATA
40]
41]! The objective function;
42][COST] MIN =
43] @SUM( INBOUND(I,J,K): INCOST(I,J,K)*VOL(I,J,K))
44] + @SUM( OUTBOUND(I,K,L): OUTCOST(I,K,L)*DEM(I,L)*YWC(K,L))
45] + @SUM( PLANTS(J): FCP(J)*PSITE(J)
46] + @SUM( COMBIK(I,K): PCOST(I,J)*VOL(I,J,K)))
47] + @SUM( WARE(K): FCW(K)*WSITE(K)
48] + WCOST(K)*@SUM( COMBIL(I,L):YWC(K,L)*DEM(I,L)));
49]! Production capacity constraints;
50] @FOR( PLANTS(J): [PCAP]
51] @SUM( COMBIK(I,K): VOL(I,J,K)) < PCAP(J)*PSITE(J));
52]
53]! Production meets demand;
54] @FOR( ITEMS(I):
55] @FOR(WARE(K): [BALANCE]
56] @SUM( PLANTS(J):
57] VOL(I,J,K)) - @SUM( COMBKL(K,L):DEM(I,L)*YWC(K,L))=0));
```

```
58]
59]! Warehouse capacity constraints;
60] @FOR( WARE(K): [WCAP]
61] @SUM( COMBIL(I,L): DEM(I,L)*YWC(K,L)) < WCAP(K)*WSITE(K));
62]
63]! Single-sourcing constraints;
64] @FOR( CUST(L): [SOURCING]
65] @SUM( WARE(K): YWC(K,L)) = 1);
66]
67] @FOR( COMBKL(K,L): [FORCE] YWC(K,L)<WSITE(K));
68]
69]! Set the binary restrictions;
70] @FOR( WARE(K):
71] @BIN(WSITE(K)));
72] @FOR( COMBKL(K,L):
73] @BIN(YWC(K,L)));
74] @FOR( PLANTS(J):
75] @BIN(PSITE(J)));
END
```

The @SUM operator replaces the $\sum$, and the @FOR defines loops through the indices. The bracketed terms in the model are labels for ease of identification of solution report details. The forcing constraints $y_{kl} \leq z_k$ have been added here to tighten the formulation. They render the LP relaxation "integer friendly," and can speed up the MIP solution process. ReVelle [994] reviews examples of such models and the history of such tightening constraints (see also [242, 454, 795, 869]). The small numerical example resulted in premature termination of the solver, while looser plant and warehouse capacity constraints allowed quick solution times using a 486 PC running at 25 mhz, and a student version of LINGO. (The model could be converted to MPS format using the SMPS command if needed for use with another solver. Also, sparse link sets could be used to model sparse network structures, and an ϵ-optimal termination criterion could be implemented.)

The main limitation of algebraic modeling languages for distribution system design is that they have not been interfaced with the highly specialized solvers needed to optimize at the level of detail needed in most real applications. When this limitation is overcome, such languages will graduate from a supplementary role to a primary one.

9.5 Conclusions

We began with a series of memos in order to illustrate analysis of a distribution system in a way that is compatible with developing intuition into the problem. We then briefly described two types of distribution system design tools, diagnostic tools and language tools. The strength of diagnostic tools is in quickly assessing the pay-off potential of change and the consequent merit of more detailed study. We see diagnostic tools as complementary to the specialized optimizers required for most practical settings. The algebraic language tools provide efficient model specification and, at the least, a first step in more detailed analysis of a company's logistics system. Applications of these tools alone may even be sufficient in some cases to establish the confidence needed for a final decision. We have purposely left our descriptions concise and rely on the following annotated bibliography as a reference for further detail.

9.6   Annotated Bibliography

9.6.1   Related to Diagnostic Tools

Two papers that are helpful for providing a general background on the use of diagnostic tools for analysis are [889, 518]. Both authors cover a number of applications and give examples of approximate models for quick analysis.

The literature on market area models is extensive and spans many disciplines. The origins of modern activity on market area models can be traced back at least 50 years to the work of German economist August Lösch (1906 - 1945). An excellent review of the literature is given in [391]. We will limit overlap by restricting mention to a few papers that help illustrate the range of application of market area models. We will also point out more recent work and identify sources for average distance formulas.

The papers [744, 451, 453] give more background on the diagnostic tool employed in memo 2. Illustrations in [453] are drawn from applications in the auto parts, consumer products, food, and mining industries. Smith [1089] describes market area models for estimating the number of service personnel who are responsible for machine repairs (e.g., photocopiers, computer systems) in assigned geographic territories. The tools are relevant for planning social services as well (public health nurses, social workers). Other public sector applications include capacity and market area decisions for solid waste transfer-stations [1212], bus garages [1175], and fire stations [1205]. In economic planning and analysis, market area models have been used to estimate the benefits due to transportation improvements [861] and are well-established in analyzing spatial pricing policies under competition [72].

Examples of recent work on market area models for distribution system design include [158, 159, 160, 161, 162, 163, 1182, 1183]. Campbell [158, 159, 161, 162] analyzes a network comprising a single plant supplying distribution centers which serve local markets. Hub location is addressed in [163] and the merits of a two-echelon distribution system with primary and secondary facilities are analyzed in [1183]. Dynamic demand is introduced in [160], for the case of deterministic increasing demand over time, and in

[1183] for a system with changing and uncertain demand.

Market area models make use of average distance expressions (see [366, 740, 1101] for formulas). Average distance measures are used to study the role of transshipment centers in many-to-many logistic networks in which vehicles make deliveries from many sources to many destinations [256]. Near optimal network geometry is modeled using a small set of parameters.

9.6.2   Related to Algorithmic Tools

Aikens [13] reviews operations research contributions to the modeling and analysis of distribution systems design that involves "determination of the best sites for intermediate stocking points." Models are classified according to:

- whether the underlying distribution network (arcs and/or nodes) is capacitated;
- the number of warehouse echelons (zero, single, or multiple);
- the number of commodities (single or multiple);
- the underlying cost structure for arcs and/or nodes (linear or nonlinear);
- whether the planning horizon is static or dynamic;
- patterns of demand (e.g., deterministic or stochastic, influence of location, etc.);
- the ability to accommodate side constraints (e.g., single-sourcing, choice of only one from a candidate subset, etc.).

Discussion begins with heuristic and branch-and-bound approaches applied to the single-commodity, uncapacitated, zero warehouse echelons, linear cost model often referred to as the simple plant location model, and ends with Benders decomposition applied to the multicommodity, capacitated, single warehouse echelon case as proposed in [454], and elaborated upon in [455]. (A numerical example of the Benders decomposition approach appears in [786]. Early recognition of the potential of Benders decomposition is discussed by Balinski [50] who cites the working paper [49].)

Work considering stochastic systems is sparse, and work directly considering nonlinear costs is practically nonexistent. Of course piece-wise linear approximations for nonlinear concave throughput cost functions were proposed long ago, as in [365]. Geoffrion [452] provides guidance on constructing optimal objective function approximations. A recent contribution considering the stochastic demand, zero warehouse echelons, capacitated, single-commodity case is [736], "for which no exact algorithm had been previously developed." A brief literature review is provided (and [771] is suggested for further background). Solution of large-scale stochastic demand problems involving multiple commodities, intermediate facilities, with various capacity constraints awaits further developments.

Beasley [66] reviews the literature of capacitated, single commodity warehouse location problems (that omit the plant-warehouse linkage), and reports on a tree search procedure capable of solving up to 500 potential warehouse locations and 1000 customer zones. Use of Lagrangian relaxation for bounding and imbedding problem reduction rules follows on a rich history of such strategies in solving facility location problems.

Structure-exploiting factorization has also contributed to the evolution of algorithms applied to logistics models as discussed in [458]. Examples of advances in factorizing embedded network structures are [151, 796], while [150] provides an example of the robustness of the Benders decomposition strategy. The latter work introduces "goal cuts" within a decomposition approach to accelerate convergence when solving multicommodity production/distribution problems for Nabisco Brands, Inc. Plant location decisions are combined with assignment of production facility types to the plants.

Most published accounts of optimizing models applied to distribution system design have been of the MIP variety where binary facility location variables are used to select from a finite set of candidate sites. In [447] a continuous model (location can take place anywhere in the plane) is paired with the standard MIP approach to decide supply points for shipping to customers of a large Belgian brewery. Tactical issues involved in implementation are addressed. A good account of the practical issues involved in successfully applying MIP models to support both long-term and short-term distribution system planning decisions is given in [483]. The development and use of a decision support system (DSS) for DowBrands, Inc., based on an MIP model of a distribution system having two warehouse echelons is described in [1011]. Geoffrion and Powers [457] discuss the need for comprehensiveness in the distribution planning system. Shapiro, Singhal, and Wagner [1064] use a case study to illustrate how integrated logistics planning advocated by logistics practitioners (and espoused in [968]) can be successfully employed through DSSs based on mathematical programming models.

10
SITING EMERGENCY SERVICES

Vladimir Marianov
Catholic University of Chile

Charles ReVelle
The Johns Hopkins University

10.1 Introduction - What are the Important Issues?

How many ambulances are needed in a particular geographical area, and where should they be deployed to assure reliable service to medical emergencies? Where should the fire-fighting companies be sited in a city so that responses to fires arrive in time to minimize damages and to save lives? When should a police car be at the corner of Elm and Main streets or, more pointedly where should it be at each moment through the day to reduce the risk of crime in the area under study? These questions, relating to the design and operation of emergency services, have been under study by a number of researchers over the last 25 years. Planners of emergency systems must answer questions like these when they confront the design or the reconfiguration of emergency medical systems or of emergency repair systems, or of police operations, or fire fighting systems. In this chapter we critically review efforts that have been made to answer these challenging questions. These efforts are reflected in a rich variety of mathematical models which have been developed to assist emergency system planners with the development of deployment strategies.

One criterion for judging the performance of emergency services is the speed at which the system reacts when an emergency call is logged. Another measure is the ability of personnel to deal effectively with the situation once the server (vehicle) is on the scene. The initial spatial allocation of servers, the principal focus of this paper, influences powerfully the efficiency of the response. To decide on a spatial allocation requires that a number of questions be answered and a number of issues be addressed:

1. The first important question is "How many servers are needed?" This question has more than one possible answer. If there was no *a priori* budget constraints, the minimum number of servers might be sought that would achieve "coverage" of the entire population (or all possible customers). This solution might or might not coincide with the solution that minimizes the cost of the system. If the budget is limited, the number of servers to be deployed must be limited as well. When the budget for deployment of servers is limited, it makes sense to do as well as possible, to seek the best possible performance with the resources at hand. For instance, the goal might be to "cover" the maximum population (or customers) with a "good quality of coverage." Quality of coverage must be defined by the particular requirements of the system under study.

2. A second crucial question is "How long can customers involved in an emergency afford to wait for service before the consequences of a lack of response become intolerable?" Different answers to this question must be offered for different emergency services. In the case of emergency repair services, a failure causes an economic loss which, in most of the cases is proportional to the time the failure lasts, and consequently, to the time it takes to correct the situation. On the other hand, shorter response times impose greater resource requirements on the system, which in turn increase its cost. Thus, response times are determined by an explicit or implicit compromise between the cost of the repair system and the cost of the failure.

In the case of most medical emergencies, the risk of loss of life is known to increase with response time. Consequently, one reasonable objective is to minimize response time, given a limited budget. The siting associated with a minimum response time solution, however, may leave some demand areas very distant from their closest responding unit. Another approach is to assure that the response to all calls is within a time standard or, equivalently, to have a server available within a distance standard or time standard of each and every customer. Application of such a standard deals directly with the issue of some demands left too distant from their nearest responding unit. When dealing with fire fighting services, intuition indicates that a shorter response time always results in less property damage, but loss of life does not follow this formula through the entire range of response time. For fire services, the Insurance Services Office (ISO), an organization of the insurance industry, rates cities according to their fire protection capability (Insurance Services Office, [622]). One of the standards used in this rating is the distance between customers and fire companies. The greater the extent of demand outside of the distance standards, the lower the rating a city achieves, indicating a higher risk to property. Thus, it is reasonable to use these distance standards in the design of fire fighting systems in urban areas. In the case of police beats, the hazard of crime may be related to the frequency of a police car passing a certain point.

3. A third essential question is "what does 'coverage', or 'good quality coverage' mean?" Early models defined coverage of a customer as the presence of one server initially stationed within a distance standard of that customer. As models have evolved, the meaning of coverage has been refined. For example, some police emergencies cannot be controlled by a single policeman. Coverage, for such emergences, then, must be defined as the response to the emergency by, say, p policemen within some time frame. If fewer than p policemen attend the call, the emergency may not be counted as covered. This raises the issue of multiple-server response.

The usual fire emergency puts both people *and* property at risk. When both are ar risk, an engine company is not enough; both engine companies (or brigades) and ladder companies (brigades) are both needed at the scene of the fire to protect property and people. Furthermore, different numbers of companies are needed in different cases. ISO defines standard response to a fire in medium size cities as a response by three engine companies and two ladder companies within a distance standard. This type of response is referred to as a multiple-server, multiple-type response. Furthermore, when more than one server is needed at the emergency scene, an optimal server location pattern ought to consider more than one server initially stationed at the same position. The issues of co-location of servers and the possibility of depot capacity limits introduce still more conceptual challenges to the modelling and solution of the emergency siting problems.

4. Another question involves what to do when servers are not available. Servers may not be available because of such factors as equipment break-down, rest breaks needed by the drivers and personnel, and, last, but far from being least, by the time needed by servers to attend other calls. Depending on the system, there may be a high probability that the nearest server is busy attending a call when a new call for service appears. This leads to the phenomenon of congestion, and the question of which unit to dispatch when the nearest server or servers are not available. An inappropriate dispatch policy could make a system non-functional, despite all location optimization efforts.

5. Still another issue is "Can the data needed for modelling be collected?" Models should be formulated in such a way that they use only data that can be collected and used, and they must be robust, in the sense that the system designed by the model should not be too sensitive to small data errors. Often, there is not much data available, and the available data may have errors, some of which are known. Also, the underlying parameters may, in fact, be random variables. As an example, travel times are random in nature, since they depend on traffic levels and on the current condition of the road network.

6. Additionally, the analysis must take note of the nature of the emergency system under study . Is it private or public? Different design criteria may be used in the two cases: profit maximization in one, cost minimization in the other. Are there any existing systems that can cooperate or will the systems be in competition with one another. Do "mutual aid pacts" exist which allow servers to cross jurisdictional boundaries to provide assistance when a server in one system is available and the nearest server in the other is not. Is there any interaction between private and public systems (ambulances, for instance)? Do fire services sometimes "fill in" for ambulance servers when a timely ambulance response is impossible or unlikely?

7. Another issue important in some systems is workload. Achieving equity of workload is one facet of redesigning an emergency system. The actual level of workload can also be crucial since "burnout" of emergency personnel can occur with excessive workloads. Workloads run in parallel with congestion; greater workload units are, of course, the most busy units and the least likely to be available for a timely response.

8. Still a further significant factor in siting is the political feasibility of closing some locations or the acceptability of re-location alternatives. Closing a fire station that has served as a focus for community activity and that provides a sense of security to residents may be difficult indeed. Siting a fire station in an otherwise quiet suburban neighborhood may also be problematic.

9. The last factor in siting to which we refer here is how to present alternatives for decision making. Undoubtedly, the decisions must involve multiple objectives, but the question is how to display and present alternatives when they are characterized by many measures of goodness.

Once all these questions and issues are addressed, a solution method has to be chosen to "solve" the emergency system design problem as it is finally characterized. Alternatives can be found using either methods based on mathematical programming or methods based on iterative improvements of an initial system design, perhaps using a descriptive model, until a system has evolved which displays adequate performance. The mathematical programming (or optimization) models are designed to find alternative superior solutions which trade off the most important objectives against one another. A solution provides the number and positions of servers, given the constraints and the definition of the objectives of the problem.

Although optimization models utilize a more simplified view of reality than descriptive models, the computer time required for optimization to obtain an alternative is often shorter than the time required with the descriptive model. Importantly, optimization models implicitly consider more alternatives than can be considered by iterative solution of descriptive models. Further, mathematical programming models almost always require less complete data to determine the necessary parameters. Optimization models can be deterministic or probabilistic, depending on the particular system parameters, and desired results. Probabilistic models generally need more data than the deterministic models. For reviews of optimization models, see [264, 993].

Iterative methods, in contrast to optimization models, on the other hand, produce successively improved versions of a system design (location pattern), beginning with an initial design and creating new designs until a system is obtained which fulfills the specifications to a desired degree. At each step, new designs of the system are investigated and their performance is evaluated in terms of the stated criteria. Iterative methods are based on improving, at each step, the performance of the created system, as the proposed locations are modified incrementally. The hope is that the iterative procedure can converge to a good solution. In order to test or evaluate the location pattern at each step, these models use either simulation or some other descriptive/evaluative model. After the user or the model proposes a trial location of the servers, the descriptive model is used to solve for the steady state distribution of all the system-wide, region-wide and server-specific parameters.

These models, most of which use queueing theory, can be very realistic and complete. To take full advantage of these characteristics though, reliable and detailed data, whose collection is a very difficult task, needs to be used in the computation of the model parameters. Such data may not, in fact, be available or even obtainable within a reasonable cost or time frame for use in models. Most of descriptive models which utilize queueing for the analysis of the performance of mobile servers on congested systems, are nonlinear models. Sometimes, models which are suitable for locating just one server, have also been used as a sub-algorithm for locating several servers. In general, a fairly large computational effort must be exerted if these models are to be used for the location of servers. Simulation is used in the same manner as queueing in such evaluative models.

In the discussion that follows, we first focus on optimization models, describing the models in the order of their evolution, an order that reflects the recognition of model needs that occurred through time. All of the optimization models deal with coverage or availability. Why do we focus the review on models that use coverage and/or availability? Other models that focus on average response time or average distance have been built (see [995] for a review of these average response time models.) We focus on coverage and availability since these concepts have their roots in the standards and specifications of real emergency systems. Coverage within distance standards comes from the arena of fire protection as well as from emergency medical services. Availability of servers within a time standard comes from the emergency medical sector. Coverage refers to the positioning for a particular demand node of a server within a distance or time standard of that node. Availability refers to the actual presence of the server within the standard at the time that a call from that demand node arrives.

We proceed from optimization models to descriptive models, principally queueing models. These models are merely evaluative of any particular pattern of deployment. When these models are coupled with heuristic procedures, however, they may be used to approach "good" solutions to the problem of server deployment.

10.2   Methods Based on Deterministic Optimization Models

A number of approaches have been developed which are based on mathematical programming/optimization constructs. The aim of the mathematical programming models is to deploy service units in an optimal fashion (minimum cost maximal covering or some other measure of merit), given the constraints and requirements that are set for the system. The first model in the sequence of emergency service covering models was the Location Set Covering Problem (LSCP) [1141, 1140]. The LSCP sought to position the minimum number of servers in such a way that each and every demand point on the network had at least one server initially positioned within some distance or time standard S. In an emergency service context, this means that everyone in the entire population has at least one emergency vehicle initially positioned within the time or distance standard. Note that in this problem statement, a demand point or demand node is considered as covered if the server is deployed within the standard of the demand node, even though this server may be busy at times, even most of the time. The formulation of the model is as follows:

$$\text{Minimize } Z = \sum_{j \in J} x_j \tag{10.1}$$

subject to

$$\sum_{j \in N_i} x_j \geq 1 \qquad \forall i \in I; x_j \in \{0, 1\}; \qquad \forall j \in J \tag{10.2}$$

where

J set of eligible emergency vehicle sites (indexed by j);

I set of demand nodes (indexed by i);

x_j 1 if a server is stationed at j and 0 otherwise;

N_i $\{j | t_{ji} \leq S\}$; with t_{ji} = shortest time from potential vehicle location j to demand node i; and

S time or distance standard for coverage; that is N_i is the set of nodes j located within the standard for demand node i. If a call for service originating at this node is answered by available servers stationed inside this neighborhood, it will be answered within the time or distance standard.

The objective (10.1) minimizes the number of facilities required. Constraints (10.2) state that the demand at each node i must be covered by at least one server located within the time or distance standard S.

This model can be easily solved by its linear programming relaxation in which the integer variables are required simply to be non-negative. [1141] reported that an occasional cut constraint (about 5% of the runs) may be needed if the solution of the linear programming problem produces a non-integer number of facilities. The problem can also be solved by the method of reductions [1140].

The p-center problem can be solved by using the LSCP recursively. The p-center problem seeks the location of the p facilities which minimizes the maximum distance separating any demand point and its nearest facility. In order to obtain a solution to this problem, a sequence of Location Set Covering Problems is solved, in which the distance standard is successively reduced by the smallest unit of measurement of distance. As distance standard is reduced, the number of required facilities may stay constant (p) in the solution of several runs. When a value of the standard distance is reached at which the number of facilities increases to $p + 1$ or higher, the preceding value of the standard distance corresponds to the smallest maximum distance for which p facilities are feasible. [850, 216] describe this method of solution.

The LSCP idea has been extended to siting problems in which coverage is required through time. Time here refers both to future time and time-of-day. The first adaptation of the covering concept to future time was provided by [1005] who projected a spatial expansion of demand through multiple periods and sited facilities in such a way as to achieve coverage with the least number of facilities. [1044] applied multi-objective programming to dynamic siting utilizing permanently placed facilities. He demonstrated how coverage in one-period can be traded off against coverage in some other period. In 1982, [507] focussed on covering demands through time with the minimum number of facility years and the restriction that a facility once in place remains in place through the rest of the horizon. [214] investigated a similar problem but with the option of phasing in and phasing out facilities with associated costs for a facility entering or leaving.

In addition to these models which design through future time, [991] recognized that significant variation in demand occurred through the course of a day. They created TIMEXCLP a model derived from Daskin's MEX-CLP, which sought ambulance positions that maximized coverage through a day of one-hour intervals which intervals displayed differing call frequencies. Repede and Bernardo allowed the number of vehicles at a site to vary through the day as demand evolved. Although little other work on location in response to time varying demands seems to have appeared, the idea is intriguing and important. Periods of low demand at night might even warrant night-time closing of units if responses could still occur within the standard. The authors of this review anticipate further work in this important area.

The Location Set Covering Problem requires coverage of all demand points, no matter how small their population, how distant they are from other demands on the network, or how small their need for service. Recognizing that the resources required to cover all points of demand, no matter how small or remote, could be excessive, [223, 1200] framed a new problem that did not require the coverage of all nodes. Church and ReVelle called it the maximal covering location problem (MCLP), the name we use here, and White and Case referred to it as the partial covering problem. This model assumes a limited budget, which is reflected as a constraint on the number of servers to be sited. Thus, the model seeks the placement of a fixed number p of servers - probably insufficient to cover the entire population within the standards - so that the population or calls for service that have a server positioned within the standard S would be maximized:

$$\text{Maximize } Z = \sum_{i \in I} a_i y_i \qquad (10.3)$$

subject to

$$y_i \le \sum_{j \in N_i} x_j \qquad \forall i \in I, \qquad (10.4)$$

$$\sum_{j \in J} x_j = p \qquad (10.5)$$

$$x_j, y_j \in \{0, 1\} \qquad \forall j \in J, i \in I,$$

where additional notation is

y_i 1 if node i is covered and 0 otherwise

p the number of servers to be deployed; and

a_i the population at demand node i.

All other variables and parameters are the same as defined before. The objective (10.3) maximizes the sum of covered demands. Constraint (10.4) state that demand i cannot be covered unless at least one server is located within the time or distance standard S. Constraint (10.5) gives the total number of facilities that can be sited. Relaxed linear programming supplemented by occasional use of branch and bound was utilized by Church and ReVelle to provide solutions to this problem. Church and ReVelle also suggested heuristic procedures for solutions, as did White and Case. The most informative use of the problem statement is to develop a tradeoff curve between population covered and the number of facilities utilized.

The maximal covering idea was generalized for fire protection by [1046]. Their model accounts for coverage by two different types of service. In addition, both the servers and the facilities which housed the servers were to be sited simultaneously. Of several model types, the most general formulation was that known as the FLEET model (for facility location, equipment emplacement technique). The FLEET model sought the locations of a limited number of engine companies (the pumper brigades) and truck companies (ladder brigades) as well as the fire stations that housed them. The goal of the FLEET model was coverage of the maximum number of people by both an engine company sited within an engine company distance standard *and* a truck company sited within the truck company distance standard. That is, coverage required the simultaneous siting of both types of service within their respective distance standards. Other objectives included in a multi-objective formulation of the FLEET model were maximum coverage of fire frequency, maximum coverage of property value, maximum coverage of population at risk (as measured by the product of fire frequency and population for each demand point), and maximum coverage of property at risk (formed in a similar fashion). The engine and truck company distance standards against which all coverage was measured were the same values for all objectives.

The maximum population coverage version can be stated mathematically as

$$\text{Maximize } Z = \sum_{i \in I} a_i y_i \tag{10.6}$$

subject to

$$y_i \leq \sum_{j \in N_i^E} x_j^E \qquad \forall i \in I \tag{10.7}$$

$$y_i \leq \sum_{j \in N_i^T} x_j^T \qquad \forall i \in I \tag{10.8}$$

$$x_j^T \leq x_j^S \qquad \forall j \in J \tag{10.9}$$

$$x_j^E \leq x_j^S \qquad \forall j \in J \tag{10.10}$$

$$\sum_{j \in J} x_j^E + \sum_{j \in J} x_j^T = p^{E \cdot T} \tag{10.11}$$

$$\sum_{j \in J} x_j^S = p^S \tag{10.12}$$

$$x_j^E, x_j^T, x_j^S, y_i \in \{0, 1\} \qquad \forall j \in J, i \in I,$$

where

x_j^E 1 if an engine company positioned in a fire house at site j; 0 otherwise;

x_j^T 1 if a truck company positioned in a fire house at site j; 0 otherwise;

x_j^S 1 if a fire station is established at site j; 0 otherwise;

N_i^E $\{j | t_{ji} \leq E\}$; set of potential engine sites j which can cover node i by virtue of being within the engine distance standard E;

N_i^T $\{j | t_{ji} \leq T\}$; set of potential truck sites j which can cover node i by virtue of being within the truck distance standard T;

$p^{E \cdot S}$ number of engine and truck companies; and

p^S number of fire stations

In this formulation, the first two constraint types define coverage as achievable only if *both* one or more engine companies are sited within the engine distance standard *and* one or more truck companies are sited within the truck distance standard. The third and fourth constraint types allow companies to be sited only at nodes that have been allocated a fire station. The fifth constraint limits the total number of companies without specifying any fixed proportion of engines and trucks, reflecting the major cost of personnel as opposed to equipment. The sixth constraint limits the number of stations. Schilling *et al* solved the linear relaxation of the problem, resolving fractional solutions with branch and bound.

The MCLP has also been applied to dynamic siting of facilities. Schilling [1044] showed how to adapt the model to planning for uncertain futures. Using multiple scenarios of future growth, he demonstrated how robust alternatives could be chosen which accommodated the uncertainty of future patterns of development.

The models presented above have been used not only for the design of emergency systems, but also for other services, as well. The set covering model has been used to allocate bus stops [477]. The maximal covering location model, and different variants of it, have been used for the location of health clinics [357] hierarchical health services [864] for the allocation of marketing resources in journals [350], optimal choice of pressure measurement points in the eye for early detection of glaucoma [709], and many other applications [219].

Sometimes, two or more servers of the same type are needed at the same time at the scene of the emergency. In the case of fire fighting systems, there are accepted rules about the number of servers attending the scene of the fire. The Insurance Services Office (ISO) defines two types of requirements for the coverage of an area by fire companies: first-due coverage and standard response coverage. First-due coverage requires each demand point to have an engine company (which includes pumpers) located within its first-due distance standard and a truck company (which carries rescue ladders) located within its first-due distance standard. The FLEET model just described is a first-due model; requiring only one of each unit type within their respective distance standards.

Standard response coverage, in contrast, is designed to bring sufficient fire fighting and rescue manpower to the scene of a fire; it requires enough engine companies and truck companies within applicable distance standards to furnish this manpower requirement, which depends in turn, on the expected size of the fire. The number of companies required for the standard response is typically three engine companies and two truck companies. The standard response requirement raises the complicating issue of allowing more than one server to be sited at the same location.

Given the assumptions of earlier models, the siting of more than one server of a particular type at a single location would have been an inefficient allocation of resources; only one server within the standard for each service type was enough for coverage. A needed response by more than one server of a particular type, however, suggests that siting more than a single unit at a site could become a good , perhaps optimal, choice. Hence, co-location of servers has to be considered. And, as a further complication, the issue of station capacity also requires addressing and resolution.

Some models have explicitly dealt with the possibility of co-location and station capacity. A model which limited the number of stations but allowed more than one server at the same site was suggested by [112]. Their hybrid FLEET model requires solution of an integer as opposed to a zero -'one program. However, their model considered only one type of vehicle and placed vehicle capacity limits at the stations. Marianov and ReVelle [801], [802] propose and solve a capacitated model for the maximal standard response problem. These models sought the placement of a fixed number of engine and truck companies so that the population or calls for service with a full standard response sited within appropriate distance standards would be maximized. The problem statement allows multiple servers of one or more types located at a single site, up to the capacity limit of that site. A simplified statement of their capacitated model is the following:

$$\text{Maximize } Z = \sum_{i \in I} a_i y_i \tag{10.13}$$

subject to:

$$y_i + w_i^E + u_i^E \leq \sum_{j \in N_i^E} \sum_{k=1}^{C_j} x_{kj}^E \qquad \forall i \in I \tag{10.14}$$

$$y_i + u_i^T \leq \sum_{j \in N_i^T} \sum_{k=1}^{C_j} x_{kj}^T \qquad \forall i \in I \tag{10.15}$$

$$y_i \leq w_i^E \qquad \forall i \in I \tag{10.16}$$

$$w_i^E \leq u_i^E \qquad \forall i \in I \tag{10.17}$$

$$y_i \leq u_i^T \qquad \forall i \in I \tag{10.18}$$

$$x_{(C_j-k+1)j}^E + x_{kj}^T \leq x_j^S \qquad \forall j \in J, \ k \leq C_j \tag{10.19}$$

$$x_{(k-1)j}^l \leq x_{kj}^l \qquad \forall j \in J \ k \leq C_j - 1, \ l = E \text{ or } T \tag{10.20}$$

$$\sum_{j \in J} \sum_{k=1}^{C_j} (x_{kj}^E + x_{kj}^T) = p^{E+T} \tag{10.21}$$

$$\sum_{j \in J} x_j^S = p^S \tag{10.22}$$

All Variables $\in \{0, 1\}$ \tag{10.23}

The new variables and parameters are:

x_{kj}^E 1 if a kth engine company is located at site j, 0 otherwise

x_{kj}^T 1 if a kth truck company is located at site j, 0 otherwise

x_j^S 1 if a station (depot) is located at site j, 0 otherwise;

y_i 1 if demand node i is covered by the standard response (3 engines, 2 trucks), 0 otherwise;

w_i^E, u_i^E, u_i^T 1 if demand node i is covered by a second engine, first engine, first truck, respectively, and 0 otherwise;

C_j capacity of site j.

While N_i^T and N_i^T have similar definitions as earlier, the respective distance standards are different in that they apply to the standard response rather than to first due coverage. The objective maximizes population coverage. The first two constraints allow coverage only if both three engines *and* two trucks are sited within distance standards of the demand node. The next three constraints avoid coverage by a second server without previous coverage by a first one, and coverage by a third server without coverage by the second server. The next two constraints, together, have three effects: first, they allow the siting of servers only at stations. Second, they preclude the location of more than C_j servers (more than $c_j - k$ engines plus k trucks) at a site. Finally, they state that a $(k+1)^{\text{th}}$ server cannot be located at a site until the kth server is sited there. The last constraints limit the number of servers and stations to be located. This model faces the issue of station capacity in such a way that when solved using its linear relaxation, most integer variables are more likely to have integer values in the solution. [802] propose also a probabilistic version of their model.

10.3   Deterministic Models Addressing Congestion

The preceding formulations considered that the servers were continuously available. None of them considered the possibility of congestion occurring. By congestion, we mean the condition in which an emergency call in a particular demand region occurs while all of the potential servers in that demand node's coverage area are engaged in service. In such a situation, it may not be sufficient for each demand to have one server sited nearby geographically (in the neighborhood). Instead, for each demand, we would seek to have one or more servers *available* with some reliability in the demand node's coverage area. One of the possible ways to assure this availability of service under congestion conditions is to site for each demand node and area a sufficient number of servers to insure that at least one server will be available for service in the neighborhood at all times; i.e. that congestion will not defeat the idea of coverage.

Two kinds of optimization models have been developed which address congestion; deterministic, or redundant coverage optimization models, and probabilistic optimization models. Redundant coverage models seek to position servers in such a way that more than one server is located in the neighborhood of each demand. In other words, redundancy in server location is sought for each demand area, in the hope that users in that area will still have an available server even in a congested situation. [82, 265, 76, 358] maximized redundant coverage given a fixed number of servers. Redundant coverage was measured as the difference between the number of servers stationed within the standard and the minimum number required for coverage. In the case of the Location Set Covering Problem, that minimum number is one. When the number of servers being sited is also the minimal number required for coverage of the system of demands, the problem reduces to finding the alternate optimum to the LSCP which maximizes the total number of redundant servers. The mathematical formulation in non-standard form is:

$$\text{Maximize } Z = \sum_{i \in I} a_i r_i \tag{10.24}$$

subject to

$$r_i = \sum_{j \in N_i} x_j - 1 \qquad \forall i \in I \tag{10.25}$$

$$\sum_{j \in J} x_j = p \tag{10.26}$$

$$x_j \in \{0, 1\} \qquad \forall j \in J, \forall i \in I,$$

$$r_i = \text{integers}$$

$$a_i = \text{a weight characterizing the importance of covering demand}$$
$$\text{node } i \text{ with redundant servers.}$$

where r_i is an integer variable that measures the number of coverers above one for demand node i. The models of Berlin and Daskin and Stern did not utilize the weight a_i in the objective; hence their models simply maximized the total of redundant coverers without regard to whether the demand nodes covered more times were or were not important. A centrally located but low population demand node could be covered many times with equal pay off in these two formulations. Benedict and Eaton corrected this problem in their formulations by weighting the excess coverage with a weight characteristic of the demand, e.g. population or call frequency. The solution methods were, again, solution of the linear programming relaxation of the models with cutting planes or branch and bound. These latter were applied as needed but, in fact, were seldom used.

This set of models still requires mandatory coverage of each demand node by a first coverer, as can be seen from the first constraint of the model. A further disadvantage of these models was that redundant coverers could concentrate on some demand nodes, leaving others with only a first coverer. In other words, adding a seventh coverer to a demand node has the same advantage as adding a second one to another demand node if both nodes have the same weight. Hogan and ReVelle [589] approached the correction of these problems by maximizing backup coverage, which they defined as the coverage of demand points by two of more servers. They stated two models, called the BACOP models for backup coverage problems. BACOP 1 is similar to the preceding model with its first constraint replaced by

$r_i \leq \sum_{j \in N_i} x_j - 1 \ \forall i \in I$, with the variable r_i set equal to one if two or

more servers cover demand node i, and zero otherwise. BACOP 1 seeks to maximize population with has at least one server in addition to the first. It de-emphasizes multiple redundant coverers of a node and focuses on the first redundant coverer of a node. The BACOP1 still requires first coverage of all the demand nodes.

To relax the first coverage requirement, Hogan and ReVelle formulated the BACOP 2 model, which trades off first coverage against backup coverage. Its formulation is

$$\text{Maximize} \qquad Z_1 = \sum_{i \in I} a_i y_i \qquad\qquad (10.27)$$

$$Z_2 = \sum_{i \in I} a_i r_i \qquad\qquad (10.28)$$

subject to

$$r_i + y_i \leq \sum_{j \in N_i} x_j \qquad \forall i \in I \qquad\qquad (10.29)$$

$$r_i \leq y_i \qquad \forall i \in I \qquad\qquad (10.30)$$

$$\sum_{j \in J} x_j = p \qquad\qquad (10.31)$$

$$x_j, r_i, y_i \in \{0, 1\} \qquad \forall j \in J, i \in I,$$

where all notation is defined as before. The objectives maximize first coverage and second coverage, respectively. The first constraint says that coverage by a first and second server is bounded by the number of servers stationed in the neighborhood of demand node i. Thus, if only one server is located in the neighborhood, first coverage only is achieved. If there are two or more servers, first and backup coverage are achieved. The second constraint indicates that backup coverage can not be fulfilled without prior achievement of first coverage. The last constraint limits the number of servers to be deployed. The authors report that marginal reductions in first coverage can strongly improve backup coverage. They solved the linear relaxation of the model with occasional branch and bound. Storbeck [1102] used the difference between two non-negative variables to maximize backup coverage. His model has the following formulation:

$$\text{Maximize} \quad Z_1 = \sum_{i \in I} a_i k_i^+ \tag{10.32}$$

$$Z_2 = -\sum_{i \in I} a_i k_i^- \tag{10.33}$$

subject to

$$k_i^+ - k_i^- = \sum_{j \in N_i} x_j - 1 \qquad \forall i \in I \tag{10.34}$$

$$\sum_{j \in J} x_j = p \tag{10.35}$$

$$x_j, \ k_i^- \in \{0, 1\}, \ k_i^+ \text{integer} \qquad \forall j \in J, i \in I$$

where

k_i^+ 1 if coverage in excess of one for node i; 0, otherwise, and

k_i^- 1 if coverage less than one for node i; 0, otherwise.

The first objective maximizes backup coverage of population, while the second one minimizes population "undercoverage".

10.4   Methods Based on Probabilistic Optimization Models

Probabilistic optimization models take explicit account of the probabilities of servers being busy to compute the amount of redundancy actually needed. In other words, they use explicit probabilistic constraints *inside* the mathematical programming model. [1100], in an unpublished work using results from queueing theory, formulated an early set of such models, most of them non-linear, which located servers so as to trade off customer's waiting cost against service cost.

Later, [191] formulated a probabilistic version of the LSCP. In their model, the probability that at least one server is available within the distance standard to each demand node is forced to be greater than or equal to some reliability α. To compute such a probability, they make use of estimates derived from simulations of the probability q of a server being busy. The probability of being busy is also referred to as the *busy fraction*. This busy fraction is assumed to be independent of the probability of other servers being busy. Location patterns derived were used to estimate new busy fractions which were used, in turn, in a repeat solution of the LSCP. Chapman and White's model, however, could not be solved to convergence.

Daskin [262] utilized the Chapman and White notion of a server busy fraction to formulate the Maximum Expected Covering Location Problem, (MEXCLP). Daskin maximized the expected value of population coverage within the time standard, given the p facilities are to be located on the network. He utilized a single system- wide busy fraction q in his model and assumed a binomial distribution of the probability of k servers being busy. Daskin computed the increase in the expected coverage of a demand when a kth server is added to its neighborhood. A call is served only if there is at least one server available to node i. The probability P_k of this event, given that there are k servers in the area, each with a busy fraction q, is equal to 1 minus the probability of all k servers of node i being busy, that is, $1-q^k$. The increase in the expected coverage of a call caused by the addition of a kth server, H_k, is just $P_k - P_{k-1}$. Thus $H_k = (1-q^k)-(1-q^{k-1}) = (1-q)q^{k-1}$. The model maximizes this expected coverage for all possible number of servers at each neighborhood, and for all demand nodes:

$$\text{Maximize} \quad Z = \sum_{i \in I} \sum_{k=1}^{n_i} a_i(1-q)q^{k-1}y_{ik} \quad (10.36)$$

subject to

$$\sum_{k=1}^{n_i} y_{ik} \leq \sum_{j \in N_i} x_j \quad \forall i \in I \quad (10.37)$$

$$\sum_{j \in J} x_j = p \quad (10.38)$$

$$y_{ik} in \{0,1\} \forall i, k; \quad x_j \text{ integers } \forall j$$

where

y_{ik} 1 if node i has at least k servers in its neighborhood, 0 otherwise;

x_j the number of servers at site j; and

n_i the maximum number of servers in N_i.

The first constraint says that the number of servers covering demand i is bounded from above by the number of servers sited in the neighborhood. The second constraint limits the number of servers to be deployed. Declining weights $(1-q)q^{k-1}$ on the variables y_{ik} make unnecessary any ordering constraints for these variables, and push the variables in the solution toward integrality if the linear relaxation of the model is solved. Daskin proposed a heuristic method of solution for MEXCLP, which provides solutions for the system, for different ranges of values of q.

Recognizing the need for including in the model the possibility of siting depots which are capacitated, [112] modified MEXCLP to consider the location of vehicles of only one type and depots with limited room for units. They referred to their model as the Multiple cover, One-unit Facility Location, Equipment Emplacement Technique (MOFLEET).

[997, 998] looked at the problem in a different way. Instead of maximizing expected coverage, they constrained the level of server availability to be greater than or equal to a preset value, while minimizing the total number of servers. Their model, the Probabilistic Location Set Covering Problem (PLSCP), utilized a region-specific busy fraction, instead of a system-wide busy fraction. They modeled reliability as a binomial process as a first approximation. The model was essentially a version of the LSCP with a probabilistic constraint. Observing that a more realistic model should make use of server-specific busy fractions and in view of the difficulty of working with such parameters, ReVelle and Hogan, introduced a *local estimate* of the busy fraction, q_i, in the coverage area around node i.

As in Chapman and White's original model, the probability in ReVelle and Hogan's model of at least one server being available within the service radius of node i must be greater than or equal to α. By analogy to Daskin's estimate of a system-wide busy fraction, the local estimate of busy fraction in the service region around demand node i was given by an estimate of the service time needed in the region, divided by the available service time in the region, that is:

$$q_i = \frac{\bar{t} \sum_{k \in M_i} f_k}{24 \sum_{j \in N_i} x_j} = \frac{\rho_i}{\sum_{j \in N_i} x_j}$$

where

$\bar{t}$ average duration of a single call, in hours;

f_k frequency of calls for service at demand node k, in calls per day;

M_i set of demand nodes located within S of node i;

N_i $\{j | t_{ji} \leq S\}$; that is N_i is the set of nodes j located within the time or distance standard of demand node i; and

ρ_i utilization ratio.

The probability that at least one server is available within time standard S when node i requests service is 1 minus the probability of all servers within S of node i being busy. Since ReVelle and Hogan assumed the binomial distribution for the probability of one or more servers being busy, this probability is

$$1 - P[\text{all servers of node } i \text{ are busy}] = 1 - \left[\frac{\rho_i}{\sum_{j \in N_i} x_j} \right]^{\sum_{j \in N_i} x_j}$$

Requiring this probability of at least one server being available to be greater than or equal to α leads to a nonlinear probabilistic constraint. This probabilistic constraint does not have an analytical linear deterministic equivalent. ReVelle and Hogan found the numerical deterministic equivalent to be

$$\sum_{j \in N_i} x_j \geq b_i \qquad (10.39)$$

where b_i is the smallest integer which satisfies $1 - \left(\frac{\rho_i}{b_i}\right)^{b_i} \geq \alpha$ and they sought the least number of servers and their positions subject to constraint (10.39) on the availability of servers. The probabilistic location set covering problem with local estimates of busy fraction has, then the same form of the LSCP with constraint (10.2) replaced by constraint (10.39).

Because the solution of the PLSCP could lead to a large number of servers, potentially larger than available funds could achieve, ReVelle and Hogan [999] formulated the Maximum Availability Location Problem, MALP. The model sought to maximize the population which had service available within a stated travel time with a specified reliability, given that only p servers are to be located. Using the same reasoning as in PLSCP, ReVelle and Hogan computed the number b_i of servers needed for α-reliable coverage of node i, and maximized the population with b_i or more servers; i.e., with α-reliable coverage. Their MALP is stated as follows:

$$\text{Maximize} \qquad Z = \sum_{i \in I} a_i y_{ib_i} \qquad (10.40)$$

subject to

$$\sum_{k=1}^{b_i} y_{ik} \leq \sum_{j \in N_i} x_j \qquad \forall i \in I \qquad (10.41)$$

$$y_{ik} \leq y_{ik-1} \qquad \forall i, \ k = 2, 3 \ldots b_i \qquad (10.42)$$

$$\sum_{j \in J} x_j = p \qquad (10.43)$$

$$x_j, y_{ik} \in \{0, 1\} \qquad \forall j \in J, i \in I \qquad (10.44)$$

The variable y_{ik} is one if k servers are potential coverers of node i. The objective maximizes the population with at least b_i coverers stationed within the standard. Constraint (10.41) states that there are at most as many potential coverers as servers sited within the neighborhood of node i. Constraint (10.42) is an ordering constraint, which forces a node to have $k - 1$ coverers before having k coverers. Constraints (10.43) and (10.44) are self-explanatory. In the fire protection arena, [1000], formulated a comparable model, the Probabilistic Facility Location, Equipment Emplacement Technique, PROFLEET. This model considered the deployment of several types of vehicles, simultaneously covering each emergency, as well as the siting of depots or stations. The model considered independence between availabilities of engines and trucks, and [803] presented a second version, in which availabilities of engines and trucks were no longer independent.

Recently, [51] formulated a new version of the PLSCP, in which a desired level of reliability is mandatory for each demand. This level is achieved through a probabilistic constraint. An upper bound of the "uncoverage probability" of each demand is constrained to be less than a predetermined value, as opposed to the PLSCP of ReVelle and Hogan, in which an estimate of this probability is constrained. Independence is assumed between probabilities of servers being busy.

Another recent model by [485], maximizes the expected number of calls reached within a set time threshold. In this nonlinear model which is based on Jarvis' [636] mean service time computation, the service time depends on call location. Again, independence is assumed between probabilities of servers being busy. The authors present a heuristic method of solution.

To summarize the preceding probabilistic models, the MEXCLP used a system-wide busy fraction, whereas, the PLSCP, MALP and the PROFLEET models utilized region-specific busy fractions. All of these models, as well as the models by Ball and Lin and by Goldberg and Paz made the simplifying assumption that the probabilities of two vehicles being busy within the same region are independent. In order to relax this assumption, the binomial distribution can no longer be used, and a queueing behavior must be assumed.

Marianov and ReVelle [804] presented the Queueing - PLSCP, or QPLSCP, in which they relax the independence assumption and model the behavior in each region as an M/M/s - loss queueing system. (Poisson arrivals, exponentially distributed service time, s servers, loss system). With this new formulation, the reliability constraints are finally put on a sound foundation by the use of queueing theory to model the arrival - departure process within the location model itself. In a recent work by Batta, Dolan and Krishnamurthy [58], the independence assumption is (approximately) relaxed through the use of Larson's approximated hypercube. However, in that model, called AMEXCLP, the busy fraction of servers is assumed to be the same over the whole system, while the QPLSCP uses a more realistic, neighborhood-specific, busy fraction.

The general formulation of the QPLSCP is the same as for the PLSCP. However, the computation of the parameter b_i is quite different. Marianov and ReVelle find the probability of all s servers in a region being busy, p_s, to be

$$p_s = \frac{\dfrac{\rho_i^s}{s!}}{1 + \rho_i + \dfrac{\rho_i^2}{2!} + \ldots + \dfrac{\rho_i^s}{s!}}$$

where ρ_i is the queueing theory utilization ration. The authors compute the number s of servers which make this probability smaller than a present value, and this number becomes the parameter b_i of the QPLSCP model.

10.5   Descriptive Models and Heuristic Solution Procedures

The most well known of the queueing descriptive models are the Hypercube [737] and the approximated hypercube [738]. Hypercube models, as well as most of queueing descriptive models for the analysis of the performance of mobile servers on congested systems, are nonlinear models. The hypercube model, which builds on previous developments [181, 741] for two servers, describes a spatially distributed queueing system with distinguishable servers. Each server may have two states: busy or free, and the state of the system of servers is given by a vector whose components are the individual states (busy or free, 1 or 0) of each server. Thus, the system has 2^p states, where p is the total number of servers. The state space for three servers may be represented by the vertices of a three-dimensional cube, and the generalization to p servers leads to a state space which is formed by the vertices of a p-dimensional cube, that is, a hypercube. In this representation, the number of edges of the hypercube that one has to pass to go from one state to a different one, represent the Hamming distances between these two states (number of positions that differ in value between two state space vectors). In this system, the initial geographical positions of the servers are known. Also known are the call rates of each demand node i, which are assumed Poisson distributed, with mean λ_i. Service times are exponentially distributed with mean $\frac{1}{\mu}$. Thus, the system behaves as an M/M/p queueing system. Servers may attend calls originating at any demand node, but every server has its own primary service area, and there is an ordered preference list of servers for each demand node, given by the dispatch policy. Each call is attended by *one and only one* server.

In order to calculate the steady state probabilities of each state and, consequently, all the information needed on how occupied the servers are, the transition rates between states must first be computed. This computation is made using the following procedure:

1. For each demand node i and for each state b_s of the system (or the hypercube), look at the preference list and determine which idle server k will attend calls from i, given the state of the system is b_s. Call $b_s + k$ the state that corresponds to the same servers busy as in state b_s, but with server k also busy. Increment the transition rate from state b_s to state $b_s + k$ in λ_i. Suppose now that the preferred server is not unique, but there are s preferred servers equally likely to attend calls from node i when the system is in state b_s. Then, each of the s servers defines a new state, which corresponds to b_s plus that particular server busy. Increment all the transition rates from state b_s to each one of these new states, in $\frac{\lambda_i}{s}$. Do the same for each node and each state of the system.

1b. For each state in which one or more servers are busy, increment the transition rate from that state to all the states resulting from making idle one of its busy servers, in μ. This procedure leads to the transition matrix of the system.

2. Once the transition matrix is determined, use an iterative procedure for computing the steady state probabilities of each state of the system. The balance equations are:

$$P_k(\lambda_k + w_k\mu) = \sum_{\substack{j=k-(1 \text{ server busy})}} P_j\lambda_k + \sum_{\substack{j=k+(1 \text{ server busy})}} P_j\mu$$

where P_k is the probability of being in state k and w_k is the weight (number of nonzero componenets of the state vector) of state k. One of the equations must be removed and replaced by an equation forcing the probabilities to sum to one. The iterative procedure suggested in [737] is:

$$P_k^n(\lambda_k + w_k\mu) = \sum_{\substack{j=k-(1 \text{ server busy})}} P_j^{n-1}\lambda_k + \sum_{\substack{j=k+(1 \text{ server busy})}} P_j^{n-1}\mu$$

where P_k^n is the value of the probability P_k at the nth iteration.

Once the steady state probabilities of each state are computed, it is also possible to compute all the system's performance characteristics, that is:

1. Region-wide mean travel time, workload imbalance, and fraction of dispatches which are interdistrict

2. Workload of each response unit

3. Mean travel times to each district and of each response unit

4. Fraction of responses into each district which are interdistrict

5. Fraction of responses of each response unit that are interdistrict

6. Fraction of all responses that result in unit n being dispatched to node j.

This analytical model gives very precise measurements of all of the relevant parameters of the emergency system, without the drawbacks of a simulation. In many cases, however, a very precise solution of the problem is not justified, either because the available data is not accurate, or because there are political, legal, spatial or administrative constraints. Also, solving 2^p simultaneous equations requires extensive computational time, which is not always available. Recognizing these facts, [738] formulated an approximated hypercube model, where a system of only p equations need to be solved.

Instead of solving 2^p equations for the state probabilities, the approximated hypercube model solves p nonlinear equations for the utilization factors of the servers in the system, once their location is known. The assumptions are the same as in the hypercube, that is, the system behaves according to an M/M/p queueing discipline. The call rates of each demand node i are Poisson distributed with a known mean λ_i. Service times of all servers are identically, exponentially distributed with mean $\frac{1}{\mu}$. Furthermore, μ is assumed to be equal to 1. Servers attend calls originating at any demand node, but every server has its own primary service area, and its own ordered list of preferred servers, given by the dispatch policy and independent of the state of the system. Each call is attended by one and only one server.

According to these assumptions, an expression can be stated for the total rate of assignments per unit time of each server, R_j^T, that is, the absolute rate at which that server is assigned to an emergency call:

$$R_j^T = \sum_k \sum_{i \in G_j^k} \lambda_i P[k-1 \text{ first preferred units of } i \text{ are busy, unit } j \text{ is free}]$$

(10.45)

In this expression, the inner summation is over the set G_j^k, of all nodes i for which unit j is the kth preferred unit in the list. For each one of these nodes, its call rate is multiplied by the probability that all the servers which are higher than j in its dispatch list are busy, and server j is the first idle in that list. This represents the only situation in which server j is assigned to a call originating in demand node i, which happens with a rate λ_i. The outer summation completes the count, for all positions k of server j in the preference lists of the demand nodes.

Expression (10.45) is exact. However, it is not possible to compute easily the probability in the inner summation without having distinguishable servers. Thus, Larson uses an approximation. In order to develop this approximation, the following relaxations are assumed:

- There is no preference list for each demand node, but on each call the servers are chosen at random, without replacement. Thus, each server is equally likely to attend a call originating at any demand node, no matter how far away it is.

- The utilization factor of all servers in the system is the same, ρ.

In this "relaxed" system, it is possible to calculate the probability that a particular call will find busy the first $k-1$ servers chosen at random, and the kth server idle:

$$P_k = Q[p, \rho, k-1]\rho^{k-1}(1-\rho) \qquad (10.46)$$

where $Q[p, \rho, k-1]$ is a correction factor which accounts for dependence between servers (note that without this correction factor, the expression is identical to that in MEXCLP, obtained assuming a binomial distribution). This factor is computed by the author to be:

$$Q[p, \rho, k-1] = \frac{\displaystyle\sum_{l=k-1}^{p-1} \left\{ \frac{(p-k-2)!(p-l)}{(l-k+1)!} \right\} \frac{p^k \rho^{l-k+1}}{p!}}{(1-\rho)\left[\displaystyle\sum_{i=0}^{p-1} \frac{p^i \rho^i}{i!}\right] + \frac{p^p \rho^p}{p!}}$$

The probability in expression (10.45) is replaced by the probability (10.46), and recalling that $R_j^T = R_j(1 - \rho_j)$, and $R_j = \frac{\rho_j}{1-\rho_j}$, equations can be obtained for the iterative computation of ρ_j:

$$1 - \rho_j = \frac{1}{1 + \displaystyle\sum_k \sum_{i \in G_j^k} \lambda_i Q[p, \rho, k-1] \prod_{l=1}^{k-1} \rho_{n_{il}}}$$

where n_{il} is the identifying number of the lth preferred server of demand node i, and ρ is the average utilization factor in the system.

Once the utilization factors of all servers are computed in the approximated form, other parameters of the system can be also computed, with an error produced by the approximation, which the author estimates is no more than 2%, when:

- p is large

- no units are markedly different from others (in workload). This is probably due to the assumptions in the approximation.

The approximated hypercube gives a fairly complete, although approximate description of a congested system. Some other descriptive models have evolved, which generalize the results in the approximated hypercube; an example is the model by [637], which extends the validity of approximated hypercube to the case of general service time distributions. A later modification of the approximated hypercube by [154], allows the analysis of the case with co-located servers and dispatch ties.

Several descriptive models of a system consisting of a single server have been presented. In some cases, these models are used as building blocks for heuristic procedures which find best locations for the server in a congested system. [95] develop an heuristic algorithm to locate optimally one server on a congested network. They formulate what they call the Stochastic Queue Median. A model by [54], considers the situation in which there might be a selective rejection of calls by the dispatcher. The model is presented together with a greedy heuristic procedure for location of the server. [60] present a model and an algorithm for locating one server when there are calls of different priorities. [55] presents a model to study the effect of using expected service time dependent queueing disciplines on optimal location of a single server.

Based on the one-server location algorithm of [95], [97] developed two heuristics for locating p servers on a congested network. At each step, both use a procedure called Mean Time Calibration [740], which in turn includes solving the exact hypercube. After the Mean Time Calibration is performed, the first heuristic uses the 1-median to improve the location of each one of the servers, while the second one uses the Stochastic Queue Median for the same purposes. The authors compare both algorithms for a 10-node, three-server situation. In these procedures, interaction between the servers is allowed, and models are nonlinear.

Some of the models have considered also the problem of districting, or determination of optimal service territories. [94] solve the problem of districting for a two-server network in the presence of queueing. Given the locations of two servers on a congested network, a nonlinear model and a heuristic algorithm are developed to determine the optimal service territories of each server. Each district behaves as a $M/G/1$ system, with FIFO queues. Optimality means, in this case, minimum average response time to a random customer. No interaction exists between districts.

Later [98] use the Stochastic Queue Median, combined with this 2-server districting algorithm, to develop a general location - districting iterative algorithm for two units, and for n-nodes, m-server networks. In the first case, the best districting is found for a first location choice, followed by a re-location of both servers given fixed districts. This procedure is iterated as needed. In the case of m servers, at each step, two servers are optimally located and their optimal service areas found, while all the remaining units stay at a fixed location. No interaction exists between districts.

In general, a fairly large computational effort is required if these models are to be used for location of servers.

10.6   Conclusions

The development of models and methodological frameworks to design or reconfigure emergency systems has taken place over the span of a quarter century. It has developed alongside and in concert with the evolution of the modern computer as it transisted from a room-filling behemoth to a desktop associate. And like the computer on which the models must rely, the models and methods are not done evolving. The shape of the next models can be predicted by simply observing how the current generation still falls short of perfectly describing reality. We will focus on five areas.

First, we should begin to see a new generation of models which deal with the issue of co-location of servers from different emergency systems. [1003] introduce this line of research in the FAST (fire and ambulance siting technique) model that examines the link between ambulance and fire company siting. In the United States, ambulance deployment has traditionally taken place either at hospitals or at fire stations or both, but rarely have ambulances been positioned at free standing ambulance stations. [1003], in a deterministic covering model, examine the consequences of allowing the ambulances to be sited free of constraints on the location of other services. These models should eventually develop all the probabilistic sophistication and nuances of the models discussed above.

Second, we can expect that the estimate of server or region-specific busy fraction will be refined. Although we have moved from deterministic to redundant to probabilistic models, and although within this last category we have moved from a system-wide busy fraction to a region-specific busy fraction using queueing concepts, we still have not precisely matched the busy fractions estimated by simulation. Unless the challenge exceeds the imaginative powers of investigators, we will soon see server-specific busy fractions or more refined region-specific busy fractions.

Third, we should see focus developing on workload issues, a topic that has largely been ignored till now and one that greatly concerns emergency system planners. The issues of busy fraction and workload are tightly connected so progress on the former should bring achievement on the latter as well.

Fourth, we should see a gradual melding of the two lines of evolution, queueing and covering. It is hard to predict how this will take place, but certainly the use of heuristics offer the descriptive queueing models an opportunity to compete with the covering/availability models in the arena of design. And the introduction of queueing concepts by Marianov and ReVelle into covering/availability models suggest movement from the other side as well.

Lastly, we have a warning of competition in emergency unit deployment from a different direction than we have discussed so far. The technique of Geographical Information Systems (GIS) combined with simulation may prove to be seductive indeed. Although we know of no reported joint use as yet, these two techniques offer non- technically trained planners the ability to visualize deployment patterns and try them out - in trial after trial. Acceptable solutions may, in fact, be found in this way, but the power of optimization, especially multiobjective optimization, will be lost. We can proceed as we have been proceeding, sharpening and refining our optimization models and methods to achieve an appropriate degree of reality. Nonetheless, we may yet loose the opportunity to design real emergency systems to those who can display deployment alternatives and their consequences on wall-sized screens - unless we can combine our models with and fold our models into a geographic information system.

11
CONTINUOUS LOCATION PROBLEMS

Frank Plastria
Vrije Universiteit Brussel

Dedicated to the memory of my regretted parents.

11.1 Continuous Location

11.1.1 Context

The central concern in location problems is determining *site(s)* for one or more new facilities with respect to a set of fixed points, often called existing facilities, markets or users, sources or *destinations* (which we will use here as a generic name), with which it (they) should interact. Examples of *interaction* are movement as for transport of goods and/or service people, a physical link, like a high voltage or pipeline, a communication link, attraction of potential customers, or an unwanted influence like (risk of) pollution, radiation or detection, etc.

Such problems are termed continuous when the *underlying space* both for facility sites and given points is a *continuous* one, i.e. all points under discussion are determined by way of one or more (according to the dimension) variables (*coordinates*) which may vary continuously. As such they were appropriately termed as "site generation" models by [786], since no a-priori knowledge of particular candidate sites is assumed, and it is the model which is supposed to generate appropriate ones. This excludes the discrete case, also termed "site selection" models by the same authors, where a given finite set of candidate sites is considered, represented by discrete variables, and also the general network case where description of a point involves either identification of the node it lies at (discrete) or both identification of the containing edge (discrete) and the (continuous) position along this edge.

The special case of a linear or circular network may be considered as both continuous and network, and will in our continuous framework be described as one-dimensional. Most continuous location problems will, however, be considered in a space of dimension at least two. *Two-dimensional* problems are the most popular for evident reasons of geographical nature (even the whole of the earth as a (topological) sphere is two-dimensional), but higher dimensional problems also appear, such as within a multiple floor building or underwater for three dimensions, whereas several more abstract settings may give rise to even more dimensions.

The continuous setting results in the use of two main mathematical tools: *geometry* and *analysis*. The models are indeed mainly of geometrical nature, while the study of their properties involve both geometry (usually of convex sets) and functional (mainly convex) analysis - see [839] for an excellent introduction. For optimisation one relies on *linear/ nonlinear programming and global optimisation*, as opposed to discrete, combinatorial and/or integer programming, and in many cases the recent field of *computational geometry* can be put advantageously into use.

11.1.2 Form

The quality of the interactions is considered to be directly related to the relative spatial position of the interacting points, and usually expressed by some notion of *distance*. Thus the concept of distance is central to continuous location. However, many different distance measures may be of interest depending on the application, and the study and choice of adequate distance concepts has almost become a research field in its own right. In view of its crucial importance we will start with an overview of distance measures in section 11.2.

Once the distance measure(s) to be used has been determined, one may start comparing different sites on the basis of the quality of the involved interactions. It is traditional to represent objectives as disutilities or costs, and therefore the aim will usually be minimisation. If closeness is valued as will be the case in our four first examples of interactions cited above, this (dis)quality will be nondecreasing with distance and the interaction is called *attracting*. For interactions of (ob)noxious or hazardous type, the (dis)quality will be nonincreasing with distance, and we will call it *repelling*. The location problem consists of the minimisation of the (dis)qualities of all interactions.

Thus the initial paradigm is multi-objective. Since the qualities directly depend on distance, and this dependence is always assumed to be monotonous, we face a *vector optimisation* problem, where each component is represented by an individual distance. Such problems are usually considered to be solved by the determination of the Pareto-optimal, or *efficient* solutions. The work in this field is discussed in section 11.3, together with related voting concepts of particular interest in competitive situations.

It is well-known that the set of efficient solutions is seldom directly useful in practice, because it usually is too large and may contain many unsatisfactory solutions. In practice one will often need a more precise description of the actual aim of the problem described by a *globalising* (or utility-) function which combines all individual distances into one global value, viewed as the overall quality of the solution. Thus the optimisation problem has become single objective and the notion of optimality well defined.

Such globalising functions may take many different forms, depending on the application context, yielding a wealth of different location problems, many of which remain as yet unexplored. As a main classification factor we will distinguish between pure *location problems* where the only aim is the determination of an optimal site(s), and *location-allocation* problems in which other aspects, such as the determination of the "active" interactions, i.e. those which directly influence the global value of the solution, are an important part of the solution to the problem at hand. Evidently this distinction is not always clear-cut, viz. the globalising function consisting of the largest among all distances may be either viewed as a pure location problem (as we will do here), when the knowledge of which distance is actually the largest is not relevant, or as a location-allocation problem when this knowledge is considered important. The second classification factor is somewhat more precise : the number of facilities to be located, primitively subdivided into one (*single facility*) and several (*multifacility*). Thus we discuss "pure" single facility location problems in section 11.4, single facility location-allocation problems in section 11.5, and their respective multifacility versions in sections 11.6 and 11.7. This naming convention is classical, except for the term single facility location-allocation problem, which is novel and proposed here to the research community. Section 11.8 groups some types of related problems that we were unable (or unwilling) to classify in either of the previous four groups.

11.1.3 Status

After some pioneering work in the first half of the century, locational analysis, as a field of scientific research, reached momentum in the early sixties, and is steadily growing since then. In particular the study of discrete and network location problems have known a tremendous period of growth. Although historically older, with roots going back to the eighteenth century, continuous location theory has known a less rapid, but steady, development. Figure 1 shows the number of publications per year since 1960 contained in a bibliography restricted to continuous problems and directly related fields, as surveyed below. It was started in 1980 and regularly updated since then, and contained around 1600 references at the time this chapter was written. The trend clearly shows that the continuous location field is in full progress, a fact we demonstrate in this chapter. That the subject has grown into a mature field of research has recently been accredited by the prestigious Mathematical Reviews and Zentralblatt für Mathematik / Mathematics Abstracts by including it in their 1991 Mathematics Subject Classification under number 90B85 - Continuous location, one of the 18 titles within the section 90Bxx Operations Research and Management Science.

Although attempted, the bibliography mentioned above is probably far from complete, and many references therein were not available to us. Even

FIGURE 1. Papers on continuous location and related topics (1960-1993)

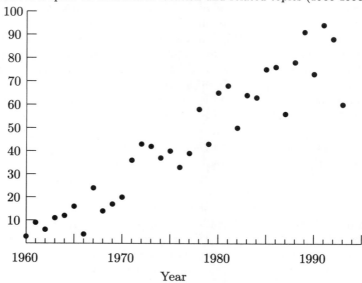

so it would have been an impossible task to attempt to describe all results we are aware of, and a heavy selection had to be made, with preferential treatment of the more recent work. Therefore this chapter must be seen as a partial and highly subjective view of the state of the art in the summer of 1994. We apologize to any fellow researcher who would feel his work was left out. Over 400 titles, some of which, unfortunately, still unpublished are cited in this chapter. We felt this mostly very recent work should not go unmentioned and feel confident that most of it soon will be available in print.

Other surveys, such as [557, 545, 385, 127] will complement this chapter. More details may be found in [786], to which we refer in general for most of the history and now classical results. For an overview of literature on applications the reader may consult [585].

11.1.4 Trends

Continuous location has long suffered from a consensus that it is merely theory without much practical use, more mathematical programming than operational research. This is no longer true, if it ever were. During the last decade the main trend has been directed towards bringing the theoretical models closer to the realities met in practical applications. Indeed, whereas traditional models were unconstrained, most recent work takes into consideration the fact that in practice feasible sites are always confined to only part of the underlying space. The expansion of the family of distance measures and/or the classes of globalising functions, still enabling actual solution of the resulting models, strongly enlarges the modelling possibilities and hence allows to fit the models more closely to the real-world.

The recent interest and research on the location of (ob)noxious facilities is not just fashionable environmental awareness, but rather reflects the fact that the annoying, if not dangerous or lethal effects are of continuous spatial nature. Continuous location theory is the instrument by eminence for modelling and supporting locational decisions in the environmental field.

At the other hand several of the classical models are still under scrutiny in order to deepen the understanding of their properties and, for some of the more notoriously difficult ones, the search is still vivid towards satisfactory solution methods.

11.2   Distance

11.2.1   Assumptions

A distance measure on a space X attributes to each pair of points x, y of X a real value $d(x, y)$. The most popular (basic) properties connected with a suitable distance measure are those required for a *metric*:

- *nonnegativity*: $d(x, y) \geq 0$ for all $x, y \in X$.
- *stationarity*: $d(x, x) = 0$ for all $x \in X$.
- *definiteness*: $d(x, y) \neq 0$ for all $x \neq y \in X$.
- *symmetry*: $d(x, y) = d(y, x)$ for all $x, y \in X$.
- *triangle inequality*: $d(x, y) + d(y, z) \geq d(x, z)$ for all $x, y, z \in X$.

The two first properties are generally accepted, and the third more relates to the discussion of what is a point: when two points are at zero-distance, redefine them as one point, thereby recovering definiteness. The last property is directly related to the idea of distance as minimal path length, and is then a direct consequence of the relaxation principle: the shortest way for going from x to z ("directly") is certainly not longer than the shortest way restricted to passing through some given y. This "evident" property does however not always hold, e.g. for cognitive distances, measuring the subjective view on distance by people (see the nice overview paper [933] on distance notions in spatial analysis). It is however the symmetry that has been most often attacked as unrealistic, since there exist many circumstances in which movement is not reversible without change, e.g. one-way streets in a city, travel towards or from a traffic-congestion point, movement up/down-hill, in the presence of currents and wind, etc. [956, 960].

For optimisation purposes some more properties of the distance measure are often required (only applicable when X is a vectorspace); e.g.

- *convexity:* $d(x,.)$ is a convex function for all $x \in X$.
- *analytic expression:* $d(x,y)$ is given by a closed analytical formula, only involving the coordinates of x and y.
- *linearity:* $d(x,y)$ is a piece-wise linear function of the coordinates of x and y.

The convexity assumption is needed in order to enable the use of convex analysis, the main analytical tool in nonlinear optimisation [839]. An analytical expression of distance is useful since it allows for application of the classical tools of calculus, while linearity leads to linear programming problems for which optimisation techniques and programs abound.

These assumptions are clearly dictated by the pragmatic wish to be able, without too much effort, to tackle problems. But in actual applications these properties are, as a rule, not met, as clearly indicated by [933]. Therefore some research has been devoted towards comparing and approximating actual distances, stemming from real-world situations like road networks, with theoretical distance measures which are (mathematically speaking) "better behaved" [784, 79, 138]. As a consequence, models substituting the actual distance measures with theoretical distances are necessarily approximations to the problems under scrutiny, thus invalidating the "optimality", strictu sensu, of the solution(s) found. This situation calls for new tools which would allow for direct use of actual distance measures, or/and for more attention to sensitivity analysis of the obtained solutions, a question we will come back to in the sequel.

With respect to the convexity assumption, its combination with the basic properties (without symmetry) was shown by [1215] to imply that d is derived from a *gauge*. This is a realvalued function g defined on a real vectorspace X satisfying:

- nonnegativity: $g(x) \geq 0$
- definiteness: $g(x) = 0$ iff $x = 0$
- positive homogeneity: $g(r.x) = r.g(x)$ when $r > 0$
- sublinearity: $g(x + y) \leq g(x) + g(y)$.

It is entirely determined by its unit ball $B = \{x \in X \mid g(x) \leq 1\}$ by the Minkowsky relation $g(x) = \inf\{r > 0 \mid x/r \in B\}$, which may be any compact convex set the interior of which contains the origin. A distance measure is derived from g by $d(x,y) = g(y - x)$, hence by translation from x to the origin. If in addition symmetry of distance is assumed, then g will satisfy $g(-x) = g(x)$ or, equivalently, its unit ball B will be centrally symmetric w.r.t. the origin, and g will be a *norm*.

11.2.2 Norms and Gauges

The most familiar and widely used distance measure is the *euclidean* (sometimes confusingly referred to as rectilinear). It is derived from the euclidean norm $g(x) = \sqrt{\sum_{i=1}^{n} x_i^2}$, where $x_i (i = 1, \ldots, n)$ is the i-th coordinate of x in $\mathcal{R}^n$. It is *isotropic*, i.e. rotation invariant, and applies when movement is allowed homogeneously in all directions. It possesses all properties mentioned except for linearity. This means that problems involving euclidean distances are usually handled analytically, but in some cases pure geometric techniques may apply: in the plane the unit ball is a circle, a well-known and easily handled geometrical object.

The second topper is the *rectangular*, city-block, taxi, Manhattan or Hamming distance (also dangerously mentioned as rectilinear, a term sometimes rather interpreted as euclidean), derived from the rectangular norm $g(x) = | x_i |$. Its use is mostly rationalized by reference to movement in a perfectly rectangular network, being the length of a shortest path consisting of pieces parallel to either one of the (arbitrarily chosen) coordinate axes. The actual reason for its popularity is in my view to be sought more in its nice linearity properties, and hence the ease with which it may be handled in optimisation, since nowhere does a road network have the perfect orthogonal structure it presupposes. Recently it has however found a very nice and quite adequate application in the field of robotics [417, 421] when movement is restricted to two directions only, generated by one motor each. If the motors are not allowed to work simultaneously, and their constant speeds are equal, the minimal time necessary to move from one position in the plane to another is given by their rectangular distance.

If the motors may work independently and simultaneously, we will obtain another time-distance measure given by the longest of the two displacements. This is the *maxi* or Tchebycheff distance derived from the norm $g(x) = \max_{i=1}^{n} | x_i |$, of which the above seems to be the first application in the location field. In the plane $\mathcal{R}^2$ the maxi-distance may be seen as rectangular after a change of coordinates (rotation over 45^0 and scale divided by $\sqrt{2}$), thereby explaining perhaps why it did not receive more attention. In higher dimensional setting this equivalence breaks down, however, and the study of maxi-distance related problems could be interesting and probably not too hard in view of the linearity property which it possesses [926].

All of the above mentioned distances are part of the family of ℓ_p-distances, derived from the ℓ_p-norms $g(x) = \ell_p(x) = \sqrt[p]{\sum_{i=1}^{n} |x_i|^p}$, for $1 \leq p < \infty$. For $p = 1$ we obtain the rectangular norm, hence usually denoted by ℓ_1. The euclidean norm is ℓ_2. Taking the limit for $p \to \infty$, the Tchebycheff norm ℓ_∞ arises. We thus have a continuous and decreasing family of norms $(\ell_p(x) \leq \ell_q(x)$ when $p \geq q$, where equality is reached only for points on the coordinate axes) which includes all the classical cases mentioned above. Due to their fairly simple analytical expression, and stimulated by approximation studies hinted at before and discussed below, much of the current research directly assumes ℓ_p-distances. A disadvantage of these measures (except ℓ_2) is however their direct dependence on the choice of the coordinate axes, which is often simply ignored and hence totally unrelated to the actual problem [605] for the ℓ_1-case).

General gauges with the linearity property are called *polyhedral* gauges, and were already introduced in location theory by [1214, p.48] being probably one of the papers most often referenced to in later continuous location work. The unit ball of these gauges is a convex polyhedron (polygon in the plane). The derived distance arises when movement is restricted to a finite set of (oriented) directions, defined by the vectors from the origin to the vertices of the unit ball, with speed proportional to the (euclidean) length of these vectors. Symmetric polyhedral gauges are often called *block norms* after [1173]. They correspond also to the A-distance defined by [1202].

Recently [960] proposed skewed norms as another family of gauges, applicable to asymmetric distance situations. For any norm N and vector p (which should not be too big, more precisely the dual norm $N^d(p)$ less than 1) the formula $g(x) = N(x) + \langle p, x \rangle$ (where $\langle ., . \rangle$ denotes scalar product) defines a gauge. Taking N as the euclidean norm or linear transformation of it (i.e. choosing a possibly skew alternate coordinate system) one recovers all ellipsoïdal gauges, i.e. those having an ellipsoid as unit ball, the origin being any interior point. This covers previously proposed applications like movement on an inclined plane [587] or flight under a steady wind.

11.2.3   Fit

The question of how to best fit real-world (mostly road-network) distances by theoretical ones has been centered mainly towards ℓ_p-distances. [782, 784] have done the pioneering work in this respect. They tried out two different accuracy measures (sum of absolute, resp. squared normalized, deviations) on several models, one being an inflated ℓ_p-norm, thereby showing that values of p other than 1 or 2 could be useful. Their results, valid for a sample in the USA, were somewhat attacked by [80] (see also [785, 81] and the related work by [874, 875] based on a similar study for Germany, and later for several regions worldwide [79], in which it appeared that simply inflating euclidean distance by a fixed factor (to be estimated), a so called *detour-factor*, gave almost as good a fit as the additional estimation of p in the inflated ℓ_p-model, except, strangely, for the USA). [79] comes to the conclusion that the accuracy improvement obtained by this additional degree of freedom cannot be predicted by way of some simple network-characteristics such as road density, area of the region, size of detour-factor or relative accuracy of the detour-factor fit.

In view of the fact that such an approximating theoretical distance differs from region to region, it will have to be re-estimated in each practical case. Therefore the simplicity of the fitting process may be more decisive than the accuracy obtained, if within acceptable bounds, and the detour-factor model has a clear advantage since it relies on simple regression, while a cumbersome grid search method was used for the more involved Love-Morris model, which [137] replaced by a more efficient but still discrete search technique. In fact, by the orientation-dependency of the general ℓ_p-norms this latter should include also the estimation of the appropriate coordinate system, leading in $\mathcal{R}^2$ to a model with five degrees of freedom (inclination and scale for each of both axes and value of p) instead of two. An empirical study in this direction is performed by [788], who also consider some types of block norm approximations, and conclude that these are less accurate than those obtainable within the family of rotated and inflated ℓ_p-norms. See also [141] for a rotation fitting technique and Chapter 1 for more details.

Other distance fitting models have been proposed. [874] tries combining two different ℓ_p-norms, [1171] combine ℓ_1 with ℓ_∞, in their "one-infinity" norm proposal and [1173] for somewhat more general block norms, and [138] combine ℓ_1 with ℓ_2, fitting of which is amenable to relatively simple regression methods [136]. The remark about rotational dependence still holds for all of these proposals and seems quite worthy of further investigation, as would be consideration of other types of theoretical distance measures.

In an attempt to model the fact that the route taken for movement between two points is not necessarily always the same, [147] propose to define distance as a random variable. More precisely they consider ℓ_p-distance with random p, and show that the expected distance may then be closely approximated by the ℓ_p-distance for some fixed p.

A relevant question is also the approximation of a given norm by a member of a given family of norms, and its quality. Worst-case analysis leads to comparison bounds as used by [956, 408]. Average (or median) case is studied by [183] for any pair of ℓ_p-norms. [140] investigate the "bias" of the ℓ_p-norm with respect to the euclidean norm in function of the point of evaluation; they conclude that any ℓ_p-norm with $p > 2$ may be approximated sufficiently closely by another one with $1 < p < 2$, after a 45^0 rotation.

As a last remark on the fitting topic, it seems quite surprising that none of these studies adresses the question of fitting the distance up to a fixed destination point which are exactly the types of distances used in location! It seems intuitively clear that better fits would be obtained, and the main access-roads to such a point (city) would give a (at least heuristic) clue to the orientation problem, or the main directions of block norms/gauges. Of course, from the viewpoint of locational modelling, such an approach will necessarily lead to *mixed-distance* models, where the distance measure differs with destination, thereby increasing the complexity of the models, which remain, however, often still solvable.

11.2.4   Paraphernalia

Several types of nonconvex distance measures have been proposed in the context of location. [934, 935] introduce metrics in the plane, suited for local situations around a center with either radial or circumferential structured network or a combination of both. [1095] looks at the Jaccard-metric, extending to the positive orthant $\mathcal{R}^n$ a dissimilarity measure for binary vectors used in cluster analysis, but indicates no direct application field. [778] find that the ℓ_p-distance formula with $p < 1$ may be useful in facility layout situations, although these "hyperrectilinear" distances are not metrics and fail to obey the triangle inequality. [859] look at the continuous version of circum-radial metric [934], by considering the "ring-radial" distance metric obtained when movement is restricted to straight lines emanating from a fixed center and circles centered at the same point. A similar but still different crane-distance, is obtained when considering the time necessary for a stationary crane to move from one position to another [1013]. In the area of robotic movement other quite complicated distance concepts abound, the study of which seems to be largely untouched in location context, notwithstanding their importance in modern automatisation.

[62] consider situations in which several subregions allow for different metrics, and links between them allow to move from one to the other. A more or less inverse distance model has been studied by [893] in which high-speed corridors (such as highways) are superposed on a rectangular distance space. Such mixed planar/network models have great application potential and deserve more attention.

Another related distance notion deserving more popularity in the location field, but already quite well-studied for motion-planning in the field of robotics, is to consider shortest paths in an heterogeneous environment, defined either by the presence of obstacles or by regions allowing for different speeds or travel costs. The location problems considered have until now been restricted to the former easier case, either using rectangular [59], or euclidean movements [989, 918]. For the latter, more general case, the problem of how to determine distance algorithmically seems to have been satisfactorily settled only recently by [748] for the rectangular and [858] for the euclidean case. The consideration of location problems in such an exciting and quite realistic environment is certainly strongly suggested.

All distance measures discussed above relate to the plane, or at least a linear space. The Earth as (approximately) a sphere is topologically quite different. The most appropriate distance measure on a sphere is the great-circle distance, if necessary improved by the geodesic distance taking the slight non-sphericity of the earth into account. The results of [782] indicate that for relatively small regions, such as Wisconsin (or even the whole of continental US) a fitted planar distance (as discussed above) gives better results than a (more simple) spherical distance fit. For wide regions, or the whole earth it is however not possible to use planar models and location problems on the sphere regularly appear in the literature [307].

A more exotic underlying space is the torus, location problems on which have recently found an application in optimal grid positioning [954] where ℓ_p-type distances may be used.

One powerful tool related to distance measures, with strong application potential for location is the Voronoi diagram of a finite point set, i.e. the partitioning of the space (usually plane) into points closest to each of the given ones, and its numerous extensions. The study of the efficient determination, coding search problems have been at the center of computational geometry [972, 362] and were recently brilliantly surveyed by [39]. Properties of Voronoi diagrams for general metrics are discussed by [699]. Chapter 6 gives an overview of their applications in continuous location.

11.3   Dominance, Efficiency and Voting

11.3.1   Pareto

For a destination a in a space X, to which distance is measured by d_a, a site $x \in X$ is *no worse* (*better*) than some other site y if $d_a(x, a) \leq$ (resp. $<$)$d_a(y, a)$. Two points are always comparable for this destination.

When however several destinations are considered, as is usually the case in location theory, comparison is not so simple: often one site is better for some while the other site is better for others. When will we say that one site is better than another? Several different answers are possible.

The first one is to ask that the comparison comes out the same for all destinations. This is the traditional Pareto view on such multi-objective problems, and leads to the following notions, where we take some possible regional constraints into account, by specifying a feasible set $S \subset X$.

With respect to some set of destinations $A \subset X$, often supposed to be finite, and for each $a \in A$ a distance measure d_a up to a, one says, comparing two sites $x, y \in S$, that

- *x dominates y*   iff x is no worse than y for all $a \in A$
- *x strictly dominates y*   iff x dominates y and is better than y for
                             at least one destination.
- *x strongly dominates y*   iff x is better than y for all $a \in A$.

Consider subsets $T \subset S$ which seem more promising to contain "good" sites, in the following way: T is a *dominator* (of S) if for any $y \in S \backslash T$ there is some x in T dominating y. Similarly T is a *strict* (resp. *strong*) *dominator* if any $y \in S \backslash T$ is strictly (resp. strongly) dominated by some $x \in T$.

At the other hand one may consider the "best possible" points in S as follows: A site $x \in S$ is *efficient* (or Pareto-optimal) if there is no other site $y \in S$ strictly dominating it. It is called *weakly* (resp. *strictly*) *efficient* if there is no other site $y \in S$ which strongly dominates (resp. dominates) it. We will denote the sets of weakly (strictly) efficient sites w.r.t. S by $E(S)$, (resp. $WE(S)$ and $SE(S)$) and omit mentioning S when no constraints are present, i.e. $S = X$. It is clear that $SE(S) \subset E(S) \subset WE(S)$.

The study of these notions in location theory was initiated by [723] showing that for A finite in $\mathcal{R}^n$, all distances euclidean, and no constraints ($S = X = \mathcal{R}^n$), the convex hull of A equals SE (which he called the set of admissible points, also called the metric hull or set of minimal points [69, 915]. Weakly efficient points were studied by [397] in the context of approximation theory and by [940, 941] in the geometry of normed spaces.

Subsequently most research has been directed towards unconstrained problems with a fixed norm and finite A. [1187] prove that the convex hull of A is always a dominator for any fixed norm in the plane and on a line, and proceeded in [1188] showing that this convex hull contains E for any ℓ_p-norm $(1 < p < \infty)$ in the plane. In fact [948] characterizes all norms for which E is equal to the convex hull of A as follows: any norm on the line, any *round norm* (i.e. with unit ball without nontrivial line segments on its boundary) on the plane, and any *ellipsoidal norm* (i.e. euclidean after change of coordinates, or, equivalently, induced by some scalar product) in dimensions higher than 2. Furthermore, all three efficiency concepts coincide when the norm is round [949]. [1131] obtain the same result in the plane, where these sets then coincide with the convex hull of A.

The situation is much different when the norm is not round, in particular for block norms. [346] show that in this case the efficient set is a union of "elementary convex" sets, i.e., in the plane, the pieces (their boundaries, etc.) - or complex - in which the space is cut when drawing lines through each destination, parallel to the travel directions. The determination of E for the ℓ_1-norm may be done in $O(n \log n)$ time by the method of [185] and was further studied by [1172] for their one-infinity norm and for a general block norm by [1131, 928, 929] the latter also considering WE and SE. It turns out that as a rule these all differ, and remarkably only depend upon the travel directions and not the speeds!

In the presence of mixed norms the exact determination of the efficient sets seems much harder, except of course when all norms are multiples of a fixed norm, which has no influence on the dominance concepts, and thus reduce to fixed norm problems. The known results are restricted to the plane and mainly concern domination, initiated by [554] who introduce the *octagonal hull* of A as the smallest convex polygon containing A with sides parallel to the axes or main diagonals, and show that it is a dominator for mixed ℓ_p-norms. [173] show that E is then always contained in the octagonal hull of A, and indicate when this bound is reached. Afterwards they generalize these results to general norms and "polygonal hulls" of A, which reduce to the octagonal hull when all norms are symmetric around the axes and the main diagonals (which includes all ℓ_p-norms) [174].

For mixed block norms (and even gauges) the results of [346], stating that the efficient sets are a union of elementary convex sets, still hold, and [343] indicates some rules to be taken into account for their determination. Since this does not lead to an efficient algorithm, [175] indicate how to construct a superset, which is sometimes exact but degenerates to the whole plane in the general case. In such a case [951]'s rather crude rectangular dominator, valid in any mixed norms case, may prove to be useful and directly extends to gauges and any dimension. Thus the determination of efficient sets for mixed norms problems in general still leaves much room for further research and even more for fixed or mixed gauges problems (not mentioning nonconvex distances), for which virtually nothing is known, except for the polyhedral case, as indicated above, and some "convex hull" type properties on the sphere with spherical distance [21, 295].

Until recently the consideration of constraints in these multi-objective problems, had been neglected. It seems that the first rather vague results ($E(S)$ is visible from E) were obtained by [948], but never published. It is only since 1992 that research on this topic really started, mainly in Sevilla, Spain. [167] show first that for the euclidean norm (or any ellipsoidal norm) and convex constraint set S, the sets $E(S)$ and $WE(S)$ coincide and equal the orthogonal projection (according to the scalar product corresponding to the norm) on S of the convex hull of A, and [172] proceed to study the nonconvex case since "assuming the feasible region to be convex amounts to ignoring geographical and legal reality" [385]. This is done using ccd's for S, i.e. coverings by closed convex sets, leading to explicit constructions for polygonal ccd's. The former result is extended to a general round norm by [179] replacing orthogonal projection by closest points (in terms of the norm considered). Clearly this avenue of research is only just starting and much remains to be done, e.g. block norms, well studied in unconstrained cases, have not yet been touched upon.

Up to this point we only discussed the notions of dominance and efficiency for *attracting destinations*: nearness to the central facility was viewed as an advantage. However, with the current trend towards environmental concern many facilities are viewed as noxious, and nearness to them is felt as a disadvantage. Therefore one should also consider *repelling destinations*, and this leads to novel problems, by inequality inversion when deciding that a site is better than another. Obviously in the plane these problems are only of interest in constrained context, since otherwise no such anti-efficient points exist, any site being dominated by even further sites. To our knowledge the only work in multi-objective noxious facility location in terms of dominance and efficiency has been done by [166] leading to some nice and instructive constructions, some of which may be more readily found in [177]. We expect that the near future will see more work in this direction, including mixed attracting and repelling situations (also called "semi"-obnoxious) of even greater practical interest, and as yet totally unexplored.

11.3.2   Condorcet, Simpson, et al.

A second way of deciding that one site is better than another stems from the theory of voting and has direct applications to situations with competition. Assigning a normalized weight to each destination, e.g. its population relative to the total population, and setting an acceptance level α ($0 < \alpha \leq 1$), we say that $x\alpha$-dominates y iff the total weight of all destinations for which x is better than y is at least α, or, in other words, x is preferred to y by a fraction of at least α: strong dominance may then be seen as 1-dominance. By analogy we may talk about α-dominators and α-

efficient points, 1-efficiency meaning weak efficiency. Clearly α-dominance implies β-dominance for any $\beta < \alpha$, and hence in this case the α-efficient set contains the β-efficient set.

A $\frac{1}{2}$-efficient point is generally known as a *Condorcet-point*, and may be seen as an optimal site in terms of simple majority voting: no other site is preferred to it by a majority of voters. In most cases however no Condorcet-points exist, and one has to choose $\alpha > \frac{1}{2}$ in order to obtain α-efficient points. Taking for α the smallest of such fractions, the corresponding α-efficient points are called *Simpson-points* after [1081]. These are sites a minimal number of voters will be against, when compared to any other site, or, viewed as a competitive situation, the points at which a first firm should locate in order to loose as little as possible to a competitor locating anywhere next, assuming nearness only determines the markets.

A limited number of researchers have addressed the question of existence of Condorcet-points and determination of Simpson-points in the continuous location context: [963, 1190, 815, 715, 1189, 296, 274, 275, 342, 168, 169, 170]. [274, 275] (and [342] in the more general setting of a distribution of users) show that Condorcet-points exist for any destination set in the plane, only when distance is measured by a block norm with exactly two directions, i.e. an ℓ_1-norm possibly after coordinate change. [274, 275] also states some conditions on the destinations in order to have a Condorcet point for ℓ_p-norms, and gives some counterexamples. [342] indicates explicit algorithms for constructing the Simpson-set for a general block norm in the plane. [168, 169, 170] investigate these questions in the presence of a closed convex locational constraint. With ℓ_1-norms constrained Condorcet points always exist and are found as closest point projections of the unconstrained ones [168]. How to determine constrained Simpson points when other distance measures are used, is described in [169, 170].

[912] take a different view on spatial voting and define a strong point as that position of a candidate which minimizes the probability to be defeated by another candidate, whose position is chosen at random. They show that under very general conditions in the plane there is a unique strong point, which may be found as the centre of gravity of the voters (destinations) with special weights. Another notion connected to voting theory is the *yolk* introduced by [836]: it is the smallest (euclidean) ball intersecting all median hyperplanes. Recent results and a literature list on this subject is found in [1143].

These voting topics leave ample room for research. Other approaches to competitive location models are hinted at in sections 11.5 and 11.7.

11.4   Single Facility Location Problems

11.4.1   General

General single facility location problems are obtained as follows: Let X denote the *space of sites* and A be a finite *family of destinations*, each having a location in X. Although in some cases we may have two destinations at the same location, we denote location by the same notation as the destination. In stochastic versions the set of destinations A will sometimes be infinite, as described by some continuous distribution.

For each destination $a \in A$, d_a denotes the *distance measure* up to a. To any site $x \in X$ a *vector of distances* $D(x)$ is then associated, with components the distances of x to each destination: $D(x) = (d_a(x, a))_{a \in A} \in \mathcal{R}_+^A$. These distances are combined by a *globalizing function* $G : \mathcal{R}^A \to \mathcal{R}$: $t = (t_a)_{a \in A} \to G(t)$ into a global objective value assumed to be a disutility, hence to be *minimized* within a given *feasible region* $S \subset X$. In short, we face the problem

$$\text{MIN } \{G(D(x)) \mid x \in S\}, \tag{11.1}$$

any optimal solution of which will be called *optimal site*.

The most studied globalising functions are of following two kinds:

1. G decomposes into a sum of one-dimensional functions, i.e.: $G(t) = \sum_{a \in A} g_a(t_a)$, in which case (11.1) is called a *minisum* problem,

2. G is a maximum of one-dimensional functions, $G(t) = \max_{a \in A} g_a(t_a)$, and we obtain a *minimax* problem.

Viewing $g_a(d_a(x, a))$ as a *functional distance* concept [933] such as travel time, transport cost, etc., the former corresponds to minimising average functional distance, a typical centralised strategy [1181] while the latter may be seen as an attempt to justice by minimising the situation of the worst-off destination, a criterion proposed by [985]. Note that in applications in economics, where functional distance is taken to be transport cost the standard assumptions on g_a is to be nondecreasing and concave, although most models studied assume linearity or convexity in order to be solvable.

Of course, many other possibilities exist, some of which will be indicated in the sequel, including section 11.5 where we artificially set apart some particular choices of G which include other aspects than the mere determination of an optimal site. For a study of axioms leading to different types of globalising functions see [176].

Usually a destination $a \in A$ is either attracting or repelling, and this will be reflected by G being nondecreasing or nonincreasing in t_a, respectively. Under suitable monotonicity and continuity conditions on G optimal site(s) exist and are (anti-)efficient points to the corresponding multi-objective location problem, by extension of the results of [951] to include repelling destinations as well.

Methods to solve (11.1) in this broad context include *(quasi) convex programming* when X is a linear space, the d_a being gauges and G a (quasi) convex function (see [194] for a subgradient method, [952] for cutting plane methods, [426, 427, 428] for an ellipsoid method, [839] for a primal-dual proximal method, [180] for a grid- approximation technique) and *global optimisation* when X is the plane, the d_a gauges, and G Lipschitz and boxwise optimisable (e.g. monotone in each component) [957]. See also Chapter 3. This latter method allows for the additional determination of a region of near-optimal sites, of particular interest in practice, at the one hand when the actual distance measure was approximated by a "nicer" theoretical distance, and further data are not necessarily very precise, and at the other hand because this is the type of sensitivity analysis decision makers call for. Most published work however concerns more restricted settings, as reviewed below.

Uncertainty may appear in location models in different ways. At the one hand the exact location of the destinations may be unknown, or there may be a huge number of them clustered in some regions, and it is not possible to take them individually into account. At the other hand the amount of interaction (e.g. demand) is often varying with time and may not be considered as fixed. Of course both types of uncertainty may simultaneously be present. This leads to various types of stochastic versions of location models, a general overview of which is given by [771].

11.4.2   Minisum problems

11.4.2.1 Fermat-Steiner-Weber

The simplest problem in this class is obtained when G is taken as linear, $G(t) = \sum_{a \in A} w_a t_a$, yielding the Fermat-Steiner-Weber problem, shortly called WP. This problem has, by far, attracted most of the research in continuous location, by setting it in various contexts. The coefficients w_a are traditionally called *weights*, and are usually assumed to be positive.

In the very general setting of a fixed metric, [1214] showed his celebrated *majority theorem*, stating that when one destination has a majority weight (at least half of the total weight of all destinations), then its location is an optimal site. Recently [955, 198] showed that this result still holds in mixed attraction and repulsion (negative weights) situations, when the majority is held by an attracting destination. The majority theorem was also adapted by [956] to asymmetric distance measures, redefining the majority concept to reflect the "amount" of asymmetry.

The original WP is in the plane, using euclidean distance, and has a history dating back to Fermat in the 17th century, see [588, 1051] for the early history and [786, 1194] for later historical developments. Even in this seemingly simple case direct analytic solution is impossible, as shown by [79], except for some particular situations: up to four destinations with equal weights [1194], up to three destinations with general weights [938], when one destination comes sufficiently close to holding (or holds) a majority [1186], and finally when the destinations are colinear, and the problem reduces to the simple one-dimensional case (see below).

One of the main difficulties of the WP is due to its nondifferentiability, occuring at least at each destination point, due to the fact that a norm is never smooth at the origin. This leads to the optimal solution often occuring at a destination, when the corresponding weight is sufficient to counteract the pulling influence of the other destinations. Exact conditions for such destination optimality were investigated by [1186, 724] for the classical WP, by [650] for mixed ℓ_p-norm problems and by [948, 949] for general mixed norms and convex nonlinear cost functions g_a.

WP may be solved by an analog method, the Varignon frame [786, p.27] and by any convex programming method. By far the most popular technique of this kind is the method of [1186], which may be viewed as a calculated step gradient method. Basically one ignores nondifferentiability, and partly separates the unknown variables out of the first order optimality conditions, yielding a fixed point interpretation, which is solved in the classical iterative way. After several independent rediscoveries, [724] pointed out a weakness in Weiszfeld's method in that it gets stuck at a destination if it arrives there, and proceeded to show how to adapt the method by constructing the steepest descent direction at such a point. A complete convergence proof of this method is given by [910, 617]. The first of these authors introduces a step-length parameter into the algorithm, which may be chosen freely between 1 and 2, possibly different at each step, without impeding convergence. How to choose good values for this parameter is investigated empirically by [310]. [201, 202] indicates how to adapt the method to the situation where the sum of powers of euclidean distances should be minimized. [189] raise some still open questions about the size of the set of points from which a destination is reached, and in [190] indicate the theoretical possibility of polynomial-time algorithms.

In case of squared euclidean distance one obtains a "minimal sum of squares" model, classical in statistics, which is directly solved by the centre of gravity, and whose level sets are circles around it [1199].

When the rectangular distance is used the Weber problem is extremely easy to solve, since the objective function decomposes into two independent one-dimensional instances. The Weber problem on a line is simply solved by any *median* point (considering weights of destinations as frequencies of observations) a property still holding for a continuous weight distribution. Discrete medians may be calculated in time linear in the number of destinations. [425] thoroughly study the shape of the objective in planar rectangular distance WP and derive its contour lines. Recently [78] considered WP with rotated axes in the calculation of rectangular distance.

Similar simplicity is obtained when polyhedral gauges (or block norms in case of symmetry) are used as distance measure, since the WP may then be formulated as a linear program, see e.g. [1173]. [345] show that the set of optimal solutions is an elementary convex set, even with mixed gauges and in any dimension, whence there always exists an optimal solution at an intersection point (being the extreme points of elementary convex sets).

Following the [782] study of approximated distance measures, WP has also been extensively studied for ℓ_p-norms [786], often with the g_a being convex power functions. Weiszfeld's algorithm is usually adapted to this case, either directly, or using some hyperboloid approximation rendering the distances everywhere differentiable, and linear rate convergence was proven for any fixed p with $1 \leq p \leq 2$, while counterexamples for convergence exist with $p > 2$ [870, 139, 142]. This method and results are extended by [432, 433] to the more general case of any globalising function which is nondecreasing, differentiable and has a property related to quasiconvexity.

Properties of Weber problems in higher dimensions and general norms have been studied by [239, 184]. Solutions may be obtained by the primal-dual proximal point algorithm of [840] and the cutting plane method [952].

One of the practical difficulties with such iterative methods is to determine the precision obtained in order to know when to stop the calculations. Therefore several lower bounds on the objective value were developed, an overview of which is given by [779]. The more generally applicable primal-dual method of [840] and cutting plane method of [952] automatically generate such lower bounds, the latter also allowing for an evaluation (in the plane) of distance to the optimal solution in spatial terms, as opposed to distance in value terms obtained from lower bound calculations.

[652] take up the WP with hyper-rectangular distance measure and prove an intersection-point optimality theorem similar to the rectangular norm case. [859] find an analogous property for ring-radial distance WP leading to an $0(n^2)$ finite solution method; the same result holds for crane-distance and Jaccard metric WP, as shown and exploited by [1013] at the one hand, and by [1095, 1176] at the other hand.

The case of asymmetric distance measures has been studied theoretically by [345] and applied by [587, 333] using a Weiszfeld-type algorithm. [204] gives an improved method for the Hodgson et al technique, thereby indicating that their destination optimality conditions are incorrect. A complete analysis of these conditions, including the Drezner-Wesolowsky proposals, who simply ignored the destination optimality case, is given in [959].

[930] study conditions under which minisum and minimax single facility problems with general cost functions have a unique optimal solution.

It appears that the cost functions claimed by economists, i.e. concave or even discontinuous, have received only scant attention, most probably due to the possible existence of local minima [239, 870]. [616] consider block norms and show that the intersection point property remains valid for concave cost functions. They propose a finite descent method yielding a local minimum, while [1151] discuss a global optimisation method. The general norm cases however have almost never been tackled, and for the moment the only solution technique available seems to be global optimisation, such as Big Square Small Square (see below), although some problems of convergence may arise in the presence of discontinuities in cost. These questions certainly call for a closer look.

11.4.2.2 Uncertainty

Uncertainty in the destinations, also called regional demand, is modelled by a continuous spatial distribution of one or more destinations. The minisum objective then corresponds to minimising the expected value of the distance of the facility to the random destinations. The analytic evaluation of such an objective function of integral type is however only possible in some simple situations like rectangular or circular regions, yielding objectives which are not always easy to optimise [22, 774, 327]. [713] therefore propose the use of approximate average distances. The most elegant and versatile approach to date is however to optimise directly without evaluating, a possibility described by [171] by way of the ellipsoid method, which calls for more readily available gradient information only. It may be mentioned that this approach is also able to handle the location of a facility of area type instead of point type, a case previously studied by [303].

Another way to handle regional demand, is to circumvent the difficulties inherent to the stochastic approach by replacing each region by some representative point, a centroid, and solve the resulting classical location problem [77]. The simplicity of this approach has made it attractive to many practitioners, but raises many novel questions about the quality of the so obtained solutions [293, 646] and Chapter 2. There is indeed an evident "aggregation error", in addition depending on the way the used centroids are obtained, which in turn should depend on the distance measure used. These questions have mainly been addressed in the context of multifacility location-allocation, and will thus be further discussed in section 11.7.

In case of uncertainty about the weights to be used in a WP, the most general question is to determine all the points which may be optimal for any choice of positive weights. Such points have been called *properly efficient* in a more general context by [448, 723] showed that for euclidean distance they consist of all destinations and all relative interior points of their convex hull; this result was extended to the constrained case by [167]. Further results are obtained by [345, 346, 343] showing a.o. that efficiency and proper efficiency coincide for polyhedral gauges. [344] overviews relationships between Kuhn's result and extensions with characterizations of inner product spaces. A related problem of finding a minimal set containing an optimal site for any choice of nonnegative weights is taken up by [315] for ℓ_p-distances in the plane and for more general norms by [347].

Instead of considering any combination of weights, one may constrain the weights by bounds or inequalities and ask for the possible optimal sites for the corresponding Weber problems. This type of question have remained largely open, and is directly connected to a sensitivity analysis of the Weber point(s) to a change in weights, which seems far from trivial [300].

Another approach to uncertainty in the weights is to assume a distribution to be known for each of them, usually independently, attempted only by a few authors. [1206] looks at the WP on the line, with a multivariate normal distribution for the weights, and derives for each destination the probability of being optimal. [328] do a similar study in the plane with ℓ_p-distances. More recently [1075] study a block norm (there called A-distance) situation with weights independently normally distributed and consider the minimisation of some given percentile of the Weber objective. In terms of applications one may consider the idea of fixed destination points with uncertain or varying weights as quite realistic. Therefore these topics clearly need closer scrutiny, opening new avenues of research, like [336] for dynamic location. A situation where the destinations are not fixed, but moving in some prescribed way, is studied by [3]. If at the other hand the facility is allowed to move in order to service random demand, questions arise about whether it is best to use stationary positions and where, or better move, but how? Such questions are investigated by [758, 25] on a linear space, concluding that in such a case stationary points are better. It is an open question whether a similar result holds in planar situations.

[319] investigate the asymptotic behavior of the optimal Weber-site when a large number of destinations are randomly and uniformly drawn in the unit ball.

In the same vein it is quite useful to be able to say something rational about the degree of nonoptimality of a given site. This type of information is important when considering a possible relocation decision of a facility. Such spatial type of sensitivity analysis, related to the shape of the objective function, seems to have been largely ignored, with some exceptions such as [519], and may be partially obtained by the BSSS method (see below). Another pertinent question is the degree of non-optimality obtained by optimisation using another distance measure. [716] propose evidence that this would be rather low, but this is somewhat disclaimed by the results of [183]. More research in this direction is clearly called for.

11.4.2.3 Constraints

It is only since the last decade that serious attention has been paid to locational constraints. General convex optimisation techniques are mostly able to handle convex constraints [616].

Many practical applications call for methods able to handle feasible regions of any shape, even disconnected ones. An early contribution from [550] develops a special method for solving WP with ℓ_p-distance within a finite union of convex polygons. This was followed up by [547] where mixed ℓ_p-norms and nonlinear cost functions are allowed in a similar environment, introducing the "Big Square Small Square" method (shortly called BSSS hereafter) of global optimisation, later extended by [957] to even more general objectives with arbitrary polygonal shaped feasible regions.

Previously the type of constraints were mainly restricted to maximal distance constraints around destinations, indicating that service is only possible or useful within a certain distance [1041, 1179]. These problems are related to bicriterion location problems in which both a minisum and minimax objective are to be minimised, as studied by [548, 556, 23].

[528, 526] consider "restricted" minisum problems, where the sought site must be outside some convex (or slightly more general) forbidden region(s), and [29] takes additional barriers to travel into account.

11.4.2.4 Sphere

The Weber problem on the sphere, using great circle distance has been studied by many researchers often attempting to adapt Weiszfeld's algorithm, but without convergence proof. A short overview may be found in [786], and a more complete summary of older results in [330]. Some more recent work is [301] giving perhaps the first theoretically convergent scheme, based on an iterative reduction to a spherical minimax problem. [1222] describes a different globally convergent method while [540] uses global optimization.

[307] shows that when the destinations are randomly drawn from a uniform distribution any site becomes optimal asymptotically, both in terms of maximisation and minimisation. Since many known results depend on whether all destinations lie in one hemisphere, the method of [1145] to check this property may prove useful.

It is a pity that only a few results were published for constrained problems on the sphere, like [281], while this topic clearly calls for more interest, especially in the worldwide multinational environment of today. See also [1223] for an application to an aircraft refueling problem, leading to a constrained spherical Weber problem with nonlinear costfunctions.

Other metrics on the sphere such as geodesic distance taking the asphericity into account and asymmetric distance for high altitude flight in presence of the jet stream could be incorporated in the models, giving rise to challenging new problems.

11.4.2.5 Repulsion

Minisum objectives have not received much attention yet for modelling noxious facilities. The only known application, [676] is to minimise the total pollution load on a set of destinations due to the location in a given region of a gas expelling plant. The objective function is obtained by considering Gaussian dispersion models for the gas plume for several main wind directions, and summing the effects on all destinations. Although this does not conform, strictly speaking, to our general model, since the effect does not only depend on distance, but also on direction, a BSSS technique may be used for optimisation. An improved model with more wind directions was described in [664, 673]. Recently the BSSS methodology was adapted by

[671], to locate a facility on a network, but with two-dimensional polluting effects. See Chapter 19 for more details.

Weber problems including both attracting and repelling destinations have been studied only recently, although mentioned occasionally as interesting since many years. It seems that [1124] is the first to study WP with positive and negative weights, limited to two attracting and one repelling destination, a study continued in [1126]. [335] consider the general case with positive total weight, and give algorithms for rectangular, squared euclidean and euclidean distances, the first two exact and the last a modified Weiszfeld method without convergence proof. The constrained case is considered by [198] who reduce the problem to concave minimisation and solve it by global optimisation. See also [955] for related results with a general metric. It may be mentioned that the BSSS technique may tackle this and even more complicated situations [957].

11.4.3 Minimax problems

11.4.3.1 Centers

Similar to minisum problems, the simplest minimax models are those obtained by choosing linear $g_a : g_a(t_a) = w_a t_a + c_a$, with positive weights w_a, yielding *weighted center* problems; the cases with $w_a = 1$ and $c_a = 0$ for all destinations $a \in A$ are usually called (unweighted) *center* problems. Such minimax models may be seen as to correspond to the social oriented notion of justice, as introduced by [985], and therefore also often called *Rawls-problems*. Applications with positive c_a are described by [379] and with negative c_a by [113].

A repelling version of this problem, i.e. locating a circle or rectangle of fixed dimension so as to cover a minimum weight of given weighted points in the plane, was studied by [337].

In case of fixed euclidean distance in the plane, the center problem asks for the center of the circle with smallest radius covering all destinations, a question posed by [1116], with clear applications to location of emergency facilities, radio emitters, etc. A first geometrical solution procedure is given by [1117], who attributes it to Peirce, and was rediscovered independently by [218]: broadly speaking it consists in starting with any covering circle and a finite iteration of radius-reductions. A more efficient method was found by [379], which increases the radius until all destinations are covered, the worst-case complexity of which was shown by [318] to be at least $0(n^2)$, but reaching an ϵ-approximate solution in $0(n)$ time. [1063] describe an $0(n \log n)$ technique based on Voronoi diagrams. The $0(n)$ techniques of [817, 352] seem to be mainly of theoretical interest, since computational results indicate that they are clearly outperformed by both first methods, even for a thousand destination points.

Planar weighted euclidean center problems may be solved by adaptations of techniques for the unweighted case, see [188, 192, 571, 818].

The higher dimensional euclidean center problem, or minimal covering (hyper)sphere problem (equivalent in $\mathcal{R}^3$ with finding the smallest radar beam to cover given targets, as noted by [743]) was shown to be solvable by a finite series of convex quadratic programs by [378]. For the weighted version see [314]. For fixed block gauges the unweighted center problems may be solved by simple geometric means [961].

For other norms, the most studied one is the rectangular norm, for which the weighted center problem may be solved analytically (see [786] for the plane, and [1094] for the unweighted version in three-dimensional space). Maxi-norm problems are equivalent to the rectangular norm ones in the plane, and [926] indicates that the simple analytical solution extends to higher dimensions in this case, contrary to the rectangular norm. Block norms (even mixed) center problems may be formulated as linear programs, although the dimension of the resulting LP's are quite large.

ℓ_p-norm center problems in any dimension may be solved by primal or dual nonlinear techniques described by [927]. Weighted center problems in the plane with ℓ_p-distance are studied by [325]. A solution technique for this problem on the sphere with great-circle distance is developed in [330].

A related problem is to cover a given convex region by a convex polygon or ellipsoid of given shape with minimal area. [1028] indicate how this may be solved by convex programming and investigate also the extended and more complicated version where rotations are allowed. Locating a linear segment of fixed length in the plane so as to minimise the maximum closest-point distance of n destinations to it gives rise to a similar problem, studied by [618] for euclidean distance. They show it may be solved in $O(n)$ in case the segment has a fixed orientation, $O(n^2)$ when it must lie on a line through some fixed point -reducible to $O(n)$ when this point lies in the convex hull of the destinations-, and $O(n^4 \log n)$ in the general case. This is related to the minimax location of a line, shown to be solvable in minimal $O(n \log n)$ time [820, 871, 749].

Probabilistic versions of the center problem, in which demand arises at random points in the plane according to some distribution, and the expected maximum distance is taken as an objective, were studied a.o. by [165, 825, 434].

Constrained minimax problems have received much less attention, although linear or convex optimisation methods are able to handle linear or convex constraints. For a quite powerful technique of this type see [614]. [526] consider a convex forbidden region and rectangular distances. More general constraints were first tackled by [123] using a computer graphical interactive technique in which the power of human vision is used as a feasibility checking subroutine, while [547] develop a totally algorithmic version in the most general minimax setting with general nondecreasing g_a. Incidentally, the same paper describes the BSSS method, there only advocated for minisum problems, which easily adapts to the minimax case as well [957]. Since any minimax problem with gauges and nondecreasing g_a's yields a quasiconvex objective, the ellipsoid method of [426] or cutting plane method of [952] may also be used when the feasible region is convex.

11.4.3.2 Anti-centers

Many installations which are either polluting or involve a risk to the environment have an effect which spreads out in all directions and therefore mainly depends on distance. As a typical example one may cite the location of storage tanks for highly inflammable and/or poisonous substances, see e.g. the study of [1158] of a peak shaving installation for liquid natural gas in the Netherlands. Therefore the site of such facilities should be chosen as far away as possible of population centres, leading to maximising minimal distance, which, e.g. by sign inversion, is easily reduced to an *anti-center* or minimax location problem with repelling destinations. Evidently such problems should be constrained in order to admit an optimal solution.

Rectangular distance versions, although not very realistic (see [385] for a critical discussion), were studied in [331, 823, 880, 32]. For euclidean distances this model was studied a.o. by [260, 326, 827, 829, 831, 182, 674, 675, 661, 388, 177] by several techniques and for different types of constraints. [1142, 758] discuss these types of problems under the name "largest empty circle, square", which correspond to taking euclidean and Maxi-distances, and extends them to "largest empty triangle, rectangle" problems which do not conform to our general model, since largest is meant in terms of area. A recent contribution in this field is [484] who indicate further references.

Another variant is the *design centering* problem asking for the largest copy of a given convex polyhedron, allowing for translations and homotheties only, which may be inscribed in a given polyhedral region (not necessarily convex) [596]. An application is the diamond cutting problem studied by [890].

11.4.4   Other objectives

11.4.4.1 Cent-dian

A combination of the economic aspect of the Weber objective and the social aspect of the Rawls objective is obtained by taking a convex combination of both minisum and minimax globalising functions. This *cent-dian* objective was used parametrically by [23] for solving the bicriterion problem mentioned in section 11.4.2. [167] give an axiomatic characterization of this objective, leading to an intuitive interpretation of the parameter involved. With linear (or convex) costfunctions this objective is convex, and may be minimised by convex optimisation [952].

11.4.4.2 Inequality

A site may be considered ideally distributionally just when all destinations are at equal distance. Clearly this is unfeasible in most cases, and should be relaxed into minimising some measure of deviation among the distances. At the one hand this leads to considering the *range* in distances, i.e. the difference between maximal and minimal (possibly weighted) distance, as the objective to be minimized, studied by [822] for rectangular distance, and [321] for euclidean and rectangular distances.

At the other hand the *variance* in the distances to the destinations may be used, as proposed by [836] for network situations, and mentioned in [957] as solvable by BSSS. These types of objectives tend, however, to yield optimality at infinity, and should normally be used in constrained environment. They have applications in metrology for estimation of out-of-roundness of round production parts, as discussed by [1224, 1162] (see also [1161] for the similar problem of fitting rectangular or square shapes to points and further references to other such important applications in automated quality monitoring). We feel that the use of other norms would also apply in similar applications involving parts of other shapes.

Other deviation type objectives like length of confidence interval, interquartile, coefficient of variation, entropy, etc. may prove to be useful. Most of these and other globalising functions may be considered as *inequality measures*, as overviewed and classified by [383]. These measures lead to even more exotic location models, most of which seem not yet to have been studied in continuous space.

11.4.4.3 Generalized minimax

Yet other proposals are to minimize the maximum of a number of simpler (e.g. minisum) objectives, studied by [309, 584] for simple distances, which generalize both the minisum and minimax objectives discussed before. [427, 428] show how these kinds of problems may be solved by an ellipsoid method, when the simpler objectives are quasiconvex functions of gauge-distances. A particular case is the *round-trip location problem* introduced for rectangular distance by [186], asking to minimize the maximum trip-distance around pairs of given destinations, and for which [611, 302] develop $O(nlogn)$ solution methods. For other distance measures see [329].

[584] investigate the multiobjective problem where each criterion is either of minisum or of minimax type with respect to fixed destinations, but with varying weights. Next to Pareto-efficient and simple minimax, they also propose minimising the lexicographic maximum, a novel objective which is related to the restricted location problems they previously studied. [977] determine the efficient set for such multi-objective problems with block norms.

11.4.4.4 Queuing

Taking congestion effects into account, leads to new types of single facility location models. Following previous study in network context, basic congestion type models are derived from the classical $M/G/1/\infty$ and $M/G/c/c$ queuing models, for which closed formulae are available. Calls occur at destinations as independent time-homogeneous Poisson processes. The servers are located, when idle, at some home location to be determined. Total service time of a call consists of random on-scene and off-scene service times and a travel-time component which is a function of distance between home location and calling destination. The $M/G/1/\infty$-based model assumes a unique server, any call occurring while the server is busy being placed in a queue. The *stochastic queue median* problem aims at finding a home location minimising expected response time. [317] show convexity of this objective function, and derive some localisation results in the euclidean distance case, also studied by [654]. [59] consider rectangular distance with barriers. [126] performs sensitivity analysis in the call rate under ℓ_p-distance, extending the results of [128] for rectangular distance.

The $M/G/c/c$-based model assumes c servers, all with the same home location, no queue being allowed: any call occurring when all servers are busy is lost. The *stochastic loss median* problem asks for the home location which minimises the fraction of rejected calls. [95] show that this model reduces to a Weber problem.

Other queuing location models, taking into account priorities or different objective functions have been addressed by [430, 431, 1228]. These studies open up a new and promising area for future location research.

11.4.4.5 Remarks

Obviously, interesting but virtually untouched globalising functions abound and we feel confident that many more will be uncovered and studied as the application fields of continuous single facility location(- allocation) expands. But this should not become a Pandora's box of untractable models, and therefore uniform solution methods of wide scope, such as BSSS, are welcome. Techniques for selecting rationally among this wealth of models, initiated by the axiomatic approach of [176], may turn out to be quite useful and should get more attention.

One "application" area, better to be viewed with a sorry smile, are attempts to determine the "centre of France, the EEC, Europe, the USA, ..., the world" [39] giving rise to much local pride and economic exploitation. These "studies" seem to make exclusive use of the centre of gravity paradigm, which can hardly be considered as based on economical or social views, but rather stems from tradition in classical physics and statistics and the misconception about the Weber site which still seems to exist with some geographers. Much more appealing definitions of central points in such a silly endeavour exist, as exemplified by this survey. But trying to convince politicians of this would only lead to overwhelming numbers of new "centres", each as (un)plausible as any other ...

Finally we want to include here two problems which do not really fall in our general scope. One consists of finding the site from which the maximum sight angle of a number of given shapes is minimised, studied by [306]. The other one concerns the determination of analytic centres of polytopes, minimising some objective related to distance to the faces instead of points (thus related by geometric duality to the problems discussed here). This topic is of particular interest in the design of interior point algorithms for linear and nonlinear programming, a flourishing research area indeed. Recent contributions to this subject are [38, 711].

11.5   Single Facility Location-Allocation Problems

11.5.1   Minisum

Location-allocation problems include other aspects to be determined than just the site of the new facility. Viewed as a location problem this means that the evaluation of the globalising function G at some site is itself an optimisation (sub)problem, which may be nontrivial.

As a first example, consider minimising the sum of interaction costs with a fraction (90% say) of the destinations, where determination of which destinations to be taken into account is part of the sought solution. Such a model could be used as an estimator of location in robust statistics [530] and a way of detecting "outliers". A slightly more involved minisum location-allocation model arises when some destinations are supply points, the others being demand points, and total available supply exceeds total demand to be met. A simple form of such a model was introduced by [681] in which evaluation of G reduces to solving a continuous knapsack problem.

More complicated models arise when recognising that location and design of a production plant are interrelated, leading to *production-location* models initiated by [872]. The book of [606] gives a thorough discussion of this topic, a review of which will not be attempted here. It seems, however, that much research remains to be done in this field. One particular aspect is optimal price-setting for products, leading to *profit-maximising location problems*, which may be equivalent under certain assumptions to the traditional locational objective of transport cost minimisation [1130]. A recent contribution in this direction is found in Chapter 15 and [922], who advocates BSSS-variants to solve the planar single facility profit-maximising Weber problem under several spatial pricing policies: spatial discriminatory, uniform delivered and uniform mill pricing. Other economical theories on the behaviour of managers depart from the classical goals of cost minimisation and/or profit maximisation, leading to new objectives in location theory, as recently indicated by [1157], opening up even wider new avenues of research.

A different type of models, classified as minisum location-allocation (according to our broad definition of this class) are *location-routing problems*, or traveling salesman location problems, in which the destinations are to be visited along a tour of minimal length, instead of each individually. Although starting to be popular in network context, in continuous context we know of [153, 320, 1079, 107, 1077].

11.5.2   Minimax

An example of a single facility location-allocation model of minimax type is to find a site minimising the distance within which a given fraction of the destinations lie, which are considered to be weighted (e.g. by population or demand). Such *minquantile location* problems are closely related to *maxcovering location* problems asking to determine the site from which a maximal number of destinations lie within a given (functional) distance. For example, the unweighted euclidean distance version asks to position a circle of given radius so as to cover a maximum number of given points in the plane. A simple $O(n)$ solution method is indicated by [824], while [294] studies the weighted version. [183] indicate how to solve both planar minquantile and maxcovering problems parametrically, i.e. for all fractions and all radii (resp.), even in the presence of convex constraints, both for inflated euclidean and mixed block norms cases. The method is based on the property that optimal solutions to the minquantile problem for any fraction are always completely determined by at most three destinations, and therefore only $0(n^3)$ candidate solutions have to be examined. Maxcovering problems in higher dimensions have applications in marketing for product-positioning, as surveyed by [1048]. A recent contribution is [247].

Quantiles are position parameters of the distribution of distances, thus in a sense similar to weighted minisum (viewed as a mean). Another possibility is the *Hurwicz* objective proposed by [176] in which a convex combination of minimal and maximal distances to some destination is considered, and may be viewed as a trade-off between attracting and repelling effects of the facility. This objective seems open for study.

11.5.3   Competition

A related class which could be called maximin location-allocation problems are *market maximising location problems*, in which a new facility is to be located in a competitive environment where other competitors have their own market-shares. The objective will be to capture, by a good choice of the site, the biggest possible market share.

When market share is deterministic, by simple allocation to the closest facility, one obtains a Hotelling [599] type model, which as such seems not to have been studied. [288] introduces the concept of unequally attractive facilities, where allocation is not necessarily to the closest facility, but still deterministic, and indicates how to optimally locate in such an environment. This is further analysed in Chapter 13 and [666].

[289] considers a probabilistic market allocation rule, the probability of patronising a facility (here called retail outlet) being inversely proportional to (euclidean) distance. The captured market share may then be expressed in closed analytical form in terms of distance, and a Weiszfeld type solution method is proposed.

Conditional location problems are somewhat similar. In minisum version, one wants to increase the number of existing (cooperating) facilities by a new one in such a way as to minimise the new obtained way of serving the clients by re-allocating them to the new facility when closer. This problem seems to have been studied only by [203] for euclidean distance. A minimax version of this problem is the 1-center case discussed more generally by [205] for euclidean and [880] for rectangular distances.

11.6   Multifacility Location Problems

11.6.1   General

Similar to the general single facility location model, we may state a multi-facility location problem formally as follows:

Let X be the space of sites, A a family of destinations (in this context often called *existing facilities*), considered as points in X. Let V denote a finite set of *new facilities* to be located in X, which will interact with destinations and among themselves. The structure of the interactions may be represented by an undirected graph $H = (A \cup V, E)$ (the *interaction graph*) where nodes indicate destinations or facilities and edges represent interactions. Note that, destinations being fixed, it is not necessary to consider edges between nodes in A. Any choice of sites for the new facilities is then an *embedding* of H in X with fixed points indicated by the destinations, to be viewed as some mapping $x : A \cup V \to X : f \to x_f$, with $x_a = a$ for all $a \in$ A. With each interaction $e = (u, v) \in E$ between nodes u and v one associates a *distance measure* d_e, used to calculate the distance between the sites x_u and x_v chosen for u and v, which we will denote as $d_e(x_e)$. This associates to any embedding x a *vector of distances* $D(x) = (d_e(x_e))_{e \in E} \in \mathcal{R}_+^E$. These distances are then combined using a *globalizing function* $G : \mathcal{R}_+^E \to \mathcal{R}$ into a global value to be minimized. Locational constraints may be considered of two main kinds : *constraints on nodes* of type $S_f \subset X (f \in A \cup V)$, saying that only embeddings with $x_f \in S_f$ are admissible (note that these are always implicitly present for destinations since we impose $x_a = a$, i.e. $S_a = \{a\}$ for all $a \in A$), and *constraints on edges* of type $T_e \subset X \times X (e = (u, v))$, saying that only embeddings with $(x_u, x_v) \in T_e$ are admissible. The most important example of constraints on edges are *distance constraints* such as $d_e(x_u, x_v) \leq (\geq) r_e$ stating that the distance over which this interaction takes place remains within (or is at least) some given value r_e. Clearly constraints on edges are also a form of interaction between facilities, and the corresponding edge should be present in the interaction graph H, even when their distance is not used in the globalising function G. All these constraints may be viewed as one global constraint on the embedding x of type $x \in S \subset X^{A \cup V}$.

We thus face the problem

$$MIN\{G(D(x)) \mid x \in S\}. \tag{11.2}$$

It is clear that the multifacility location problem is much more complicated than the single facility one it generalizes. Clearly the interaction graph H may be assumed to be connected, since otherwise the problem decomposes in independent subproblems corresponding to its connected components. Actually, destinations being fixed, it may be assumed that the subgraph of H induced by V, i.e. the interaction graph among new facilities only, is connected.

In view of the higher complexity it is not surprising that multifacility location problems have only been studied for some of the "simpler" situations described for single facility problems. Most of the literature is concerned with minisum problems, while some attention went to minimax versions. No technique seems to have been developed to attack (11.2) in the broad setting described above, the primal-dual method (see below) being for the moment the most versatile, as long as the problems remain convex. An extension of BSSS to multi-facility problems, mimicking its success on planar single facility models, seems out of question due to the much higher dimension (in terms of variables).

11.6.2 Minisum

When G is taken as linear, with positive weights, and distances are measured by norms, we obtain the extension of the Weber problem to multiple facilities, by far the most studied multifacility problem. This model, introduced by [872], has a high degree of nondifferentiability, appearing at least whenever two or more facilities *coincide*, i.e. have the same location. It is considered a hard nonlinear optimisation problem and used as test problem for nondifferential optimisation techniques.

The easiest version is obtained for the rectangular norm, for which [554] established that there always exists an optimal solution with new facilities at intersection points (cfr. section 11.4.2) inside the convex hull of the destination points like in the single-facility case. The problem decomposes into one-dimensional ones like in the single-facility case, solvable as a network flow problem, [155], a minimum cut method, [208], a linear program, [648], or a relaxation method, [269].

For fixed block norm problems in the plane the intersection point property, ubiquitous in the single facility model, holds whenever the norm has a unit ball with four extreme points, but may fail in general [837]. Mixed block norm problems may be transformed into linear programs, and are thus relatively easy to solve, although the LP may have a high number of variables [786]. This approach easily accommodates linear constraints.

Other localization theorems have been obtained: [420] show that in the fixed euclidean norm case all new facilities may be restricted to lie within the convex hull of the destinations, without loss of optimality, a result which directly extends to any fixed norm in the plane, as shown by [554]. In this latter paper a similar result is obtained for mixed ℓ_p-norms, replacing convex hull by octagonal hull. The former result was improved by [837] replacing convex hull by the often smaller metric hull.

In general, the simple question to check optimality of a given embedding is far from trivial, due to the nondifferentiability in presences of coincidence, a situation often occurring at the optimal solution, and therefore to be studied in detail.

Sufficient conditions for coincidence at optimum involving the weights only, similar to (and derived from) the majority theorem, have been derived by [649] in the general setting of distances measured by metrics, and extended by [751] introducing attraction trees, which, when present in H, imply coincidence of all their nodes. Detection of attraction trees in general graphs was, however, shown to be NP-complete [858], but may be achieved by a boolean program, as described in [752].

Necessary conditions for optimality have been obtained by several authors, mainly in the fixed euclidean norm case [270]. Exact optimality conditions were obtained by [647] for coincidences involving at most two facilities, and for the general case by [937] for fixed euclidean norm and, more generally, [958] for mixed norms. It turns out that checking optimality is easy when the subgraph of H, induced by the coincidences at the embedding under scrutiny, is a tree, but presence of cycles leads to a nontrivial flow-existence problem on this subgraph under nonlinear constraints.

It should also be noted that, contrary to the single facility Weber problem, the euclidean multifacility location problem may have several optimal solutions as demonstrated by [1017].

Solution methods of many different kinds have been applied, mainly to the euclidean norm case and fixed ℓ_p-norm case. [872] proposed to adapt Weiszfeld's method to the multifacility problem, and [909] showed that this method yields decreasing objective values as long as coincidences do not appear. [1016], however, construct a counterexample showing that the method may converge to an suboptimal (nondifferentiable) solution, but subsequently [1018] show that working with a slightly (hyperboloid) perturbed and differentiable "distance measure" leads to convergence. Convergence of the unperturbed version was obtained when H is a tree by [979] using an amended method involving a matrix inversion at each step, and by [936] by a recursive-type Weiszfeld method. [429] extend the work of [1018] (and simplify their proof) by showing convergence with linear rate of Miehle's algorithm adapted to much more general situations: a differentiable perturbation of a fixed ℓ_p-distance measure with $1 \leq p \leq 2$ and a fairly general differentiable globalising function.

Other techniques, sometimes allowing for linear constraints on nodes, have been tried, such as approximation methods [393, 218], a trajectory method [322], (sub)gradient methods [194], but with limited success. Much more powerful methods were developed more recently by [911, 157] which include special treatment in case of coincidences. All these methods being iterative ones, using local information as derived from (sub)gradients, should include some stopping rule. In order to measure the quality of the solution reached the lower bound given by [780], applicable to fixed ℓ_p-norm models may prove useful.

Up to date the most comprehensive method for solving minisum multifacility location problems seems the primal-dual technique of [613] which applies to arbitrary mixed norms problems including arbitrary convex constraints, both on nodes and edges. The main advantage of this method is that it avoids all problems of nondifferentiability and automatically generates lower bounds enabling rational stopping rules, while its main disadvantage is that convergence may be rather slow, and therefore it may not be competitive with the last of previously mentioned techniques, when applicable, which exhibit quadratic convergence.

Similar multifacility problems have been investigated on the sphere with great-circle distance by [280], proposing a Weiszfeld-like method without convergence proof. [44] consider also the problem on the sphere (or spheroid) but use euclidean distance in $\mathcal{R}^3$, (or *chordal distance*) which seems hardly applicable except when all destinations are sufficiently close to each other, in which case the planar assumption will probably suffice.

The planar version with squared euclidean distance has been investigated by [392], showing it to be solvable as a set of linear equalities, and used for the solution of floor layout problems by [286].

It is clear that minisum multifacility location problems still offer great opportunities for research, extending what has been and still may be done for single facility models. The difficulties involved are, however, much larger, in particular when repelling destinations would be considered, since these imply loss of convexity, thus all of the techniques above become inoperable.

11.6.3 Minimax multifacility problems

Much less attention has been paid to the minimax multifacility problem, obtained by taking for $G(t) = \max_{e \in E}(a_e t_e + b_e)$. Note that we consider only linear functions on each individual interaction distance, this being the only type which seems to have been studied, and then only for positive a_e and nonnegative b_e (often taken to be zero).

The rectangular norm version may be reduced to a linear program [1191, 380, 868] or solved by a network flow method [272]. Euclidean or fixed ℓ_p-norm problems have mainly been attacked by several nonlinear programming approaches [789, 381, 323]. The subgradient technique of [194] may handle convex constraints. The most widely applicable technique, accounting for mixed norms, fixed costs ($b_e \neq 0$) and any type of convex constraints, is an adapted version of the primal-dual method, as described by [615].

A different type of minimax problem was studied by [1088] in which the aim is to locate a given number of independent detection stations, detecting the occurrence of some event with a probability that decreases with (euclidean) distance, so as to maximise the probability of detecting an event occurring anywhere within a given region.

A version with repelling destinations was described by [332], but the proposed solution method only applies to one dimension. A similar planar model with rectangular distances is studied by [146].

11.7   Multifacility Location-Allocation Problems

In location-allocation problems for several facilities, the aim is to find at the one hand a site for each facility (*location*) and the way the destinations are to be served by these facilities (*allocation*). Objectives of several types are possible as discussed below.

11.7.1   Minisum

The traditional minisum version is usually called the (continuous) *p-median* or multi-Weber problem: each destination will be served by one of the p new facilities (to be located), and the sum of costs (assumed to be linear in distance) of all these services is to be minimised. Introduced by [237], and widely applied in very different settings as comprehensively reviewed by [585], this problem turns out to be very hard to solve, consisting simultaneously of a nontrivial combinatorial part (allocation, i.e. a partitioning of the set of destinations into groups to be served by each facility) and a nonlinear part (location, i.e. solving a Weber problem for each group). The objective is neither concave nor convex and not everywhere differentiable [238]. In fact twenty years later it was shown to be NP-hard by [819].

[238] (continuous setting) and [799] (discrete setting) developed the popular *alternating* heuristic (also often called sequential location and allocation, or shortly SLA), which consists of alternating between an allocation phase in which sites of facilities are kept fixed (each destination is served by the nearest facility) and a location phase in which allocation (the partition into groups) is kept fixed and the single facility problems within each group are solved separately, the whole process stopping when no new improvement is found. It turns out that this technique, if restarted several times from different, usually randomly chosen starting locations, quickly finds good solutions, and quite often an optimal one, although this may not be established by the method itself. See [1093] for experiments with this method for euclidean, squared euclidean and rectangular distances. [130] attempt a statistical estimation of the global optimal solution after repeated SLA. The heuristic has been used by [1023] to investigate the

optimal choice of the number of new facilities, a problem of great interest in cluster analysis. See [623, 712] for other types of heuristics.

The one-dimensional multi-Weber problem may be solved by dynamic programming [776]. In the plane with rectangular distances we have again an intersection point property, where attention may be restricted to those intersection points inside the convex hull of the destinations, and even some of these may often be excluded [783] thereby reducing the problem to a discrete p-median problem for which several solution methods exist.

The euclidean norm case has attracted most attention. Recently [1021] indicated a nice application of locating steam generators in heavy oil fields. The first exact solution method was developed by [722] as a branch and bound method with rather weak lower bound calculations. Recognising that the convex hulls of the groups of destinations in the optimal partition must be disjoint (the groups are separated by the mediatrix between the serving facilities' sites) led [905, 908] to a very efficient solution of the 2-facility case (only $n(n-1)$ partitions by a line should be considered), earlier also developed by [561] but never published (see also [299]). [1025] prove that the number of disjoint convex hull partitions of n points in the plane into p groups is polynomial in n for fixed p, which in theory would allow an extension of Ostresh's method to more facilities, but does not seem to be easily implementable in practice. A more recent approach using the disjoint convex hull property, developed by [1020, 1022] consists of generating the list of all possible convex hulls, together with the value of the corresponding Weber problem, and use this as data of a discrete set-covering problem. All these exact solution methods are able to solve rather small-scale problems: the largest reported in the last paper has 25 destinations and 7 facilities.

Different techniques based on nonlinear programming have been developed by [200, 881]. A similar but much improved technique, applicable to ℓ_p-norms problems, is described by [119]. These methods find local minima which often seem to be global ones on small-scale problems. Although these are in fact heuristic methods, the main advantage is that much larger problems may be handled than by exact methods, and the space may be more than two-dimensional: [119] report on 50 destinations and 2 facilities in $\mathcal{R}^{10}$!

As mentioned in section 11.4 about the Weber problem, regional clusters of destinations are often approximated by representative points. In location-allocation problems this is often necessary to strongly reduce the number of destinations, in order to obtain a solvable problem. This leads however to different types of "aggregation errors" (see [584] for an overview). Although seemingly standard, these effects of aggregation were insufficiently investigated, especially in continuous setting. First steps are found in [422] who derive a tight bound on the error in the objective function, and [423] proposing a special aggregation scheme for the rectangular distance case which minimizes such a bound, and [584] for an iterative technique based on Voronoi diagrams to eliminate errors due to misallocations of unaggregated destinations. Clearly this topic deserves much more attention, which it already received in discrete setting.

A block norm version involving concave cost functions described by [836, 838] with applications to siting of social community centres, has the intersection point property, and a stepwise improvement method yielding local optima is applied.

11.7.2   Minimax

The minimax version, better known under the name (continuous) *p-center problem*, may be viewed, in euclidean distances, as covering all destinations by p equal identical circles of minimum radius. [810, 819] prove its NP-hardness, both for euclidean and rectangular distance, but the results of [1025] show that for euclidean distance, the planar problem, in principle, is polynomially solvable for fixed p.

Exact solution methods for the rectangular distance p-center problem were described by [1178] (valid for any ℓ_p-norm), [304, 28, 705, 926]. Drezner's approach yields extremely efficient methods for 2 ($0(n)$) and 3 ($0(n \log n)$) facilities in the plane, while Pelegrin extends Aneja et al's method to maxinorm problems in any dimension, with applications to cluster analysis. [61] investigate the loss in objective value observed when only destination points are allowed as sites for the facilities (i.e. the discrete version), and look also at rectangular distance with barriers.

For euclidean distance p-center problems exact solution techniques are provided by [200, 298, 1166, 206] all involving rather inefficient branch and bound schemes. Several heuristic methods exist, applicable to general norms in any dimension, provided the single facility problem may be efficiently solved [298, 369, 354, 925, 183]. [394] combine a relaxation technique and interaction with a user providing some decisions through a graphical support.

11.7.3   Variants

An alternative to the p-median problem arises when the facilities to be located have limited production or supply capacity. The allocation part of the problem then involves the solution of a transportation problem. This interesting type of problem was investigated by [1120] and seems to have been neglected since then. [143] recently propose a new variant, which may be considered as a continuous version of the simple plant location problem, well known from discrete location theory, but including economies of scale.

One of the difficulties of the p-center problem, especially in branch and bound approaches, is that usually many optimal solutions exist: it is the size (radius) of the largest cluster which alone determines the objective value; this leaves much freedom to the choice of the other clusters. One possible way to overcome this problem is to minimize the sum of maximal within-group distance instead, yielding the *p-center-sum* problem, investigated by [953], who develops a branch and bound method, based on the fact that the one-dimensional version is extremely easy to solve.

Conditional p-median and *p-center* problems are studied by [203] extending his [200] nonlinear local minimisation method for the unconditional versions. For conditional *p*-center problems exact methods have been developed by [308], who proposes a method based on a sequence of $O(\log n)$ unconditional *p*-center problems, and [205] giving an extension of their [206] method for the unconstrained case with the same complexity, which solves instances with 200 destinations, 3 fixed facilities and 5 new facilities.

[1177] describes several heuristic procedures for the multiple facility version of the maxcovering location problem (section 11.5). [824] show that it is sufficient to consider a finite number of candidate solutions and indicate how to exploit this fact. In [826] they extend this result to a more general objective consisting of maximising the sum over all destinations a function which is non-decreasing in the number of facilities by which it is covered.

Theories related to location-allocation but including economies of scale effects for the facilities are given by *market area models*. An excellent overview of this literature, together with a model of very general type and a study of its properties is given by [391]. See also [534, 535, 537] for extensive studies of market area shapes under many different behavioral assumptions, and [1030, 1128] which show, following a suggestion of [361] and the earlier study [1167], that market areas are quite sensitive to the distance measure used. This topic has very close connections with *Voronoi-diagrams* discussed in Chapter 6 and exhaustively surveyed by [39] and exposed in all details in the magnificent book [897], with a chapter devoted to location problems.

Other types of competitive location of multiple facilities, e.g. connected to Nash-equilibria, have been studied quite intensively in dimension 1, either on the line or on a circle, but not enough in more general continuous settings. An overview of the competitive field may be found in [374], while two examples of recent contributions are [121, 878].

As far as we know no work on repelling location-allocation seems to have been done, except perhaps a recent work [720] on how to place n empty squares of maximum size among n given points in the plane, each point being a cornerpoint of one square.

11.8   Other Related Problems

Out of a large family of problems in some way related to location in continuous environment, we select three examples we think are particularly interesting.

11.8.1   Steiner trees

A widely investigated problem related to minisum multifacility problems, but including a combinatorial aspect, is the *Steiner tree problem*, asking for a spanning tree for a set of (n given) points of minimal total edge-length, where it is allowed to use additional points (Steiner points) as nodes of the tree. Once a choice has been made on the number of Steiner points (at most $n-2$) and the interconnections to be considered (called the topology) one faces a multifacility location problem of the simplest unweighted type, which is quite easily solved (see [832] for the euclidean distance case). The difficulty resides in the high number of possible topologies, making the problem NP-complete both for euclidean and rectangular distance [443, 444]. [608] surveys the literature on euclidean Steiner tree problems. A more complete coverage of this subject may be found in [609]. Some of the most recent contributions are [341] with good polynomial time heuristics in the euclidean case, [610] about probabilistic results for the rectilinear case, and [1211] for situations with obstacles. Steiner trees are an example of minimal length networks connecting given points, a subject more generally treated by [873]. See also [259] for a related hierarchical network design problem.

11.8.2   Hubs

We also mention here the related *hub-(multi)facility* location problem, introduced by [901] consisting of minimising global transport costs between given source-destination pairs, by way of a given number of "hubs", i.e. transit points to be located [43]. Recent contributions to the planar version are [903, 45], the latter also considering the version on the sphere.

11.8.3   Routing

Routing problems are usually considered in network setting. However in many cases they may also have a clear and nontrivial continuous character, like the strategic routing of ocean liners taking the (unfortunately constantly changing) currents into account as studied by [769], which is in fact a continuous shortest path problem related to the motion in a heterogeneous environment mentioned in section 11.2. Hazardous and/or noxious material transport is a longstanding problem in which interest has, for evident reasons, recently strongly risen. [768] give an overview of work in this field, with relations to location. [334] indicate how to construct a route connecting two given line segments and maximising the minimum weighted euclidean distance to a given set of points. [665, 1067] look for a route within a permitted region and between two given points, which minimises the accident probability due to the presence of a finite number of given damage sources, where the individual damage probability depends decreasingly on euclidean distance.

11.8.4 Concluding remarks

In the quickly growing field of computational geometry many more problems related to continuous location abound [972, 362]. We also want to mention again that several novel and quite hard location problems, directly connected with Voronoi diagrams, are being studied, mainly in Japan. See [897] for details and references.

One might also argue that e.g. regression analysis in statistics and some topics in approximation theory in functional analysis are parts of continuous location theory, where not points, routes or networks are located, but rather lines, (hyper)planes, or even affine subspaces of any other fixed dimension [711, 807]. But this topic would evidently lead us too far ..., and we'd better stop this overview at this point.

12
GLOBAL MANUFACTURING STRATEGY

Vedat Verter
University of Alberta

M. Cemal Dincer ˙
Bilkent University

12.1 Introduction

In response to the intensifying rivalry in many industries, firms are restructuring themselves to operate on a global basis. Globalization of a firm constitutes diversification of its operations to different countries so as to take advantage of the inefficiencies in the international product, factor, and capital markets. Global configuration of operations provides access to cheaper labor and raw materials, subsidized financing opportunities, and larger product markets. A global firm however, can achieve competitive advantages only if its geographically dispersed activities are effectively coordinated. That is global firms "must learn to operate as if the world were one large market - ignoring superficial regional and national differences" [759]. Although multinational companies have a long history, globalization of operations is a recent phenomenon. A typical multinational corporation consists of several *strategic business units* (SBUs) each functioning in a particular industry and serving a well-defined market segment. Each SBU, in turn, consists of several functional units such as purchasing, marketing, finance, personnel, R & D, and manufacturing. Such a three-level hierarchy is depicted in Figure 1. Traditionally, multinational companies were operated as *multidomestic* corporations where each SBU is run as a domestic firm. What we are currently witnessing however, is an internationalization at the SBU level via the location of functional units at different countries. As a result of these developments, managers are facing new challenges such as, the need to incorporate the differentiating features of the international environment in decision making, and the difficulties of dealing with the increased organizational complexity due to globalization.

In this chapter, we will confine ourselves to the strategy planning problems at a specific functional unit in the global firm, i.e. manufacturing. The long term goals for manufacturing performance and the policies adopted

FIGURE 1. A three-level corporate hierarchy

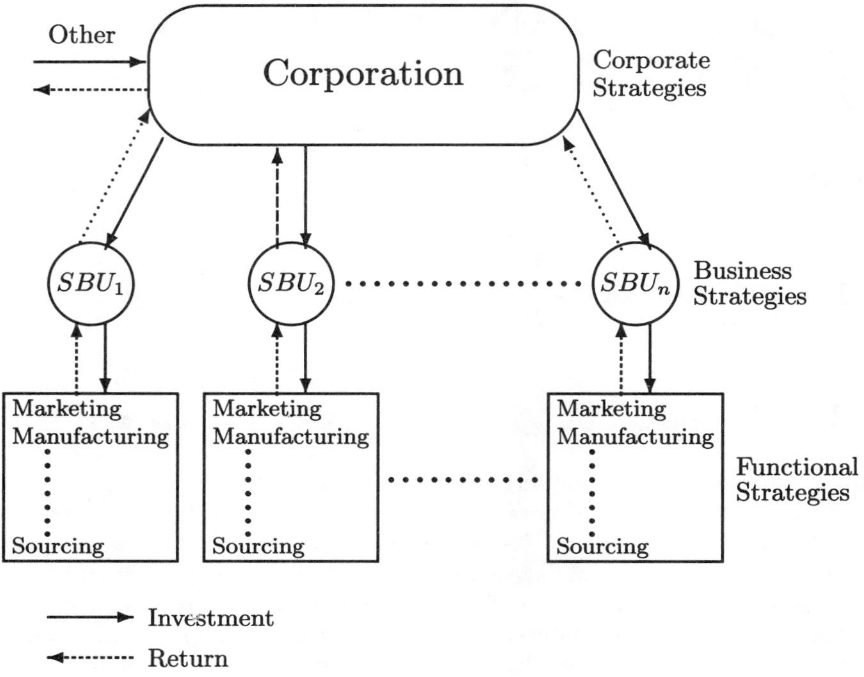

to achieve those goals constitute *manufacturing strategy* of a firm. *Cost, quality, delivery performance*, and *flexibility* are the most common criteria to evaluate performance of a manufacturing system. It should be noted that firms are not in a position to choose between cost and quality or dependability and flexibility objectives. Empirical studies did not justify the existence of such tradeoffs suggested in early conceptual work on manufacturing strategy. In accordance with the intensive rivalry in global industries, *innovativeness* and *time-based competition* are also emerging as important manufacturing objectives. Policies that enable a firm to meet its long term goals comprise a collection of strategic decisions. Leong et al. [756] pointed out the consensus among several authors about the strategic decision areas for manufacturing. Manufacturing strategy decisions can be categorized as *structural* decisions that are associated with configuration of the manufacturing facilities, and *infrastructural decisions* that address the people and systems that run the manufacturing activity. Structural decisions include location of the manufacturing facilities, and the manufacturing technology to be adopted as well as the capacity to be built-in at each facility. The decisions concerning the linkages among the facilities that perform different stages of the production process, and the interactions of manufacturing facilities with suppliers and customers are also in this group. Infrastructural decisions are associated with production planning and control, quality control, workforce management, new product development, and performance measurement systems. Note that, manufacturing is an integral part of the firm, and therefore manufacturing strategy cannot be formed without considering its interactions with the business and corporate strategies as well as with other functional strategies as pointed out by Fine and Hax [403].

Production-distribution networks provide an effective tool in modeling manufacturing structure of a firm. A typical production-distribution network is depicted in Figure 2. Nodes of the network represent vendors, manufacturing facilities, distribution centers, warehouses, and customer zones whereas arcs represent the flow of items. In a global firm, nodes of the production-distribution network are located in different countries, and items flow across national boundaries. Each layer of the network in Figure 2 is called an echelon, and geographical dispersion of facilities usually results in a multi-echelon configuration. Another common feature of global manufacturing is the production and distribution of a variety of products to benefit from economies of scope. The multicommodity nature of international production-distribution networks is fostered by the different needs of each national product market. Manufacturing strategy planning involves decisions regarding the long term changes in configuration of the production-distribution network of the firm. Facility location decisions are crucial in strategy planning, especially in globalization of the manufacturing activity. Availability and cost of production factors such as, qualified labor, reliable raw material and component supply, and manufacturing technology vary significantly among countries. Thus, within the international context, location of a facility interacts with its size, technology content, and product range in facilitating the achievement of strategy goals. Ignoring these interactions in locational decisions will presumably lead to a suboptimal manufacturing configuration in a global firm. We suggest that global manufacturing strategy of the firm provides a framework for the facility location decisions.

The aim of this chapter is to provide a survey of the analytical models that are relevant to the facility location decisions in a global firm. This is an emerging field of research and most of the reviewed studies are far from capturing all of the characteristics of global manufacturing. Our criterion for including an analytical approach in this survey is that it should incorporate at least one of the characteristics of global manufacturing such as, multiple products, multiple echelons of manufacturing facilities, exchange rates, or price uncertainties. The remainder of this chapter is organized as follows. The second section briefly describes the requirements of the global manufacturing strategy planning process in setting the framework for locational decisions. The third section reviews the analytical models for the multicommodity, multi-echelon production-distribution system design

FIGURE 2. Production-Distribution network

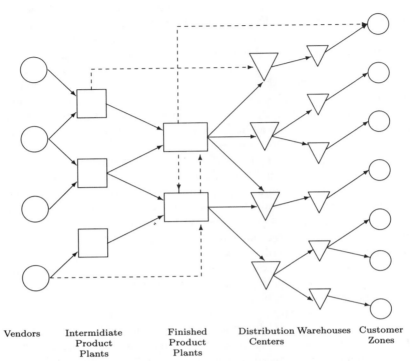

| Vendors | Intermidiate Product Plants | Finished Product Plants | Distribution Centers | Warehouses | Customer Zones |

problem. Note that, although these models can also be used for strategy planning in domestic firms, they provide a sound basis for incorporating various features of the international environment. Thus, designing global manufacturing networks is usually perceived as an extension of the basic multicommodity, multi-echelon network design problem, as presented in the fourth section.

12.2   Global Manufacturing Strategy Planning Process

Access to low cost production input factors constitutes the earliest and most common motivation for global manufacturing. The international differentials in manufacturing costs however, are diminishing as pointed out by Flaherty [407]. Improvements in the transportation and communication industries however, foster the emergence of global products that can simultaneously be marketed in several countries. This enables the global firms to achieve scale economies in constructing their manufacturing facilities, and hence in their manufacturing technology investments. Proximity to customer zones, use of local technological resources, and pre-emption of competition via early-mover advantages are further strategic reasons for global manufacturing as observed by Ferdows [399]. Hitt et al. [575] pointed out that there is a positive correlation between global diversification of operations and firm performance.

There are several distinguishing features of the international environment which have to be taken into consideration for global manufacturing strategy planning. According to classical economic theory, the *law of comparative advantages* provides a basis for international transactions. That is, trade will be mutually advantageous if countries are relatively more efficient in producing different goods. Ideally, an international equilibrium will be attained when exports and imports of individual countries reach a balance in their own currency. Exchange rates among currencies provide the means for adjustment of the international disequilibrium. This is apparent from the following theorems on the relationships between exchange rates, interest rates and price levels: *Purchasing Power Parity* (PPP) states that exchange rates off-set the differences between national price levels of freely tradeable goods, and *International Fisher Effect* (IFE) states that exchange rates off-set the differences between interest rates for different currencies. These fundamental exchange rate relationships however, are based on some assumptions such as; perfect product markets (no transportation cost or time, no barriers to trade), perfect financial markets (all relevant information is reflected in prices, no taxes, no transaction costs, no controls), and future certainty. These assumptions are quite unrealistic in many cases, since product and capital markets are not "perfect" and the future is not known with certainty. Deviations from PPP exist due to lags in market responses, transportation costs, national differences in the price ratios of internationally traded goods to domestically traded goods, government interventions and risk expectations. Furhermore, deviations from IFE exist due to the availability of subsidized financing and differentials in corporate tax rates. In addition to the various arbitrage opportunities provided by these imperfections, uncertainties regarding future exchange rates, interest rates and price levels cause various types of risks the firms have to undertake when designing their manufacturing strategy. Distinguishing features of the international environment and their potential impacts on firm behavior are depicted in Table 12.1.

Despite growing importance of the globalization phenomenon, the review by Adam and Swamidass [9] showed that the international content of manufacturing activity is among the missing themes in manufacturing strategy literature. Recently, Miller and Roth [844] provided a taxonomy of the manufacturing strategies identified by a survey of 164 manufacturing companies in the United States. De Meyer [276] presented a similar empirical study for European manufacturers. In an earlier study, Ferdows et

TABLE 12.1. Distinguishing Features of the International Environment

Phenomena	Effect
Deviations from PPP	goods arbitrage opportunity
Deviations from IFE	financial arbitrage opportunity
Price uncertainty	market risk
Exchange rate uncertainty	currency risk
Government interventions	political risk
Quotas, local content rules	constrain flow of items
Tariffs, duties	increase transaction costs
Cultural, taste differences	product tailoring
Language, skill differences	human factors management

al. [400] compared the evolving global manufacturing strategies in United States, Europe and Japan. Although these empirical studies provide some insight about the current practice in global manufacturing, Skinner [1085] pointed out the lack of methodologies to identify the most appropriate manufacturing structure in achieving the manufacturing strategy objectives. A prevailing approach is due to Kotha and Orne [714] who extended the work of Hayes and Wheelwright [568, 569]. They perceived the manufacturing activity as a three dimensional structure, the dimensions being process structure complexity, product line complexity, and organizational scope. Factors leading to the high or low level of complexity at each of these dimensions are further identified, and analysed. In this context, manufacturing strategy constitutes a movement in the three dimensional space from a point that designates the current manufacturing structure to a target point implied by the strategy goals. Their work revealed the need that the product range and organizational structure of each manufacturing facility should be consistent with the target market segment and geographic scope of operations. Further, each plant should select and adopt the technology that is most appropriate in terms of its strategic role in the company. Ferdows [399] presented a scheme for matching the technology content of a manufacturing facility with the primary reason for its location in a foreign country. Although the work of Kotha and Orne [714] provided valuable guidance on how to achieve the fit between strategy and structure, their representation of the manufacturing activity remains quite descriptive.

The aim in manufacturing strategy planning is achievement of long-term sustainable *competitive advantages* over the rivals via configuration and coordination of the firm's production-distribution system. Global firms have facilities located in different countries. Note that there are differences between countries in terms of cost and quality of the available production factors as well as institutional and cultural infrastructure. Thus, in many industries, it is possible to identify some countries that provide *compara-*

tive advantages for performing certain stages of the manufacturing activity. Kogut [706] described the interplay of the firm specific competitive advantages and the location specific comparative advantages for global firms. One of the primary requirements of global manufacturing strategy planning process is the ability to take advantage of the interrelations between competitive and comparative advantages. A critical factor in global competitiveness is the comparative advantages of the home base for global firm's activities in the relevant industry. A home base is the country which reaps the profits and which is usually where the majority of production and management takes place. Porter [970] provided a paradigm to assess the attractiveness of a nation as a home base for an industry, and identified the determinants of national advantage: factor conditions; demand conditions; related and supporting industries; and firm strategy, structure, and rivalry. Government intervention and uncertainty are perceived as the factors influencing these determinants. This theory is implemented through the analysis of a selection of industries in each of the following countries: Denmark, Germany, Italy, Japan, Korea, Singapore, Sweden, Switzerland, United Kingdom and the United States. Kogut [707] pointed out the importance of creating *operational flexibility* within the corporation in order to benefit from being global. Operational flexibility provides the capability to explore arbitrage and leverage opportunities. Arbitrage opportunities are associated with global configuration and include; production shifting in response to changing circumstances in factor markets, information arbitrage, tax minimization via transfer pricing, and financial arbitrage via subsidized loans. Leverage opportunities arise from global coordination and may be a hedge against the political risk associated with international investments.

Global coordination of the manufacturing activity is equally important with the configurational decisions in global manufacturing strategy planning. The inbound and outbound logistics activities assure the coordination of material flow. Fawcett [396] stressed the importance of strategic logistics in global manufacturing success, and presented a review of the literature on logistics. The sharing and use of information throughout the global network is essential in managing global manufacturing. De Meyer and Ferdows [277] presented empirical evidence suggesting that integration of information systems within manufacturing, and between manufacturing and other functions is a growing concern in Europe, North America and Japan. Flaherty [406] described the role of support activities in coordination of the manufacturing and technology development taking place in internationally dispersed facilities. Global coordination of the manufacturing activity also involves allocating responsibility to the dispersed facilities and aligning their technology development efforts which need to be supported by the organizational culture as pointed out by Misterek et al. [857]. In order to assure success at the firm level however, the manufacturing activity has to be coordinated with the other functional units. The cross-functional coordination of manufacturing and marketing is emphasized in the work of Hill

[573].

This section is intended to be a brief account of the factors that differentiate global manufacturing from the manufacturing activity in a domestic firm. Naturally, these factors should be incorporated in the manufacturing strategy of a global firm. Thus, requirements of the global manufacturing strategy planning process constitutes a framework for facility location decisions. An analytical approach for locational decisions can facilitate the achievement of strategy objectives to the extent that it incorporates the interaction of configurational decisions and the characteristics of the international environment. Having set the criteria to evaluate the relevance of an analytical model for facility location decisions in a global firm, we now turn to review the state-of-the-art. It should be noted that the above discussion is confined to the effects of globalization on the manufacturing activity. The changes in firm structure due to globalization, such as joint ventures, strategic alliances, franchising and licensing are out of the scope of this chapter. The interested reader however, is referred to the recent bibliography in [742] and to the books edited by Porter, [969] Ferdows, [398] and Sheth and Eshghi [1072].

12.3   The Production-Distribution System Design Problem

Given a set of markets to be served, the sets of alternative facility locations and technologies for each stage of the production and distribution process, and a set of alternative vendors *the production-distribution system design problem* involves decisions regarding the number and location of facilities in each echelon, the product range, capacity and technology content of each facility as well as the flow of items through the network so as to minimize the total cost of serving the clients. The cost items involved in designing a production-distribution network are the fixed costs of facility construction, the variable costs of technology acquisition and operation as a function of the capacity to be built-in at each facility, and the inbound and outbound logistics costs. Considerable progress has been achieved in the development of tools for solving the relatively simpler single commodity single-echelon production-distribution system design problems as presented by Verter and Dincer [1163]. The design problem of a global firm however, would normally have a multicommodity, multi-echelon structure on which the literature is rather sparse. An earlier review of the facility location models for distribution planning problems is due to Aikens [13]. In this section we review the literature pertaining to the development of analytical techniques for solving the production-distribution system design problem stated above in its most general form.

The uncapacitated facility location problem (UFLP) that involves locating an undetermined number of facilities to minimize the fixed setup costs plus linear variable costs of serving clients constitutes a building block in solving the multicommodity, multi-echelon design problems. The multicommodity uncapacitated plant location problem (MUPLP) is a generalization of the UFLP where multiple products are required by the clients, and an additional fixed cost is incurred if an open plant is equipped to manufacture a particular product. One of the earliest works on multicommodity problems is due to Warszawski [1174] where each plant is restricted to produce a single commodity. Warszawski [1174] devised a heuristic algorithm whereas, Neebe and Khumawala [885] suggested a branch and bound procedure for this class of problems. Klincewicz, Luss and Rosenberg [704] provided an optimal and various heuristic branch-and-bound procedures for solving the MUPLP. They suggested decomposition of the nodal problems into separate UFLPs each associated with a product, for the calculation of the lower bounds. Akinc [14] analyzed the capacitated version of the problem where the size of a plant to be established is bounded above. He presented a branch-and-bound algorithm which constitutes an extension of the Akinc and Khumawala [15] algorithm for the capacitated facility location problem to the multicommodity setting. Klincewicz and Luss [703] developed a dual-based algorithm for solving the MUPLP. Their algorithm is inspired by the dual-based approach of Erlenkotter [390] for the UFLP. Erlenkotter's dual ascent and dual adjustment procedures are extended to generate a good feasible solution to the dual of the linear programming relaxation of MUPLP. These procedures are incorporated in a branch-and-bound algorithm for providing an optimal solution to the MUPLP. Klincewicz and Luss [703] reported solving extensions of the 25 plant locations 50 markets Kuehn and Hamburger [721] problem as well as a set of random MUPLP's. The computational performance of their algorithm is encouraging since the set of sixteen MUPLPs (based on the Kuehn and Hamburger problem) consisting of 3, 5, and 10 product problems, required only 18.42 seconds on the average on an Amdahl 470/V8 computer.

The two-echelon uncapacitated facility location problem (TUFLP) is a generalization of the UFLP where the commodity is processed at two echelons of facilities before being served to the clients. Kaufman, Eede and Hansen [680] suggested a generalization of Efroymson and Ray's [365] UFLP algorithm, for simultaneous location of uncapacitated single-commodity plants and warehouses. It is also possible to utilize their model for locating two echelons of warehouses given the locations of production facilities. Ro and Tcha [1009] provided a branch-and-bound procedure for solving the TUFLP with some side constraints. Their work extends the Efroymson and Ray [365] and Khumawala [693] algorithms to the TUFLP. Tcha and Lee [1122] suggested a branch-and-bound approach for the multiechelon UFLP in which the dual ascent procedure of Erlenkotter [390] is utilized. Their computational experiments however, focus on the TUFLP. Gao and Robinson [442] devised a dual-based optimization procedure for solving the TUFLP. Erlenkotter's dual ascent and dual adjustment procedures are extended to generate a good feasible solution to the dual of the linear programming relaxation of TUFLP. This solution provides a lower bound on the value of the optimal solution to TUFLP. Further, based on the dual solution, a feasible primal solution is constructed using the complementary slackness conditions. As in Erlenkotter [390] these procedures are incorporated in a branch-and-bound algorithm for providing an optimal solution to the TUFLP. Gao and Robinson [442] reported solving TUFLPs with 25 facility locations at each echelon and 35 markets in 2.4 seconds on a CDC Cyber 170/855 computer.

Capacitated, multicommodity, multi-echelon formulations are mostly focused on location of a single echelon of warehouses on the basis of existing production facilities. Elson [377] presented one of the earliest studies where the availability of management options to expand existing distribution centers (DCs) in addition to opening new ones is also incorporated. Proposed optimization procedure however, decomposes the commodity flows into the plant-to-DC and the DC-to-customer portions. Geoffrion and Graves [454] provided the most influential work on the multicommodity production-distribution system design problem. Given a set of production plants each with known capacity, the authors were concerned with locating a single echelon of DCs and assigning those DCs to customer zones in order to satisfy the demand. The model formulation is as follows:

$$\text{Minimize} \quad z = \sum_{p,i,d,j} c_{pidj} X_{pidj} + \sum_d [F_d Y_d + v_d \sum_{p,j} D_{pj} Z_{dj}] \quad (12.1)$$

$$\text{subject to} \quad \sum_{d,j} X_{pidj} \leq S_{pi}, \quad \forall p, i, \quad (12.2)$$

$$\sum_i X_{pidj} = D_{pj} Z_{dj}, \quad \forall p, d, j, \quad (12.3)$$

$$\sum_d Z_{dj} = 1, \quad \forall j, \quad (12.4)$$

$$\underline{V}_d Y_d \leq \sum_{p,j} D_{pj} Z_{dj} \leq \overline{V}_d Y_d, \quad \forall d, \quad (12.5)$$

$$\text{Linear configuration constraints on Y and/or Z,} \quad (12.6)$$

$$Y_d, \ Z_{dj}, \in \{0,1\}, \quad \forall d, j, \quad (12.7)$$

$$X_{pidj} \geq 0, \quad \forall p, i, d, j, \quad (12.8)$$

where

p = index for commodities,

i = index for the existing production plants,

d = index for potential DC sites,

j = index for customer zones (CZs),

S_{pi} = production capacity of plant i for commodity p,

D_{pj} = demand for commodity p in CZ j,

$\underline{V}_d, \overline{V}_d$ = minimum, maximum allowed annual throughput for DC d,

F_d = annualized fixed setup cost of opening DC d,

v_d = variable unit cost of throughput for DC d

c_{pidj} = unit cost of producing and shipping commodity p from plant i to CZ j through DC d.

The decision variables are:

X_{pidj} =amount of commodity p produced and shipped from plant i to CZ j through DC d,

$Y_d = 1$ if DC d is opened, 0 otherwise,

$Z_{dj} = 1$ if DC d serves CZ j, 0 otherwise.

Constraints (12.2) are supply constraints and (12.3) ensure that demand from each CZ will only be satisfied by a DC assigned to serve that CZ. The *single-sourcing* of CZs by DCs is imposed by constraints (12.4). That is the model suggests construction of a *dominant* DC for each CZ which fully serves the demand. Constraints (12.5) keep the total annual throughput of each DC between the required limits. They also enforce that a closed DC cannot be assigned to serve a CZ. Linear configuration constraints (12.6) allow representation of managerial requirements about the selection of DC sites and the DC-CZ assignments in the model. The objective is to minimize the sum of total production, transportation, DC construction and operation costs. Geoffrion and Graves [454] adopted a variant of the Benders [75] decomposition that solves the master problem as a feasibility problem. This is primarily in order not to waste effort solving a master problem to optimality when there are only a few Benders cuts to represent the subproblem at the earlier iterations. Their algorithm also describes how to synthesize the dual solutions to the single-commodity transportation sub-subproblems to obtain dual solutions to the multicommodity transportation subproblem. The authors reported application of the solution technique to a real problem for a major food firm with 17 commodity classes, 14 plants, 45 possible DC sites, and 121 CZs. They also mentioned another large scale application for a major manufacturer of hospital supplies.

Moon [863] extended the problem formulation to incorporate the non-linearities in DC throughput costs due to economies of scale. He presented an application of the Generalized Benders Decomposition devised by Geoffrion [449] to the nonlinear production-distribution system design problem. Approximate dual prices are generated by solving linear (instead of concave) subproblems which are then adjusted to better represent the concavity in throughput costs. These adjusted dual prices are incorporated in the Benders cuts. The computational results are reported to be encouraging. Van Roy [1159] presented an extended application of the Geoffrion and Graves [454] model for multi-level production-distribution planning and transportation fleet optimization. The problem belongs to a liquified petroleum gas company with 2 commodities (propane and butane), 2 refineries, 10 potential bottling plant locations, 40 potential depot locations, 40 potential breakpoints (transporters' home sites), and 200 customer regions. Location and capacity expansion decisions associated with the bottling plants, depots and breakpoints are given together with the decisions concerning the transportation fleet size, and the transportation shift systems and schedules. Note that, this problem requires optimization of the location decisions regarding three echelons of the production-distribution system compared to Geoffrion and Graves [454] concerned with the location of a single-echelon of facilities. The problem was solved using a matrix generator for network-like problems and MPSARX [1160], a general-purpose mathematical programming software system augmented with automatic reformulation and cut generation features.

Cohen, Lee and Moon [230] presented an integrated model for production-distribution system design as an implementation of the manufacturing strategy paradigm suggested in Cohen and Lee [228]. Production plant and DC locations, DC-CZ assignments, and flow of raw materials, intermediate and finished products through the system are simultaneously provided so as to minimize the sum of plant / DC construction and operation costs, raw material purchase costs, and transportation costs. Constraints ensure that the number of open plants, production volume at each plant, and DC throughput levels are within their upper and lower limits, and certain production plants and DCs are fixed open as a managerial policy. Customer demand for each product need to be fully satisfied taking into account the production capacity and the raw material supply constraints. Cohen, Lee and Moon [230] made a special effort to capture the scale and scope economies in production costs. The base level production costs at each plant are adjusted via a production cost multiplier which is a function of the capacity utilization rate and the number of products produced. Thus, the model is a large scale mixed-integer mathematical program with a nonlinear objective function. The authors devised an iterative solution procedure controlled by a model hierarchy. The algorithm requires an initial plant configuration to be provided. Then, a DROP/ADD heuristic is utilized to generate a new plant configuration with either one less or one more plant. The current DC configuration is taken as input and initial DC-CZ assignments are either carried out by an assignment heuristic or provided by management. The first submodel is a linearly constrained nonlinear mathematical program that deals with the material flow through the network, and the product range at each facility. A simplex based algorithm is suggested for solving this subproblem. Submodel 1 provides plant production capacities as an input to the submodel 2 which is a distribution system design problem solved by the Geoffrion and Graves [454] procedure. Submodel 2 provides new DC configuration and DC-CZ assignments for the next iteration of submodel 1. The DROP/ADD heuristic is activated for a new plant configuration upon convergence of the subproblem iterations. Cohen and Lee [228] reported the use of their model in the consulting practice of Booz Allen and Hamilton Inc.

Cohen and Moon [231] employed the model described above to investigate the impact of production scale economies, manufacturing complexity and transportation costs on production-distribution systems. They analyzed the behavior of optimal solutions in response to variations in the input parameters of the production-distribution system design problem. It has been observed that economies of scale and scope as well as transportation costs can significantly affect the system structure. Recently, Cohen and Moon [232] presented a plant loading model with economies of scale and scope. The plant loading problem takes the configuration of plants and DCs, and the DC-CZ assignments given, in order to optimize the product range at each facility and the flow of materials through the production-distribution system. Note that, this problem corresponds to the first subproblem of Cohen, Lee and Moon [230]. In [232] cost of complexity is captured via a fixed cost of assigning a product line to a plant They represented economies of scale by the aid of a piecewise linear concave production cost function compared to the nonlinear representation in Cohen, Lee and Moon [230]. A variant of Benders decomposition is suggested for solving this plant loading problem.

Defining the production-distribution system design problem at the outset enables us to provide an assessment of the state-of-the-art on multi-commodity, multi-echelon networks. It is observed that an overwhelming majority of the prevailing methods for designing such systems focus on the optimization of location and allocation decisions ignoring other dimensions of manufacturing structure i.e. capacity, technology content, and product range of the manufacturing facilities and their vertical integration. The capacity acquisition costs are not incorporated in the existing models which implies an implicit assumption that they would be the same at all sites. Since this assumption is not valid in the international context, the information provided by these models regarding the size of a new facility might be far from optimal. Although, it is possible to include the capacity acquisition costs in fixed setup costs in the capacitated models, it is highly likely that such predetermined sizes for the new facilities would be suboptimal. Further, there exist a variety of manufacturing technologies for each product that can be adopted in providing the capacity to be built-in a new plant. There is a vast literature on the analysis of investments in manufacturing technology such as [402, 974]. Incorporation of the technology selection decisions in production-distribution system design models would lead to a significant improvement in their capability to assist strategic decision making. Modeling technology content of each facility in the network would also allow for addressing production flexibility, quality, and delivery performance, as well as costs in the network design process.

12.4 Designing International Production-Distribution Systems

The previous section enables the reader to trace the development of methods for production-distribution system design and their validation through real life applications. These models allow for the incorporation of multiple echelons of facilities, multiple commodities, and the nonlinearities due to economies of scale and scope that are inherent in international networks, and hence provide valuable insights in designing such systems. Note that however, the analytical approaches mentioned above are confined to a cost minimization objective. This creates a deficiency in dealing with the uncertainties associated with product markets which are crucial in the international context. A significant line of research focused on the *international plant location problem* (IPLP) to remedy this weakness. Unfortunately, the improvements in the incorporation of uncertainty are offset by the fact that IPLP addresses a very simplistic (single commodity, single echelon) production-distribution system.

IPLP constitutes a challenging version of the UFLP within the international context, and is stochastic by nature due to the randomness in price and exchange rate movements. National governments provide subsidized financing (as well as low tax rates) to attract multinational companies to locate production plants in their country. Multinational companies on the other hand, use foreign financing packages to hedge against international price and exchange rate fluctuations. Thus, financing decisions are an integral part of IPLP due to risk reduction strategies as well as locational incentives via subsidized interest rates. The pioneering work in modeling the interaction between international location and financing decisions is due to Hodder and Jucker [577]. That model however, is restricted to a deterministic setting. Hodder and Jucker [578] extended their previous work to incorporate uncertainty, ignoring financing decisions. They presented a single period model where a multinational company is assumed to be a mean-variance decision maker in terms of the after-tax profits. They modeled the random deviations from PPP via a single factor price generating mechanism as follows:

$$P_j e_{1j} = b_j(P_1 + \epsilon_j), \tag{12.9}$$

where

P_j = the price at market j,

e_{1j} = the units of the numeraire currency per unit of currency j,

b_j = market adjustment parameter

P_1 = the random price in the home country with mean $\overline{P_1}$ and variance σ_p^2, and

$$\epsilon_j \sim N(0, \sigma_j^2), \quad cov(\epsilon_j, \epsilon_k) = 0 \;\; \forall j, \forall k \neq j.$$

Note that, although price uncertainty is explicitly taken into account, the incorporation of exchange rate uncertainty is rather implicit. Hodder and Dincer [576] developed a model for simultaneous analysis of the international location and financing decisions. The mixed integer program has a quadratic objective function due to adoption of the mean-variance framework. The authors suggested a multifactor approach to diagonalize the variance-covariance matrix in the objective function. This results in a considerable reduction in the computational difficulty of solving IPLP. Recently, Min [848] suggested a chance-constrained goal programming model in order to incorporate the presence of dynamism and multiple objectives in the locational decisions of multinational firms. Although far from capturing the total complexity, international plant location methods provide a viable building block in designing international production-distribution systems.

Pomper [964] provided one of the earliest studies on international investment planning. He proposed a model to assist management in the evaluation of alternative manufacturing policies on a *global* basis. The model prescribes the optimal time-phasing of the location, technology and capacity investments as well as the optimal flow of materials throughout the future network. Pomper [964] analyzed the *single-commodity, single production stage* firms. He assumed that the multicommodity, multi-echelon structures can be decomposed into these easier to handle type of elements. Uncertainty in the environment is modeled by an *uncertainty tree* to represent the time-phased relationships among the environmental scenarios each occur with a certain probability conditional to the previous state of the environment. The expected present value of consolidated cash flow is maximized. Financial decisions are not considered although, Pomper [964] accepted that the international financial markets are not perfect. Economies of scale in production costs and in investment costs associated with the alternative technologies are approximated via fixed-charge linear functions. Dynamic programming is used to model and solve the international investment problem where a *manufacturing state* is defined to be the number of plants of each technology in each country. An alternative mixed-integer programming formulation of the problem is also presented which is claimed to be superior in large scale applications. Pomper [964] reported an application of his model to a mature agricultural chemical product of a US-based multinational chemical company.

Kendrick and Stoutjesdijk [692] devised an investment project evaluation model. The *single-country* based firms are analyzed taking into account their international activities such as imports and exports. Their model can be conceived as a manufacturing strategy planning tool since the chosen investments constitute means to implement the manufacturing policies. Decisions prescribed by the *multiproduct, multiperiod* model are; increments to the capacities of production units, shipments from plants to markets and among plants, exports, imports, domestic purchases of production factors, and by-product sales. The only set of integer variables in the mathematical program represents the capacity expansion decisions. Economies of scale in capacity acquisition is represented via a fixed-charge linear approximation. A *two-stage* production structure is incorporated and the future is assumed to be known with certainty. Net present costs are minimized to satisfy the demand by upgrading the current system via capacity expansion investments. Kendrick and Stoutjesdijk [692] suggested usage of the general-purpose integer programming softwares for solving their model.

Cohen, Fisher and Jaikumar [227] proposed a normative framework for strategic management of the international production-distribution systems. The firm's product range, production plant locations, capacities and production technologies are taken as given and the raw material sourcing, production, and market supply decisions are optimized. The *multicommodity, multiperiod* model seeks to maximize the net present value of the after-tax profits in the numeraire currency of the firm. Many of the international issues are incorporated such as duties and tariffs, currency exchange rates, differences in corporate tax rates in each country, market penetration strategies, and local content rules. Economies of scale in raw material purchasing is represented by the availability of a set of vendor contract options. There are fixed costs of plant loading which can be interpreted as costs of *complexity*. Furthermore, the fixed plant loading costs enable representation of the economies of scale in production via several "pseudo-products" corresponding to the various cost rates associated with different levels of production. Production is assumed to have a *single-stage* structure and production plants are modeled to have capacity limits both in terms of the overall product range and on a per product basis. The Cohen, Fisher and Jaikumar [227] model is a mixed-integer nonlinear program. Nonlinearity in the objective function is caused by the co-existence of the *financial decisions* namely, transfer prices and overhead allocations together with the *operational decisions*. Hence, the authors suggested a hierarchical solution procedure. First step involves optimization of the operational variables concerning vendor contract selection, plant loading, purchasing, production, and market supply decisions on the basis of a fixed level of the financial variables. Second step solves for the optimal values of the financial variables, given optimal levels of the operational variables provided by step 1. This provides input for the next iteration of the first step. Cohen, Fisher and Jaikumar [227] suggested adoption of a mean-variance framework for incorporation of the price and exchange rate uncertainty in the international markets. The authors also suggested utilization of transfer prices for tax minimization purposes as well as for country-decomposed implementation of global manufacturing strategies.

Cohen and Lee [229] reported application of a variant of the Cohen, Fisher and Jaikumar [227] model in a multinational company manufacturing personal computers. In the Cohen and Lee [229] study the product structure is modeled to include major components, subassemblies, and finished products. Obviously this increases the tradeoff capability. However, it should be noted that the model is *deterministic* and *single-period* which may partly offset the above enhancement. Cohen and Lee [229] perceived global manufacturing strategy as a collection of *component strategies* which are designed at various echelons of the production-distribution system. Component strategies are associated with the raw material sourcing, plant charter, and distribution/market supply activities. Firms have *policy options* for each of the component strategies. Various combinations of these policy options constitute the *global policy options* for the firm. The Cohen and Lee [229] model is essentially a *strategy evaluation model* since provision of the set of available global policy options is required by the solution procedure. A particular global policy option is translated into the *structural decisions* of the model in an interactive manner. That is values of the indicator variables are fixed in order to evaluate the global policy option under consideration. Cohen and Lee [229] specified the following zero-one decision variables:

- Assignment of finished products and subassemblies to plants,

- Assignment of vendors to plants for each major component and subassembly,

- Assignment of vendors to DCs for each finished product that is sourced directly from vendors to DCs,

- Assignment of supply links from plants to DCs and from DCs to CZs,

- Transfer pricing policies for assigning transportation costs for intermediate and finished goods from one plant to another,

- Transfer pricing policies for assigning transportation costs of finished goods from plants to DCs,

- Determination of whether demands from a market region are to be satisfied.

Then, the remaining problem is a large scale linear program which involves material flow decision variables that denote the quantities of major components, subassemblies and finished products transported through the production-distribution system, and production decision variables that denote the quantities of items manufactured at the plants.

Recently, Huchzermeier [602] presented a model for global manufacturing strategy planning under exchange rate uncertainty. He suggested a multinomial approximation to the stochastic exchange rate process. A stochastic dynamic programming formulation is developed for evaluation of the global manufacturing strategy options. An *option* O_t defines all operational and financial policies together with the structure of the production-distribution system at period t. State of the firm at the beginning of period t is determined by the current realization i of the exchange rates denoted by vector e_{0t}^i, and O_{t-1}. Hence, the recursion function is:

$$V_t(e_{0t}^i, O_{t-1}) = \max_{O_t \in \Omega_t} \{P_t(e_{0t}^i, O_{t-1}, O_t) + \Phi_t \sum_j \pi_{ij} V_{t+1}(e_{0,t+1}^j, O_t)\},$$

(12.10)

where

$V_t(.)$ = discounted value of the firm at period t, given the adoption of O_{t-1} and the realization of e_{0t}^i,

Ω_t = set of available global manufacturing strategy options at period t,

Φ_t = risk adjusted discount factor for period t,

π_{ij} = the stationary transition probability from exchange rate realization vector i to exchange rate realization vector j.

The profit at period t is:

$$P_t(e_{0t}^i, O_{t-1}, O_t) = SP_t(e_{0t}^i, O_t) - \delta(O_{t-1}, O_t),$$ (12.11)

where

$SP_t(.)$ = expected global after-tax profits for operating under O_t and e_{0t}^i,

$\delta(.)$ = switching cost from O_{t-1} to O_t.

At each period, the subproblem SP_t is formulated as a stochastic program with recourse in order to also incorporate the demand uncertainty. The subproblem is essentially a constrained resource allocation problem and constitutes a variant of the Cohen and Lee [229] model. Huchzermeier [602] suggested a hierarchical procedure for solving the integrated model. Computational tractability however, decreases as the number of exchange rate processes and the number of demand scenarios increase.

Although, designing multicommodity, multi-echelon production-distribution networks attracted the attention of researchers for two decades, the studies that extend the basic design problem to incorporate the distinguishing features of international environment are not that many. It is observed that the most recent analytical methods for international production-distribution system design are developed for evaluation of alternative manufacturing configurations. Thus, the strategy design process is confined to the strategy options envisaged by management. This means that identification of the optimum configuration is conditional to its provision by the management as a viable strategy alternative. Given the size and complexity of the arising mathematical programs for designing international networks, the loss of computational tractability when such a model is treated in its entirety is quite natural. However, the alternative manufacturing configurations for a global firm proliferate, and hence generation of strategy options for manufacturing structure constitutes a problem by itself. Therefore, research is needed on the development of solution techniques that will enhance the optimization capability of international production-distribution system design methods in terms of the structural decision variables.

12.5   Concluding Comments

The prevailing analytical approaches for location of manufacturing facilities in a global firm can be classified into two broad categories:

i) *Strategy Evaluation Models:* These models are comprehensive in nature, and usually incorporate some of the features of international environment. Specification of a set of alternative manufacturing configurations is required. Each element of this set is represented by fixing values of the associated structural (integer) variables in the model. The remaining model in the (continuous) flow variables can then be used to evaluate the manufacturing configuration under consideration.

ii) *Strategy Generation Models:* These models focus on the relatively simpler basic design problems and the IPLP. The structural and material flow decisions are provided simultaneously. It is however, still quite cumbersome to provide exact solutions to many of the arising mathematical programs which necessitates development of heuristic algorithms in many cases.

At this stage, the methodology for international production-distribution system design is in need of several enhancements to better capture the dynamics of global manufacturing strategy planning. First, the design techniques should be improved to address all the relevant competitive priorities such as quality, delivery performance and flexibility rather than focusing on cost as a predominant objective. Incorporation of the technology content of each facility in the multicommodity, multi-echelon network models would be a significant contribution toward that direction. Second, the design process should also address the policies regarding coordination of internationally dispersed facilities that would lead to global competitive advantage, not just the configurational decisions alone. Third, it should be realized that a production-distribution network represents only the manufacturing activity of the firm and its linkages with the buyers and suppliers. A comprehensive model of global competition however, requires incorporation of actions of the firm's current and potential competitors as well as government interventions which may require a game-theoretic setting. Finally, strategy generation capability of the models for international production-distribution system design needs to be improved via the development of new approaches to optimize the structural decisions.

Part III

COMPETITIVE FACILITY LOCATION

13
COMPETITIVE FACILITY LOCATION IN THE PLANE

Tammy Drezner
California State University, Dominguez Hills

13.1 Introduction

Facility location models deal mainly with the location of plants, warehouses and other industrial facilities. One branch of location theory deals with the location of retail and other commercial facilities which operate in a competitive environment. The facilities compete for customers and market share, with a profit maximization objective. The customary objective function to be maximized is the market share captured by the facilities. All competitive location models attempt to estimate the market share captured by each competing facility in order to optimize its location. The best location for a new facility is at the point at which its market share is maximized. For a survey of various competitive facility location models see [288, 289, 290, 291, 374, 469].

The first modern paper on competitive facility location is generally agreed to be Hotelling's paper on duopoly in a linear market [599]. Hotelling considered the location of two competing facilities on a segment (two ice-cream vendors along a beach strip). The distribution of buying power along the segment is assumed uniform and customers patronize the closest facility. When one facility is located and there is no competition, all customers patronize the existing facility. However, when a competing facility is introduced and is located at a different point on the segment, the customers on one side of the midpoint between the two facilities patronize one facility and the customers on the other side of the midpoint patronize the second facility. If one facility is held fixed in place, the best location for the second is either immediately left or right of the fixed one, depending on which segment - left or right of the existing facility - is longer. In extensions of Hotelling's model [296, 438, 496, 511, 512, 513, 757, 898, 973, 992, 1090] it is assumed that customers patronize the closest facility.

The assumption that customers patronize the facility closest to them implies that the competing facilities are equally attractive. For equally attractive facilities, the plane is partitioned by a Voronoi diagram [897, 898]. It is implicitly assumed that all customers located at a demand point patronize the same facility. This, in turn, implies an "all or nothing" property. The combined buying power at a demand point is assigned entirely to one facility and none is assigned to other facilities, unless two or more facilities are equidistant. Such an assumption is used to simplify the models. In reality, customers at a demand point divide their buying power among several facilities.

An extension to Hotelling's approach is the location-allocation model for the selection of sites for facilities that service a spatially dispersed population. Both the facilities' locations and the allocation of customers to them are determined simultaneously. The allocation of customers to facilities is made using Hotelling's proximity assumption - each facility attracts the consumers closest to it. The market share attracted by each facility is calculated and the best locations for the new facilities are then found. Multifacility location-allocation models analyze the system-wide interactions among all facilities. Goodchild [496] suggested the location-allocation market share model (MSM). A retail firm is planning to open a chain of outlets in a market in which a competing chain already exists. The entering firm's goal is to maximize the total market share captured by the entire chain. Most location-allocation solution methods rely on heuristic approaches that do not guarantee an optimal solution, rather they provide good solutions for implementation. The best locations are selected from a user-provided set of potential sites. A book edited by Ghosh and Rushton [471] provides a collection of papers on the subject. See in particular the chapters by [68, 220, 902]. A comprehensive review of location-allocation models can be found in [466].

When the facilities are not equally attractive, the proximity premise for allocation is no longer valid. To account for variations in facility attractiveness, a deterministic utility model for competitive facility location is introduced by Drezner [288]. Hotelling's approach is extended by relaxing the proximity assumption. Consumers are known to make their choice of a facility based on factors other than distance alone. Therefore, it is assumed that customers base their choice of a facility on facility attractiveness which is represented by a utility function. This utility function is a composite of facility attributes and the distance to the facility. The utility function represents the expected satisfaction. A distance differential is calculated based on the attractiveness difference between the competing facilities. It is suggested that a customer will patronize a better and farther facility as long as the extra distance to it does not exceed its attractiveness advantage, i.e., the calculated distance differential. For example, paramedics transporting a motor vehicle accident victim will by-pass a nearby hospital in favor of a farther, better equipped trauma center as long as the difference in quality of care exceeds the adverse effect caused by the extra distance. A break-even distance is defined based on the distance differential. At the break-even distance the attractiveness of two competing facilities is equal. This break-even distance, therefore, is the maximum distance that a customer will be willing to travel to a new facility based on his perception of its attractiveness and advantage, or disadvantage, relative to existing facilities. All customers at that demand point will patronize the new facility if it is located within the break-even distance. While customers are no longer assumed to patronize the closest facility, customers at a certain demand point are assumed to apply the same utility function. Therefore, they all patronize the same facility. The "all or nothing" property is maintained in this extension.

To address the "all or nothing" assumption of the deterministic utility model, a random utility model is introduced by Drezner [290]. The deterministic utility model is extended by assuming that each customer draws his utility from a random distribution of utility functions. This assumption eliminates the "all or nothing" property since a probability that a customer will patronize a particular facility can be established and is no longer either 0% or 100%. Both the deterministic and random utility models are discussed later in this chapter.

An alternative approach to the location of competing facilities, based on the gravity model, was introduced by Huff [603, 604] and is extensively used by marketers. According to the gravity model [236, 990], two cities attract retail trade from an intermediate town in direct proportion to the populations of the two cities and in inverse proportion to a power of the distances from them to the intermediate town. Huff proposed that the probability that a customer patronizes a retail facility is proportional to its size (floor area) and inversely proportional to a power of the distance to it. Facility size, or square footage, is a surrogate for facility attractiveness. Huff depicted equi-probability lines. A customer located on such a line between two facilities patronizes the two facilities with equal probability. These equi-probability lines divide the region into catchment areas, each dominated by a facility, in a manner similar to the Voronoi diagram [897]. See also Chapter 6 in this book. These lines do not define an "all or nothing" assignment of customers to facilities, rather, at any demand point, the proportion of consumers attracted to each facility is a function of its square footage (attractiveness) and distance. Other relevant papers based on the gravity model are [67, 581, 733].

In the original Huff formulation, facility floor area serves as a surrogate for attractiveness. A major improvement on Huff's approach was suggested by Nakanishi and Cooper [884]. They introduced the multiplicative competitive interaction (MCI) model. The MCI coefficient replaces the floor area with a product of factors, each a component of attractiveness. Each factor in the product is raised to a power. Thus, the attractiveness of a facility is a composite of a set of attributes rather than the floor area alone. Nakanishi and Cooper's idea was elaborated on and applied by Jain and Mahajan [635] to food retailing using specific attractiveness attributes. All these papers suggest the evaluation of a user provided discrete set of potential sites for the location of a new facility.

Huff's and Nakanishi and Cooper's models were extended to the location of multiple facilities by [8, 465]. Achabal et al. [8] extended the MCI model to the location of multiple facilities which belong to the same chain. They modeled the problem as a nonlinear integer programming problem and employed a random search procedure combined with an interchange heuristic to identify optimal and near-optimal sets of locations. Ghosh and Craig [465] proposed a franchise distribution model. An expanding franchise seeks to maximize sales while minimizing cannibalization of franchise outlets. They too formulated the model as a nonlinear integer programming problem but included additional factors in the model such as advertising. These two models also select the best locations from a user-provided set of alternative sites.

Other papers [581, 1209] suggest variations on Huff's formulation by replacing the distance raised to a power with an exponent of the distance. This formulation accelerates the distance decay. It does not change the nature of the solution.

Finding the best location for a new facility (or multiple facilities) in a continuous space using the gravity model objective is discussed in Drezner [289] for the single facility case, and in Drezner et al. [291] for the location of multiple facilities. These extensions of the gravity model are also discussed later in this chapter.

Central Place Theory [217, 562] is yet another related theory. It provides a framework for analyzing the size and spacing of retail centers. The hierarchy of service centers represents differences in availability of goods and services of varying order. Customers are assumed to travel to the closest facility that offers the service or goods sought. Losch [770] examined the interplay between range and threshold. The range is the maximal distance travelled to a facility, a spatial extent of centers. Christaller [217] defined range as market area delineation and spatial coverage. The range is similar in concept to the break-even distance used in Drezner [288]. Threshold is the total effective demand, or "critical mass", required to support a particular facility. The ratio between the total demand and the threshold level determines the maximum number of facilities that can be profitably located in an area. While Christaller assumed that any place that offers a higher-order good will also offer all lower order goods, Losch relaxed this constraint. A more recent paper on central place theory is [68]. A review is provided in [471].

The remainder of this chapter is devoted to a discussion of 1) the deterministic utility model [288], 2) the random utility model [290], and 3) gravity based models [289, 291]. These models deal with finding the optimal location for one or more new competing facilities anywhere in a continuous plane using various assumptions.

13.2   The Deterministic Utility Model

The deterministic utility model in competitive facility location was introduced by Drezner [288]. Non-equally attractive facilities have been extensively studied in the context of consumer spatial choice behavior. Various behavioral models are employed in an effort to discern preferences and their components as well as to predict consumer choice. It is generally agreed that customers, through a decision-making process, make a choice of the facility which is expected to maximize their satisfaction. This choice is determined by some formula according to which customers evaluate alternative facilities' attributes weighted by their personal salience to arrive at an overall facility attractiveness. A trade-off between distance and attractiveness takes place. Based on this premise researchers of spatial choice behavior employ a utility function in order to measure facility attractiveness or utility. This utility function represents the degree of expected satisfaction with each alternative as a function of the relevant characteristics of the facility [287, 495, 975, 1052, 1053, 1096, 1134, 1135].

The utility function establishes a relationship of the type

$$U = F(x_1, x_2, ..., x_m) \tag{13.1}$$

In most models it is assumed that the function F is additive in the independent variables leading to the general additive function:

$$U = \sum_{p=1}^{m} w_p f_p(x_p) \tag{13.2}$$

where m attributes x_p for $p = 1, ..., m$ are considered, each with an associated weight w_p. A function $f_p(x_p)$, is assumed. These models assume that customers patronize the facility with the highest utility. Spatial choice behavior is taken a step further [288], to be used in facility location. Based on aggregated utility values for existing facilities and a utility function for a new facility, the best location is found for the new one. This best location is found using the "break-even" distance concept. At the break-even distance the utilities of the existing facilities and the new facility are equal. The maximum distance that a customer is willing to travel to a new facility can be calculated by applying any of the spatial choice behavior models. Suppose there are k existing facilities in the area. Consider an existing facility j which is compared, by demand point i, to a new facility to be located in the area. The attributes of the existing facility (including the distance d_j) are known, therefore its utility U_j (calculated by (13.1)) can be established. In calculating the utility U of the new facility, all independent variables (attractiveness attributes) in the functional relationship are known except the distance d from demand point i. The distance d is regarded as the first attribute x_1. The utility U for the new facility is reduced to a function $U = F(d, x_2, ..., x_m)$. Since all attributes $x_2, ..., x_m$ are known, U is a function of the distance d only, and can be expressed as $U(d)$ of the distance to the new facility. A customer prefers the new facility to existing facility j if $U(d) > U_j$. The maximum distance for which the customer prefers the new facility over existing facility j is the break-even distance. The break-even distance is calculated by solving $U(d) = U_j$. Since $U(d)$ is a decreasing function of d, this equation has a unique solution. The solution is defined as the break-even distance, Δ_{ij}, relative to facility j. Δ_{ij} is the maximum distance that customer i will travel from his base to the new facility rather than patronize existing facility j. The break-even distance relative to all existing facilities is the minimum among the break-even distances. The break-even distance D_i from demand point i to the new facility is therefore:

$$D_i = \min_{1 \le j \le k} \{\Delta_{ij}\} \qquad (13.3)$$

If the distance from demand point i to the new facility is less than the break-even distance D_i, a customer located at demand point i will patronize the new facility. This is analogous to saying that if the new facility is located inside a circle centered at demand point i with a radius of D_i, demand point i will patronize the new facility.

The distance between demand point i and the new facility is defined as $d_i(X)$ where X is the location of the new facility. The market share, $M(X)$, attracted by the new facility located at X is:

$$M(X) = \sum_{d_i(X) < D_i} B_i \qquad (13.4)$$

where the break-even distance D_i is defined by (13.3), and B_i is the buying power associated with demand point i. The buying power is the aggregated discretionary spending power at demand point i for the product(s) in competition. Note that if $d_i(X) = D_i$, the buying power at demand point i is equally divided among all facilities with the same maximal utility (including the new one). This case represents a "tie". Equation (13.4) states that the market share captured by the new facility is the sum of the buying power (B_i) at all demand points that are attracted to it. The market share $M(X)$ is to be maximized by the selection of the best X. The detailed algorithm for solving this problem is presented in [288]. Note that the optimal location for the new facility is sensitive to the attractiveness of the facility. Different attractiveness measures may yield different optimal locations.

13.3   The Random Utility Model

As mentioned above, the deterministic utility model employs an "all or nothing" assignment of customers to facilities. To account for variations in individual utility functions, the deterministic utility model is extended to the random utility model. It assumes that the utility function varies among customers located at a demand point. In [290] it is assumed that the utility function has the form

$$U(Q, d) = \sum_{i=1}^{M} w_i Q_i - d \qquad (13.5)$$

where Q is a vector of M quality measures and d is the perceived distance between the customer and the facility. The vector w represents the weights adjusted such that the coefficient for the distance is "-1". It is assumed that all the variables: the quality measures Q_i for $i = 1, ..., M$, the weights w_i for $i = 1, ..., M$, and the distance d are drawn from a probability distributions with known means and variances. Since the utility function is a sum of distributions, for a sufficiently large M $U(Q, d)$ is approximately normally distributed by the central limit theorem. In the computational experiments that follow constant weights w_i's, normally distributed quality measures Q_i's, and normally distributed distance d were employed. Hence, $U(Q, d)$ is normally distributed for any M.

The probability that a customer will prefer a certain facility to all other facilities is calculated by applying the multivariate normal distribution [312, 643]. Calculating multivariate normal probabilities involves a [k-1]-dimensional integral which is very difficult to calculate for large k's [312]. Reasonable run times can be achieved for dimensionality not exceeding six. That means that calculations involving up to seven facilities (including a new one if one exists) can be calculated in reasonable time.

For a relatively large k, when the calculation of the multivariate distribution takes too long, a reasonable simplifying assumption is that the variances of the utility functions are the same. The detailed calculations are discussed in [290]. Once the probabilities are calculated, the market share captured by a particular facility (new or existing) can be calculated as a weighted sum of the buying power at all demand points. For each demand point, the buying power is multiplied by the probability that a customer will patronize that particular facility.

13.3.1   Finding the best location

It was empirically shown in [290] that the market share can be approximated by an exponential decay function. This exponential decay function is very well fitted by a function of the type: $f(d) = f(0) \cdot e^{-d(\alpha + \beta d)}$ where α and β are positive constants. A method for finding α and β for each demand point is presented in [290]. Finding the best location is then transformed to maximizing the function $f(x, y)$: $f(x, y) = \sum_{i=1}^{n} w_i e^{-d_i(\alpha_i + \beta_i d_i)}$ where d_i is the distance between demand point i and the prospective location (x, y).

This objective function is an approximation of the market share captured by the random utility method. Maximizing $f(x, y)$ yields an approximation to the optimal location. Such location problems are successfully solved by the Weiszfeld algorithm [786, 1186]. The Weiszfeld algorithm is an iterative procedure. The algorithm is started at an initial arbitrary solution, and the solution is successively improved each iteration [290]. The issue of finding the best location that maximizes the random utility function is further discussed in [340].

13.4   Gravity Models

The gravity based model is based on the assumption that "The probability that a customer patronizes a facility is proportional to its attractiveness and inversely proportional to a power of the distance to it". Huff's [603, 604] facility floor area and Nakanishi and Cooper's [884] MCI coefficient are measures of attractiveness, or in short attractiveness.

The following definitions are used in the mathematical formulation [291]:

n The number of demand points (each demand point represents a small area around it)

(a_i, b_i) The location of demand point i for $i = 1, ..., n$

B_i The available buying power at demand point i for $i = 1, ..., n$ (The total buying power for all customers assigned to the demand point)

k The number of existing competing facilities

c The number of existing facilities which are part of one's own chain. The first c out of k facilities are assumed in this category. c can be as low as zero and is bounded by k

p The number of new facilities to be located in the area

(x_m, y_m) The location of new facility m, for $m = 1, ..., p$

d_{ij} The distance between demand point i and existing facility j. $i = 1, ..., n; \ j = 1, ..., k$

$d_i(x_m, y_m) = \sqrt{(x_m - a_i)^2 + (y_m - b_i)^2}$ The distance between demand point i and new facility m. $i = 1, ..., n; \ m = 1, ..., p$.

E_j The measure of attractiveness of existing facility j for $j = 1, ..., k$

A_m The measure of attractiveness of new facility m for $m = 1, ..., p$.

λ The power to which the distance is raised.

13.4.1   The model

The model finds the best location for new facilities whose individual measures of attractiveness are known. When p new facilities are opened in an area, the total market share T attracted by these facilities and by those already part of the franchise [289, 291, 603, 604, 884] is:

$$
T = \sum_{i=1}^{n} B_i \frac{\displaystyle\sum_{m=1}^{p} \frac{A_m}{d_i^\lambda(x_m, y_m)} + \sum_{j=1}^{c} \frac{E_j}{d_{ij}^\lambda}}{\displaystyle\sum_{m=1}^{p} \frac{A_m}{d_i^\lambda(x_m, y_m)} + \sum_{j=1}^{k} \frac{E_j}{d_{ij}^\lambda}}
\tag{13.6}
$$

The objective is to find the best location in the plane that maximizes the total market share captured T using equation (13.6).

A sequence of algebraic manipulations [289, 291] leads to a somewhat simpler minimization problem, one of minimizing the total buying power *not attracted* by the chain.

$$
F = \sum_{i=1}^{n} B_i \frac{\displaystyle\sum_{j=c+1}^{k} \frac{E_j}{d_{ij}^\lambda}}{\displaystyle\sum_{m=1}^{p} \frac{A_m}{d_i^\lambda(x_m, y_m)} + \sum_{j=1}^{k} \frac{E_j}{d_{ij}^\lambda}}
\tag{13.7}
$$

The two sums for existing facilities are constants. Let

$$
G_i = B_i \sum_{j=c+1}^{k} \frac{E_j}{d_{ij}^\lambda}; \quad H_i = \sum_{j=1}^{k} \frac{E_j}{d_{ij}^\lambda}
\tag{13.8}
$$

These definitions lead to:

$$
\text{minimize} : \left\{ F = \sum_{i=1}^{n} \frac{G_i}{H_i + \displaystyle\sum_{m=1}^{p} \frac{A_m}{d_i^\lambda(x_m, y_m)}} \right\}
\tag{13.9}
$$

13.4.2   The Minimization of F

The concepts of convexity and concavity of functions are very important in finding the optimal solution of a function [1226]. Concave functions have a unique maximal point whereas convex functions have a unique minimal point. If a function is concave, then its three-dimensional plot will look like one "hill". If it is convex, it looks like one "crater". Functions are either concave or convex or neither. A search for the maximal point of a concave function can be started at an arbitrary point, and as long as the procedure "climbs up the hill", the best location must be found because the top of the hill will be reached. If the function is not concave, the function may contain several "hills". The highest "hill" represents the global maximum. All the hilltops are local maxima (one of which is the global one). If a maximization procedure is started at an arbitrary point and proceeds by going up the hill, a local maximum will be reached, which may or may not be the optimal location. The customary way to handle these situations is to repeat the search from many starting points ending at several different local maxima, and selecting the best one found. If the process is repeated enough times, the global maximum is likely to be reached. The function F in (13.9) is not convex, therefore there may be more than one local minimum. A more detailed discussion of this issue can be found in [289].

In [289] a procedure based on Weiszfeld's algorithm [786, 1186] is proposed to solve the single facility location problem. This algorithm can be used in a univariate search. A univariate search is a heuristic method in which each variable is optimized while the other variables are held constant. An iteration consists of optimizing the first variable while holding the others constant, then optimizing the second variable while holding the others (including the first one) constant, optimizing the third variable, then the fourth etc. until all the variables are optimized once, one at a time. Each iteration improves the objective function. The algorithm is terminated when the improvement in the objective function value is smaller than a pre-specified tolerance level for a full round of optimizations on all the variables. For our particular problem it means finding the best location for one facility while holding all other facilities rooted in their places. Once this is accomplished, the optimized facility is rooted in its location and another facility's location is optimized. One iteration consists of finding the best location for all p facilities, one facility at a time. The algorithm is terminated when the changes in location for all new facilities are small.

Two ways for employing the univariate search are proposed in [291]. The one which provided better solutions is presented here.

Algorithm 1

1. Randomly generate p sites for the p new facilities.

2. Perform one Weiszfeld iteration for each of the facilities while holding the others rooted in their places. (See [291] for details.)

3. After all p facilities were relocated one by one in Step 2, calculate the changes in their locations. If the location changes are less than a prespecified ϵ, terminate the algorithm.

The following are two practical approaches to solving (13.9) [291]:

The AMPL Approach

Instead of pursuing a specialized algorithm for the solution of these problems, standard non-linear programming codes available for the solution of non-linear programming problems like (13.9) can be used. The student version of AMPL [418] was used to solve these problems. Since these standard programs assume that the objective function is convex, one still needs to re-solve the problem repeatedly with various starting points and select the best solution for implementation. Also, the non convexity of the objective function may cause difficulties in the solution procedure itself. The AMPL modeling program is quite easy to write, is very short and compact, and is easy to follow. The full program is given in [291].

The Excel Approach

Spreadsheet software now incorporate optimization capabilities. The "solver" option in Excel can be used to solve (13.9). In Excel 4.0 this option is found under "formula". A spread sheet is built to calculate the objective function, and the solver provides the optimal solution with the calculated market share at that point. Since the problem is not convex, the procedure has to be repeated using many randomly generated starting locations and the best solution is selected for implementation.

13.5   Computational Results

Extensive computational results are reported in [288, 289, 290, 291]. The results were obtained on a 486 PC compatible computer. This section summarizes the main results.

For illustrating the various models an example problem is used. It consists of $k = 7$ existing facilities and $n = 100$ demand points uniformly distributed in a square area of 10 by 10 miles. Each demand point is assigned a buying power $B_i = 1$ for a total buying power of 100. The problem is depicted in Figure 1.

The best location of an eighth facility was found using the various methods. It is interesting to note the variability in the location solutions and market share captured as found by the different models, each using a different set of assumptions. The four location solutions are depicted in Figure 1.

Equal attractiveness: Assuming equal attractiveness of facilities for the deterministic utility model [288] each demand point selects the closest facility. The optimal solution is at $x = 3$, $y = 2.25$, with captured market share of 15.00.

Deterministic Utility: Assuming non-equal attractiveness for the deter-

FIGURE 1. Locations of Existing Facilities and Optimal Location for a New Facility by Various Methods

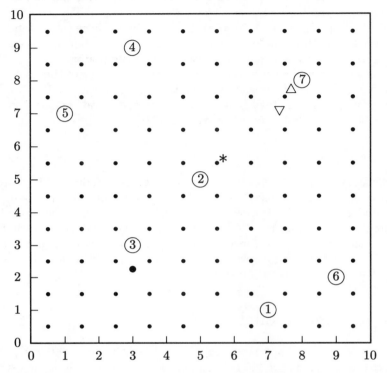

ⓙ Facility j; • Demand Point; ● Equal Attractiveness (Proximity); △ Deterministic Utility; ▽ Random Utility; ∗ Gravity Model.

ministic utility model [289], the utility function used is:

$$U = \frac{0.80Q_1 + 0.73Q_2 + 0.72Q_3 + 0.68Q_4 + 0.66Q_5}{1.5} - d \quad (13.10)$$

with the quality measures given in Table 13.1. The location is at $x = 7.56$, $y = 7.66$, and market share of 22.00.

Random utility: Assuming a random utility model [290], the utility function (13.10) was used with the means of the Q's given in Table 13.1, the standard deviations of the Q's being "1", and standard deviation of the distance being 0.5. The location was at $x = 7.22$; $y = 7.06$ with a captured market share of 15.09.

Gravity model: The location that maximizes the market share using gravity models and assuming equal attractiveness for the existing

TABLE 13.1. Average Quality Measures for Facilities

Quality Measure	Rating of Quality Measure by Facility Number							
	1	2	3	4	5	6	7	New
1	6.8	7.8	7.6	8.0	8.1	6.1	6.1	7.8
2	7.2	6.4	7.2	7.2	6.5	5.8	6.7	6.4
3	6.9	8.9	7.1	6.1	7.6	8.4	7.4	7.1
4	8.1	7.2	6.0	6.7	8.4	8.0	6.5	8.0
5	8.7	6.3	7.2	7.3	7.1	7.6	7.9	6.9

TABLE 13.2. Optimal Locations and Market Shares by Various Methods

Method	Location		Market Share Captured (by Method)			
	x	y	Proximity	Deter. Utility	Random Utility	Gravity Model
Proximity	3.00	2.25	15.00	16.00	12.67	12.26
Determ. Utility	7.56	7.66	11.00	22.00	14.71	11.86
Random Utility	7.22	7.06	10.00	16.00	15.09	12.31
Gravity Model	5.59	5.54	9.00	11.00	12.67	12.93

facilities and the new one [289] is at $x = 5.59$, $y = 5.54$ with a market share of 12.93.

These optimal locations for an eighth facility and the market share captured employing the various methods are reported in Table 13.2 and depicted in Figure 1. All optimal locations appear to be arranged around the area diagonal. This is partly because the problem is close to symmetry around that diagonal. Usually such locations are placed in "holes" containing customers distant from an existing facility. Three of the four location solutions are located in the "hole" between facilities #2 and #7, while one solution is situated to take advantage of the "hole" under facility #3. Optimal locations tend to be close to an existing facility such that the new facility captures a large share of that existing facility's market share.

It appears that the deterministic utility model yields the largest market share. However, the best location is obtained using the assumption that is *closest* to actual customer behavior.

The multiple facility case applying gravity models [291] was run using the Weiszfeld approach, AMPL, and Excel. Problems which require finding the location for p new facilities, each with attractiveness of $\frac{1}{p}$ for $p = 1, ..., 10$, were solved. This particular attractiveness distribution provides for the same *total* attractiveness for all problems.

For $p = 1$ there are 11 local optima, therefore the algorithms were run from 100 different starting solutions. The three methods provided comparable results. The Weiszfeld algorithm was the fastest, a little over one second per starting solution; AMPL took about 5-6 seconds per starting solution and Excel was the slowest, taking about 40-50 seconds per starting solution.

The problem for $p > 1$ was also solved by Algorithm 1, by AMPL, and by Excel starting at 100 different starting solutions. The number of local optima is very large and for $p \geq 4$ all 100 solutions were different. The number of local optima is evidently much larger than 100 for a moderately large p. The Weiszfeld procedure (Algorithm 1) provided the best solutions, and AMPL provided the second best solutions. Run times for Algorithm 1 are comparable to AMPL, however Excel was much slower.

Various simulations and sensitivity analyses were also reported in [288, 289, 290, 291]. It is illustrated that in gravity based models, an increase in the attractiveness of a facility increases its market share but at a lower than linear rate. A thirty-fold increase in attractiveness results in a nine-fold increase in market share captured. These simulations can be used to analyze the impact of the introduction of a new facility or changing the attractiveness of existing facilities on the market share captured by an existing facility.

13.6   Conclusions

Models for finding the best location for competing facilities have been used by researchers and practitioners for some time. Most models, however, are useful for locating facilities which are equally attractive. In reality, facilities differ in the total "bundle of benefits" they offer customers. A survey of models for the location of competing facilities which are not equally attractive is presented here. Three models are analyzed: the deterministic utility model, the random utility model, and the gravity model. The utility models are predicated on consumer spatial choice models as well as on the premise that facilities of the same type are not necessarily comparable. The facilities vary in one or more of the qualities which make up their total attractiveness to customers. Furthermore, varying importance assigned to each of these variables by different customers will result in a selective set of consumers patronizing each.

In all models discussed in this chapter it is assumed that the new facility (or facilities) can be located in the continuous plane rather than at a pre-specified set of possible points. This formulation distinguishes the models discussed in this chapter from most of the models in the literature that discuss selection of sites from a discrete set of points (see, for example, Chapter 16 in this book).

In the deterministic utility model it is assumed that all customers at a demand point use the same utility function and select the facility to patronize by that utility function. This leads to the "all or nothing" property. All customers associated with a particular demand point patronize the same facility.

In the random utility model it is also assumed that customers select the facility that maximizes their expected utility. However, customers' ratings of the utility components are stochastic by some statistical distribution. This assumption remedies the "all or nothing" deficiency since customers at a certain demand point are no longer assumed to patronize the same facility. Customers located at a particular demand point divide their buying power among several existing facilities. The random utility model incorporates some of the features of the gravity based models and some of the features of spatial choice models.

In gravity models, a "customer selection rule" which depends on the attractiveness of the facility and the distance to it is used. The selection rule is probabilistic implying that the buying power of consumers located at a demand point is divided among the facilities and that the "all or nothing" property does not apply. Three solution approaches are suggested for finding the best location according to gravity models. The first applies a well-known algorithm taken from the location literature - the Weiszfeld algorithm. A special algorithm that solves these problems is constructed. The other two approaches suggest the use of standard mathematical programming solution codes. The readily available AMPL program or the Excel spread sheet are suggested. While the Weiszfeld approach requires the coding of a special program, using AMPL or Excel is very simple. In most cases, the specialized program yields the best results.

There are two main applications for the models discussed in this chapter. The first application is the location analysis of a new facility. If the quality of the new facility is pre-determined (i.e., by fund allocation whereby the decision maker has no control over this parameter), the model is directly applied. The best location for the new facility along with the market share captured at that location are found. If the decision maker is flexible in selecting the appropriate overall quality for the new facility, then he/she should repeat the analysis for various combinations of quality measures. The combination which yields the largest market share and the highest expected profit is selected for implementation. A cost/benefit analysis is always recommended for an affirmative decision.

The second application is an analysis of the impact of changes in quality in existing facilities on the market share captured by one's facility. These changes may involve either changes in one's own facility or changes in other facilities or a combination of both. Once changes in market share captured by one's facility are assessed, a cost/benefit analysis will assess the profitability of expanding that facility (the threshold concept in central place theory). A cost/benefit analysis will also indicate the profitability of improving one's facility in order to counteract improvements made by competitors.

An additional contribution of the models is that a decision maker will be able to perform a "what-if analysis" and anticipate the impact on his/her facility of either improvements in other existing facility(ies) or of the introduction of a new facility (either superior or inferior to his/her own). In this case he/she needs to know the overall attractiveness of the proposed new facility or the difference in overall attractiveness pre-post improvements in an existing one.

Using the models, a decision maker can assess:

1. the impact on location of changes in attractiveness for his new facility;

2. the impact on market share of change in location for his new facility;

3. the impact on market share of changes in attractiveness at his existing facility(ies);

4. the impact on his facility of changes in other facilities or the introduction of a new facility.

These models afford the anticipation of the impact of future actions. In a highly competitive market such as exists domestically, and in the face of increasing global competition, the ability to optimize location in terms of market share provides a strategic advantage for decision makers.

14
MULTIFACILITY RETAIL NETWORKS

Avijit Ghosh
New York University

Sara McLafferty
Hunter College

C. Samuel Craig
New York University

14.1   Introduction

Of all the elements of retail strategy, few are as important as the choice of locations from which to sell goods and services. While the location decision has always been critical, the rapid growth and expansion of multifacility networks in retailing has heightened the importance and complexity of the location decision. Multifacility networks involve a single firm operating more than one retail outlet in the same geographic market. Typical examples of multifacility networks include chain stores and franchise systems. For these firms, opening new outlets is an important avenue for growth. By locating new outlets, whether in existing markets or in new ones, the firm expands the geographic area it serves and increases the potential for sales and profits. At the same time, over-expansion will adversely affect outlet performance and eventually the viability of the network.

A number of considerations make the location decision extremely critical. It is a relatively permanent decision and not easily reversible. Store closures impose financial costs and also run the risk of tarnishing the firm's image. Store locations play a major role in determining the type and number of people the firm will attract. Good locations can provide ready access to a large number of target customers and increase sales. In a highly competitive retail environment, even a slight difference in location can have a significant impact on market share and profitability. A poor location is a liability that is difficult to overcome [635].

Any firm that operates retail outlets must implement a location policy that is consistent with its overall marketing strategy. The location policy translates the firm's marketing strategy into concrete actions that are sensitive to the spatial pattern of demand, competitive conditions and the availability of desirable sites. In today's highly competitive retail environment, firms can create competitive advantage by achieving a strong market presence by locating multiple stores in the same market. Locating multiple units in one market has a number of advantages. For one, it creates market presence so that all consumers in the market area have relatively easy access to a firm's stores. Second, it allows for managerial efficiencies as well as scale economies in distribution, warehousing and transportation costs. Third, it increases the efficiency of advertising and promotional expenditures in the local market. In sum, concentrating outlets in an area can create synergy, improving the performance of individual outlets that are a part of a larger network.

Organizing such networks, however, is a difficult task. Traditional methods of site selection, such as the *analog* model [233, 33, 34] are not well suited for this purpose since they are limited to analyzing single-store locations. These single store procedures ignore the impact that an individual store may have on other outlets in the market area operated by the same firm. Establishing a network of two or more outlets, on the other hand, requires systematic evaluation of the impact of each store on the entire network of outlets operated by the firm and the consideration of the "system-wide store-location interactions" [8, p.8].

A number of reviews of the literature on retail site selection now exist. See, for example, [246, 267, 469, 1218, 255, 1156] among others. This paper focuses on the use of location-allocation models for systematically evaluating and planning store networks and illustrates the application of such models in retailing. Our purpose is not to present an exhaustive review of all models that have been proposed. Instead, our objective is to identify different classes of multifacility location models, and compare and contrast their application. We also identify important issues in broadening the use of location-allocation models in retail location planning.

The rest of the chapter is organized into four major sections. The first section discusses the basic features and components of location-allocation models for retail location planning. In the second section five classes of location-allocation models −p-median, covering, p-choice, consumer preference, and franchise models − are briefly described. The third section presents two case studies which apply location-allocation models to design a network of service centers. Next, we discuss avenues for extending retail location models. The paper then ends with an agenda for the future.

14.2   Location-Allocation Models

The growth of multi-store networks has prompted the development of location-allocation models for planning retail networks. In the context of retail site selection, location-allocation models determine the best *locations* for new retail outlets based on stated corporate objectives and *allocations* of consumers to those locations based on the expected pattern of consumer travel. The allocations depict the flows of consumers to stores and thus define each store's trade area. They are used to forecast sales and market share of outlets at different locations. Both the optimal location and the allocations must be determined simultaneously, since the optimal locations depend on consumer travel patterns (the allocations) and the travel patterns, in turn, depend on store locations.

In addition to determining store locations, firms must decide on the number of outlets to open in a market area. By opening a larger number of outlets the firm can increase its accessibility to customers and increase its market share. But, opening more stores increases the cost of establishing and maintaining the network. Moreover, as the number of stores increases, the new stores cannibalize the sales of the existing stores, reducing average sales per store. In deciding how many stores to open, the firm must trade off the likely revenue generated by opening new stores with the cost of establishing and operating them. The marginal benefits of opening a new store must be weighed against the incremental cost. The decision on how many stores to open is not independent of where they are located. A smaller number of well located stores may be preferable to a larger number of poorly located outlets. For any given number of stores, then, it is essential to know which locations within the market area are most desirable.

Location-allocation models are a useful tool for retail locational planning because they can analyze both the location and allocation aspects of siting and the system-wide impact of individual store locations. These models offer procedures for systematically evaluating store locations and finding sites that maximize corporate goals such as market share or profits. Location-allocation models can be used to develop a network of stores in a new market area, to expand an existing network or to relocate or close existing stores.

The earliest attempts at formulating location-allocation problems date back to the beginning of the 20th Century when Alfred Weber attempted to find the most efficient point of production for an industrial plant given the raw material sites and market locations [435]. It was not until the early 1960s, however, that researchers came up with a mathematical solution to the generalized Weber problem [725, 237]. Since then, research on location-allocation modeling has increased rapidly, fueled by advances in computing technology and the growing awareness of the applicability of these models to real world planning problems (for a review of the development of location-allocation models see [1001, 579, 754, 551, 471, 127, 466].

The application of location-allocation models for retail planning is a relatively recent phenomenon. In applying the models to the retail sector, researchers must address certain features of retail locations. First, the overall objective in locating retail stores is typically a measure of profitability that relates to larger corporate goals and objectives. These measures must incorporate both cost and revenue projections and consider the impacts of new retail outlets on the performance of the firm as a whole. Second, the demand for retail goods and services is often complex and difficult to model. It depends not only on locational factors, but also on non-locational factors such as price, assortment, and image. The relationships among these factors may be non-linear and vary from place to place. Third, in the retail sector competition is a major factor. In choosing locations, it is essential that firms analyze and anticipate the locations and corporate strategies of competitors. These distinctive features make retailing a challenging new area for the application of location-allocation models.

14.3   The Components of Retail Location-Allocation Models

Location-allocation models utilize two basic types of information. The first consists of a set of mathematical rules that describe the overall objectives to be achieved in locating facilities and the spatial behavior of consumers in the retail system. These are expressed in the *objective function* and the *allocation rule*. The second comprises the geographic database that describes the area in which the model is to be applied. The database must include information about the spatial distribution of demand for the retail goods and services, location of existing stores, feasible sites for new stores, and the travel times or distances between demand areas and feasible sites. These various components of location-allocation models are described below.

14.3.1   Objective Function

The objective function specifies the criteria to be optimized in choosing store locations and thus serves as the yardstick against which alternative location plans are evaluated and compared. In retail applications the objective typically is to maximize consumer accessibility to stores [496], maximize estimated market share [467, 289], or maximize profitability [8, 464]. Multiobjective models have also been developed to incorporate several alternative measures in the objective function [847, 497, 465].

14.3.2   Allocation Rule

In location-allocation models it is critically important to predict accurately how consumers choose among alternative stores in an area, since the consumer choice process determines the trade areas and market shares of outlets. In some situations, for example, choice may depend simply on distance or travel time; while, in other cases, consumers may make complex trade offs between distance and nondistance factors, such as price, assortment and quality. The allocation rule must also indicate whether consumers patron-ize a single store, or shop at several stores with differing frequencies. Since consumer choice varies from one situation to another, there is no single al-location rule that is appropriate for all situations. One of the advantages of location-allocation models is their flexibility in handling different allocation rules depending on the application.

14.3.3   Spatial Pattern of Demand

An essential component of location-allocation models is information on the geographic pattern of demand for retail goods and services. Typically, the demand within a small geographic area is estimated from secondary sources and assigned to a point within the area. It should reflect not only population, but also income, demographic characteristics and other factors known to affect the demand for goods and services offered by the retailer.

14.3.4   Feasible Sites

A second type of geographic information required in location-allocation models is the list of feasible sites for new stores. These are sites that are considered appropriate for building new outlets, because they meet certain requirements regarding land, access and infrastructure. In *network location models*, a limited number of feasible sites are identified. *Planar location models*, on the other hand, assume that any location in an area is a feasible site. Typically, planar models are used to identify the optimal locations and then an attempt is made to find available sites near those locations. It may be the case, however, that no site is available near an optimal location, or the location may be infeasible due to zoning regulations, topography or some other factor that prohibits commercial development. Another problem is that by not first identifying the set of feasible sites, the cost of real estate at different sites cannot be considered in choosing the optimal locations.

14.3.5   Distance or Time Matrix

One of the fundamental factors affecting consumer shopping behavior is the distance or travel time to stores. In location-allocation modeling, this information is stored in a matrix which shows the distance or travel time from each demand point to each feasible site. Ideally, this matrix is gener-ated by calculating the shortest travel time between each pair of points on the transportation network. Alternatively, distances can be used in lieu of travel times. For distances accurately to reflect travel times, they should be adjusted for physical or social barriers to travel, congestion and road conditions. When detailed information on actual travel times and distances are unavailable, they are often approximated by using the Euclidean or the city-block metric to calculate the distance between two spatial coordinates.

These distances are easy to compute from maps using geographic information systems (GIS). For convenience, in the following discussion the terms shortest distance and shortest travel time are used synonymously.

14.3.6   Store Locations and Characteristics

The final type of information consists of the location of existing stores operated by the chain and its competitors and the characteristics of those stores. The list of characteristics should include all factors hypothesized to influence consumer store choice decisions, for example, price, quality, service and merchandise assortments.

14.4   Five Types of Location-Allocation Models

Location-allocation models incorporate a variety of allocation rules and objective functions. For the purpose of discussing their application to retail location planning they can be grouped into five categories: (i) p-median models, (ii) covering models, (iii) p-choice models, (iv) consumer preference based models, and (v) franchise models. The models differ in their assumptions about consumer behavior and the environment in which the stores operate. In this section we briefly discuss the features of these models as a backdrop to the case studies and extensions discussed in later sections.

14.4.1   p-Median Models

The extension of the Weber problem to multiple supply points led, in the early 1960s, to the formulation of the well-known p-median model, which has been critical to the development of location-allocation models. The objective of the classic p-median problem [237, 509] is to find the locations for a given number (p) of facilities that minimize the average distance that separate consumers from their nearest facility [68]. As is well known, the solution to this problem is the p weighted medians of the demand points represented on a graph [509]. The p-median model has been applied to a wide range of facility location problems, including food distribution facilities, public libraries, swimming pools, and health facilities, to name just a few. It also provided the impetus for the development of various other types of location-allocation models.

Insofar as distance affects access, the p-median solution maximizes consumer accessibility to service facilities. It is clearly applicable in those retail contexts where maximizing consumer access to stores is a major objective and it is reasonable to assume that consumers visit the nearest outlet. This is likely to be the case for convenience stores, fast food outlets, and services such as banks and post offices.

To state the network version of the p-median problem mathematically, define w_i as the demand for retail goods in the ith demand zone, F as the set of feasible sites, and d_{ij} as the distance between i and feasible site j. The objective function of the p-median problem can then be written as:

$$MIN \left\{ \sum_i \sum_{j \in F} X_{ij} w_i d_{ij} \right\} \tag{14.1}$$

where, $X_{ij} = \begin{cases} 1 & \text{if } d_{ij} = \min\{d_{ik}|k \in F\} \\ 0 & \text{otherwise} \end{cases}$

To determine the optimal locations for p facilities, the objective function (14.1) is solved subject to the following constraints:

$$\sum_{j \in F} X_{jj} = p \text{ for all } j$$

$$X_{ij} \leq X_{jj} \text{ for all } i, j$$

$$\sum_{j \in F} X_{ij} = 1 \text{ for all } i$$

The constraints ensure that only p facilities are located, that no consumer is allocated to a site that has no outlets, and that all consumers are allocated to at least one outlet. X_{ij} operationalizes the allocation rule, since it defines the set of demand points served by each outlet. The variable takes the value of one when the outlet at j is closest to i; otherwise it takes the value of zero. Thus only the distance of each demand point to its nearest outlet is considered in the objective function. The objective function minimizes the distance separating consumers from their nearest facility.

The p-median problem is the foundation for two retail location models proposed by Goodchild [496]. Suppose that a chain is planning to add new outlets in a market area in which it already operates some stores and that the firm's objective is to maximize consumer access to its outlets, both new and old. To deal with this type of situation, Goodchild proposed a variation of the p-median model, called the *competition-ignoring model* (CIM). The objective function of the CIM is:

$$MIN \left\{ \sum_i \sum_{j \in F \cup E} X_{ij} w_i d_{ij} \right\} \tag{14.2}$$

where, E is the set of existing store locations, and

$$X_{ij} = \begin{cases} 1 & \text{if } d_{ij} = \min\{d_{ik}|k \in F \cup E\} \\ 0 & \text{otherwise} \end{cases}$$

As the name suggests, CIM ignores the location of competing stores in determining optimal sites. The assumption is that consumers first choose a particular chain that they wish to patronize and then select the outlet of that chain nearest to them. This may be the case, for example, when consumers have a strong preference for particular chains and strong loyalties to those firms. Implicit, too, is the assumption that the greater the network accessibility to a chain's outlets, the higher the likelihood that potential customers will choose it. Since it does not consider the location of competing stores, CIM represents an aggressive corporate strategy that seeks to maximize market penetration.

Of course, ignoring competitive locations is often problematic. If there is little difference in the image and offerings of competing stores, consumers have less loyalty to one chain, and therefore patronize the closest outlet irrespective of which chain operates it. In this case, the firm's objective should be to maximize the total demand within the proximal market areas of its own outlets. To determine the best locations, the allocation rule in the previous problem must be changed to:

$$X_{ij} = \begin{cases} 1 & \text{if } d_{ij} = \min\{d_{ik}|k \in F \cup E \cup C\} \\ 0 & \text{otherwise} \end{cases}$$

where C is the set of competing outlets. This is the *market share model* (MSM) proposed by Goodchild [496]. The model searches for gaps in the existing coverage of the market and locates outlets in areas distant from competing stores. In choosing new locations the model avoids areas served by competitors and tries to establish the firm's own geographical niche.

Retail location models based on the p-median are generally useful in determining sites that maximize the population's accessibility to retail services. These models are relatively simple, and different variations of the model can be formulated by modifying the general p-median problem (see, for example, [574]). Since the allocation rule is based on the assumption that consumers travel to their nearest facility, the models can be used without gathering detailed information on consumer shopping patterns.

Underlying the p-median objective of minimizing average (or total) distance to facilities, is the assumption of a linear relationship between accessibility and demand (or utilization). But a linear relationship between accessibility and demand may not be valid in all cases. Hakimi [513] proposed a generalization of the p-median problem that incorporates a non-linear relationship between accessibility and demand. Consider n demand nodes located on the vertices of a graph. Let X_p represent the p sites ($p < n$) for locating facilities, and w_i be the potential demand at i. Now let the portion of this potential demand which is actually spent at an outlet be a nondecreasing convex function of the distance from i to the nearest facility in X_p, $f_i(d(i, X_p))$, where $d(i, X_p)$ is the shortest distance between i and the closest facility in X_p. The p locations which maximize demand can then be found by optimizing the following objective function:

$$MAX \left\{ \sum_{i=1}^{n} w_i f_i(d(i, X_p)) \right\} \qquad (14.3)$$

Hakimi shows that, if the function $f_i(.)$ is specified as $f_i(.) = 1/(1 + d(i, X_p)^\alpha)$, then like the p-median problem, the network version of this problem has an optimal solution in which all the p locations are on the vertices of the graph. (A number of other specifications of $f(.)$ are possible. One possibility, e.g., is to consider the average distance to all facilities in the area. Another is to consider a step function relating demand to distance.)

14.4.2 Covering Models

A second type of location-allocation model of interest in retailing is the covering model. Covering models were originally developed for public sector location problems, such as the location of emergency medical and fire services; but the potential application of these techniques is much broader. They are important in designing multiunit networks for service oriented retail firms where access is a major determinant of patronage. The objective of covering models is to identify locations that provide potential users access to service facilities within a specified distance or travel time. This is important in situations where access plays a key role in determining the level of service utilization or the quality of the service delivered. For many convenience oriented retail facilities, such as movie theaters, banks, fast food restaurants, and ice cream parlors, most customers live near the outlet. Customers may obtain substitute products or services or completely forgo consumption, if an outlet is not accessible to them. A measure of the level of service provided by these outlets, therefore, is the proportion of total population that is located within the immediate neighborhood of the outlets. By using a covering model the planner can identify locations that maximize the number of people within a specified "reservation" distance or travel time constraint.

One of the first types of covering models to be proposed was the *set covering model*. This model assumes that consumers residing beyond a specified maximum distance or travel time (S) from an outlet are not adequately served and, therefore, do not use the service. The objective of the set covering model is to find the minimum number and locations of facilities needed to serve all potential consumers within the specified distance or travel time [1139]. But the goal of providing universal service may not be feasible because of the cost of operating a large number of outlets, making it necessary to tradeoff the cost of locating additional outlets with the potential revenue generated from incremental coverage. Funds saved by constructing fewer outlets can be used to improve other aspects of service quality such as the number of service personnel or operating hours. Instead of aiming for universal coverage, the retail manager may seek to maximize the amount of potential demand covered by a fixed number of service centers. This is the *maximal covering location problem* (MCLP) proposed by Church and ReVelle [223]. The MCLP can be written mathematically as follows:

$$MAX \left\{ \sum_i w_i Y_i \right\} \tag{14.4}$$

subject to:
$$\sum_{j \in N_i} X_j \geq Y_i \text{ for all } i$$

$$\sum_j X_j = p$$

$$X_j = \begin{cases} 1 & \text{if facility at } j \\ 0 & \text{otherwise} \end{cases}$$

$$Y_i \in \{0, 1\}$$

The critical operational variable of the model is the set N_i, defined for each demand point. This set designates the set of outlets within the specified distance or time from demand point i – the set of outlets that are considered accessible to that demand point. The objective of maximizing the demand that is covered is operationalized through the definitions of Y_i and the first constraint. This constraint dictates that Y_i is equal to 0 if N_i is empty. An uncovered zone does not contribute to the objective function since the corresponding Y_i is 0. (In the set covering problem no N_i is allowed to be empty). The second constraint limits the number of outlets to a specified number, p. The objective function maximizes the amount of demand that is covered by p facilities within the specified accessibility criterion.

14.4.3 p-choice Models

Like many business organizations, retail firms often use market share to measure the effectiveness of their location strategies. To estimate market share for a set of locations, retail analysts must predict the amount that consumers spend at each of the chain's outlets. In location- allocation models, the allocation rule simulates consumers' shopping patterns. The implicit assumption underlying p-median models, for example, is that consumers travel to their nearest outlet and purchase their entire requirement of goods and services from that store. A store's market share is then given by the ratio of the demand within the store's proximal market area to the total demand. The assumption of nearest-center travel is also made in covering models, with the additional stipulation that customers utilize the service only if the closest outlet is within a reservation distance.

It is now well established that the nearest center hypothesis used in almost all classic location models is too restrictive. Customers of retail stores do not always patronize the nearest facility and they often patronize more than one facility. There are a number of reasons for such behavior. First, as Ghosh and Rushton [471] note, consumers may have a set of activities they wish to perform within a given space and time constraint. This situation gives rise to the phenomenon of multipurpose trips, where consumers fulfill diverse needs on a single trip. In organizing multipurpose trips consumers often bypass their nearest facility and choose one that is convenient given the other needs to be fulfilled on that trip. (There is now a growing literature which looks at the implication of multipurpose shopping for retail location [900, 876, 47, 468, 1127, 621].)

Implicit in the nearest center allocation rule is the assumption that all facilities provide the same assortment and level of service and charge similar prices. Thus facilities are distinguished only by their locations. In practice, however, even facilities providing similar types of services have varying assortments, prices and quality. This is another reason for consumers to visit facilities other than the nearest one. Moreover, the stochastic nature of prices and quality may prompt consumers to visit a portfolio of stores rather than a single one.

The lack of empirical support for the nearest center hypothesis has led researchers to propose stochastic models of choice that incorporate distance (or travel time) as well as *store* characteristics such as service level and facility size. (Numerous empirical studies of consumer behavior conducted in diverse geographical locales refute the nearest center hypotheisis for consumer-oriented services [494, 34, 224, 635]. For a review see [601]). Two early proponents of this approach were Huff [603] and Lakshmanan and Hansen [733], who proposed probabilistic choice rules based on the spatial interaction model. These choice rules can be written as:

$$P_{ij} = \frac{\frac{f(A_{ij})}{g(d_{ij})}}{\sum_{k \in K_i} \frac{f(A_{ik})}{g(d_{ik})}} \tag{14.5}$$

where A_{ij} is a measure of attractiveness of facility j to consumer i, d_{ij} is the distance or travel time separating consumer i and facility j and K_i is the set of stores that are in consumer i's choice set. In particular applications the allocation rule can take various forms depending on the specification of the functions $f(.)$ and $g(.)$, which map nonnegative real numbers into positive numbers with $f(0) > 0$ and $g(0) > 0$. Readers will recognize the similarity of this allocation rule to the spatial interaction models widely used in geography to study migration, trade flows and consumer travel in general [570]. Two specific forms of Equation (14.5) that are commonly used in retail research are the *multiplicative competitive interaction (MCI)* suggested by Nakanishi and Cooper [884] and the *multinomial logit (MNL)* model [813, 1206].

To consider the development of spatial interaction based location-allocation models, consider the following objective function which is a generalization of the p-median objective:

$$MAX \left\{ \sum_i \sum_{j \in F} w_i P_{ij} X_j \right\} \qquad (14.6)$$

where P_{ij} $[0 < P_{ij} < 1; \sum_j P_{ij} = 1, \forall i]$, is the probability that customers at i patronizes facility j, w_i is the number of customers at i, and X_j is one if a store is located at j. This is the p-choice model [466]. This is a generalized, unconstrained version of the p-median problem, because the all-or-nothing nature of consumer choice in the p-median problem is relaxed [68, 902]. Optimizing objective function (14.6) leads to a set of facility locations that maximize the number of customers served. The locations maximize the firm's expected market share. The p-choice problem has been the basis of a number of theoretical [1207, 67, 855, 755] and applied facility location problems [1121, 580, 8, 467, 682, 344].

Probabilistic choice rules like that in Equation (14.5) are now extensively used to predict the market share of facilities at different locations. The set of store attributes, A_{ij}, should include all relevant store and site characteristics hypothesized to influence consumer choice. For retail facilities, for example, the important store attributes include objective as well as subjective measures of price, quality of service, quality of goods, convenience of hours of operation, and proximity to home and/or work (see, for example, [266, 226, 988, 460, 414]). In their study of food retailing, Jain and Mahajan [635], for example, used such factors as store size, the availability of credit card services, the number of checkout counters, whether the store was located at an intersection, in addition to distance, to explain the market share obtained by individual outlets. In addition to retail stores, choice models have been used to predict the market shares of shopping centers [446, 1185], retail banks [538], and hospitals [411].

To apply the p-choice model in retail settings, the retail analyst first needs to analyze the pattern of consumer travel to existing facilities, and then apply the results from the empirical study in the allocation phase of location-allocation algorithms (for a case study see [469], see [289] for an example of the p-choice problem on the plane). Although the p-choice and p-median problems are similar in structure, they generally produce very different solutions. The most important difference is that in the former, facilities no longer have well-defined geographic market areas. Each store's market area is a probabilistic surface that shows the probability of a customer from a given area patronizing that facility. The exact nature of this probability surface depends on the parameters of the spatial interaction model. The greater the impact of distance on choice, the higher the likelihood that consumers patronize their nearest store, and less the overlap in market areas. This typically gives rise to optimal location patterns in which stores are dispersed. On the other hand, when the impact of distance is relatively low compared to that of facility attraction, facilities tend to gravitate toward areas of high demand, leading to a clustered location pattern.

14.4.4   Consumer Preference Based Models

The fourth class of retail location-allocation models consists of models that directly incorporate consumer preferences. In these models, the allocation rule is based on consumer evaluation of hypothetical choice experiments, rather than observed choices. Consumers are asked to evaluate various choice scenarios, and their evaluations are used to predict choices in the allocation rule. Consumer preference based models overcome one major problem of the p-choice models just described – context dependency. At the root of the context dependency problem is the finding that the observed parameter values of the functions $f(.)$ and $g(.)$ are highly sensitive to the set of alternatives over which choices are observed [1029, 416, 413]. Consequently, past behavior can be a poor predictor of future choice when new alternatives become available. If new facilities represent major departures from existing norms of facility design or service level, then the utilization of new facilities is unlikely to conform to past choices. Yet, ironically location-allocation models are often applied in precisely these situations.

Including measures of locational structure in the calibrated choice functions is one remedy to the context dependency problem [355, 460]. But such measures are difficult to implement without systematic empirical research on the impact of locational structure on choice. Another approach is to assess consumer choice directly using structured choice experiments. Conjoint and information integration methods are typically used for this purpose. Conjoint analysis has been widely used in marketing studies to assess the impact of brand and service characteristics on consumer preferences [1213]. Instead of analyzing past choices, these methods utilize consumer evaluations of hypothetical choice descriptions to calibrate the utility function. Properly conducted experimental studies can provide considerable insight into consumer behavior and patronage decisions while removing the effects of the spatial context from the choices themselves.

One of the earliest consumer preference based location models was proposed by Parker and Srinivasan [917]. They used conjoint analysis to determine how consumers evaluate different attributes of a primary care facility, including travel time. The utility function calibrated through the conjoint procedure was then used in the allocation phase of a location-allocation model to determine the optimal location and attributes for a primary health care facility. Since the conjoint method used preferences for hypothetical choice tasks in estimating the utility function, the parameters were not biased by the location and characteristics of existing outlets.

To implement the conjoint method the first step is to identify the store characteristics (or attributes) that influence consumer preferences, and the levels that these attributes can potentially take. The impact of different attributes on preference is then assessed by calibrating a linear-additive function relating outlet characteristics to utility. Assume, for example, that there are S service attributes of interest and an attribute s can take M_s possible values and that there are A accessibility classes. We can then write:

$$U = \sum_{s=1}^{S} \sum_{m=1}^{M_s} \beta_{sm} Z_{sm} + \sum_{a \in A} \alpha_a V_a \qquad (14.7)$$

where Z_{sm} is 1 if the outlet has the m'th level of attribute s and V_a is one if the outlet is in the a'th accessibility class, and U is the utility. The β parameters show how utility changes as a result of change in service characteristics and α measures the impact of each accessibility class on utility. Consumer choice can then be simulated by assuming either that they choose the outlet that gives the maximum utility, or that they patronize multiple outlets with frequencies proportional to their utilities. The choice function can then be incorporated in the allocation phase of the location-allocation model. For application of conjoint methods to retail choice see [772, 464, 880], and the case study presented later in this chapter.

14.4.5   Franchise Models

Franchising is a form of business in which the parent company (the franchisor) grants individual franchisees a license to engage in commerce using its business practices, goods and services, and trademark in return for predetermined fees and royalties. Although the operations of franchised outlets are in many ways similar to other kinds of retail stores, a number of special considerations arise in making location decisions for franchise stores. Most importantly, the dispersed nature of ownership makes it difficult for the franchisor to focus solely on systemwide revenue or profitability criteria. The goal of maximizing systemwide revenue may be at odds with the individual franchisees' goals of maximizing their own revenues and profits [1227, 682, 1097]. In locating outlets, both the franchisor's and the franchisees' goals must be considered simultaneously.

Pirkul, Narashimhan and De [945] proposed a model for franchise location decisions that allowed for the possibility of opening both company owned and franchisee-owned outlets. The objective is to maximize expected returns from both types of stores to the franchisor, subject to the constraints of available budget and warehouse capacity. A notable feature of this model is the link between the locations of warehouse and retail stores, a consideration ignored in most other models. However, the potential profitability of outlets at different locations is calculated exogenously, rather than being derived as an output of the model.

Ghosh and Craig [465] focus on a different type of tradeoff that is required in making franchise expansion decisions. In expanding a franchise system the location planner is faced with two potentially conflicting goals. First, the decision maker wants to add additional outlets in order to maximize total revenues of the franchise network. By doing so, the franchisor's income is maximized (through licensing and royalty fees) and its competitive position is strengthened. But adding new outlets can potentially cannibalize the sales of existing franchisors, creating the potential for conflict in the system. The traditional method for dealing with this type of spatial conflict has been to impose territorial restrictions to prohibit the location of new franchise outlets within a specified minimum distance from existing outlets. Territorial restrictions, however, ignore the positive impact that a new outlet may have on existing stores (through increased advertising, name recognition etc.) and the negative impact of forcing new outlets to locate in undesirable sites to avoid proximity to existing stores.

To consider the needs of both new and existing outlets in franchise expansion Ghosh and Craig [465] propose a multiobjective approach to generating potential location strategies. They define X_j $(j = 1, 2, \ldots, J)$ to be a zero-one variable which takes the value of one if an outlet is opened at a feasible site j. The list of feasible sites is ordered such that $j = 1, \ldots, j^*$ are locations of competing outlets, $j^* + 1, \ldots j^{**}$ are locations of the franchisor's existing stores, and $j^{**} + 1, \ldots, J$ are potential sites for new outlets. Let S_j be the potential revenue at site j, calculated based on a spatial interaction model. The locations that maximize system wide revenue can be found by solving the following objective function:

$$MAX \left\{ \sum_{j=j^*+1}^{J} X_j S_j \right\} \tag{14.8}$$

Note that the objective (14.8) is the sum of the revenues of existing stores and the revenues of the new stores.

In maximizing systemwide revenues objective (14.8) does not consider how total revenue is distributed between these two sets of stores. As stated earlier, the locations that maximize the revenue of new franchisees may actually cannibalize the sales of existing ones. The authors argue that it is necessary to find compromise solutions that optimize the franchisor's goal of maximizing systemwide revenue while protecting the interests of the existing franchisees. They use the constraint method proposed by Cohon [234] to generate such multiobjective solutions. The first constraint imposed is: $[S_j X_j \geq S_j^* X_j \mid j = j^* + 1, \ldots, j^{**}]$ where S_j^* is the revenue of store j prior to system expansion. This constraint ensures that the sales of existing outlets do not decline when a new outlet is added to the system (it should be noted that in their model the addition of new outlets may increase the sales of existing stores due to the increased competitiveness of the entire chain). In addition to protecting the interests of the existing franchisees, following Balakrishnan and Storbeck [48], the authors also impose the constraint: $[(S_j - T)X_j \geq 0 \mid j = j^{**} + 1, \ldots, J]$ where T is the minimum expected revenue of new franchisees. This ensures that only locations that produce a positive return on investments are chosen for the new franchisees.

14.5   Applying Covering Models for Service Center Location

This section presents two case studies to illustrate the application of location-allocation models for planning a network of retail service centers. The case studies involve a proposed network of "carry-in" service centers for repair and maintenance of micro computers. The service centers are designed to serve small businesses owners –the primary target market– who will travel to these centers and drop off equipment for repair. Such a service is attractive to small businesses in rural/suburban areas, where "on site" service is typically lacking or costly. Since users must travel to the centers, utilization will depend on consumer accessibility to the network of service centers.

The first case study focuses on the use of maximal covering model to find the number of outlets and the best configuration of locations to provide a given level of service. The second case study extends the basic covering model to incorporate nondistance factors in the allocation rule and assess the profitability of alternative networks. These nondistance factors broaden the application of the covering model to allow managers to evaluate many aspects of the service offering through the use of conjoint methods.

To solve the location problem a geographic information system is created describing the area in which the service centers are to be located. The framework for this information system is a geocoded data base consisting of the spatial coordinates of 64 demand zones in the region. Each of these points is considered the center of demand as well as a feasible site for locating new outlets. Because of the grid-like road network in the area, distances between demand points are measured from spatial coordinates using the city block metric. Due to the large size of the study area, the focus is on finding general areas in which service centers are to be located, rather than identifying precise sites.

A critical element in applying location models is to estimate the geographic distribution of potential demand for the service. To estimate demand, information on the number and type of small businesses in each demand zone was obtained from secondary sources. Information on the pattern of microcomputer use by different types of small businesses, and the likelihood of using the repair service was obtained through a survey of small businesses in the area. The survey revealed that accessibility is a critical determinant of service utilization – the probability that a small business owner will use a carry-in service center is inversely related to distance to the center. Survey responses indicated that the expected rate of utilization falls sharply when the distance to a center exceeds six miles. A six mile limit is, therefore, used to define the accessibility constraint in the covering model. For expositional clarity we assume that the same distance criterion applies for the entire target population; however, this assumption may be relaxed to allow different segments of the population to have different sensitivities to distance.

14.5.1   Network Size and Accessibility

The initial set of analyses examined the number of outlets required to cover all small businesses in the area within the six mile constraint. The set covering model was used to determine the number and location of facilities required to provide total coverage. The solution indicated that 17 outlets are needed to cover the total population within 6 miles. However, if all 17 outlets are opened, the level of utilization at several of the centers is too low to make them economically viable.

To analyze the relationship between investment in service outlets and accessibility, the maximal covering location problem (MCLP) is solved for values of p (number of outlets) ranging from 1 to 17, with S fixed at 6 miles. The results from this series of analyses showed the relationship between the number of outlets and the percentage of small businesses covered within the distance criterion. The incremental demand covered decreases as more outlets are added. Just 9 outlets can cover 90% of the businesses within 6 miles, but an additional 8 outlets are required to cover the remaining 10% of the businesses. When the number of outlets is small, the optimal locations are in high density areas and the chosen locations have many potential customers within their service areas. For example, a single service center can cover 31% of the businesses in the region. Adding another outlet increases coverage to about 45% and five outlets covers nearly 76%.

In formulating a location strategy, a critical decision is the number of outlets to open. Examining the relationship between the number of outlets and coverage allows an assessment of the consequences of different location strategies. Even if funds are available to cover the entire region, it may not be profitable to do so because of the required investment. Several demand points in the region are relatively remote and difficult to serve. Providing service to these areas is expensive, while the incremental revenue gained is small and likely to be less than the cost of providing the service. Moreover, the resources freed by opening fewer outlets may be utilized to improve other aspects of service provision to increase patronage.

14.5.2   Optimal Location Plans

For this study area five well-placed service centers can cover 75% of small businesses within the stated six mile criterion. This level of coverage, however, will be assured only if the centers are placed at their optimal locations. As stated earlier, the covering model not only provides guidelines for the

TABLE 14.1. Comparison of Two Coverage Models

	Percent Demand in Distance Class	
Distance Class	MCL Plan	Modified Plan
Less than 2 miles	13.28	16.15
Between 2 and 3 miles	8.52	11.82
Between 3 and 4 miles	16.17	12.17
Between 4 and 5 miles	12.05	12.02
Between 5 and 6 miles	25.81	23.67
Total 6 miles or less	75.83	75.83

number of centers to open, but also specifies the locations of the outlets. One feature of covering models is that they typically generate many alternate optimal solutions. That is, other combinations of five locations exist which achieve the same level of coverage as the optimal solution. Some of these alternate solutions may be preferable when criteria other than coverage are also considered, and they should be examined in choosing the final location network.

One way to generate alternate solutions is to modify the MCLP to maximize coverage as well as minimize the distance from covered demand points to their closest outlets. This modified problem can be solved in two stages. First the MCLP is solved for the given combination of critical distance and number of centers (S and p). Next, it is solved again with a stricter distance constraint. Thus, if the first problem was solved with S equal to 6 miles, S may be set equal to 3 miles in the second. Further, an additional constraint is included in the second problem which requires that while maximizing coverage within the shorter distance, coverage within the larger distance cannot be less than that achieved in the previous problem. The second problem thus seeks alternate optimal solutions to the first which maximize coverage within a shorter distance ($S = 3$) without sacrificing the coverage obtained in the initial solution ($S = 6$). The solution to the second problem will locate facilities closer to users without sacrificing the level of overall coverage.

Table 14.1 compares the results of two MCLP solutions for the study area. It shows the percent of demand within different distance levels from their closest facility for two different MCLP solutions. The first column is the result of solving the problem with $S = 6$ and $p = 5$. The second column shows the distance distribution for the modified MCLP with $S = 3$, $p = 5$, and the constraint that the level of demand covered within 6 miles cannot be less than in the original problem. Comparing this column with the first reveals that in the optimal solution to this problem locations are closer to the demand points than in the original MCLP solution. The proportion of demand within 3 miles of a center is 27.97 in the second case, compared to only 21.80 originally. This improvement is achieved without sacrificing coverage within 6 miles: 75.83% of demand is covered within 6 miles in both cases. The second solution may be preferable because of its greater accessibility to nearby centers.

Covering models, like the one just discussed, are simple and useful aids for designing service location networks. They can, however, be modified and improved depending on the needs of the particular application. Three general areas of modification include differential weighting of demand points, incorporating the impact of non-distance factors on service utilization and optimizing profitability of networks. We now present a second case study to illustrate these modifications. This case study, too, involves the location of carry-in computer repair centers targeted to small business owners, but it considers a larger geographical region consisting of 164 demand nodes. Based on land availability, infrastructure and other factors, 124 of these were designated as feasible locations for opening service centers.

14.5.3   Maximizing Utilization

Covering models have a simple allocation rule which assumes that consumers utilize the service fully as long as the distance to their nearest outlet is less than the critical distance constraint. In many cases the relationship between distance and utilization is more complex and decreases with increasing distance. One way to represent this is a step function relating utilization to distance. Utilization decreases in a stepwise manner out to the critical distance limit, beyond which it falls to zero. Another assumption of the covering model is that utilization depends only on distance. But service characteristics such as speed of service, operating hours, quality of services, and price, are likely to influence service utilization rates, and must be considered along with accessibility.

One can modify the basic covering model both to incorporate a stepwise relationship between utilization and accessibility and to consider non-distance factors. To do so, the objective function is changed to maximize the expected level of utilization, where utilization depends on both distance and non-distance factors. In estimating utilization, several distance or travel-time classes can be considered and the demand covered within each accessibility class weighted according to the expected level of utilization. The model incorporates non-distance factors by considering the impact of outlet characteristics on utilization. The result is a weighted covering model which maximizes expected utilization.

Consider a planning region consisting of a set of discrete demand nodes I. The level of demand at each zone i is w_i. Let Q be the set of possible designs for service centers and A the set of accessibility classes. Individual elements of Q and A are designated as q and a respectively. Define S_{iqa} as the probability that a potential user at i will utilize an outlet that has design q and is in accessibility class a from i. The firm's objective is to choose from the set of feasible locations F and set of designs Q, the locations and designs

of "p" outlets that maximize the total level of utilization. The solution to
this problem can be found by solving the following combinatorial problem:

$$MAX \left\{ \sum_{i \in I} \sum_{q \in Q} \sum_{a \in A} w_i S_{iqa} Y_{iqa} \right\} \qquad (14.9)$$

subject to: $\sum_{j \in N_{ia}} X_{jq} \geq Y_{iqa}$ for all $i \in I, a \in A$

$\sum_{a \in A} Y_{iqa} \leq 1$ for all $i \in I, q \in Q$

$\sum_{q \in Q} X_{jq} \leq 1$ for all $j \in J$

$\sum_{j \in F} \sum_{q \in Q} X_{jq} = p$ for all $j \in J, q \in Q$

$X_{jq}, Y_{iqa} \in \{0, 1\}$

Like the earlier models, this one, too, is concerned with accessibility to the
service network. In this case, however, the weighting system (S_{iqa}) translates accessibility to expected utilization rates and the objective function
achieves the goal of maximizing utilization.

The first constraint in the model defines for each demand zone the set
of centers that can provide coverage within each accessibility class. Note
that in contrast to the MCLP a number of N_i variables are defined here
—one for each accessibility class. Thus, N_{ia} defines the set of outlets within
the a'th accessibility class of demand zone i. As in the MCLP model, the
coverage vector for each demand zone (the N_{ia}) defines the corresponding
Y_{iqa} variables used in the objective function. Y_{iqa} takes on a value of 1 if
node i is within the a'th accessibility class of an outlet. On the other hand,
if N_{ia} is a null set, Y_{iqa} is equal to zero. The next constraint ensures that
only one of the A coverage variables associated with a node is assigned a
value of 1. As the objective is to maximize expected utilization, only the
Y_{iqa} corresponding to the highest S_{iqa} is allowed to equal one. The next
constraint rules out the possibility that the same outlet is assigned more
than one design. The final constraint in the model restricts the number of
outlets to a specified number p.

To implement the above combinatorial problem, conjoint analysis is used
to estimate the impact of accessibility and service center characteristics on
utilization, S_{iqa}. The first step in the conjoint procedure is to identify the
attributes that affect consumer propensity to use the service centers. These
include accessibility as well as non-distance factors. For the computer repair
centers four important factors were identified through personal interviews
and surveys of potential users. The identified factors were: (1) travel time,
(2) hours of operation, (3) speed of service (turnaround time), and (4)

annual fee for a service contract. The fee would cover the cost of all repairs during the year. Each of these factors can be controlled by the company in designing the service network. The firm can easily change operating hours or the annual fee. Speed of service can also be increased by hiring extra service personnel to reduce potential service backlog. Finally, travel time can be shortened by having more service centers.

Once the relevant service attributes are identified, the next step is to define the range of possible values of these attributes. Based on interviews with potential users and company personnel four possible levels were identified for each of these attributes. To determine the optimal service design, consumer sensitivity to these attributes must be ascertained in order to estimate their impact on service utilization. By how much, for example, will service utilization increase if turnaround time reduces from 24 hours to 2 hours; or if travel time is reduced from 40 minutes to 20 minutes? To empirically assess the impact of different service characteristics on utilization, we assume a linear-additive relationship between service attributes and utilization as in Equation (14.7). To estimate the parameters in the equation, hypothetical service outlet descriptions were created following a full-profile fractional design. A sample of 400 potential consumers was then asked to indicate their likelihood of using the service, if offered, for each of these descriptions. For more details see [464].

The responses to conjoint measurements can be analyzed at several levels of aggregation. At one extreme, individual responses can be analyzed separately to determine parameter values specific to that individual. The other extreme is to estimate a single set of parameters that represent the aggregate preferences of all individuals in the sample. An intermediate possibility is to divide the respondents into groups and estimate a set of parameters for each segment. In this study, the respondents were divided into two segments based on population density, since preliminary results indicated a significant difference among responses from these two groups. OLS regression with dummy variables was used to estimate the model parameters.

Respondents from the high density segment were most sensitive to operating hours followed by turnaround time, travel time and annual fee, respectively. In the low density segment, on the other hand, respondents were most sensitive to the annual fee and least sensitive to speed of service. For example, in the low density segment, decreasing the annual fee from $700 to $400 increases expected utilization by 34%. Similarly, reducing travel time from 40 minutes to less than 10 minutes increases expected utilization by 20%.

14.5.4 Calculating Profitability

The focus of the objective functions of the models discussed so far has been on increasing accessibility and utilization. But in the final analysis, retail planners have to be concerned with network profitability. To evaluate the profitability of different networks, the revenues and costs of different strategies must be evaluated. The revenue and cost of a network depend on the location of the centers and also on the design of the outlets. For example, although faster turnaround time and longer operating hours increase revenue because of higher utilization, they also increase operating costs. Similarly, utilization can be increased by opening more outlets, but this increases costs too. The optimal design, therefore, must reflect the tradeoffs between the costs and revenues associated with different strategies.

Two types of costs must be considered in calculating profits for carry in service centers. First are the annual operating costs, including the costs of manpower, supplies, working capital and site (mortgage or rent). The second type of cost is the variable cost of servicing contracts. This depends on expected failure rates, types of repairs needed and cost of replacement parts. Such information can be obtained from engineering estimates of product reliability and expected failure rates. The difference between the annual fee charged for each contract and the cost of servicing it is the net amount the firm earns per contract.

An additional complication in designing service networks for computer repair centers is the fact that the demand for computers, and thus computer repairs, is not static but growing with time. Moreover, because the service concept is novel, only a fraction of the potential demand may be realized in the initial phases of the project. With time, however, this proportion is expected to increase. Therefore, to assess overall profitability the model must look at profitability over a multiperiod planning horizon rather than a single period. To represent the temporal growth in demand, the potential demand at time t, w_{it}, is calculated as $w_i(1 - e^{-\Theta t})$.

The profitability of a network over a planning horizon of T years can now be written as:

$$\frac{\sum_{t \in T}[\sum_{j \in F}[(\sum_{i \in I} \sum_{q \in Q} \sum_{a \in A} w_{it} S_{iqa} Y_{iqa} C_q) - F_{jqt}]]}{(1+r)^t}$$

where F_{jqt} is the fixed cost at time t of an outlet with service design q located at site j, C is the contribution per customer, and r is the discount rate used by the firm to calculate the net present value (NPV) of future revenues. This equation is used in the covering model, replacing the more general objective function presented before. As discussed earlier consumer stated preferences are used to estimate the values of S_{iqa}.

14.5.5   The Optimal Network

How many computer repair centers should the company open? The answer to this important planning question can be found directly from Table 14.2 which reports results from applying the network optimization model. It shows the number of small businesses covered by networks of different

TABLE 14.2. Coverage and Profits for Networks of Different Sizes

No. of Centers	Percent Demand Covered	Marginal Coverage	Net Present Value†	Marginal NPV
1	29.7	29.7	48.31	48.31
2	49.6	19.9	78.71	30.40
3	65.1	15.5	94.42	15.71
4	75.4	10.3	99.51	5.09
5	84.7	9.3	100.00	0.49
6	91.5	6.8	95.56	-4.44
7	93.4	1.9	77.88	-17.68
8	95.1	1.7	60.00	-17.88
9	96.6	1.5	41.52	-18.48
10	97.8	1.2	22.43	-19.09
11	98.7	0.9	3.31	-19.13
12	99.5	0.8	-19.38	-22.68

† NPV is expressed as a ratio of the NPV of the optimal five center network.

sizes, the marginal increase in coverage gained by increasing the number of centers, the relative net present value and the marginal net present value of each network. To be covered, a small business must be less than 40 minutes away from its nearest service center. As might be expected, coverage increases as more centers are opened. However, the marginal coverage decreases as more centers are opened. Beyond six outlets, each additional center adds less than 2% to total coverage. Although not shown in Table 14.2, 16 centers are needed to provide universal coverage within 40 minutes.

The profitability of each network depends on the cost of establishing and operating the outlets and the expected revenue. The marginal profitability of additional outlets decreases rapidly because the marginal coverage provided by the new outlets gets progressively smaller. After five centers are opened, additional centers decrease the profitability of the network. The NPV of the six center network, for example, is 4.44% less than that of the five-center network. Although the NPV of the six-center network is substantial, the incremental revenue generated by the sixth center is not enough to compensate for the additional cost of establishing and operating the sixth center. Based on the criterion of return on marginal investment, therefore, only five centers should be opened in the region.

Table 14.3 shows the performance of the individual centers in the optimal five-center network in terms of coverage, expected number of service contracts, and profit. There is considerable variation in the expected level of activity at each center and, consequently, also in the level of profit. Outlet 3 is expected to serve 1,537 service contracts and contribute 31.2% of total network profit. Outlet 5, on the other hand, is expected to have only 780 contracts and account for 4.4% of system profit. Outlets at the other

TABLE 14.3. Performance of Individual Outlets in 5 and 6 Center Networks

Center	Expected Number of Service Contracts	Expected NPV	Percentage of Total NetworkProfit
Five Center Network			
1	1221	20.0	20.0
2	1206	19.4	19.4
3	1537	31.2	31.2
4	1364	25.0	25.0
5	780	4.4	4.4
Total	6108	100.0	100.0
Six Center Network			
1	1221	20.0	20.9
2	1085	15.2	15.9
3	799	4.3	4.5
4	1537	31.2	32.6
5	1237	20.5	21.5
6	780	4.4	4.6
Total	6659	95.56	100.0

sites contribute approximately 20 to 25% of total profit. Information on expected number of service contracts is important in planning workload distribution and manpower, equipment and working capital requirements.

As noted earlier, expanding from a five-center to a six-center network decreases the total profitability of the network. The reason for this drop in profitability can be seen from comparing the performance of the sites in the five-center network to the performance of sites in the six-center network shown in the bottom part of Table 14.3. The two networks have four sites in common, but outlet 2 in the five-center network is replaced by two new outlets in the six- center network. The combined profit of these two centers (outlets 3 and 6 in the bottom half of the Table) is greater than that of the outlet 2 in the five-center network. However, the two locations draw many of their customers from the service area of outlet 4, decreasing the profit of this center by about 20%. Thus, as more and more outlets are added to the network, the new centers cannibalize sales from existing ones. Although the new centers may be profitable when viewed independently, they reduce revenues of existing centers, decreasing overall profitability. Decisions on the optimal number of centers to open, therefore, must be made by considering the profitability of the total system and not that of the individual outlets separately, as is often the case in practice. The network optimization model determines not only the optimal locations of the centers but also the best design for these centers. For the five-center network the best design is to operate 9-5 Monday to Saturday, with a turnaround time of 24 hours, and an annual fee of $700. Deviations from this design reduce the profitability of the network. For example, utilization is highest when the annual fee is $400, turnaround time is less than 20 minutes and the centers are open 9-8 Monday to Saturday. This design increases utilization by 12%, compared to the optimal design, but profitability decreases. Revenue per customer decreases because of the lower fees, and operating costs increase due to longer operating hours and the need for more staff to complete repairs more rapidly. Similarly, although operating costs can be decreased by reducing operating hours and increasing turnaround time, utilization and profits will drop. The design chosen by the model reflects the best tradeoff between cost and revenue and thus maximizes profitability.

14.6   Extension to Basic Models

The previous sections provided an overview of the different types of location-allocation models and case studies of their application to retail location planning. The different types of models discussed earlier provide building blocks for formulating retail location models for different market scenarios. Each of these models is based on explicit and implicit assumptions regarding consumer behavior, the demand environment and supplier behavior. In the final analysis, it is the reasonableness of these assumptions that determines the applicability of the models in different planning contexts.

The expanded use of location-allocation models for retail planning depends critically on the development of models that can capture more complex behavior patterns of both producers and consumers. In this section we focus on developments that have brought greater realism to retail location models by better representing the uncertainties in the market environment. The environments in which firms compete are rarely static. They change as a result of changes in consumer behavior, residential patterns, and competitive locations. Retail location models have tended to neglect the impact of such changes on the profitability of chosen locations. Most models have been myopic, selecting optimal locations for a single point of time based on existing spatial distributions of population and competing stores [467]. Given that store locations are long-term decisions that are costly to change, retail location models must help the planner identify locations that are robust to different environmental conditions.

Ghosh and McLafferty [467] point to a number of sources of change in the retail environment that can affect the firm's choice of locations. For one, the spatial distribution of demand can change due to population growth or decline caused by migration and housing developments. The growth of new subdivisions can cause a fundamental shift in the geographical pattern of demand in an area. Second, alterations to the road network and traffic patterns are another source of change that affects retail locations. Finally, the addition or abandonment of stores by competitors can alter the competitive environment. A change in the competitive environment will affect the performance of stores whose locations were chosen without such knowledge.

14.6.1   Scenario Planning

One way of dealing with uncertainties in location planning is through scenario planning. Scenario planning provides flexibility on strategic decisions, an integration of future environments, and a broader perspective for the decision maker [348]. At the heart of scenario planning is the realization that a single long range forecast is typically inadequate in capturing how the future will unfold. Instead a set of alternative scenarios are developed by identifying events that can potentially alter the environment. The planner can then consider the different contingencies and select locations that hedge against the risk of uncertainty. Two examples of scenario planning in location decisions are [1045, 467].

Recognizing the uncertainty surrounding forecasts of future demand patterns, Schilling proposed that "If a decision maker can delay choosing, he or she is more likely (through the reassessment of prior probabilities afforded by new information) to select a future locational configuration that will more closely match the actual future distribution of demand than if the selection were required immediately [1045, p.3]." Noting that not all facilities are built at the same time, Schilling argued that the network design decision can be viewed as a series of irreversible decisions made at discrete points over the planning horizon. Every time a decision is made to locate a facility it forecloses future options. To maintain flexibility the decision maker should postpone the closing of options as long as possible. This is done by identifying locations that are desirable in all future scenarios and opening facilities in those locations first.

Ghosh and McLafferty [467] propose a multiobjective programming approach to dealing with uncertainty. Suppose, for example, that S future scenarios are identified. If $Z_k(X_1, \ldots, X_n)$ is a measure of the performance (market share or profits) in scenario k, the best possible strategy is given by the solution to the following multiobjective problem:

$$MAX[Z_k(X_1, \ldots, X_n), k = 1, 2, \ldots, S] \qquad (14.10)$$
$$(X_1, \ldots, X_n) \in F$$

where $X_1, \ldots X_n$ is the set of decision variables and F defines the set of feasible strategies. There may be no unique solution to this problem, however, since no single strategy performs best in all scenarios.

When there are no unique solutions to a multiobjective problem, a crucial task is to identify the set of noninferior solutions. One procedure for generating this set of noninferior solutions is the weighting method [800, 234]. The objective function (14.10) can be transformed into a single objective problem by the following linear combination:

$$MAX \left\{ \sum_{k=1}^{S} w_k[Z_k(X_1, \ldots, X_n)] \right\} \tag{14.11}$$

Cohon has shown that the solution to (14.11) is a noninferior solution to the multiobjective problem, and suggested the generation of the entire set of noninferior solutions by solving (14.11) repeatedly with various sets of weights [234, p.104].

14.6.2   Competitive Location Models

In addition to changes in consumer behavior and residential patterns, competition also creates uncertainty in locational planning. Unlike their counterparts in the public sector, retail organizations are rarely monopolists; competition is the rule rather than the exception. In such environments retail firms must anticipate and plan for possible future changes – changes that can occur as competitors attempt to improve their own performance by seeking their own preferred strategies. It is necessary, therefore, to find facility locations that not only improve performance in the short run, but also protect that performance from future changes in the competitive environment.

Although the scenario planning approaches discussed earlier are helpful in dealing with certain kinds of uncertainty, the large number of potential contingencies to be considered makes them less effective in dealing with future competitive environments. In recent years there has been considerable interest in developing competitive location models for network applications. These generally assume a duopolistic market in which both firms operate multiple outlets and the planner's objective is to find locations that are immune from the possible impact of new competitive facilities. (For a review of competitive models see Chapters 13 and 16 in this book). The application of these models in the retail context will greatly advance the state-of-art in retail location planning.

Fundamental to most competitive location models are the concepts of medianoids and centroids proposed by Hakimi [511]. To describe these concepts we adopt Hakimi's notations. A network is a weighted graph $G(V, E)$, where V and E are the sets of vertices and edges, respectively with a nonnegative weight $w(v)$ associated with each vertex v in V. Let $X_p = \{x_1, \ldots, x_p\}$ denote the set of p existing facilities located on G, and $Y_r = \{y_1, \ldots, y_r\}$ be the set of r new facilities to be located on G. Define $W(Y_r|X_p)$ as the cumulative weight of all vertices that are closer to a facility in Y_r than to one in X_p. If Y_r is such that $W(Y_r|X_p) > W(Y_r|X_p)$ for all Y_r on G, then Y_r is called the $(r|X_p)$ medianoid.

Suppose now that a firm is planning to locate p facilities in a market in which no competitive facilities currently exist, but the firm foresees the possibility of a competitor opening r new facilities. Then the objective should be to choose the p facilities such that the firm suffers minimum loss when competitors' facilities are opened. This is the $(r|p)$ centroid of G, which is given by the X_p that minimizes $W(Y_r|X_p)$. Thus the centroid is the solution to the following minimax location:

$$\min_{x_p} \max_{Y_r} W(Y_r(X_p) \mid X_p) = W(Y_r^*(X_p) \mid X_p^*) \qquad (14.12)$$

Ghosh and Craig [463] present an application of a competitive location model for locating retail outlets for a chain of convenience stores. The chain (chain A) is planning to enter a new market and open a number of new outlets. The managers, however, have reason to believe that a competing chain (chain B) also plans to open new outlets in the same market. The model proposed by these authors determines the optimal number of stores and their locations for chain A, anticipating the possible entry of competitors.

In this model, the expected revenue at a location is a function of the spatial distribution of demand and the distance separating consumers from their nearest outlet:

$$Q_{it} = \sum_{j=1}^{N} (P_{jt} a_{jt} d_{ij}^{-u}) X_{ij}$$

where Q_{it} is the expected revenue at location i; P_{jt} is the size of the target population at place j at time t; a_{jt} is the budget allocated for the purchase of goods sold at convenience stores; and d_{ij} is the distance from i to j. X_{ij} is equal to one if i is the closest store from j. The expected profit can then be calculated as:

$$R_i = \sum_{t=1}^{T} \frac{Q_{it} C_i - F_{it}}{(1+r)^t}$$

where C_i is the profit margin as a percent of revenue, F_{it} is the fixed cost of operating store at i during period t including the cost of capital, r is the discount rate and T is the length of the planning horizon. The authors propose a heuristic algorithm to determine the minimax locations using this objective function.

An interesting result regarding competitive locations was noted by Ghosh and Craig [463] and later expanded by Ghosh and Buchanan [461]. It is intuitive to expect that the first firm to enter into a market will have an advantage due to its ability to preempt the best locations. Ghosh and Buchanan [461], however, present examples of markets in which no such first entry advantage exists. The second firm can always gain the same if not a larger share of the market than the first. Eiselt and Laporte [372] demonstrate that this paradox occurs when there is no unique Nash equilibrium solution to the competitive location game. Although the reasons for the occurrence of the first entry paradox are not clear, it seems to depend on the relation between the spatial distribution of consumers and the location of potential sites [359]. See [111] for more on the first entry paradox.

14.6.3   Optimal Timing of Facility Locations

Although location models have become increasingly sophisticated in how they deal with uncertainties in the environment, the *timing of entry* decision has been rather neglected. Most models assume that decisions are made at the start of the planning horizon and that all facilities are located at that point in time. Ideally what is required, however, is to determine jointly the optimal timetable of facility openings and their locations. Two early attempts at solving the timetable problem were provided by Scott [1057] and Sheppard [1065]. They proposed location-allocation models that consider the dynamic evolution of demand and the trade-off between the cost of establishing facilities and the discounted future returns from providing the service. By incorporating the temporal pattern of demand and the cost of service in making location decisions, these models provide a valuable framework for retail location planning. The models, however, do not consider the competitive environment.

In competitive markets, the timing decision is more complex, since competitors may preempt desirable sites and preempt entry. Competitive entry reduces the firm's options for future locations. Eaton and Lipsey [360], for example, have shown that established firms may preempt to deter entry by competitors. (See [1197] for an empirical test). This means that in addition to demand evolution, firms must try to anticipate when competitors are likely to establish their own facilities and the specific sites at which these facilities are likely to be opened. Rao and Rutenberg [982] propose an approach to determining the optimal timing of market entry by specifying a set of decision rules for the firm. The decision rules are dynamic in that the outcomes change over time depending on the actions of the firm and the competitor in the previous time periods. The strategies of all firms in the market are determined by the application of these decision rules [472]. Selecting the optimal time table for facility openings is similar, in many respects, to determining the optimal timing for capacity expansion, R&D investments, switching from exports to foregin direct investment, and technology adoption. These problems, however, are typically devoid of any locational components.

14.7   Conclusions

This chapter reviewed location-allocation models for designing multifacility retail networks. These models provide comprehensive planning tools that encompass, within a single decision-making framework, questions about the number, locations and design of retail facilities as well as forecasts of sales and profits. Recent developments have extended these models to a wider range of objective functions, increasingly complex allocation rules, and more complex competitive environments, and have greatly enhanced the applicability of the models to retail planning. At the same time, the increasing availability of machine-readable databases and microcomputer software has made the models easier, cheaper and more efficient to use. This should lead to wider adoption of these methods by retail location analysts.

Although the past decade has witnessed many important developments in retail location modeling, several areas remain important topics for further research. There is a continuing need to develop new objective functions that more realistically reflect the retail environment and firms' long term goals and objectives. These objective functions will typically be multiobjective in nature and incorporate complex measures of the profitability and costs of retail outlets and the short- and long-term effects of retail competition. The objective functions might also differ depending on the organizational structure of the retail firm, as is the case with franchise systems. Research is also needed on integrating market segmentation principles with location-allocation models. Different market segments have distinct demand patterns for retail goods and services and are likely to make different spatial choices and have differing allocation rules. If location- allocation models are to provide accurate forecasts of store sales and profits, they must reflect the inherent segmentation that exists in retail markets. It is also important to refine and improve the allocation rules used in location-allocation models. Recent developments in spatial choice modeling should be incorporated, and researchers should consider critically the effects of specification and aggregation errors on the results of the modeling process.

Finally, the basic concept of location-allocation modeling should be broadened to include more input from and interaction with retail decision makers. Clearly, models cannot substitute for intelligent decision making, nor can they incorporate the full complexity of real world problems. Effective locational analysis requires the insights and judgments of retail planners and their detailed knowledge of local market conditions and the firm's needs and priorities.

15
ECONOMIC MODELS OF FIRM LOCATION

Dominique Peeters
Université Catholique de Louvain

Jacques-François Thisse
Université de Paris I-Sorbonne
CERAS-ENPC

15.1   Introduction

The economic theory of the location of the firm has been very much dominated by the Weber model. In its simplest form, this model can be formulated as follows. A firm purchases two inputs, available each at a given place, and uses them to produce a single output which is sold on a given market. It is supposed that the quantities of the goods purchased and sold are given a priori, as well as the prices on their respective markets, and that the production costs are the same across locations. The sole factor on which the firm has some control is the place where it will settle. The Weber problem then consists in choosing a location that minimizes the firm's total transportation cost. This model has been successful in explaining the observed locations of several industries [1007]. It also turned out to be flexible enough to accommodate several extensions, including locational constraints, generalized transportation and production cost functions, dynamic and stochastic elements, while its scope has been broadened to deal with the location of public services [1194].

The Weber model has had a lasting influence on the subsequent developments in location theory, notably by imposing transportation cost minimization as a paradigm. The apparent discrepancy between this criterion and the criterion of profit maximization considered in the economic theory of the firm has puzzled economists and geographers for a long time. For instance, Lösch [770, p. 28] sharply criticized Weber for eliciting a location that minimizes transportation costs:

> Weber's solution for the problem of location proves to be incorrect as soon as not only cost but also sales possibilities are considered. His fundamental error consists in seeking the place of lowest cost. This is as absurd as to consider the point of largest sales as the proper

location. Every such one-sided orientation is wrong. *Only search for the place of greatest profit is right* (our italics).

Various attempts have been proposed for reconciling the two approaches by determining the conditions under which the firm maximizes its profit by locating at the point of minimum transportation cost. A first answer has been provided by Isard [625] who showed that this occurs when the production function is with fixed technical coefficients. Soon after, Moses [872] proved that this condition was also necessary. The implication of these results is clear: besides the special case of a Leontiev production function, a more general approach than the Weber model is needed when substitution between inputs is possible. The book by Hurter and Martinich [606] gives an account of the tentatives made to integrate production theory together with the Weberian framework. Moses' paper gave rise to some later work by Alonso [19] who also accounted for demand considerations. Alonso's basic idea was to follow Lösch and to propose a firm's location model no longer driven by transportation cost minimization but instead by profit maximization. However, as shown by [891, 1130], Alonso's analysis was incomplete: there is an underlying connection between the two objective functions. Specifically, when the quantities transported are not known beforehand but only after the resolution of the location-production-distribution problem faced by the firm, it is possible to define an *ex post* transportation cost function by using the so-obtained quantities. The profit-maximizing location than minimizes this function if the firm pays the transportation of all its inputs and outputs. This result gives a posteriori a justification to Weber's approach and proves the depth of his intuition. Nevertheless, the optimal quantities to be hauled must still be determined and the Weber model definitely needs to be generalized. Another consequence of this result is that it confirms Lösch's suspicion about actual locations analyses. Since the solution of the location problem is at the same time the solution of an *ex ante* profit maximizing and an *ex post* cost minimizing problems, inductions about the firm's objective function from an observed situation can be misleading for they are based on elements that are consequences and not causes of the decision process (see the reference to Riley above).

Surprisingly enough, the pricing problem has generally been left outside of facility location analysis, even if many early authors were well aware of it [592, 770, 501]. A crucial observation is that space brings in difficulties in the perfectly competitive framework [1056]. Indeed, in a geographical setting where production and consumption activities do not occur at the same place(s), the number of firms that can supply a given market is generally small, thus endowing each of them with some market power. As a consequence, the economic environment of the firm is determinant for its location choice. Studies of the relationships between location and pricing have been confined to the analysis of spatial competition. Here, the basic framework has been provided by Hotelling [599] in his famous 'Stability in

Competition'. This family of models is very much studied and various ramifications have been developed outside the field of spatial economics [440]. The Weber and the Hotelling traditions are two branches of location theory that offer an interesting contrast. Whereas cost-oriented location models generally assume fixed requirements for the firm's output, spatial competition models postulate that firms have location-independent production costs. Thus, the impact of space either on demand or on production costs is neutralized. Introducing simultaneously spatial heterogeneity in both sides of the problem is thus a question in order.

In this chapter, it will be assumed that the firm operates as a *spatial monopoly*. This situation occurs when the firm is the sole producer of the commodity considered, or when its potential competitors are settled far away and/or transportation costs are high enough to prevent them from invading the firm's market area, or when there exists an agreement between firms allowing each of them a given territory that will not be encroached on by the others. Consequently, the firm is free to choose at which price it wants to sell its product on each market. By contrast, spatial competition resorts to game theory and involves the computation of Nash equilibria where several firms choose their prices strategically [441].

Another important observation concerning the formation of prices in the space-economy is that space separates markets in a natural way, which gives the firm the possibility to discriminate among them. Unless external factors intervene (such as antitrust laws), they are no reasons why a firm would renounce to this opportunity. More generally, discrimination in a spatial setting corresponds to the way transportation costs are passed on to the clients, a particular specification defining a particular *spatial price policy*. No discrimination occurs when each client pays its own transportation cost, while some form of discrimination exists when the firm absorbs some transportation costs or charges clients with phantom freights. A number of policies have been proposed in the literature: spatial discriminatory pricing, uniform delivered pricing, zone pricing, mill pricing, pick-up or delivery, etc. The reader is referred to [942, 502] for real-world examples of the main pricing schemes used by firms.

The problem we want to study here can then be stated as follows: *Given a spatial system of demand functions and a spatial price policy, find the location and size of the firm, the prices set on the different markets, and the distribution system that maximize the firm's profit.* Building a theory providing an answer to this problem acknowledges a fact which is left out in most of the literature on facility location: the pricing and location decisions are basically interdependent [501, 19, 1035]. Indeed, the optimal location can only be found once we know the quantities to be distributed, hence the price charged in each market. On the other hand, it is clear that the optimal price(s) depends on the location selected. To quote Greenhut [501, p.267]:

The inclusion of cost and demand factors in one model points out the need for a broader statement of the determinants of plant location than one which concludes that firms seek the location of least cost, or one which holds that firms seek the location offering the largest market area. This need is fulfilled by the concept of the maximum-profit location, which by definition may be referred to as that site from which a given number of buyers (whose purchases are required for the greatest possible profits) can be served at the lowest total cost. While the lowest level of average production cost at this site may be higher than that existing at alternative ones, the monopolistic control gained over larger numbers of buyers (spread over a market area) makes it the maximum-profit location at the optimum output.

The attempts made to integrate locational decisions and spatial pricing into *operational models* are very few (see [545] for a survey). The objective of this paper is to partially fill up in that gap by comparing the impact of different spatial price policies upon the firm's location, both in the continuous and discrete frameworks of location theory. We first define the five price policies considered in this paper and study some of their properties in a simple economic model; we then sketch the solution method that allows one to determine the optimal price(s) in more realistic environments. In the subsequent section, we introduce the five location problems for *a single plant firm in the case of a continuous space* and for *a multiplant firm in the case of a discrete space*. Finally, using algorithms developed in separated papers [536, 553], we determine and compare the profit-maximizing price(s) and location(s) for typical examples.

15.2   Spatial Pricing Policies

We first present the general problem of the spatial monopolist located at a given fixed location. It is supposed that the clients are uniformly spread with unit density over a straight half line whose origin is at the firm's location. Each client is endowed with a linear demand function $D(\tilde{\pi}) = [a - b\tilde{\pi}]^+$, where $a > 0$, $b > 0$, $[x]^+ = \max(x, 0)$, and $\tilde{\pi} = \tilde{\pi}(\tilde{r})$ is the *delivered price* at location $\tilde{r} \geq 0$. For every client, there is also a *mill (or f.o.b.) price* $\tilde{p}(\tilde{r})$ linked to $\tilde{\pi}(\tilde{r})$ by the relationship

$$\tilde{\pi}(\tilde{r}) = \tilde{p}(\tilde{r}) + t\tilde{r}. \tag{15.1}$$

The mill price is thus equal to the delivered price minus the corresponding transportation cost. Note that neither $\tilde{p}(\tilde{r})$ nor $\tilde{\pi}(\tilde{r})$ need to be the same for all clients. The particular shape of $\tilde{\pi}(\tilde{r})$ as a function of $\tilde{r}$ defines the *spatial price policy* of the firm, that is to say, the way in which transportation costs are passed on to the clients. The firm has a constant marginal production cost v and the transportation cost per unit of distance and volume is a constant t. Let $\alpha = a/b$ denote the maximum willingness to pay of the clients. To simplify notation, Beckmann [70] has suggested the following change of variables: $\pi = (\tilde{\pi} - v)/(\alpha - v)$; $r = t\tilde{r}/(\alpha - v)$ which merely consists in choosing an appropriate unit of distance for the transportation rate to be equal to one, normalizing the marginal production cost to zero by translating the axis system, and rescaling the monetary and output units for the intercept and slope of the demand function to be both equal to one. In other words, we have $a = b = t = 1$ and $v = 0$. The size R of the firm's market area is determined by the set of points where the firm finds it profitable to sell and each client wants to buy. The profit function of the firm is given by

$$P_R(\pi) = \int_0^R [1 - \pi(r)]^+ [\pi(r) - r]^+ dr = \int_0^R B_r(\pi(r)) dr \qquad (15.2)$$

where $B_r(\pi)$ is the benefit (or profit) obtained when supplying the client located at r at the delivered price π. Thus the problem of the firm amounts to choosing the function $\pi = \pi(r)$ such that:

$$P_R = \max_\pi P_R(\pi) \qquad (15.3)$$

subject to the constraint

$$\pi \in \Pi \qquad (15.4)$$

where Π is the set of admissible spatial price policies that are defined below. The last stage of the optimization problem consists in determining the optimal size of the market area:

$$P^* = \max_R P_R \qquad (15.5)$$

This problem is conditional to the constraint (15.4) and that the optimal size of the market area depends on the spatial price policy considered.

In what follows, we consider the main geographical pricing systems that are adopted by firms and, therefore, specify the constraint (15.4). Following a particular spatial price policy has an impact on the economic choices of the firm. Results comparing the prices at different locations and the levels of profit and production are also given. The results for the three first policies are based on [70]. Those for the fifth policy are borrowed from [439].

15.2.1   Mill pricing

In this system, the firm sets a mill price p^M which is identical for all clients, regardless of their location. This occurs when the clients travel to the firm and take care of the transport of the commodity, or when the firm delivers the commodity and charges its clients their respective transportation cost. The particular form of (15.4) for this policy is then:

$$\pi(r) = p^M + r. \qquad (15.6)$$

This pricing policy is the only one that does not involve discrimination among clients and, for that reason, has been advocated by antitrust authorities. The following results can be easily established by standard calculus after inserting (15.6) into (15.2). The optimal price is: $p^M = \frac{1}{2} - \frac{R}{4}$. Hence p^M decreases as the market area extends. Moreover, this area cannot expand beyond $R = \frac{2}{3}$ for the delivered price corresponding to clients located further away would exceed their maximum willingness to pay. The level of production is: $Q^M = \frac{R}{2}(1 - \frac{R}{2})$. This is a quadratic function in R whose maximum value is attained when $R = \frac{1}{2}$. Finally, the profit function is given by $P^M = \frac{R}{4}(1 - R + \frac{1}{4}R^2)$. This function reaches its maximum at $R^* = \frac{2}{3}$, which corresponds to the optimal size of the market area. The optimal price is then $p^{M*} = \frac{1}{3}$, the volume of production $Q^{M*} = \frac{2}{9}$, and the profit $P^{M*} = \frac{2}{27}$.

15.2.2   Uniform delivered pricing

It is now supposed that the firm delivers its output and charges every client in $[0, R]$ the same delivered price π^U. Constraint (15.4) is then:

$$\pi(r) = \pi^U. \tag{15.7}$$

Under uniform delivered pricing, the clients located close to the seller are discriminated against more remote clients because the mill price $\pi^U - r$ is larger for the former than for the latter. This form of discrimination is far and away the most widespread. The optimal price is $\pi^U = \frac{1}{2} + \frac{R}{4}$. It increases with the size of the market area. Observe that the firm refuses to sell beyond $R = \frac{2}{3}$. Hence the optimal size of the market area is identical to that obtained under mill pricing. However, the reasons to exclude remote clients are different: here it stems from a decision of the firm not to supply them, whereas in the previous case the decision is made by the clients not to buy. This may have some importance when dealing with regulated monopolies, because in such cases public authorities may impose to the firm to meet all the demands on a given territory. Interestingly, the respective levels of production and profit are the same as under mill pricing: $Q^U = \frac{R}{2}(1 - \frac{R}{2})$ and $P^U = \frac{R}{4}(1 - R + \frac{1}{4}R^2)$. It should be emphasized that this equivalence is associated with different consumer benefits since those located near the firm get a higher surplus under mill pricing while the opposite holds for the remote clients. Since the optimal size of the market area is $\frac{2}{3}$, we have $\pi^{U*} = \frac{2}{3}$, $Q^{U*} = \frac{2}{9}$, and $P^{U*} = \frac{2}{27}$.

15.2.3   Perfect spatial discrimination

The hypothesis of constant marginal production cost allows the firm to fix at each point of its market area a price that maximizes its local benefit $B_r(\pi(r))$ independently of the prices set at other points. A simple calculation shows that $\pi^D(r) = \frac{1}{2} + \frac{r}{2}$, that is to say, each client is charged half of the transportation cost. It is worth noting that this price depends only upon the location of the client and not on the size of the market area. In the absence of arbitrage among clients, the firm has a natural way to separate the local markets. Somewhat surprisingly, the level of production is the same as in the preceding two cases: $Q^D = \frac{R}{2}(1 - \frac{R}{2})$. However, the profit is higher: $P^D = \frac{R}{4}(1 - R + \frac{1}{3}R^2)$, since the firm can earn as much as possible from every local market, independently of the others. We see that the firm now chooses to supply the clients in the interval $[\frac{2}{3}, 1]$. Applying perfect discrimination leads to a larger market area. The total quantity produced is $Q^{D*} = \frac{1}{4}$ and the profit $P^{D*} = \frac{1}{12}$.

An interesting implication of these results is that *perfect spatial discrimination allows more clients to be supplied*. However, this price policy has also some major drawbacks: the list of prices is far from being transparent to the clients and bookkeeping may turn out to be very burdensome when the number of clients is high. Therefore, it is interesting to envisage substitutes that avoid some of these undesirable side-effects.

15.2.4 Zone pricing

The difficulties that arise with perfect spatial discrimination have led to the implementation of zone pricing. The clients of the firm are clustered into different zones and the firm sets a uniform delivered price in each one of them. Two cases may then arise: in the first one, the zones are endogenous, while in the second they are given a priori (for instance, they correspond to administrative districts). We present here the results corresponding to the former case only since, in the latter, they depend on the specificities of the partition considered. Suppose that there are Z zones (this number is exogenous to the problem) and that π_z^Z is the price prevailing in zone z. The client located at r is allocated to the zone z where the benefit that the firm can earn by placing him there exceeds any other allocation:

$$\pi(r) = \pi^z \quad \text{if } B_r(\pi_z^Z) = \max(B_r(\pi_{z'}^Z) \mid z' \in Z). \tag{15.8}$$

The first difficulty is to determine the boundaries of the different zones. After some calculations, it can be shown that the profit-maximizing zone z is the segment $[\frac{z-1}{Z}R, \frac{z}{Z}R]$, $z = 1, \ldots, Z$, which corresponds to a subdivision of the market area into Z *equal subregions*. The price in the zone z is $\pi_z^Z = \frac{1}{2} + (2z - 1)\frac{R}{4Z}$. The total quantity produced is the same as in the preceding three cases: $Q^Z = \frac{R}{2}(1 - \frac{R}{2})$. The profit ranges between those obtained under uniform delivered pricing and perfect discriminatory pricing: $P^Z = \frac{R}{4}(1 - R + \frac{4Z^2-1}{12Z^2}R^2)$. The optimal size of the market area equals $\frac{2Z}{2Z+1}$. The outer limits of the tariff-zones are then: $R_z^{Z*} = \frac{2z}{2Z+1}$, $z = 1, \ldots, Z$. The optimal price in zone z is $\pi_z^{Z*} = \frac{Z+z}{2Z+1}$, $z = 1, \ldots, Z$. The total quantity produced is $Q^{Z*} = \frac{Z(Z+1)}{(2Z+1)^2}$, while the optimal profit is given by $P^{Z*} = \frac{Z(Z+1)}{3(2Z+1)^2}$. Obviously, letting $Z = 1$ we fall back on the results obtained uniform delivered pricing, while perfect spatial discrimination prevails when Z becomes arbitrarily large.

The geography of prices for the 2- and 3-zone case is as follows. When $Z = 2$, the two zones correspond to the segments $[0, \frac{2}{5}]$ and $[\frac{2}{5}, \frac{4}{5}]$ while the delivered prices are respectively $\frac{3}{5}$ and $\frac{4}{5}$. When $Z = 3$, the zones are then given by $[0, \frac{2}{7}]$, $[\frac{2}{7}, \frac{4}{7}]$, and $[\frac{4}{7}, \frac{6}{7}]$ with $\pi_1^3 = \frac{4}{7}$, $\pi_2^3 = \frac{5}{7}$, and $\pi_3^3 = \frac{6}{7}$. Clearly, the firm's market area expands rapidly to approach its maximum size obtained under perfect spatial discrimination. Similarly, the delivered price at each location also tends towards the corresponding value under perfect spatial discrimination when Z increases. This confirms our claim that *zone pricing is a close substitute for perfect spatial discrimination*. Note, finally, that this system is viable only if the firm can prevent arbitrage among clients residing in different zones. Otherwise the clients located near the border of a zone could profitably resale to clients of the neighboring zone at a lower price than the delivered price charged by the firm.

15.2.5 Mixed pricing

A fifth system arises when the clients are free to choose between travel and pick-up at the firm at a mill price p^m or home-delivery at a uniform delivered price π^m. This is referred to as mixed pricing. The mathematical expression of a client's choice of is more complex:

$$\pi(r) = \begin{cases} p^m + r & \text{if } B_r(p^m + r) > 0 \text{ and either } B_r(\pi^m) = 0 \\ & \text{or } p^m + r \leq \pi^m \\ \pi^m & \text{if } B_r(\pi^m) > 0 \text{ and either } B_r(p^m + r) = 0 \\ & \text{or } p^m + r \geq \pi^m \\ \infty & \text{otherwise} \end{cases} \quad (15.9)$$

In words, this expression states that the client located at r will choose *pick-up* if this service is available (p^m is high enough to yield a positive profit to the firm and the delivered price $p^m + r$ does not exceed the maximum willingness to pay 1) and either home-delivery is not provided at r or is possible but at a price π^m that exceeds the full price $p^m + r$. The conditions for choosing *delivery* are similar. Finally, when none of these possibilities is available to a client, we conventionally set the price to infinity.

For a given R, it is readily verified that the clients located in the interval $[0, \frac{R}{2}]$ choose pick-up, while those in the interval $[\frac{R}{2}, R]$ prefer home-delivery. The mill price is $p^m = \frac{1}{2} - \frac{R}{8}$ and the uniform delivered price is $\pi^m = \frac{1}{2} + \frac{3R}{8}$. We immediately see that $p^m > p^M$ and $\pi^m > \pi^U$: under mixed pricing, prices are higher than under mill or uniform delivered pricing. This follows from the fact that the firm is able to discriminate against nearby clients in favor of more remote ones. Comparing with the zone system when $Z = 2$, we observe that $\pi^m = \pi_2^2$. Furthermore, the quantity produced is still the same:

FIGURE 1. Comparison of delivered prices under M, U, and D

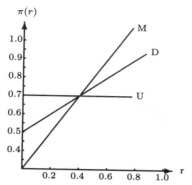

$Q^m = \frac{R}{2}(1 - \frac{R}{2})$, while the profit is now given by: $P^m = \frac{R}{4}(1 - R + \frac{5}{16}R^2)$. Again, we observe that $P^m = P^Z$ for $Z = 2$.

When the firm chooses its optimal market area, the clients in the interval $[0, \frac{2}{5}]$ choose pick-up while those located in $[\frac{2}{5}, \frac{4}{5}]$ prefer delivery. The optimal mill and delivered prices are respectively $p^{m*} = \frac{2}{5}$ and $\pi^{m*} = \frac{4}{5}$. The total production is $Q^{m*} = \frac{6}{25}$ and the profit equals $P^{m*} = \frac{2}{25}$.

15.2.6 Comparison of the price policies

It is worth summarizing our results under the form of a comparison of the different price policies analyzed. First, we report in Figure 1 the geographical schedule of prices for mill pricing (denoted by M), uniform delivered pricing (U), and perfect spatial discrimination (D). On the first half of the interval $[0, \frac{2}{3}]$, the clients are better-off under M, while those located in the second half prefer U. D is intermediate (in fact $\pi^D(r)$ is the average of the delivered prices under U and M at r). The market area is 50% larger under D. In summary, the size of the market areas rank as follows: $R^{D*} > R^{M*} = R^{U*}$, while for the profits we have $P^{D*} > P^{M*} = P^{U*}$. The total production is similarly ranked: $Q^{D*} > Q^{M*} = Q^{U*}$. Next, we compare U and D with zone pricing (in short Z). Figure 2 shows the evolution of the geography of prices when passing from a unique zone with U, through $Z = 2$ and $Z = 3$, to D. The most striking feature is how *the multiplication of zones permits the invasion of more remote markets*.

Our last comparison is between M, U, and mixed pricing denoted by m. As shown by Figure 3, this policy leads to an extension of the market area (in fact the same length and the same subdivision as under zone pricing for $Z = 2$). This is obtained by an increase of both the mill and the delivered prices. In words, the firm separates nearby and remote clients and gains on both sides. Not surprisingly, we have: $P^{m*} > P^{M*} = P^{U*}$

FIGURE 2. Comparison of delivered prices under U, $Z = 2$, $Z = 3$, and D

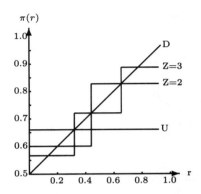

FIGURE 3. Comparison of delivered prices under M, U, and m

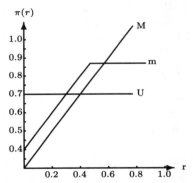

and $Q^{m*} > Q^{M*} = Q^{U*}$.

One may wonder if the properties obtained so far still hold when heterogeneity in the spatial distribution of demand is considered. Of course, such a generalization makes it difficult to establish clear-cut results and often the best we can do is to verify the robustness of these properties by means of numerical experiments. For that, one has to construct algorithms for determining the optimal price(s) for the five policies in a more complex structure than the one considered here.

15.3   Finding the Optimal Price(s)

In this section, we propose algorithms to find the optimal price(s) under alternative policies for a firm whose location is given. We first present the operational models which slightly differ from those studied in the previous section. Instead of a continuum of clients, we suppose here a finite number of markets; moreover, the demand functions, though linear, may vary from one market to the other.

15.3.1   The model

We consider a single-plant firm producing a single commodity and located in a given point of the transportation space. Clients are concentrated in a finite number of markets represented by the points $m_i, (i = 1, \ldots, I)$ of the same space. Each market is characterized by a linear demand function $D_i(\pi_i)$ for the firm's output, where π_i is the delivered price at location m_i; formally, the demand functions are given by $D_i(\pi_i) = [a_i - b_i \pi_i]^+$, where $a_i > 0$, $b_i > 0$, and $\alpha_i \equiv a_i/b_i$. Let v denote the (constant) marginal production cost function of the firm and t_i the unit transportation cost of the output from the firm to m_i. Given a price policy, the problem of the firm is to choose its price(s) in order to maximize its profit.

Let $\pi = (\pi_1, \ldots, \pi_I)$ be the vector of delivered prices on the various markets, and $\underline{q} = (q_1, \ldots, q_I)$ the corresponding vector of quantities supplied. The profit function of the firm is:

$$P(\pi, \underline{q}) = \sum_{i=1}^{I} (\pi_i - v - t_i) \, q_i. \tag{15.10}$$

Several constraints must be considered. First, the quantity shipped to m_i must be nonnegative and cannot exceed the local demand:

$$0 \le q_i \le D_i(\pi_i). \tag{15.11}$$

As in Section 15.2, the spatial price policy of the firm can be expressed through some constraints:

$$\pi \in \Pi \tag{15.12}$$

that are made more precise below. The price problem can then be formulated as follows: $P^* = \max P(\pi, \underline{q})$ subject to (15.11) and (15.12).

An alternative formulation of this problem will be useful. First, note that the firm incurs a loss when supplying the market m_i when $\pi_i \le v + t_i$; in this case, we set $q_i = 0$. On the contrary, since there is no capacity constraint the firm will meet the whole demand when $\pi_i > v + t_i$: $q_i = D_i(\pi_i)$. Let $B_i(\pi_i) = [\pi_i - v - t_i]^+ D_i(\pi_i)$ be the benefit obtained from supplying the market m_i at the delivered price π_i. Hence the objective function above can be rewritten as: $P^* = \max P(\pi) = \max \sum_{i=1}^{I} B_i(\pi_i)$ subject to (15.12).

The constraints (15.12) have already been specified in Section 15.2. The first one corresponds to perfect spatial discrimination: $\pi_i^D = \arg\max_{\pi_i} B_i(\pi_i)$. The second one arises when the firm uses uniform delivered pricing: $\pi_i = \pi^U$, $i = 1, \ldots, I$. π^U is then the only variable to be determined. Third, when the firm sets a mill price p^M, (15.12) can be stated as: $\pi_i = p^M + t_i$, $i = 1, \ldots, I$. Fourth, under zone pricing with Z zones, each being endowed with a single delivered price π^z, every market is assigned to the zone which yields the largest profit:

$$\pi_i = \arg \max_{z=1,\ldots,Z} B_i(\pi^z). \tag{15.13}$$

The optimization problem is now carried out in the $\mathbf{R}^Z$ space of the price vectors π. Finally, when clients are offered a choice between pick-up and delivery, the constraints $\pi \in \Pi$ are as follows:

$$\pi_i = \begin{cases} p^m + t_i & \text{if } B_i(p^m + t_i) > 0 \text{ and either } B_i(\pi^m) = 0 \\ & \quad \text{or } B_i(\pi^m) > 0 \text{ and } p^m + t_i \leq \pi^m \\ \pi^m & \text{if } B_i(\pi^m) > 0 \text{ and either } B_i(p^m + t_i) = 0 \\ & \quad \text{or } B_i(p^m + t_i) > 0 \text{ and } \pi^m \leq p^m + t_i \\ \infty & \text{if } B_i(p^m + t_i) = B_i(\pi^m) = 0 \end{cases} \tag{15.14}$$

15.3.2   Algorithms

15.3.2.1 Under perfect discriminatory pricing, it is readily verified that

$$\pi_i^D = \begin{cases} (\alpha_i + v + t_i)/2 & \text{if } \alpha_i > v + t_i \\ \infty & \text{otherwise} \end{cases}$$

which means that the optimal benefit is:

$$B_i^D = \begin{cases} (\alpha_i - v - t_i)^2/4 & \text{if } \alpha_i > v + t_i \\ 0 & \text{otherwise} \end{cases}$$

so that the optimal profit is given by $P^D = \sum_{i=1}^I B_i^D$. This is illustrated on the following simple example. There are four markets located along a line segment at distances 0, $\frac{1}{3}$, $\frac{2}{3}$, and 1 from the firm. The demand functions are $D_1(\pi_1) = [2 - 1.5\pi_1]^+$, $D_2(\pi_2) = [1 - \pi_2]^+$, $D_3(\pi_3) = [1 - 0.75\pi_3]^+$, and $D_4(\pi_4) = [2 - 0.75\pi_4]^+$. The marginal production cost is taken to be zero while the transportation rate is normalized to one. The benefit functions are as follows: $B_1(\pi_1) = [2\pi_1 - 1.5\pi_1^2]^+$; $B_2(\pi_2) = [-\frac{1}{3} + \frac{4}{3}\pi_2 - \pi_2^2]^+$; $B_3(\pi_3) = [-\frac{2}{3} + 1.5\pi_3 - 0.75\pi_3^2]^+$; $B_4(\pi_4) = [-2 + 2.75\pi_4 - 0.75\pi_4^2]^+$. Their respective maxima are reached at $\pi_1^{D*} = \frac{2}{3}, \pi_2^{D*} = \frac{2}{3}, \pi_3^{D*} = 1$, and $\pi_4^{D*} = \frac{11}{6}$, whereas the benefits are $B_1^{D*} = \frac{2}{3}, B_2^{D*} = \frac{1}{9}, B_3^{D*} = \frac{1}{12}$, and $B_4^{D*} = \frac{25}{48}$. The profit P^{D*} equals $\frac{199}{144}$.

This simple procedure has been implemented as a subroutine of a more general program to find the optimal prices under the five systems considered in this section.

15.3.2.2 The case of uniform delivered pricing is less simple. Figure 4 shows the profit function under uniform delivered pricing for the example provided above. It reveals the main difficulties we have to face: the

FIGURE 4. Profit function for uniform delivered pricing

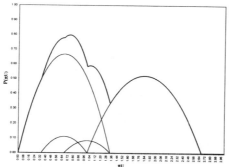

profit function is not concave (not even unimodal) and possesses points of nondifferentiability. The problem is thus to find the global maximum of a univariate function. The methods of resolution depends on the properties of the demand functions. In the general case where they are Lipschitzian, we can apply one of the methods discussed in [543, 544]. More precisely, when they are linear and when the number of markets is small, the search can be done by scanning the range of admissible prices. To this purpose, we merge the list of prices $v + t_i$ for $i = 1, \ldots, I$, and the list of the α_i. Clearly, any market such that $\alpha_i \leq v + t_i$ can be discarded. The resulting list is sorted and we obtain $\pi^{(1)} \leq \ldots \leq \pi^{(k)} \leq \ldots \leq \pi^{(K)}$. where $K \leq 2I$. We consider the interval $[\pi^{(k)}, \pi^{(k+1)}]$ and we define

$$I^{(k)} = \left\{ i \mid \pi^{(k)} \geq v + t_i \text{ and } \pi^{(k+1)} \leq \alpha_i \right\},$$

where $k = 1, \ldots, K - 1$. Then, the function $P^U(\pi^U)$ over this interval is a parabola whose equation is given by: $P^U(\pi^U) = -A^{(k)}(\pi^U)^2 + B^{(k)}\pi^U - C^{(k)}$ where

$$A^{(k)} = \sum_{i \in I^{(k)}} b_i \; ; \quad B^{(k)} = \sum_{i \in I^{(k)}} b_i (\alpha_i + v + t_i) \; ; \quad C^{(k)} = \sum_{i \in I^{(k)}} a_i (v + t_i).$$

The basic idea of the procedure is then very simple. For each interval of the list, we determine the equation of the corresponding parabola; if the maximum of the parabola is attained inside the interval, it is a candidate for the solution; otherwise the interior of the interval cannot contain the global optimum and the profit function is evaluated at the extremity of the interval where it is increasing. It can be shown that this algorithm solves the price problem in $O(I)$. We illustrate it on the example provided above. The explanations can be followed on Figure 4. The benefit functions determine five intervals reproduced in Table 15.1, as well as the markets that make up the profit function and the equation of that function within the corresponding interval.

The maximum of the first parabola is reached at $\frac{2}{3}$, which lies outside the interval under consideration. The provisional best value is obtained at the border $\frac{1}{3}$ of the interval where $P^U = 0.5$. On the second interval, we find

TABLE 15.1. Algorithmic details for solving the uniform delivered pricing problem

k	$[\pi^{(k)}, \pi^{(k+1)}]$	$I^{(k)}$	$P^U(\pi)$
1	$[0, \frac{1}{3}]$	$\{1\}$	$2\pi - 1.5\pi^2$
2	$[\frac{1}{3}, \frac{2}{3}]$	$\{1, 2\}$	$-\frac{1}{3} + \frac{10}{3}\pi - 2.5\pi^2$
3	$[\frac{2}{3}, 1]$	$\{1, 2, 3\}$	$-1 + \frac{29}{6}\pi - 3.25\pi^2$
4	$[1, \frac{4}{3}]$	$\{1, 3, 4\}$	$-\frac{8}{3} + 6.25\pi - 3\pi^2$
5	$[\frac{4}{3}, \frac{8}{3}]$	$\{4\}$	$-2 + 2.75\pi - 0.75\pi^2$

$\tilde{\pi} = \frac{2}{3}$, which belongs to the interval. The value of the profit function at $\frac{2}{3}$ is $\frac{7}{9}$, which improves on P^U. In the third interval, $\tilde{\pi} = \frac{29}{39} \approx 0.744 \in [\frac{2}{3}, 1]$, which gives another improvement $P^U = \frac{373}{468} \approx 0.797$. The maximum in the fourth interval is obtained at the interior point $\tilde{\pi} = \frac{25}{24}$, where the value of the profit function is $\frac{113}{192} \approx 0.589$, which brings no improvement on P^U. Finally, in the fifth interval we have $\tilde{\pi} = \frac{11}{6} \in [\frac{4}{3}, \frac{8}{3}]$, but $P(\frac{11}{6}) \approx 0.521 < P^U$. In conclusion, the optimal price is $\pi^{U*} = \frac{29}{39}$ and the optimal profit $P^{U*} = \frac{373}{468}$. At this solution, the firm supplies only the first three markets. The quantities delivered are respectively $q_1^{U*} = 0.885$, $q_2^{U*} = 0.256$, and $q_3^{U*} = 0.442$.

Despite its simplicity the method presented works very well for a small number of markets. However, when this number is large, one may prefer an alternative procedure proposed in [536] which cannot be described here because of the lack of space.

15.3.2.3 The case of mill pricing can be solved by using a similar method in which the lists are given by v and by the $\alpha_i - t_i$'s. Of course, every market such that $v \geq \alpha_i - t_i$ must be eliminated. After sorting the corresponding list, we obtain $v = p^{(0)} \leq \ldots \leq p^{(k)} \leq \ldots \leq p^{(K)}$. Next, we compose the index sets: $I^{(k)} = \{i \mid p^{(k-1)} < \alpha_i - t_i\}$, where $k = 1, \ldots, K$ with $K \leq I$, and the coefficients of the profit function $P^M(p^M) = -A^{(k)}(p^M)^2 + B^{(k)}p^M - C^{(k)}$ are now:

$$A^{(k)} = \sum_{i \in I^{(k)}} b_i \; ; \quad B^{(k)} = \sum_{i \in I^{(k)}} b_i \left(\alpha_i + v - t_i\right); \quad C^{(k)} = \sum_{i \in I^{(k)}} b_i \, v \left(\alpha_i - t_i\right).$$

The scanning procedure is somewhat simplified because passing from an interval $[p^{(k-1)}, p^{(k)}]$ to the next $[p^{(k)}, p^{(k+1)}]$ entails only the deletion of elements when constructing $I^{(k+1)}$ from $I^{(k)}$.

For the example, we see that there are three intervals to be explored. These intervals, the relevant sets $I^{(k)}$, and the equations of the parabola are given in Table 15.2.

TABLE 15.2. Algorithmic details for solving the uniform mill pricing problem

k	$[p^{(k-1)}, p^{(k)}]$	$I^{(k)}$	$P^M(p)$
1	$[0, \frac{2}{3}]$	$\{1, 2, 3, 4\}$	$\frac{53}{12}p - 4p^2$
2	$[\frac{2}{3}, \frac{4}{3}]$	$\{1, 4\}$	$\frac{13}{4}p - \frac{9}{4}p^2$
3	$[\frac{4}{3}, \frac{5}{3}]$	$\{4\}$	$\frac{5}{4}p - \frac{3}{4}p^2$

FIGURE 5. Profit function for zone pricing with $Z = 2$

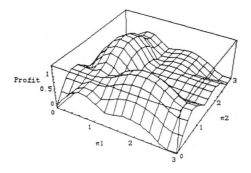

The optimal solution is $p^{M*} = \frac{53}{96} \approx 0.552$ and the maximum profit is $P^{M*} = 1.219$. The quantities purchased on the different markets are $q_1^{M*} = 1.172$, $q_2^{M*} = 0.115$, $q_3^{M*} = 0.086$, and $q_4^{M*} = 0.836$.

As in the above, another method can be used when the number of markets is large [536].

15.3.2.4 Under zone pricing, the problem of finding the optimal prices is more complicated than in the cases above because we now have to solve a multivariate global optimization problem. The profit function for the example with two zones is illustrated in Figure 5. The allocation of every market to one of the zones is made following the rule expressed in (15.13). Given the structure of the problem, we may think of using an extension of the method Big Square-Small Square developed by [547], except that the space in which we work is no more the geographical space but the price space which is included in $\mathbf{R}^Z$, where Z is the number of zones. This space is covered by hypercubes called "boxes".

The range of admissible values for the price π^z in zone z is $[\min_{i \in I} v + t_i, \max_{i \in I} \alpha_i]$. Moreover, Figure 5 reveals a symmetry in the problem about the main bisector. Using this observation, we can narrow down the search for the optimal prices by assuming that the zones are ordered in such a way that $\pi^1 \leq \ldots \leq \pi^Z$. Consequently, the set of price vectors to be explored is:

$$\Pi^Z = \{(\pi^1, \ldots, \pi^Z) \mid \min_{i \in I} v + t_i \leq \pi^1 \leq \ldots \leq \pi^Z \leq \max_{i \in I} \alpha_i\}. \quad (15.15)$$

The procedure starts with a box $[\min_{i \in I} v + t_i, \max_{i \in I} \alpha_i]^Z$ and proceeds by dividing this box into 2^Z equal subboxes, evaluating an upperbound on the value of the profit function inside each of them, eliminating those for which the upperbound is lower than the currently known best solution, and computing the value of the profit function at the center of the box with the largest upperbound in order to check if this would not improve the incumbent best solution. Moreover, the constraints (15.15) allows one to discard the boxes that do not overlap the feasible region. Since the rules are similar to those of BSSS, they will not be formally stated here but rather illustrated on the data set with $Z = 2$.

The initial step is to find a good starting solution that would reduce the search. As a first guess, one may consider the vector of the Z greatest discriminatory prices. In our example, it is $(1, \frac{11}{6})$. Market 1 is allocated to the first zone and yields a profit of $\frac{1}{2}$. Market 2 is not supplied, while market 3 is also placed in the first zone and gives a profit of $\frac{1}{12}$. Finally, market 4 constitutes the second zone by itself with a delivered price equal to $\frac{11}{6}$. This solution yields a profit of $\frac{53}{48}$. It can be improved in the following way. Consider each zone just defined and determine the optimal delivered price for the markets located inside the zone using the method presented earlier. The price vector is then $(\frac{7}{9}, \frac{11}{6})$; the corresponding profit is 1.314042. Market 2 is now supplied and belongs to the first zone. Applying the same procedure, one obtains the vector $(\frac{29}{39}, \frac{11}{6})$ which rises the objective to 1.317841. Since the zones no longer change, no further improvement can be obtained at this stage.

The adapted BSSS algorithm starts with an initial box $[0, \frac{8}{3}]^2$. The center of the box is the vector $(\frac{4}{3}, \frac{4}{3})$ which does not provide a better solution. The box is divided into four equal subboxes: $[0, \frac{4}{3}]^2$, $[0, \frac{4}{3}] \times [\frac{4}{3}, \frac{8}{3}]$, $[\frac{4}{3}, \frac{8}{3}] \times [0, \frac{4}{3}]$, and $[\frac{4}{3}, \frac{8}{3}]^2$. The third subbox can be discarded for it does not respect the constraints (15.15), except for the point $(\frac{4}{3}, \frac{4}{3})$ which has already been considered. It remains to propose an upperbound for the remaining boxes. Consider the box $Q = \prod_{z=1}^{Z} [\pi_z^-, \pi_z^+]$. The solution of the following problem

$$\overline{P}(Q) = \sum_{i=1}^{I} \max_{z=1,\ldots,Z} \max_{\pi \in [\pi_z^-, \pi_z^+]} B_i(\pi)$$ dominates the function P on the box Q and is thus a candidate upperbound. Note that the computation of this bound can be viewed as solving a "restricted" discriminatory pricing problem where the prices are limited to the specified intervals. In our example, one easily obtains 1.194, 1.381, and 0.521 for the three remaining boxes. The first and the last can be discarded, so that only the box $[0, \frac{4}{3}] \times [\frac{4}{3}, \frac{8}{3}]$ is to be considered. The value of the profit function at the center of this box gives only 1.278. The box is subsequently subdi-

FIGURE 6. Profit function for mixed pricing

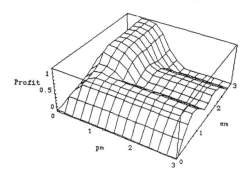

vided into the four subboxes $[0, \frac{2}{3}] \times [\frac{4}{3}, 2]$, $[\frac{2}{3}, \frac{4}{3}] \times [\frac{4}{3}, 2]$, $[0, \frac{2}{3}] \times [2, \frac{8}{3}]$, and $[\frac{2}{3}, \frac{4}{3}] \times [2, \frac{8}{3}]$, with respective bounds 1.298, 1.382, 1.2777, and 1.361. The procedure is carried on until some ending criterion is met. The possible criteria are similar to those proposed for the original BSSS algorithm: the length of the edge of the selected box does not exceed some prespecified value; the absolute or relative value of the gap between the best known solution and the upperbound is smaller than a given tolerance; the number of boxes remaining in the list is larger than the maximum allowed value. In our case, computation were discontinued when the length of that list reached 200. The heuristic solution $(\frac{29}{39}, \frac{11}{6})$ is still the best solution. The bound on the objective is 1.31981. Observe that the first zone corresponds to the markets supplied under uniform delivered pricing. Not surprisingly, the prevailing price in the most remote market is the discriminatory price since the second zone contains only this market.

15.3.2.5 Considering again our example under mixed pricing, the profit function is represented in Figure 6, where one of the axis corresponds to the mill price p^m and the other to the delivered price π^m. The clients choose pick-up or delivery according to the rule expressed in (15.14). The problem presents many similarities with the two-zone pricing case and can be solved by a similar procedure. The space to be searched in is the part of $\mathbf{R}^2$ delineated as follows. The mill price ranges between v and $\max_{i \in I}(\alpha_i - t_i)$, whereas the delivered price belongs to the interval $[\min_{i \in I}(v + t_i), \max_{i \in I} \alpha_i]$. Furthermore, pick-up is never considered by any client when $p^m + \min_{i \in I} t_i > \pi^m$. Similarly, delivery is not called upon when $p^m + \max_{i \in I} t_i < \pi^m$.

The adapted BSSS algorithm follows the guidelines exposed in case 4, except that the boxes are no more hypercubes but rectangles homothetic to the initial box $[v, \max_{i \in I}(\alpha_i - t_i)] \times [\min_{i \in I}(v + t_i), \max_{i \in I} \alpha_i]$. For our example, this box is $[0, \frac{5}{3}] \times [0, \frac{8}{3}]$. The usefulness of a good starting solution has already been noticed. A natural candidate is (p^M, π^U), that is to say, the solutions of the mill and uniform delivered pricing cases. For the example, the candidate is $(0.552, 0.744)$. For this price schedule, pick-up is chosen by the clients of markets 1 and 4, while delivery is preferred by those located in markets 2 and 3. The profit reaches a value equal to 1.2476. A higher value is obtained by finding the optimal mill price for the cluster of clients having chosen pick-up and the optimal delivered price for those using delivery. It gives the solution $(0.722, 0.810)$, corresponding to a profit of 1.32044. Since we do not observe modifications in the choices of the client clusters, we stop the heuristic search here.

Next, we enter into the core of the adapted BSSS algorithm. The center of the initial box, let $(\frac{5}{6}, \frac{4}{3})$, yields a profit equal to 0.9583, which is not better than 1.32044. The box is divided into four equal homothetic subboxes for which we have to evaluate an upperbound. That upperbound is defined in a fashion similar to the one used for zone pricing. Consider the box $Q = [p^-, p^+] \times [\pi^-, \pi^+]$. The solution of the following problem

$$\overline{P}(Q) = \sum_{i=1}^{I} \max(\max_{\pi \in [p^- + t_i, p^+ + t_i]} B_i(\pi), \max_{\pi \in [\pi^-, \pi^+]} B_i(\pi))$$

dominates the function P on the box Q and is thus a candidate upperbound. Back to the example, the four subboxes are respectively $[0, \frac{5}{6}] \times [0, \frac{4}{3}]$, $[0, \frac{5}{6}] \times [\frac{4}{3}, \frac{8}{3}]$, $[\frac{5}{6}, \frac{5}{3}] \times [0, \frac{4}{3}]$, and $[\frac{5}{6}, \frac{5}{3}] \times [\frac{4}{3}, \frac{8}{3}]$. The bound for the first three boxes is 1.38194, while the fourth, with a bound of 1.1458, is eliminated. The first box is subdivided in turn. The bound for the smaller boxes $[0, \frac{5}{12}] \times [0, \frac{2}{3}]$ and $[0, \frac{5}{12}] \times [\frac{2}{3}, 4]$ is 1.2517, which allows one to discard them. The other two boxes $[\frac{5}{12}, \frac{5}{6}] \times [0, \frac{2}{3}]$ and $[\frac{5}{12}, \frac{5}{6}] \times [\frac{2}{3}, 4]$ have the respective bounds 1.3767 and 1.3819 and are placed into the list of boxes that remain to be examined. The procedure goes on as usual. After 113 iterations, one terminates with $p^{m*} = 0.722$, $\pi^{m*} = 0.744$, and $P^{m*} = 1.32044$, i.e., the solution produced by the heuristic.

Examining the geography of prices, we see that the clients of market 1 choose pick-up, while the more remote clients of markets 2 and 3 prefer delivery. The fact that clients at market 4 select pick-up comes as a surprise. The reason is that the prevailing delivered price is too low to yield a positive profit if delivery were to be used. Nevertheless, the willingness to pay of the corresponding clients is so high that they are still ready to travel to the firm. The quantities supplied at the four markets are respectively 0.917, 0.190, 0.393, and 0.708.

15.3.2.6 Having shown how to solve the pricing problem under the five different spatial pricing policies, it may seem useful to take stock and to think about the consequences of the adoption of one particular policy. Even if the selected data set is but an example among others, it reveals some interesting features.

We begin with the delivered prices whose values are summarized in Table 15.3. The spatial heterogeneity of demand is responsible for one of the most surprising findings: the clients in the most remote market are better-

TABLE 15.3. Comparison of the delivered prices under the five policies

Market	M	U	D	Z	m
1	0.552	0.744	0.667	0.744	0.722
2	0.885	0.744	0.667	0.744	0.810
3	1.219	0.744	1.000	0.744	0.810
4	1.552	-	1.833	1.833	1.722

off under mill pricing than under any other policy, an observation which is at variance with the theoretical results of Section 15.2. The reason seems to be that this market is located far away from the firm, but is characterized by a high willingness to pay for the commodity. For the firm, the transportation cost to that market is very high. Therefore, the only way to supply it under delivered pricing is to charge the discriminatory price. Otherwise, the commodity is so desired by the clients of market 4 that they are ready to travel to the firm even when the mill price is high. We also notice that $p^m > p^M$ and $\pi^m > \pi^U$, confirming the result obtained in Section 15.2. Furthermore, the zones where one of the pricing policies is adopted are not connected.

TABLE 15.4. Comparison of the sales under the five policies

Market	M	U	D	Z	m
1	1.172	0.884	1.000	0.884	0.917
2	0.115	0.256	0.333	0.256	0.190
3	0.086	0.442	0.250	0.442	0.393
4	0.836	0.000	0.625	0.625	0.709
Production	2.209	1.582	2.208	2.207	2.208

The geographical distribution of sales reflects the spatial patterns of prices. The values can be found in Table 15.4. The importance of the "extreme" markets is evidenced, as well as the weakness of the "intermediate" markets. The loss of market 4 under uniform delivered pricing entails a dramatic fall in the total production, which is about the same for the other policies.

The final comparisons deal with the benefits that can be made from the various markets. Table 15.5 shows that *the largest part of the profits are drawn from the two extreme markets whose weight is dominant in the formation of prices under the different policies.* By definition, perfect spatial discrimination provides the firm with the highest profit at each site. The ranking of the total profits presents a slight difference with that uncovered in theory: mill pricing does better than uniform delivered pricing. Except

TABLE 15.5. Comparison of the benefits under the five policies

Market	M	U	D	Z	m
1	0.647	0.658	0.667	0.658	0.662
2	0.063	0.105	0.111	0.105	0.091
3	0.047	0.034	0.083	0.034	0.056
4	0.461	0.000	0.521	0.521	0.512
Profit	1.219	0.797	1.382	1.318	1.320

for this point, the profit ranking confirms the findings of Section 15.2.

15.4   The Price-Continuous Facility Location Problem

In this section, we consider a generalization of the Weber problem in which substitution between inputs is allowed and quantities demanded by the clients vary with the price(s) chosen by the firm [553].

15.4.1   Formulation

The transportation space is the two-dimensional Euclidean space. A single-plant firm operates as a monopolist within a given region. The markets located in this region are represented by the points $m_i, i = 1, \ldots, I$ of the Euclidean plane. Each market is characterized by a linear demand function $D_i(\pi_i) = [a_i - b_i\pi_i]^+$ for the firm's output, where π_i is the delivered price at location m_i. The (unknown) location of the firm is denoted by s. Let q^0 be the firm's total production and q_i the quantity shipped to market m_i. If the firm does not export outside the region and does not carry inventories, we have $q^0 = \sum_{i=1}^{I} q_i$. We suppose that the transportation costs are linear in the quantities shipped, but that economies-of-scales in distance may exist. Define $t_i(s)$ as the unit transportation cost of the firm's output from s to m_i, where $t_i(s)$ can be any function increasing in the distance between the two points. The firm uses the inputs $k = 1, \ldots, K$ and has a production function $q^0 = F(q^1, \ldots, q^K)$; denote by $\bar{q} = (q^1, \ldots, q^K)$ the vector of input quantities. We assume that F is independent of the firm location and that the inputs are available in unlimited quantities at given prices. Let M^k be the set of markets where input k can be purchased. Let p_m^k be the f.o.b. price of this input at $m \in M^k$ and $t_m^k(s)$ its unit transportation cost from m to s. Accordingly, the firm purchases k at the market for which the delivered price $\pi^k(s) = \min\ (p_m^k + t_m^k(s)\,|\,m \in M^k)$ is minimum.

The problem of the firm is to choose simultaneously its *location*, its *price(s)*, and its *input mix* so as to maximize its profit. This problem is called hereafter the *price-location problem of the spatial monopolist* (PLP). Let $\pi = (\pi_1, \ldots, \pi_I)$ be the vector of delivered prices at the various markets and $\underline{q} = (q_1, \ldots, q_I)$ be the corresponding vector of quantities sold. The profit function of the firm is:

$$P(s, \pi, \underline{q}, \overline{q}) = \sum_{i=1}^{I} (\pi_i - t_i(s)) \, q_i - \sum_{k=1}^{K} \pi^k(s) \, q^k. \tag{15.16}$$

Several constraints are to be considered. First, the quantity shipped to m_i cannot exceed the local demand at the prevailing local price:

$$q_i \le D_i(\pi_i). \tag{15.17}$$

Second, the production possibilities impose that:

$$\sum_{i=1}^{I} q_i \le F(q^1, \ldots, q^K). \tag{15.18}$$

Third, there may be restrictions on the set of feasible locations for the firm:

$$s \in S. \tag{15.19}$$

Fourth, as above, the spatial price policy followed by the firm is expressed by the constraints:

$$\pi \in \Pi. \tag{15.20}$$

Last, we have the usual nonnegativity constraints:

$$\underline{q} \ge 0 \quad \text{and} \quad \overline{q} \ge 0. \tag{15.21}$$

The PLP can then be formulated as follows: $P^* = \max \; P(s, \pi, \underline{q}, \overline{q})$ subject to (15.16)–(15.21).

When some standard assumptions on the production function are made, we can obtain a more tractable form for the PLP. Consider the production cost function at s: $c(q^0, s) = \min \sum_{k=1}^{K} \pi^k(s) \, q^k$ subject to $q^0 = F(q^1, \ldots, q^K)$. If F is linear homogeneous, we have:

$$c(q^0, s) = v(s) \, q^0 \tag{15.22}$$

where $v(s)$ is the site-dependent marginal production cost [606, p.86]. Several examples of such functions are now proposed. First, we have the linear production function: $F(q^1, \ldots, q^K) = \sum_{k=1}^{K} a^k \, q^k$, where the a^k are site-independent positive constants, and the marginal production cost has the form $v(s) = \min_{k=1,\ldots,K} \frac{\pi^k(s)}{a^k}$. A second possibility is a Leontief production function: $q^k = a^k \, q^0$, where $a^k > 0$ for all k, leading to: $v(s) =$

$\sum_{k=1}^{K} \pi^k(s) a^k$. We also consider the Cobb-Douglas production function: $q^0 = \prod_{k=1}^{K} (q^k)^{a^k}$ with $\sum_{k=1}^{K} a^k = 1$, which yields:

$$v(s) = \prod_{k=1}^{K} \left[\frac{\pi^k(s)}{a^k} \right]^{a^k}.$$

A final example is the constant elasticity of substitution (or CES) production function:

$$q^0 = \left[\sum_{k=1}^{K} a^k (q^k)^\rho \right]^{\frac{1}{\rho}}$$

with $a^k > 0$ for all k and $\rho < 1$. It is easy to show that

$$v(s) = \left[\sum_{k=1}^{K} (a^k)^{1-r} \pi^k(s)^r \right]^{\frac{1}{r}}$$

where $r = \frac{\rho}{\rho - 1}$. Using (15.22), it is then possible to eliminate the vector of input quantities from the PLP, which can then be rewritten as follows: $P^* = \max P(s, \pi, \underline{q}) = \sum_{i=1}^{I} [\pi_i - v(s) - t_i(s)] q_i$ subject to (15.17), (15.19), (15.20), and (15.21). Letting $B_i(\pi_i, s) = [\pi_i - v(s) - t_i(s)]^+ D_i(\pi_i)$, we finally obtain a simplified expression for the PLP: $P^* = \max P(s, \pi) = \sum_{i=1}^{I} B_i(\pi_i, s)$ subject to (15.19) and (15.20). We can also consider the profit function at any fixed location: $P_s(\pi) = P(s, \pi)$ whose optimal value is denoted: $P(s) = \max (P_s(\pi) \mid \pi \in \Pi)$. Note that this is exactly the problem considered in Section 15.3. This leads to a final formulation of the PLP: $P^* = \max P(s)$ subject to (15.19). Formally, this approach corresponds to projecting the profit function onto the geographical space and choosing the best solution among the projections. This is precisely the strategy of the algorithm presented in [553] which is a variant of BSSS. Finally, the different spatial price policies are those presented in Section 15.2.

15.4.2 An example

We provide a typical example and discuss the main economic consequences of the choice of a particular spatial price policy. Twenty points corresponding to the clients' locations have been randomly generated in a square with edge length equal to 100. The coordinates of these points are given in Table 15.6. Each client has a demand function given by $D(\pi) = [150 - \pi]^+$ and the transportation rate is 1. The firm produces this commodity using two inputs available respectively at $(45, 9)$ and $(93, 53)$. The price of each input at its source and their transportation rate are equal to 1. We assume a Leontief production function whose coefficients are equal to 1. This

TABLE 15.6. Data for the example

i	1	2	3	4	5	6	7	8	9	10
m_i^1	33	25	36	70	83	59	84	86	69	13
m_i^2	13	6	54	7	17	39	79	35	60	94

i	11	12	13	14	15	16	17	18	19	20
m_i^1	83	21	0	33	7	66	68	48	86	72
m_i^2	9	56	89	96	56	82	46	94	5	96

TABLE 15.7. Results of the example

Policy	D	U	M	$Z = 2$	$Z = 3$	m
Best location	(69.8, 34.9)	(68.0, 32.7)	(68.0, 32.7)	(70.1, 34.1)	(69.6, 33.8)	(70.0, 34.1)
Optimal prices	-	$\pi^U = 124.98$	$p^M = 92.25$	$\pi^1 = 119.60$	$\pi^1 = 119.62$	$p^m = 97.59$
				$\pi^2 = 135.07$	$\pi^2 = 132.20$	$\pi^m = 135.08$
					$\pi^3 = 141.72$	
Profit	9730.14	8761.45	8761.45	9401.93	9561.71	9400.93
Production	384	350	350	378	383	378
Time (seconds)	2.82	501.82	270.32	11.17	9.47	3154.23

TABLE 15.8. Delivered prices under the alternative policies

i	D	U	M	$Z = 2$	$Z = 3$	m
1	130.07	124.98	132.42	135.07	132.20	135.08
2	135.31	124.98	142.87	135.07	132.20	135.08
3	128.06	124.98	130.70	135.07	132.20	135.08
4	122.60	124.98	118.02	119.60	119.62	124.70
5	119.76	124.98	113.95	119.60	119.62	119.04
6	114.43	124.98	103.24	119.60	119.62	109.68
7	131.80	124.98	141.23	135.07	132.20	135.08
8	116.73	124.98	110.38	119.60	119.62	113.56
9	121.19	124.98	119.56	119.60	119.62	123.50
10	149.63	-	-	-	-	-
11	123.17	124.98	120.29	119.60	119.62	125.84
12	135.24	124.98	144.72	135.07	132.20	135.08
13	-	-	-	-	-	-
14	144.31	-	-	-	141.72	-
15	141.78	-	-	135.07	141.72	135.08
16	132.26	124.98	141.59	135.07	132.20	135.08
17	114.26	124.98	105.54	119.60	119.62	109.66
18	140.14	-	-	135.07	141.72	135.08
19	125.64	124.98	125.27	119.60	119.62	130.78
20	139.21	-	-	135.07	141.72	135.08

TABLE 15.9. Purchased quantities under the alternative policies

i	D	U	M	$Z=2$	$Z=3$	m
1	19.93	25.02	17.58	14.93	17.80	14.92
2	14.69	25.02	7.13	14.93	17.80	14.92
3	21.94	25.02	19.30	14.93	17.80	14.92
4	27.40	25.02	31.98	30.40	30.38	25.30
5	30.24	25.02	36.05	30.40	30.38	30.96
6	35.57	25.02	46.76	30.40	30.38	40.32
7	18.20	25.02	8.77	14.93	17.80	14.92
8	33.28	25.02	39.62	30.40	30.38	36.44
9	28.81	25.02	30.44	30.40	30.38	26.50
10	0.37	0.00	0.00	0.00	0.00	0.00
11	26.83	25.02	29.71	30.40	30.38	24.16
12	14.76	25.02	5.29	14.93	17.80	14.92
13	0.00	0.00	0.00	0.00	0.00	0.00
14	5.69	0.00	0.00	0.00	8.28	0.00
15	8.22	0.00	0.00	14.93	8.28	14.92
16	17.74	25.02	8.42	14.93	17.80	14.92
17	35.74	25.02	44.46	30.40	30.38	40.34
18	9.86	0.00	0.00	14.93	8.28	14.92
19	24.36	25.02	24.73	30.40	30.38	19.22
20	10.79	0.00	0.00	14.93	8.28	14.92

FIGURE 7. Market area for spatial discriminatory pricing

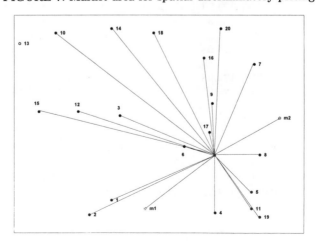

FIGURE 8. Market area for uniform mill and delivered pricing

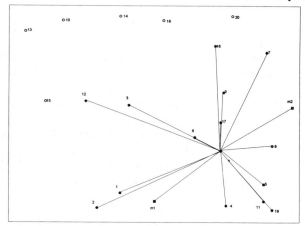

FIGURE 9. Market area for mixed and zone pricing with $Z = 2$

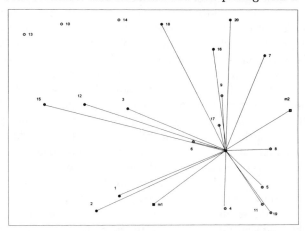

FIGURE 10. Market area for zone pricing with $Z = 3$

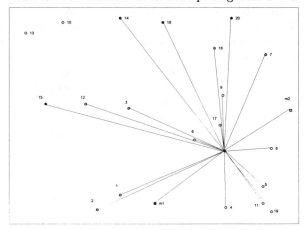

price-location problem has been solved for the five different price policies. Two and three zones were chosen for zone pricing. The results are reported on in Table 15.7, where we give successively the optimal location found by the algorithm, the corresponding optimal price(s) and profit, the total production. Finally, we give the computation time in seconds on a Macintosh Quadra 800. The optimal locations of the plant as well as the firm's market areas under the six cases are depicted in Figures 7 to 10. Figure 7 corresponds to spatial discriminatory pricing, Figure 8 to mill and uniform delivered pricing, Figure 9 to zone pricing with two zones and mixed pricing, and Figure 10 to zone pricing with three zones. In Table 15.8 and 15.9, we give the respective delivered prices and quantities purchased by the clients.

A first observation is that *the optimal locations do not differ very much.* This finding is somewhat different from what we obtain in the case of a multiplant firm and a discrete locational space. However, the choice of a price policy has a tremendous impact on the delivered prices and, hence, on the distribution of the firm's output. Spatial discriminatory pricing yields the highest profit, as well as the largest production level and market area. Uniform delivered and mill pricing have the same market areas, profits, and production, but the patterns of delivered prices inside the market area are completely different. Another interesting observation is the evolution of the variables starting from uniform delivered pricing, passing through the two- and three-zone systems, and ending with discriminatory pricing. The enlarging of the market area is remarkable and we see that three zones gives a close approximation of the discriminatory pricing results. Mixed pricing has the same market area as the two-zone system, and clients located inside the zone with the lower delivered price are those that would choose pickup in the mixed system. The delivered price for those who choose delivery is equal to the larger delivered price in the two-zone pricing case. Note also that the mill and delivered price in the mixed system are respectively higher than their correspondent under mill and delivered pricing.

We can say that this example, which is representative of most our experiments, confirms the findings of Section 15.3. This is not surprising since the optimal locations are very close of each other, while the location was given and fixed in Section 15.3. However, this conclusion is not likely to be general: discriminatory pricing being deemed to favor the invasion of remote markets, it may occur that the optimal solution under that policy (or the zone pricing system, which is its more frequent implementation) is affected by the inclusion into the market area of the firm of "heavy weighted" distant markets not supplied under mill or uniform delivered pricing.

15.5   The Price-Discrete Facility Location Problem

In order to evaluate the impact of the assumptions of the Euclidean plane and of a single plant, we suppose here that the transportation space is discrete and the firm multiplant.

15.5.1   Formulation

Our departure is the Uncapacitated Facility Location Problem, in short UFLP [243]. A fundamental feature of this problem is that each plant opened by the firm carries out the whole production process. Moreover, it is possible to determine at each potential site the optimal production technology and the sources of inputs [555]. An immediate consequence is that the production side of the firm's problem can be left unspecified, being summarized by a marginal production cost value at each possible location, whereas it was necessary to handle it explicitly in the continuous case. In this section, we consider an extension the UFLP where demands are linear. The now usual five spatial price policies are considered. This extended model has been proposed and solved by [536].

A firm produces a single commodity under monopolistic conditions and supplies I markets. Each market i has a linear demand function $D_i(\pi_i)$. The firm can operate up to J plants located at sites j. Plants and sites are similarly identified by j. The production cost function at j is:

$$C_j(q_j) = \begin{cases} 0 & \text{if } q_j = 0 \\ f_j + v_j q_j & \text{if } q_j > 0 \end{cases}$$

where q_j is the quantity produced by plant j, v_j is the local constant marginal production cost, and f_j is a fixed cost incurred if some production is undertaken at j. Let t_{ij} denote the unit transportation cost for supplying market i from site j. The problem of the firm is to determine the price charged on each market, the level of production in each plant, and the distribution pattern so as to maximize its profit.

We introduce the following variables for the price policies M, U, and D. x_{ij} is set equal to 1 if market i belongs to the market area of plant j and 0 otherwise. y_j is set equal to 1 if plant j operates, and hence the fixed cost f_j is incurred; y_j is set equal to 0 if the plant is idle. π_{ij} is the delivered price at market i when the commodity is shipped from plant j; p_{ij} is the corresponding mill price. Both prices are related through $\pi_{ij} = p_{ij} + t_{ij}$.

The profit of the firm is equal to its revenue, minus the total production cost, minus the total transportation cost. It is given by:

$$P(\pi, y, X) = \sum_{i=1}^{I} \sum_{j=1}^{J} \pi_{ij} D_i(\pi_{ij}) x_{ij} - \sum_{j=1}^{J} (f_j y_j + v_j \sum_{i=1}^{I} D_i(\pi_{ij}) x_{ij})$$

$$- \sum_{i=1}^{I}\sum_{j=1}^{J} t_{ij} D_i(\pi_{ij}) x_{ij} = \sum_{i=1}^{I}\sum_{j=1}^{J} B_{ij}(\pi_{ij}) x_{ij} - \sum_{j=1}^{J} f_j y_j$$

where $B_{ij}(\pi_{ij}) = D_i(\pi_{ij})(\pi_{ij} - t_{ij} - v_j) = D_i(\pi_{ij})(p_{ij} - v_j)$ is the benefit earned from supplying market i from plant j at the delivered price π_{ij}.

The problem of the firm can then be expressed as follows:

$$P^* = \max_{\pi, y, X} P(\pi, y, X) = \max_{\pi, y, X} \left[\sum_{i=1}^{I}\sum_{j=1}^{J} B_{ij}(\pi_{ij}) x_{ij} - \sum_{j=1}^{J} f_j y_j \right] \quad (15.23)$$

subject to:

$$\sum_{j=1}^{J} x_{ij} \le 1, \qquad i = 1, \ldots, I, \quad (15.24)$$

$$0 \le x_{ij} \le y_j, \qquad i = 1, \ldots, I; \; j = 1, \ldots, J \quad (15.25)$$

$$y_j = 0 \text{ or } 1, \qquad j = 1, \ldots, J \quad (15.26)$$

Constraints (15.24) prevent the firm from supplying a client with more than its demand. (Note that the firm is allowed not to supply a client if this entails a loss). Constraints (15.25) imply that the fixed costs are incurred for all the operating plants.

Under perfect discriminatory pricing, Hansen and Thisse [555] have shown that the problem (15.23)–(15.26) is equivalent to a UFLP. Note, indeed, that the optimal values of the π_{ij} can be obtained independently of each other because of the assumption of constant marginal production costs. Let π_{ij}^* be defined by $B_{ij}(\pi_{ij}^*) = Max(B_{ij}(\pi_{ij}) \mid \pi_{ij} \ge 0)$ and $B_{ij}^* = [B_{ij}(\pi_{ij}^*)]^+$. Since $D_i(\pi) = [a_i - b_i\pi]^+$, we have: $\pi_{ij}^* = \frac{\alpha_i + v_j + t_{ij}}{2}$ where $\alpha_i = a_i/b_i$, and

$$B_{ij}^* = \begin{cases} b_i[\alpha_i^2 - (v_j + t_{ij})^2]/4 & \text{if } \alpha_i > v_j + t_{ij} \\ 0 & \text{otherwise.} \end{cases} \quad (15.27)$$

Substituting (15.27) into (15.23)–(15.26) yields a UFLP that can be solved efficiently. In what follows, this problem is denoted UFLP-D.

Consider now uniform delivered pricing. To this end, we append the constraints

$$\pi_{ij} = \pi, \; i = 1, \ldots, I; \; j = 1, \ldots, J \quad (15.28)$$

to problem (15.23)–(15.26). The resulting model is denoted UFLP-U. An algorithm has been proposed by [558]. It relies on the observation that the subproblem that results from fixing π to a given value is a UFLP. The procedure uses a binary search on the axis of positive real numbers for determining the optimal price.

The third spatial price policy is mill pricing: the commodity is made available at each operating plant at the same mill price and each client bears its own transportation cost. Hence, we must add the constraints:

$$\pi_{ij} = p + t_{ij}, \ i = 1, \ldots, I; \ j = 1, \ldots, J \tag{15.29}$$

to problem (15.23)–(15.26). However, if the clients are free to choose the plant they want to patronize, a discrepancy may arise between their preferences and those of the firm. Each client prefers to buy from the nearest operating plant, i.e., the plant j for which

$$t_{ij} = \min(t_{ik} \mid y_k = 1). \tag{15.30}$$

On the other hand, it is most profitable for the firm to supply the clients situated at i from the plant with the lowest production and transportation cost, i.e., the plant j such that

$$v_j + t_{ij} = \min(v_k + t_{ik} \mid y_k = 1). \tag{15.31}$$

Obviously, unless the marginal production cost is the same across all plants, the allocation decisions (15.30) and (15.31) do not generally coincide. Since mill pricing typically assumes that transportation is under the clients' control, we must add new constraints expressing their behavior:

$$\sum_{\{k \mid t_{ik} \leq t_{ij}\}} x_{ik} \geq y_j, \ i = 1, \ldots, I; \ j = 1, \ldots, J \text{ and } D_i(p + t_{ij}) > 0. \tag{15.32}$$

Constraints (15.32) imply that if facility j operates and lies within the range of potential positive demand from market i at mill price p, then the clients at i either patronize facility j or another operating facility which is closer (in terms of transportation costs) to them than j. (In order to avoid ties, we assume without loss of generality that $t_{ij} \neq t_{ik}$ for $i = 1, \ldots, I$, j and $k = 1, \ldots, J$, $j \neq k$.) Problem (15.23)–(15.26)+(15.29)+(15.32) is denoted UFLP-M. For each value of p we have to solve a modified UFLP which accounts for constraints (15.32). It is denoted UFLP-O.

We now assume that the firm follows a zone price policy, i.e., the markets are clustered into Z zones and in each zone $z = 1, \ldots, Z$ a uniform delivered price π_z is charged to the clients. We also suppose that the distribution system is totally under the control of the firm, so that resales among clients located in different zones are not possible. The number of zones Z is exogenous. The following new variables are needed: x_{ijz} is set equal to 1 if market i is supplied by plant j and allocated to the zone z; otherwise it is set equal to 0.

The profit of the plant can be expressed by:

$$P(\pi, x, y) = \sum_{i=1}^{I} \sum_{j=1}^{J} \sum_{z=1}^{Z} \pi_z D_i(\pi_z) x_{ijz} - \sum_{j=1}^{J} [f_j y_j - v_j \sum_{i=1}^{I} \sum_{z=1}^{Z} D_i(\pi_z) x_{ijz}]$$

$$- \sum_{i=1}^{I} \sum_{j=1}^{J} \sum_{z=1}^{Z} t_{ij} \, D_i(\pi_z) \, x_{ijz}$$

$$= \sum_{i=1}^{I} \sum_{j=1}^{J} \sum_{z=1}^{Z} B_{ijz}(\pi) \, x_{ijz} - \sum_{j=1}^{J} f_j \, y_j \qquad (15.33)$$

where $B_{ijz}(\pi_z) = (\pi_z - v_j - t_{ij}) \, D_i(\pi_z)$ is the benefit obtained when supplying market i from plant j and assigning it to the zone z wherein the uniform delivered price π_z is charged. The problem of the firm is then expressed as follows: $P^* = \max_{\pi, x, y} P(\pi, x, y)$ subject to constraints similar to (15.23)–(15.26).

Two special cases of this problem have already been dealt with. When $Z = 1$ we have the UFLP-U; when $Z = I$, we get the UFLP-D. It is worth noting that, for a given vector of prices π, the problem boils down to an UFLP. The range of price vectors to be explored can be reduced on the following grounds. First, we can always relabel the zones such that $\pi_1 \geq \pi_2 \geq \ldots \geq \pi_Z$. Second, if $\alpha_i = \min\{\pi \mid D_i(\pi) = 0\}$ is the maximum price that clients at market i are willing to pay, then $\pi^{\max} = \max\{\alpha_i \mid i = 1, \ldots, I\}$ is an upper bound on the range of admissible prices. Third, it is never profitable for the firm to supply client i from site j when the delivered price is smaller than $v_j + t_{ij}$; so $\pi^{\min} = \min\{v_j + t_{ij} \mid i = 1, \ldots, I; \, j = 1, \ldots, J\}$ is a lower bound on the feasible prices. Consequently, the set of admissible prices in $\mathbf{R}^Z$ is $\mathcal{P} = \{\pi = (\pi_1, \ldots, \pi_Z) \mid \pi^{\max} \geq \pi_1 \geq \ldots \geq \pi_Z \geq \pi^{\min}\}$. Of course, for the problem to make sense we need $\pi^{\max} > \pi^{\min}$.

Under mixed pricing, the formulation of the problem is similar to the UFLP-M. However, the additional constraints expressing the spatial price policy are more complicated. Suppose first that π^m and p^m are the uniform delivered and mill prices set by the firm. Two cases may arise. In the first one, once a plant operates, the firm cannot refuse to supply a client showing at the plant door even if this entails a loss, i.e., if the common mill price p^m is lower than the local marginal production cost v_j. This may occur if delivery is profitable for the prevailing price π^m and the benefits drawn from this activity balances the loss caused by a price p^m which is too low at this particular location. In the second, the firm may refuse to sell under such circumstances. In both cases, we assume that the firm delivers the commodity to a client only if it is profitable for it to do so. In the first situation, π_{ij} is given by the following expression:

$$\pi_{ij} = \begin{cases} p^m + t_{ij} & \text{if } D_i(p^m + t_{ij}) > 0 \\ & \text{and either } p^m + t_{ij} < \pi^m \text{ or } \pi^m \leq v_j + t_{ij} \\ \pi^m & \text{if } D_i(\pi^m) > 0, \, \pi^m > v_j + t_{ij}, \text{ and } p^m + t_{ij} \geq \pi^m \\ \infty & \text{otherwise,} \end{cases}$$

$$(15.34)$$

FIGURE 11. Selected sites and allocations under spatial discriminatory pricing

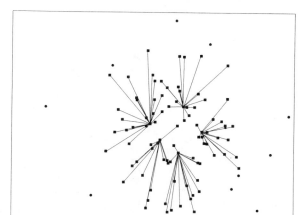

whereas in the second case we have:

$$
\pi_{ij} = \begin{cases} p^m + t_{ij} & \text{if } D_i(p^m + t_{ij}) > 0,\ p^m > v_j, \\ & \text{and either } p^m + t_{ij} < \pi^m \text{ or } \pi^m \leq v_j + t_{ij} \\ \pi^m & \text{if } D_i(\pi^m) > 0,\ \pi^m > v_j + t_{ij}, \\ & \text{and either } p^m + t_{ij} \geq \pi^m \text{ or } p^m \leq v_j \\ \infty & \text{otherwise.} \end{cases} \qquad (15.35)
$$

Furthermore, we must also consider additional constraints expressing the allocation rules of the clients to the operating plants. When client i chooses delivery, he lets the firm decide which plant will actually supply him. When he chooses pick-up, he will patronize the nearest plant as in the UFLP-O. Any couple of prices (p^m, π^m) endows each client at market i with a preference ordering on the plants to patronize: $j \preceq_i k$ if $\pi_{ij} \leq \pi_{ik}$. Therefore, we add the constraints: $\sum_{\{k | \pi_{ik} \leq \pi_{ij}\}} x_{ik} \geq y_j$, for all i and j such that $B_{ij}(\pi_{ij}) > 0$.

15.5.2 An example

We provide an example to compare the five price policies. The data can be found in Table 15.10. One hundred points were chosen according to the generation scheme explained in the previous section. Fifty among them were selected as eligible sites for plant location; they are indicated in the second column of the table. Next, we give the coordinates of the points. Each point represents a market; we report on in the column Pop the local population. The fixed costs are equal to 10,000 at every site and the marginal production cost is chosen randomly between 0 and 50.

The results are given in Table 15.11. It is seen that U and M give about

TABLE 15.10. Data for the example

i	j	$x^{(1)}$	$x^{(2)}$	Pop	i	j	$x^{(1)}$	$x^{(2)}$	Pop
1	1	4.83	35.79	384.90	51	20	-7.18	49.73	481.04
2		15.09	16.83	148.35	52		-38.18	-38.13	630.65
3	2	37.92	0.52	108.23	53		-8.1	-65.98	338.87
4		-23.38	37.52	941.53	54		4.13	116.57	273.65
5	3	70.1	38.65	379.55	55		71.53	-2.06	110.05
6		-23.08	56.4	299.79	56	21	24.24	-61.02	511.52
7		-51.73	-28.94	189.51	57		-25.67	-11.38	733.02
8		-70.42	27.22	531.86	58		40.46	-6	218.43
9		28.02	30.04	212.03	59		-62.8	-44.92	143.84
10		-18.53	-5.97	835.74	60	22	-25.97	62.26	175.73
11		-33.94	19.15	172.58	61	23	149.39	21.34	441.16
12	4	33.68	0.52	239.81	62	24	14.3	80.5	180.44
13		58.16	-0.19	487.37	63	25	-74.6	4.63	230.96
14		-13.67	41.79	530.05	64	26	-78.06	71.71	970.47
15		-22.3	15.06	948.78	65	27	85.71	-30.4	399.95
16		47.11	93.43	426.20	66		-5.06	-25.92	117.59
17		-24.96	23.5	380.79	67	28	77.28	11.62	618.89
18		67.53	11.8	654.33	68	29	85.4	-37.58	639.04
19		-27.42	-7.48	483.15	69	30	-32.6	86.67	613.67
20		-2.23	-24.43	106.83	70	31	-15.84	-2.73	165.19
21	5	-162.79	29.64	704.43	71	32	-4.86	30.73	479.82
22		35	-53.62	844.36	72	33	2.09	-8.16	467.15
23		8.24	81.19	820.70	73	34	28.96	-73.79	985.20
24		109.84	-72.21	889.07	74	35	82.3	-19.2	407.70
25	6	35.48	-25.96	559.54	75	36	7.8	-15.48	998.56
26	7	26.98	-57.48	298.42	76	37	-46.31	37.1	120.49
27		18.09	30.05	934.04	77		6.33	11.29	672.30
28		76.47	-51.01	249.40	78	38	32.52	43.59	835.91
29		-46.96	39.31	400.38	79	39	-85.8	-14.35	785.56
30		-64.88	17.46	977.87	80	40	19.07	-48.09	542.31
31		44.09	-6.42	348.18	81	41	67.12	-16.49	593.56
32		62.67	16.06	189.38	82		31.12	12.42	127.51
33		-45.83	-24.8	600.34	83		-5.83	-74.29	503.68
34	8	-33.91	10.96	734.66	84	42	-55.89	-6.94	896.17
35	9	44.18	5.33	804.83	85		3.29	-43.69	796.10
36		-0.35	-72.36	766.24	86		-33.81	-70.33	773.49
37		32.98	-25.28	793.53	87		-27.59	-7.56	325.01
38	10	41.18	37.87	839.09	88		-27.11	-52.55	348.39
39	11	37.6	5.36	676.06	89	43	-104.29	-58.08	180.41
40		-57.98	60.22	723.67	90		-59.5	6.76	214.95
41	12	21.04	21.78	620.29	91		32.79	-42.59	652.08
42	13	-14.88	56.26	322.80	92	44	-28.88	13.26	223.66
43		-10.55	-22.41	276.67	93	45	8.88	46.43	503.49
44	14	-30.81	-62.87	169.46	94	46	26.95	-20.55	789.61
45	15	35.81	81.65	999.19	95	47	13.69	30.52	980.46
46	16	-13.51	25.57	151.45	96		6.97	50.57	968.29
47		-6.42	-23.09	722.80	97	48	51.69	18.21	988.46
48	17	-51.1	9.51	450.15	98	49	53.68	-17.52	742.94
49	18	126.33	-16.93	610.33	99		70.02	72.17	956.43
50	19	11.25	-68.18	955.36	100	50	-81.76	61.79	348.52

TABLE 15.11. Global results

	D	U	M	Z = 2	m
Profit	181636	160394	158976	172814	170571
Prod.	9286	8471	8866	8898	8856
n_1	90	67	69	34	35
n_2				40	39
		$\pi^U = 75.20$	$p^M = 56.87$	$\pi_1^Z = 83.60$ $\pi_2^Z = 70.84$	$\pi^m = 83.30$ $p^m = 59.20$
Sites			Capacities		
11	1487	1558	1560	1454	1666
31	1698	2101	862	2143	1157
32			843		
36	1618		1496		1793
42			1141		
44	2196	2079	955	2005	1959
46		829		1092	
47	2287	1911	2007	2202	2281

FIGURE 12. Selected sites and allocations under uniform delivered pricing

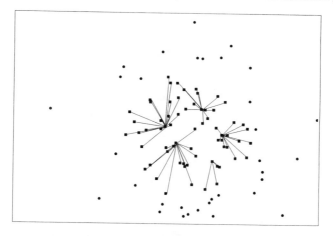

the same levels of profit; the same holds with Z and m, the latter being about half-way between the U (or M) and D. Comparing total production, D yields the highest one and U the smallest, while the three others lead to almost the same production. In the next two rows, we give the number of markets supplied under the first three policies, the number of markets in each zone for Z, while n_1 and n_2 are the number of markets where the clients choose respectively delivery or pick-up under m. The optimal prices follow. Again, we find that $p^M < p^m$ and $\pi^U < \pi^m$, whereas $\pi_2^Z \approx \pi^m$. Next, we report the list of the selected sites and their relative capacities.

FIGURE 13. Selected sites and allocations under uniform mill pricing

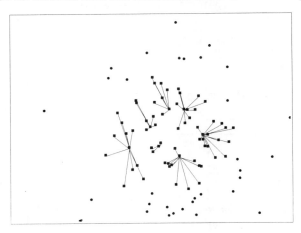

FIGURE 14. Selected sites and allocations under zone pricing

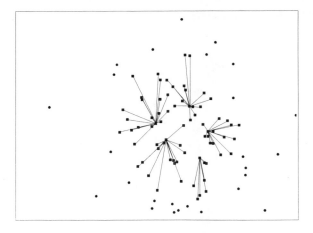

A glance shows that *there are noticeable differences in the locations chosen and the plant sizes*; in particular M leads to a larger number of smaller plants (thus entailing lower transportation costs and higher fixed costs).

We then examine the shapes of the market areas; they are depicted in Figures 11 to 15. Discriminatory pricing exhibits the largest market areas and most of the remote points are supplied. Surprisingly, the plants are not centrally located in their market areas; this is partly due to the uneven distribution of clients. We also notice in Figure 11 that some markets are not allocated to the closest plant; for instance, the point 77 of coordinates $(6.33, 11.29)$ is supplied from site 31 at $(-15.84, -2.73)$ rather than from the closer plant 47 situated at $(13.69, 30.52)$. The reason is to be found in the

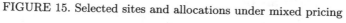

FIGURE 15. Selected sites and allocations under mixed pricing

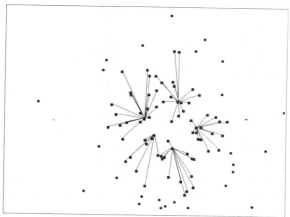

disparities of the marginal production costs: the firm supplies market i from the plant j such that $v_j + t_{ij}$ is minimum; indeed, in our example $v_{31} = 23$ and $v_{47} = 29$ and a straightforward calculation confirms that $v_{47} + t_{77,47} < v_{31} + t_{77,31}$. Comparing with Figure 12, we see that the firm practicing uniform delivered pricing looses a significant part of the peripheral markets; as a consequence, the plants appear to be more centrally established in their market areas. Four plants are located at the same place as under D, while the plant at 36 shifts to the neighboring site 46. The picture displayed in Figure 13 reveals a completely different spatial organization: there are more plants of smaller size. An explanation of this phenomenon is as follows: since the choice of the plant to patronize is under the control of the clients, the firm cannot internalize the transportation cost of its output; it must therefore adopt a particular locational pattern in order to match as well as possible the clients' preferences and its own preference. This least cost assignment rule produces market areas that are Voronoi-like. Turning now to zone pricing and comparing Figures 12 and 14 where the market areas of the two zones are represented respectively in gray and in black, we observe the same locational pattern and an extension of the market areas corresponding to the supply of seven more markets. But the most interesting point lies in the observation of Figure 15, where the clients choosing pick-up are mapped in gray and those who prefer delivery are in black. The central part of the map has many similarities with Figure 13 since the clients residing in that region follow the same behavior. At the outskirts of the region, mixed and zone pricing offer many analogies.

Finally, regarding the optimal price(s), it can be shown that three distinct spatial patterns emerge: D; U and M; Z and m. This is in accordance with our former results, given that Z is small with respect to the number of markets.

15.6   Conclusions

It is never easy to draw conclusions from an analysis based on simulations, especially when several parameters are involved as in our model. Clearly, the main difficulty here is the robustness of the results obtained. Being aware of this, we have studied several variants of our data set. They all yield results that point to the same direction. Of course, this does not preclude the possibility of getting opposite results with completely different data sets.

However, a few principles seem to emerge from our analysis.

(i) Our results show that, on average, the firm's price policy has a significant impact on its location decision, thus confirming Greenhut's [501] thesis. This is somewhat at variance with the findings of Section 4 and seems to stem from the differences in the population distributions.

(ii) Several properties established in Section 2 are supported in a model which allows for discontinuous space, irregular demand distributions, location-specific production costs and actual distances. This is comforting in that the simplifying assumptions made in theoretical location models may not be so bad after all.

(iii) The transport network configuration has an impact on firms' locations. This is interesting because shaping the transport network proves to be a rather powerful instrument in spatial planning, especially in countries with limited infrastructures like the less developed countries (see [947] for some empirical evidence and [924] for extensive experimentations assessing the impact of the transportation network on locational choices).

As a final remark, we would like to say that facility location analysis has been developed mainly to help the decision-maker in assessing the (social or private) benefit of different location systems. Besides that, we have seen in the above analysis an example of the potential offered by this body of literature for testing the generality of many theoretical properties established in simple analytical models.

16
COMPETITIVE LOCATION IN DISCRETE SPACE

Daniel Serra
Universitat Pompeu Fabra

Charles ReVelle
The Johns Hopkins University

16.1 Introduction

Competitive location has been a mainstream topic in the field of industrial organization, spatial economics, regional science and operations research. The literature is rich in papers that addresses the issue of locating firms that compete for clients in space. Beginning with the seminal work of Hotelling [599], the most traditional spatial representation of competition has been a continuous line, since it easily yields mathematical results related to price setting, production levels and locational strategies.

In the last decades, new works following the same approach have been developed using other geometrical shapes such as circles or triangles, to try to loosen the strong assumption of a linear market and to allow new strategies such as sequential entry in the market [757, 567, 1026, 973]. Although conclusions on the behavior and final strategies of the competitors are relatively easy to obtain, the spatial representation is, in most situations, still distant from reality. On the other hand, the literature on location in Operations Research and in Regional Science has concentrated on non-competitive location-allocation models, even though planar and network representations have been used.

In this sense, the competitive location models can be categorized in three spatial representations. These are (1) continuous planar space -where the potential location of the facilities is anywhere in the plane; (2) a network - where facilities are allowed to locate at the extremities of or at the intermediate points on the arcs; and (3) discrete space - where there is only a finite set of possible locations on the network. Therefore, discrete facility location can be considered as a special case of network facility location. In general, for all these spatial representations consumers are located at specific points, such as the vertices of the network. Obviously, it is not possible to determine which spatial setting is best, since the representation depends on the purpose of the model and the type of facilities to be located. An extensive bibliographic survey with over 100 citations on competitive location can be found in Eiselt et al. [374]. Several authors have started to incorporate the spatial representation of non-competitive location models into the strategies and behaviors of competitive siting situations.

Parallel to the spatial definition of the market, it is necessary to define the term "competition". Friesz el al. [438, p.48] define a competitive facility location model as "any facility location model which explicitly recognizes that a firm's location may affect its market share; and, hence, that a location must be chosen so that the objectives of the firm are optimized with respect to market share". This wide definition of competition in facility siting includes models where there is no interaction between firms, and therefore includes, for example, plant location models. For our purposes, a narrower definition will be used stemming from game-theoretical concepts. A competitive facility location model is such that there is more than one firm competing in the spatial market, and that there is interaction between them; the location decision of a firm will affect not only its market share, but also its competitor(s) market share. A similar definition has been offered by Hakimi [511].

The purpose of this chapter is threefold. First, an overview of the emerging literature on discrete competitive models as defined above will be presented. Second, some formulations on the location of competitive facilities in an oligopolistic market represented by a discrete network will be addressed, with the focus on entry barriers that can be applied to check the entry of newcomers to the market and with focus on the uncertainty in the information used to formulate strategies to maximize market share. Finally, an new model based on early works of ReVelle and Serra will be presented.

16.2   Discrete Competitive Location Models: An Overview

Hotelling's work on two firms competing in a linear market with consumers distributed uniformly along the line (also known as the ice–cream vendor problem) set the foundations of what is today the burgeoning field of competitive location. Competitive location is not only important from a practical point of view. It is also intriguing, [371, p.237], that by modifying the assumptions of the model the results obtained may differ significantly from Hotelling's conclusions: "relatively small changes in the model assumptions result in dramatic changes in the outcome".

During the late thirties and early forties, several papers using the same spatial representation as Hotelling but modifying some of the economic assumptions appeared in the economic literature [591, 757, 1090]. There followed several decades of stagnation in the contribution of new insights in the field of competitive location in linear markets Since the late seventies however, a myriad of different models have appeared in the literature of spatial economics and industrial organization. Useful reviews can be found in [965, 499, 441, 371].

Much of the effort in the new evolving literature has aimed at developing insights concerning the equilibrium pattern of competitive location and pricing. Nevertheless, "although these models (linear models) are rich in their theoretical insights about spatial competition and have greatly enhanced our understanding of locational interdependence, they provide very little guidance for developing practical approaches to facility location in competitive environments." [463].

Parallel to the development of this body of literature, a new field on location modeling was growing in the late sixties and seventies at a fast rate, namely facility location analysis. This field of research, coming basically from the fields of operations research, regional science and geography, dealt with the problem of locating new facilities in a spatial market in order to optimize one or several geographical and/or economic criteria. These criteria included overall distance minimization and transport and manufacturing cost minimization. The literature in facility location analysis is prolific: good references can be found in [284], who identified over 1500 references on the subject, [786], and [209]. Although these models used more realistic spatial representations, such as networks and planes, most of them dealt exclusively with non-competitive situations, and little attention was paid to the characterization of market equilibria.

From the late seventies, considerations on the interaction between competing facilities in discrete space have been developed following several approaches. One of the first of the questions that is addressed by several authors is the existence or (not) of a set of locations on the vertices of a network that will ensure a Nash equilibrium, that is, a position where neither firms have incentives to move. Wendell and McKelvey [1189] considered the location of two competitive firms with one server each and tried to find a situation where a firm would capture at least 50% of the market regardless of the location of its competitor. Results showed that there was not a general strategy for the firm that would ensure this capture if locating at vertices of the network. They did not develop a generic algorithm for finding solutions, but they looked at the possible locational strategies. They also examined the problem in a tree. Hakimi [512] also analyzed extensively the problem of competitive location on vertices and proved that, under certain mathematical conditions such as concave transportation costs functions, that there exists a set of optimal locations on the vertices of the network.

A similar problem was studied by [746]. Their problem not only looked at the specification of a site but also at the setting of a delivered price. They formulated the problem as a 2-stage game, where in the first stage both firms choose locations and in the second stage they simultaneously set delivery prices schedules, and the result is that there is sub-game perfect Nash equilibrium. As Hakimi did, they proved that if firm's transport costs are strictly concave, then the set of locational choices of the firm is reduced to the vertices of the network. As a consequence, the location problem can be reduced to a 2-median problem if social costs are minimized.

A similar result was obtained by [731]. They developed a two-stage game in which two firms with one server each first select their location and then the quantities they will offer to each market. They proved that a sub-game perfect Nash equilibrium exists and that the locations occur on the vertices if transport costs are concave.

The problem of two firms competing in a spatial market has also been studied in the case where the market is represented by a tree. Eiselt [368] proved that in such case there is not a sub-game perfect Nash equilibrium if both prices and locations are to be determined. Eiselt and Laporte [373] extended the problem to the location of 3 facilities in a tree. They found that the existence of equilibria depended on the distribution of weights. In both models, firms were allowed to locate on the edges of the network.

The game-theoretical models presented so far restrict themselves to the location of firms with one facility each that compete against each other. Tobin and Friesz [1137] examined the case of a profit-maximizing firm that entered a market with several plants. They considered price and production effects on the market, since the increase in the overall production level from the opening of new plants in a spatial market stimulates reactions in the competitors. These reactions might affect not only production levels, but also prices and locations.

Tobin and Friesz developed two models: (1) a spatial price equilibrium model which determines equilibria in prices and production levels for a given number of firms, and (2) a Cournot-Nash oligopolistic model in which a few profit maximizing firms compete in spatially separated markets. They used both models to analyze the case of an entering firm that is going to open several new plants in spatially separated markets, and knows that its policy will have impact on market prices. Since profits depend on location and price-levels and these depend on the reaction of the competitors, it is not possible to use a standard plant location model. To tackle the problem, they used sensitivity analysis on variational inequalities to relate changes in production to changes in price to obtain optimal locations. The model was solved using a heuristic procedure where in the first step a spatial competitive equilibrium model was obtained and, in the second step, sensitivity analysis of profit to production changes was done using an integer non-linear program to select locations and production levels likely to maximize total profits. This model was generalized by Friesz et al. [437] to allow the entering firm to determine not only production levels and the sites of its plants, but also its shipping patterns, and to examine different market strategies that can occur in the market [846]. Due to the mathematical complexity of these models, Miller et al. [845] developed several heuristic methods to tackle the problem using the approach of variational inequalities [560, 882].

Another body of literature on competitive location deals with the siting of retail convenience stores. These types of stores are characterized by (1) a limited and very similar product offering across outlets, (2) similar store image across firms, and (3) similar prices. Therefore, location is a major determinant of success.

Ghosh and Craig [463] considered the location of several retail facilities by two servers. The problem is to locate retail facilities in a competitive market knowing that a competing firm will also enter this market. They used a minimax approach, where the entering firm maximizes its profit given the best location of the competitor. Potential locations were restricted to the vertices of the network. The firm's objective is to maximize the net present value of its investment over a long-term planning horizon. The model did not allow location at the same site for both firms and did not examine the issue of ties. Ghosh and Craig used a heuristic algorithm to obtain solutions. The algorithm is as follows: for each possible set of locations of firm A, the best siting strategy is found for firm B. The final result is the set of locations where Firm A's objective is maximum given the best reactive location strategy of its competitor. A Teitz and Bart hill-climbing heuristic was used to determine the sites for both firms. The model is adapted to examine other strategies such as preemption, i.e., the identification of locations that are robust against competitive action. Other modifications included the relaxation of the number of stores that could be opened by each firm, and collusion by both servers.

In a similar model, Dobson and Karmarkar [283] introduced the notion of *stability* in the location of retail outlets by two profit maximizing firms in a competitive market. Several integer programming models were developed to identify stable locations such that no competitors can enter the market and have profits given some rules on the competitive strategies. The models were solved using enumeration algorithms.

Most competitive location models assume that consumers patronize the closest shop. Karkazis [662] considered two criteria that customers may use to decide which shop to patronize: a *level criterion* based on the preferences of a customer on the size of the facility and a *distance* criterion based on closeness to the store. He developed a model that would determine the location and number of servers to enter the market when there are other firms already operating in the market by maximizing the profit subject to a budget constraint. The problem was solved in a dynamic fashion since there is a trade-off between both criteria.

Another model that examines competition among retail stores in a spatial market was developed by ReVelle [992]. The Maximum Capture Problem (MAXCAP) has formed the foundation of a series of models. These models include issues such as the strategies that competing firms may adopt or the uncertainty that characterizes some situations. In the following sections both the MAXCAP Problem and its extensions will be examined in detail.

16.3   A Review of The Maximum Capture Problem

In a excellent review paper on competitive location, Friesz et al. [438] pointed out that ReVelle's Maximum Capture Problem was one of the three competitive network facility location models that were "likely to serve as foundations for future models" together with the ones of Lederer [745] and Tobin and Friesz [1137]. In fact, The Maximum Capture Problem initiated a series of studies on the location of retail facilities in discrete space. The purpose of this section and the following ones is to review both the original formulation and the extensions of the MAXCAP model and to give new insights on this problem.

The MAXCAP model makes the following assumptions: (1) the product sold is homogeneous, in the sense that it is difficult to differentiate it among stores belonging to different firms, (2) the consumer's decision on patronizing the store is based on distance and not on price (i.e., price is considered the same in all stores and does not have any role in the purchasing decision) and (3) unit costs are the same in all stores regardless of ownership. Examples of this type of facilities can be found in the fast food sector, in convenience stores and in the banking sector(bank branches or ATM machines), among others.

In essence, the MAXCAP problem seeks the location of a fixed number of servers (p stores) belonging to a firm in a spatial market where there are other servers from other firms already competing for clients. The spatial market is represented by a network. Each node of the network represents a local market with a fixed demand which is given. The location of the servers is limited to the nodes of the network. Competition is based on distance: a market is "captured" by a given server if there is no other server closer to it. If two servers from competing firms have a local market at the same distance, then they divide in equal part the capture, as in Hotelling's game. The objective of the entering firm is to maximize its market capture. This objective, given the assumptions on the characteristics of the retail stores, is almost equivalent to maximizing profits [545].

In this setting, the concept of market capture by an outlet is very similar to the concept of *security center* proposed by Slater [1086]. A security center is a location j which minimizes the maximum for any other location k of the difference between the demand nodes closer to j and closer to k. The MAXCAP problem extends this concept to the location of several servers. Therefore, the p servers of the entering firm located using the MAXCAP problem can be considered as security centers, since they minimize the capture that Firm B can achieve.

The formulation of the MAXCAP problem is based on the Maximal Covering Location Problem (MCLP) of Church and ReVelle [223] and can also be formulated as a P-median-like model. As mentioned before, the model assumes that a new firm (from now on Firm A) wants to enter a market in order to obtain the maximum capture, given the location of q competitor servers, that can belong to one or more firms. Without loss of generality it is assumed that there is only one competing firm (Firm B) operating in the market. The basic question is to find a set of p locations for Firm A so as to maximize its capture. The formulation of the MAXCAP model is as follows:

$$\text{Max } Z^A = \sum_{i \in I} a_i y_i^A + \sum_{i \in I} \frac{a_i}{2} z_i$$

subject to:

$$y_i^A \leq \sum_{j \in N_i(b_i)} x_j^A \quad \forall i \in I \tag{16.1}$$

$$z_i \leq \sum_{j \in O_i(b_i)} x_j^A \quad \forall i \in I \tag{16.2}$$

$$y_i^A + z_i \leq 1 \quad \forall i \in I \tag{16.3}$$

$$\sum_{j=1}^{n} x_j^A = p \tag{16.4}$$

$$y_i^A, z_i, x_j^A = (0,1) \quad \forall i \in I, \forall j \in J$$

where parameters are:

$$
\begin{array}{rcl}
i, I &=& \text{index and set of demand areas} \\
j, J &=& \text{index and set of potential locations} \\
a_i &=& \text{demand at node } i \\
d_{ij} &=& \text{distance from node } i \text{ to node } j \\
b_i &=& \text{Firm B's server closest to demand node } i \\
d_{ib_i} &=& \text{distance from demand node } i \text{ to the closest B server} \\
N_i(b_i) &=& \{\forall j \in J, d_{ij} < d_{ib_i}\} \\
O_i(b_i) &=& \{\forall j \in J, d_{ij} = d_{ib_i}\}
\end{array}
$$

and variables are defined as follows:

$$
y_i^A = \begin{cases} 1, \text{ if Firm A captures demand node } i \\ 0, \text{ otherwise} \end{cases}
$$

$$
z_i = \begin{cases} 1, \text{ if demand node } i \text{ is divided between A and B} \\ 0, \text{ otherwise} \end{cases}
$$

$$
x_j^A = \begin{cases} 1, \text{ if Firm A locates a server at node } j \\ 0, \text{ otherwise} \end{cases}
$$

The first set of constraints (16.1) depends on the set $N_i(b_i)$, which is known a priori. Each one of the demand nodes i has an associated set $N_i(b_i)$ which contains all the potential nodes at which Firm A can locate a server and capture node i. Therefore, if one of the variables x_j^A belonging to the corresponding constraint is equal to 1 (i.e., a facility is located within the capture area of node i), then capture variable y_i^A is allowed to be 1, which indicates that node i has been captured by Firm A. The second set of constraints is very similar, but this time the set O_i is used. This set includes all nodes where, if Firm A locates a server, she will divide the demand captured with her competitor ($z_i = 1$). The third group of constraints takes into account the three states that a node can have: either it is fully captured by Firm A ($y_i^A = 1$ and therefore $z_i = 0$), or it is divided between both firms, ($z_i = 1$ and therefore $y_i^A = 0$), or it is lost to Firm B ($y_i^A = 0$ and $z_i = 0$). Finally, the last constraint sets the number of servers that Firm A is going to locate. The objective defines the total capture that Firm A can achieve with the siting of its p servers. For each node, there is demand a_i to be captured. If $y_i^A = 1$, then a_i is added to the objective. On the other hand, if $z_i = 1$, then only $a_i/2$ is added to the objective, since the other half of the demand goes to Firm B.

The original MAXCAP problem considered that the demand at each node is fixed. If the demand depends on the distance to the server, then the MAXCAP problem can be reformulated using a p-median-like approach. Nevertheless, the number of variables and constraints significantly increases due to the use of assignment variables x_{ij}, where $x_{ij} = 1$ if node i is captured by a facility j belonging to Firm A.

Even though the problem of market capture was shown to be NP-hard for a general network (see Hakimi[511] on the $(r|X_p)$-Medianoid problem), ReVelle's mathematical formulation provided an optimal solution method to solve it. ReVelle used linear programming and branch and bound when necessary to solve the problem in a relatively small network (30 nodes). In most cases, the problem needed little or no branch and bound. This is probably due to the strong relationship between the MAXCAP problem and the maximal covering location problem (Church and ReVelle[223]). For large problems, the Algorithm for Market Capture (AMACA), a one-opt heuristic based on the well known Tietz and Bart method [1133] can be used.

The AMACA proceeds as follows: Given an initial location of the p Firm A servers and knowing the location of the competitors, the capture that Firm A achieves in the market can be obtained. Then, from a list of Firm A's servers, the first member of the list is picked and its location is moved to an empty node. The new capture can be computed and compared to the market capture achieved before the one-opt trade. If the objective has improved, the new set of locations is kept as the current solution. On the other hand, if the capture obtained before the trade is higher, this exchange is not considered. These one-opt trades are executed for all pairs of empty nodes and facilities. If at the end of all trades the objective has improved, then the procedure is repeated. The heuristic procedure ends whenever there is no improvement in the objective on completion of all exchanges possible. A more formal statement of the heuristic follows:

1. Locate p Firm A servers using any method.

2. Compute the capture obtained by Firm A.

3. Choose the first Firm A server off a list of its servers and trade its location to an empty node.

4. Compute the new market capture. If the value of the objective has improved, store the new locational solution. If not, restore the old solution.

5. Repeat steps 4 and 5 until all potential empty locations have been evaluated one at a time for each server.

6. If Firm A has improved its market share to a value greater than in step 2, go to step 3 and restart the procedure.

7. When no improvement is achieved on a complete set of one-at-a-time trades, stop.

The AMACA heuristic procedure is very efficient in terms of computing time but does not guarantee optimality. As it shall be seen later in this section, the AMACA procedure is very useful when the MAXCAP problem has to be solved repeatedly.

16.4   Extensions of the Maximum Capture Problem

The MAXCAP model is characterized by the following main assumptions:

1. all outlets have equal weights,

2. there is no reaction from the competitors to the entrance of new outlets,

3. there is no uncertainty in the parameters of the model.

The MAXCAP problem has been modified to relax these assumptions. In the following pages these relaxations will be examined.

Eiselt and Laporte [371] modified the MAXCAP problem to include outlets that have different weights. Their model not only optimally locates an outlet in a competitive region, but also defines its optimal weight, which is loosely defined by the authors as a measure of "the size of the facility, its relative price advantage, the courteousness of its staff, etc." (p.434). Their approach derives from gravity models (Huff, [603]) and closest point models, also known as Voronoi diagrams (see Eiselt and Pederzoli, [376]) or Thiessen polygons. The problem was cast as a non-linear integer program where the objective was to maximize profits by the entering firm. One of the caveats of the formulation is that, even though from a mathematical standpoint the formulation can be adapted to the location of several new servers, the solution method proposed becomes computationally burdensome, since it is necessary to compute for each candidate node the profits that the firm would obtain.

The issue of different weights in the location of outlets in a competitive spatial market has also been examined by Serra et al. [1060]. They consider that each firm has different servers organized in a hierarchical fashion. There is competition among outlets belonging to different Firms in each level of the hierarchy. The servers are nested, in the sense that upper-level outlets also sell products offered in the lower-level ones. The sphere of influence that an outlet has also depends on its hierarchical position. The higher in the hierarchy, the larger the sphere of influence. These factors lead to the situation were there is competition among all levels between each firm. A lower level outlet from Firm A may compete on the basis of distance for a given demand node with an upper level outlet from the competing firm that is be located further. Depending on the sphere of influence, an upper level outlet might capture a demand node even if it is located further from the demand node than a lower level outlet from the competing firm. The Maximum-Capture Hierarchical Location Problem is as follows: a Firm wants to enter a market with p_k servers in each hierarchical level k. In the market there are competitor servers that are already located. The main objective is to maximize total market capture. The demand parameter associated with the demand node at each level is defined as potential sales in monetary terms, so the objective is stated as the maximization of total market sales. The model was cast as an integer linear program and was solved using linear programming together with branch and bound when needed. In the examples used, in most cases little or no branch and bound was necessary.

The MAXCAP problem has been modified to take into account the existence of uncertainty in some of the model parameters (Serra and ReVelle, [1061]). There might be situations in which the demand or population at a node that the entering firm is setting out to capture is not a known quantity but can assume different values depending on community growth or economic vitality. Furthermore, different number and locations of competitor outlets might occur depending on market expansion and corporate strategies. In this sense, Serra and ReVelle tackle the problem using the classic scenario approach. That is, demands and/or competitor locations are different in each possible scenario. The entering firm can then use at least two different criteria to locate servers despite uncertainty about which scenario will actually occur. These criteria are:

1. To maximize the minimum capture over all scenarios (Maximin criterion)

2. To minimize the worst deviation from the maximum capture that could be obtained in a given scenario if this one would come true (Regret criterion), i.e., to minimize the maximum difference between what is achieved and what might have been achieved.

If the first objective is used, then it is necessary to obtain a set of locations that will give the largest minimum capture possible over all scenarios. It is possible to compute, for a given set of locations for firm A servers, the final capture Z_k that the entering firm would achieve in each scenario k, since $m_k = \sum_{i \in I} a_{ik} y_{ik} + \sum_{i \in I} (a_{ik}/2) z_{ik}$, where a_{ik} is the demand in node i for scenario k and y_{ik} and z_{ik} are set to 1 if node i is captured in scenario k. The model will maximize the capture in the scenario with the lowest m_k. Observe that each scenario will have different sets N_i and O_i, since these sets depend on the location of the competitors. Therefore, there will be one for each scenario (new sets N_{ik} and O_{ik}). This implies that there will be a constraint (16.6) and (16.7) for each demand node and each scenario. So the problem is formulated as follows:

$$\max Z = m$$

subject to:

$$\sum_{i \in I} a_{ik} y_{ik} + \sum_{i \in I} (\frac{a_{ik}}{2}) z_{ik} \ \geq \ m \quad \forall k = 1, \ldots, s \qquad (16.5)$$

$$y_{ik} \ \leq \ \sum_{j \in N_{ik}} x_j \quad \forall i \in I, \forall k = 1, \ldots, s \qquad (16.6)$$

$$z_{ik} \ \leq \ \sum_{j \in O_{ik}} x_j \quad \forall i \in I, \forall k = 1, \ldots, s \qquad (16.7)$$

$$y_{ik} + z_{ik} \ \leq \ 1 \quad \forall i \in I, \forall k = 1, \ldots, s \qquad (16.8)$$

$$\sum_{j \in J} x_j \ = \ p \qquad (16.9)$$

$$y_{ik}, z_{ik}, x_j \ = \ (0,1) \quad \forall i \in I, \forall j \in J, \forall k = 1, \ldots, s$$

where s is the total number of scenarios.

If the Regret objective is used, then the problem is rewritten as follows:

$$\min Z = U$$

subject to: $\quad Z_k - \sum_{i \in I} a_{ik} y_{ik} - \sum_{i \in I} (\frac{a_{ik}}{2}) z_{ik} \leq U \quad \forall k = 1, \ldots, s \qquad (16.10)$

$$+ \text{ constraints (16.6) (16.7) (16.8) (16.9)}$$

where parameter Z_k represents the maximum capture that the entering firm would obtain when optimally locating p servers in scenario k, i.e., it is necessary to compute s MAXCAP problems, one for each scenario. The left hand side of constraint (16.10) calculates the deviation from optimality in each scenario given a set of Firm A locations. The objective will try to minimize the largest deviation from optimality.

Observe that if only demands differ in each scenario, then it is not necessary to re-define sets N_i and O_i and capturing variables y_i and z_i. The new formulation is then obtained by adding only constraints (16.5) or (16.10) depending on the model to be solved (Maximin or Regret, respectively) and their corresponding objectives.

The use of linear programming relaxation and branch and bound when necessary could require large computer times if the functional form of constraints (16.5) or (16.10) does not favor 0,1 solutions. Serra and ReVelle proposed a heuristic algorithm based on the Teitz and Bart one-opt heuristic with two phases. In the first phase the MAXCAP problem is solved for each scenario individually and the solution of one of the scenarios is chosen as the initial solution depending on the objective used (Maximin or Regret). The second phase tries to improve the solution by relocating one facility at a time to an empty node as in the AMACA heuristic until no improvement is obtained. Both LP+BB and the heuristic were solved on Swain's 55-node network.

The original MAXCAP problem considered that Firm A was new in the market, i.e., did not have any server already positioned. If Firm A is already operating in the market and wants to open new outlets, then it is necessary to exclude from the network those demand nodes that were already under the the area of influence of Firm A. But a better capture can be obtained by the expanding Firm A if some of the existing servers are allowed to relocate. The new problem can be stated as follows: suppose that Firm A has p outlets in the spatial market and wants to relocate r outlets ($r \leq p$) and to open s new ones. Which outlets have to be relocated and where to locate them together with the new ones to maximize market capture? This problem - The Maximum Capture Problem with Relocation (MAXRELOC) - was studied by ReVelle and Serra [1002] and further reformulated by Serra et al. [1060]. This new problem can be easily formulated by replacing in the MAXCAP model the constraint that fixes to p the number of servers to locate (constraint 16.9) by the following ones: $\sum_{j \in J} x_j^A = p + s$; $\sum_{j \in J_A} x_j^A = p - r$ where J_A is the set of the existing p Firm A locations.

The relocation of servers is an important component of competitive facility siting because the entrance of competitors after the initial siting of outlets causes a change in the marketing landscape. ReVelle and Serra [1002] used the MAXRELOC formulation to examine different sequential strategies that two firms (a duopoly) can use when competing for clients in a spatial market with several outlets each, based on the classic economic market games suggested by Cournot and Stackelberg. At each time interval, firms are allowed to relocate some of its servers to improve their positioning. The first strategy -the Cournot Strategy - is such that a firm will locate and relocate its servers by maximizing newly captured population, regardless of the location decisions of its competitors in the same time interval. That is, in each period, each firm used the MAXRELOC formulation to relocate some of its servers using the locations obtained by its competitor in the previous period.

The second strategy - the Stackelberg strategy- is based on the assumption that one firm (Firm A) acts as a leader and positions its servers using the known locations of its competitor, and Firm B, once Firm A has relocated its outlets, uses the MAXRELOC model to relocate its servers. Therefore, Firm B acts as a follower, and it has an advantage in its strategy since it knows where Firm A has relocated its servers. These two strategies were tested repeatedly on a 55-node network and no locational equilibrium was achieved, that is, no Nash equilibrium was obtained, where both firms are not able to improve their market share by relocating some servers.

A similar problem, the Pre-emptive Capture Problem (PRECAP), was studied by Serra and ReVelle [1062]. The PRECAP analyzes the situation of a spatial market where there are as yet no competitors. A firm (Firm A) wants to locate p servers but knows that *after* their outlets' siting, one or several competitors will enter the market with several outlets too. The only information that Firm A has is that there will be q competitor servers being located in the future (say Firm B again). Therefore, Firm A wishes to pre-empt Firm B in its bid to capture market share to the maximum extent possible. The notion of pre-emptive competition has been widely analyzed in the literature of industrial organization (see, for example, Gilbert [475]). Hakimi [511] defined this problem as the $(r|p)$-centroid. Pre-emptive competition has also an equivalent concept in voting theory. A *Simpson point* (Simpson [1081]) is such that the maximum number of users closer to another point is minimum. The main goal of Firm A is to find a set of pre-emptive locations, i.e. Simpson points, that will minimize the maximum possible future capture by Firm B.

It has to be noted that if both firms wish to locate the same number of servers, then Firm B will be always able to capture at least 50% of the market by locating on top of Firm A's servers. Therefore, the best strategy for Firm A would be to obtain a set of locations such that B would have no other option than locating its servers in such a manner. As Serra and ReVelle show, in the general case, Firm B will obtain more than 50% of the market. This conclusion follows the one of Wendell and McKelvey [1189] for the special case where there is only one server. The locations obtained by Firm B once it has entered the market can be considered as *plurality points*, as defined by Wendell and McKelvey [1189] or Condorcet points, in terms of voting theory, with respect to Firm A locations. The equivalence of this type of points has been proved by Hansen el al. [559] for a general network.

The mathematical formulation of the PRECAP problem is exactly the same as the one of the MAXCAP problem, except that constraints (16.1) and (16.2) cannot be written in extensive form. The problem is that Firm A does not know the future locations of its competitors and therefore the sets $N_i(b_i)$ and $O_i(b_i)$ cannot be defined, since b_i, the closest competitor to node i, is not known a priori. Since the sets $N_i(b_i)$ and $O_i(b_i)$ contain all candidate nodes that are closer to i and at the same distance from competitor outlet b_i respectively, and since b_i is unknown, the members of the sets cannot be written down.

In order to tackle the problem, Serra and ReVelle proposed a one opt heuristic that was tested on a 55-node network. Briefly, the PRe-EMptive capture heuristic ALgorithm (PREMAL) starts with the positioning of Firm A's servers using any method (for example, p-median or covering approaches). Then, the MAXCAP problem is used to find Firm B locations. Now, the market shares of Firm A and Firm B are known. Once the initial locations for both competitors are obtained, Firm A moves one of its outlets to an empty node. Again, the MAXCAP problem is solved to obtain Firm B's response. If Firm A's market share improves after the one-opt trade, then it will keep its new outlets' locations as the current solution. Otherwise, Firm A will ignore the relocation and restore the previous solution. The procedure is repeated for all nodes and servers until no improvement is achieved on a complete set of one-at-a-time trades.

This procedure could become computationally burdensome if linear programming relaxation supplemented by branch and bound is used to solve the MAXCAP problem, since a MAXCAP problem has to be solved at each one–opt trade. If large problems were to be solved using the PREMAL algorithm, solutions to the MAXCAP problem can be obtained using the AMACA heuristic. As Serra and ReVelle show in the computational experience of the PRECAP problem, the use of the AMACA heuristic for Firm B yield solutions for Firm A which allowed for Firm B to do slightly better than if the LP version of the MAXCAP problem were solved.

16.5   Extensions of the Pre-emptive Capture Problem

The Pre-emptive Capture Problem assumes that Firm A does not know the future locations of its competitors, but does know with certainty the number of competitor servers that will enter in the spatial market. In this section the PRECAP model is modified to relax this last assumption. A new formulation is presented, together with a heuristic solution method and an example.

Now the final capture of Firm A depends not only on the location of Firm B's outlets, but also on the number of servers it throws into the competition. An objective that Firm A might consider is to minimize the average maximum capture that Firm B can achieve. In this case, the problem can be formulated as follows:

$$\max Z^A = \min \sum_{k=1}^{q} \frac{1}{k} \left(\sum_{i=1}^{n} a_i y_{ik}^B + \sum_{i=1}^{n} \frac{a_i}{2} z_{ik} \right)$$

subject to:

$$y_{ik}^B \leq \sum_{j \in N_i} x_{jk}^B \quad \forall i \in I, k = 1, ..., q \qquad (16.11)$$

$$z_{ik} \leq \sum_{j \in O_i} x_{jk}^B \quad \forall i \in I, k = 1, ..., q \qquad (16.12)$$

$$y_{ik}^B + z_{ik} \leq 1 \quad \forall i \in I, k = 1, ..., q \qquad (16.13)$$

$$\sum_{j=1}^{n} x_{jk}^B = k \quad k = 1, ..., q \qquad (16.14)$$

$$y_{ik}^B, z_{ik}, x_{jk}^B = (0, 1) \quad \forall i \in I, \forall j \in J, k = 1, ..., q$$

where additional notation is:

$$
\begin{array}{rcl}
k & = & \text{index of the number of servers that Firm B can locate} \\
q & = & \text{maximum number of competitor servers} \\
b_i^A & = & \text{closest Firm A's server to node } i \\
d_{ib_i^A} & = & \text{distance from node } i \text{ to Firm A's closest server} \\
N_i & = & \{\forall j \in J, d_{ij} < d_{ib_i^A}\} \\
O_i & = & \{\forall j \in J, d_{ij} = d_{ib_i^A}\}
\end{array}
$$

$$y_{ik}^B = \begin{cases} 1, & \text{if Firm B captures } i \text{ when locating } k \text{ servers} \\ 0, & \text{otherwise} \end{cases}$$

$$z_{ik} = \begin{cases} 1, & \text{if node } i\text{'s capture is divided between A and B} \\ & \quad \text{when B locates } k \text{ servers} \\ 0, & \text{otherwise} \end{cases}$$

$$x_{jk}^B = \begin{cases} 1, & \text{if Firm B locates a server at } j \text{ when locating } k \text{ servers} \\ 0, & \text{otherwise} \end{cases}$$

The problem has been formulated with respect to the location of Firm B's servers. Since locations and final capture for both Firms depend on the k servers that Firm B might locate, it has been necessary to redefine both the capture variables and location variables of the model. Otherwise, the formulation is very similar to the Pre-emptive Location Problem. Observe that the capture that Firm B will obtain if locating k servers is $Z_k^B = \left(\sum_{i=1}^{n} a_i y_{ik}^B + \sum_{i=1}^{n} \frac{a_i}{2} z_{ik} \right)$. Therefore, if Z_k^B is divided by k, then the capture per server is obtained. Total average capture is obtained by adding the average capture for $k = 1, \ldots, q$. The objective of Firm A is to locate first its services so as to minimize the total average capture that Firm B can achieve afterwards. If Firm A knows the probability α_k associated with the location of k servers by Firm B, then the objective can be modified by replacing $\frac{1}{k}$ by α_k. In this case, Firm A is minimizing the expected maximum capture. If $(1/k)$ is used the implied assumption is that the number of servers Firm B sites are equally likely. If q is now the maximum number of servers that Firm B might locate, it follows that $\sum_{k=1}^{q} \alpha_k = 1$.

As in the Pre-emptive Capture Problem, this model cannot be solved optimally since the locations of Firm A servers are not known and therefore constraint sets (16.11) and (16.12) cannot be written in extensive form. A modified PREMAL heuristic can be used to tackle the problem. The first step is to locate the p Firm A servers using any method. Then the MAXCAP problem is used q times to locate $1, 2, \ldots, k, \ldots, q$ Firm B servers. Now the initial objective can be computed. In the second phase, Firm A will try to improve the solution by relocating its servers using a one-opt procedure. At each iteration, Firm A will relocate one of its servers and then, as in the initial phase, use the MAXCAP problem q times to see the average market capture that B can obtain when locating $1, \ldots, k, \ldots, q$ servers and then compute the objective. If the relocation has provided a set of positions that is better than before the one-opt trade, it will keep the new set of locations as best so far. Otherwise, Firm A will ignore the relocation and will restore the previous solution. The one-opt trade will be done for all nodes and Firm A servers. If, after a complete set of one opt-trades, Firm A has improved its objective, then the procedure is restarted. When no improvement is achieved, the procedure stops and final locations are obtained. As in the PREMAL procedure, the MAXCAP problem can be solved using integer programming or the AMACA heuristic procedure, depending on the computing time and storage available. Of course, multiple starting solutions for the Firm A servers could be used. A more formal statement of the heuristic follows:

1. Locate Firm A's p facilities using any method.

2. Locate 1 to q Firm B servers using the MAXCAP Problem.

3. Compute the average market share per server for each number of servers located by firm B. Sum them. This is our initial solution.

4. Trade the location of one of the p servers of Firm A.

5. Locate 1 to q Firm B servers using the MAXCAP Problem.

6. Compute the average market share per server for each number of servers located by firm B and add them. If the objective is smaller that before, store the new locations of Firm A's servers. If not, restore old locations.

7. Repeat steps 4 to 6 until all of Firm A's p facilities and nodes have been traded.

TABLE 16.1. 55-node network demands and coordinates

node	pop	coord		node	pop	coord		node	pop	coord	
		x	y			x	y			x	y
1	120	32	31	20	77	25	14	39	47	46	51
2	114	29	32	21	76	29	12	40	44	50	40
3	110	27	36	22	74	24	48	41	43	23	22
4	108	29	29	23	72	17	42	42	42	27	30
5	105	32	29	24	70	6	26	43	41	38	39
6	103	26	25	25	69	19	21	44	40	36	32
7	100	24	33	26	69	10	32	45	39	32	41
8	94	30	35	27	64	34	56	46	37	42	36
9	91	29	27	28	63	12	47	47	35	36	26
10	90	29	21	29	62	19	38	48	34	15	19
11	88	33	28	30	61	27	41	49	33	19	14
12	87	17	53	31	60	21	35	50	33	45	19
13	87	34	30	32	58	32	45	51	32	27	5
14	85	25	60	33	57	27	45	52	26	52	24
15	83	21	28	34	55	32	38	53	25	40	22
16	82	30	51	35	54	8	22	54	24	40	52
17	80	19	47	36	53	15	25	55	21	42	42
18	79	17	33	37	51	35	16				
19	79	22	40	38	49	36	47				

TABLE 16.2. Initial locations and final results for Firm A

		CPU time
Initial Locations	5,11,12,54,55	
Final Locations		
Heuristic 1	4,13,16,31,41	20:58
Heuristic 2	4,7,13,22,49	3:55:23

8. If Firm A has reduced the objective in steps 4 to 6, go to step 4 and restart the procedure. When no improvement is achieved on a complete set of one-at-a-time trades, stop.

This new heuristic has been tested on Swain's 55-node network (1974, see Table 16.1, Figure 1)[1115] using both the AMACA procedure and linear programming and branch and bound when necessary (LP+B&B), to solve the MAXCAP subproblems (heuristic 1 and heuristic 2, respectively). Firm A wishes to locate 5 servers and knows that Firm B is considering to locate after A's initial move from 1 to 9 outlets. In the first phase of the solution method, Firm A's initial locations were found randomly and the MAXCAP problem was solved using both approaches. Results are presented in Table 16.2. Only two locations, nodes 4 and 13, are obtained in both solutions.

Table 16.3 presents the initial and final market captures for Firm B using both heuristics if finally locating k servers after Firm A's entrance in the market with p servers. For example, if the decision of Firm B were initially

FIGURE 1. Swain's 55-node network

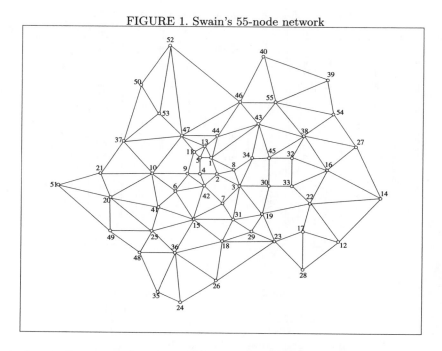

to locate 5 servers, then it would obtain 79.7% of the market. After the second phase of the heuristic, the one-opt relocation of Firm A's servers, Firm A's final sites would reduce Firm B's market capture to 50.7% if the AMACA procedure were used and to 53.9% if LP+BB were used. It was expected that Firm B would obtain better results using LP+BB since at each time that Firm A would relocate one of its servers, the MAXCAP problem for Firm B would be optimally solved. Nevertheless, the heuristic procedure proposed significantly improves the market capture that Firm A can achieve. Multiple starting positions might correct this anomaly. Finally, initial and final locations for Firm B are presented in Table 16.4.

As was expected, the heuristic that used the AMACA procedure (Heuristic 1) needed much less CPU time than the heuristic that used LP+BB to solve the MAXCAP problems (see Table 16.2).

16.6   Conclusions

We began in this chapter with an overview of the different approaches that have been proposed to analyze locational strategies in a competitive setting. Special focus was on those models where both the demand and the potential locations for servers are discrete. We then reviewed the Maxi-

TABLE 16.3. Market capture results for Firm B

# of servers	Initial Locations Capture		Final Locations			
			Heuristic 1 Capture		Heuristic 2 Capture	
1	1941.0	51.8%	473.0	12.6%	505.0	13.5%
2	2400.0	64.0%	872.0	23.3%	981.0	26.2%
3	2615.0	69.7%	1248.0	33.3%	1375.0	36.7%
4	2818.5	75.2%	1588.0	42.3%	1747.0	46.6%
5	2988.5	79.7%	1900.5	50.7%	2020.0	53.9%
6	3066.5	81.8%	2204.5	58.8%	2268.0	60.5%
7	3126.5	83.4%	2409.5	64.3%	2480.0	66.1%
8	3179.0	84.8%	2566.0	68.4%	2677.0	71.4%
9	3223.0	85.9%	2715.0	72.4%	2800.5	74.7%

TABLE 16.4. Locations for Firm B

Initial Locations — Number of servers									Final Locations — Heuristic 1 — Number of servers								
1	2	3	4	5	6	7	8	9	1	2	3	4	5	6	7	8	9
4	4	4	1	1	1	1	1	1	2	2	2	2	2	2	9	5	5
	16	16	16	4	4	4	4	4		41	17	11	11	9	9	13	9
		23	16	16	17	17	5	5			41	17	13	13	17	13	13
			23	23	27	27	17	17				41	17	16	31	17	17
				47	43	43	27	27					49	29	32	31	27
					47	47	43	40						49	42	32	31
						54	47	43							49	42	32
							54	47								49	42
								54									49

									Heuristic 2								
									1	2	3	4	5	6	7	8	9
									30	22	1	1	13	5	5	5	5
										42	22	16	16	8	8	8	8
											41	29	25	25	16	16	9
												41	29	29	25	17	13
													42	37	29	25	16
														42	37	31	17
															42	37	20
																42	25
																	31

mum Capture problem, a model that has been developed and adapted for different situations -where there is uncertainty in model parameters, where reactions from competitors to the entering firm might occur, and where outlets might have different areas of influence. Finally, a new model was proposed to tackle the Pre-Emptive Capture problem (a variant of the MAXCAP problem) in which there is little information available to the entering firm about the number and locations of future entrants in the spatial market. A heuristic method was proposed and tested on Swain's 55-node network. This network has been the basis for testing most of the models based on the MAXCAP problem, and the heuristics developed are based on the Teitz and Bart one-opt heuristic.

A common characteristic of all MAXCAP-based model is that the entering firm knows exactly the number of servers that it is going to locate. On the other hand, there is no consideration of opening costs that might differ substantially depending on the location chosen. If c_j is defined as the opening cost at node j, then the total opening cost is equal to $\sum_{j \in J} c_j x_j$. In order to obtain a trade-off between total capture and total costs, costs can be included as a second objective which is to be minimized. Then constraint (16.9), which sets the number of servers to locate, is eliminated. For the bi-objective problem, the weighting method or the constraint method can be used to obtain the trade-off curve between the two objectives. (Cohon [234]).

The bi-objective problem can be transformed into a single objective formulation if a_i is defined as potential sales if node i is captured. Then the problem is similar to the maximization of profits, even though no account is taken of variable costs, which might depend on the size of the area served (the area of influence) by each outlet. This situation can be overcome by defining a_i equal to (unit price) less (unit variable costs) times (number of potential units sold). The new objective would be defined as follows:

$$\max \pi = \sum_{i \in I} a_i y_i + \sum_{i \in I} \frac{a_i}{2} z_i - \sum_{j \in J} c_j x_j.$$

These modifications would make it difficult to use any of the one-opt heuristics since to apply them it is necessary to know the number of outlets that the firm wishes to locate. On the other hand, if the opening cost of the outlet depends on the size of the market it will serve, the problems would need to be reformulated using a P-median like approach, as proposed by ReVelle in the original MAXCAP problem.

Finally, it would be interesting to modify the MAXCAP problem to account for the location of production plants, where decisions on prices and production levels are also relevant.

We conclude by recalling the theoretical richness of the concept of Hotelling's ice cream vendors on a linear beach. We can observe that the wealth of opportunities that this original problem provided is now being paralleled by a large variety of problems based on the concept of competitive location on a network.

Part IV

ROUTING AND LOCATION

17
FLOW-INTERCEPTION PROBLEMS

Oded Berman
University of Toronto

M. John Hodgson
University of Alberta

Dmitry Krass
University of Toronto

17.1   Introduction

In traditional network location theory [509, 510] demand for service is assumed to originate from nodes of the transportation network. Customers residing at nodes of the network that wish to obtain service travel from the nodes to the service facilities to consume the service. The main purpose of the travel is to obtain the service. Alternatively, the service provider may travel to the customers at the nodes to provide the service requested. Examples of customers travelling to the facility are service activities such as weekly grocery shopping, going to the movie theatre and bringing the car to the bodyshop for repair. Example of servers travelling to the customers include ambulance, police and repair services. The median problem which minimizes customers' average travel time and the center problem where the worst case travel is minimized are known criteria for locating the service facilities [509, 510].

In this chapter we focus on models where demand for service originates not from the nodes of the network but from flow (customers) travelling on various paths of the network. Examples for such customer behaviour include in particular discretionary services (e.g. automatic teller machines, gasoline stations and convenience stores). Here significant fraction of customers consuming the service do so on an otherwise pre-planned tours (e.g. on the daily commute to work). The purpose of the travel is <u>not</u> to obtain service, but if there is a facility on the pre-planned tour the customers may choose to obtain the service. Therefore, the objective here is not to minimize the average travel time or the worst case travel time but to locate the facilities so as to maximize the total flow of customers that are 'intercepted' during their travel. Other examples of applications where maximizing the intercepted flow of customers is the objective include locating police radar traps, highway traffic monitions, billboards, etc.

This chapter is divided into two parts. The first part (Section 17.2) deals with deterministic flow interception problems. Here it is assumed that there is complete knowledge of all the paths that carry non-zero flow in the network. The actual flow rates are also assumed to be known. In the second part of the chapter (Section 17.3) we deal with probabilistic flow interception problems, where the information about paths and flows is not readily available. Here it is assumed that only information on the fraction of customers originating their travel from any node and the fraction of customers that travel from any node to all adjacent nodes is given. Clearly the information required for probabilistic flow interception problems is much easier to obtain in comparison to deterministic flow-interception models.

The presentation of both the deterministic and probabilistic models follows the same general structure. We start with a basic model that includes a variety of (sometimes restrictive) assumptions. We then present various generalizations of this basic model that relax some of the assumptions in the basic model. Examples of such generalizations include models with congestion (queueing), models where customers are allowed to deviate from their pre-planned trips to visit a facility, models where customers may be intercepted by more than one facility, etc.

A significant feature of Flow Interception Models is that greedy-type heuristic algorithms seem to perform very efficiently for both the deterministic and probabilistic models. This allows for efficient computation of accurate approximate solutions to realistic-size problems. The results on the worst-case bounds and computational experiments for greedy-type algorithms are presented below.

17.2   Deterministic Flow Interception Problems

The problem of locating m service facilities so as to maximize the total flow of potential customers who pass (are intercepted) by at least one of the facilities on their pre-planned trips was introduced independently in [96, 583]. Here we focus on deterministic versions of the problem where all paths between origin-destination pairs that carry non-zero traffic flow and the actual flow rates (e.g. number of customers per unit of time) are known.

Additional assumptions of the model discussed in [96, 583] are:

1. Customers make no deviation, no matter how small, from their pre-planned tours to visit the facilities.

2. Waiting time of customers at facility sites are negligible and do not affect the decision of the customer whether or not to visit the facility.

3. Customers are intercepted by at most one facility encountered along their route even if there are several facilities located at that route. In other words double counting of the intercepted flow is not allowed.

We present eight flow-interception problems. The first problem discussed in Section 17.2.1 is the basic one, which includes all the assumptions stated above. Also, in the same section, we discuss a related problem of finding the smallest number of facilities that will provide a pre-specified amount of flow that is required to be intercepted.

In section 17.2.2 the assumption that deviation from pre-planned trips is not allowed is relaxed. Three generalizations of the basic problem when deviations are allowed are presented. In the first generalization, a maximum deviation of up to Δ is allowed. Thus, as long as the distance from the customers' route, while travelling on their pre-planned tour, to a closest facility does not exceed Δ, these customers are counted as being intercepted. In the second generalization the fraction of customers that visit a facility is assumed to be a decreasing function of the extra distance travelled. In the third generalization the total extra distance incurred by all customers is the main focus.

In section 17.2.3 it is recognized that for service facilities with limited capacity the assumptions that waiting time of customers at the facilities is ignored, is unrealistic. In the problem considered deviations from pre-planned tours are allowed and the framework of the problem is identical to that of the second generalization discussed above. However, now we incorporate into the extra distance travelled by customers as a result of a deviation also waiting time at the facility. Therefore, it is assumed here that the fraction of customers that visit a facility is a decreasing function of the sum of the extra distance incurred and the mean travelled time.

In section 17.2.4 we recognize that for many marketing applications (e.g. billboards) the assumption that at most one interception (exposure) is considered is not realistic. Therefore, we relax the assumption of no double counting. It is assumed that the fraction of customers intercepted along a given path (with non-zero flow) is a concave nondecreasing function of the number of facilities located on that path. In other words each additional facility encountered may intercept more and more flow but at a marginal decreasing rate. Two problems are introduced. In the first one given at most m facilities the objective is to maximize the total amount of flow intercepted. In the second one it is desired to find the minimum number of facilities required to intercept the total amount of flow possible.

Finally, in section 17.2.5 we describe testing of the basic problem on a real-world example - the street grid of Edmonton, Canada.

FIGURE 1. A simple 7-node network

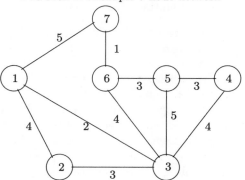

17.2.1   The Basic Model

We consider a network $G = (N, A)$ where N is the set of nodes with cardinality n and A is the set of arcs. We denote by P the set of non-zero flow paths in the network and by f_p the rate (per unit of time) of flow along path $p \in P$. As an example, Figure 1 depicts a network with $n = 7$ nodes (for our current discussion, ignore the numbers next to the arcs).

Table 17.1 below includes all the possible non-zero flow paths in P, and the corresponding flow rates. For example, the first path is between the origin node 1 to the destination node 6 along nodes 2 and 3 and the corresponding flow is 30 per day. We denote the total flow by f, $f = \sum_{p \in P} f_p$ (455 for the example in Figure 1).

Suppose there are $m \geq 1$ facilities to be located in the network. The objective is to locate the facilities so as to maximize the total flow intercepted. Let X be a set of m points from G and $I(X, p)$ be an indicator variable that assumes the value 1 if at least one point in X is included in the path p and 0 otherwise. The problem is

$$max_{X \subseteq G, \, |X|=m} \quad F(X) = \sum_{p \in P} f_p I(X, p) \qquad (P_1)$$

The search for X can be limited to the set of nodes N and thus all interior points on arcs can be excluded from consideration. Informally, the reason is that any facility that is located on an interior point of an arc can be moved to either one of the two nodes connecting that arc without any loss of flow and possibly with a gain of additional flow. Therefore $F(X)$ the set function in (P_1) is now defined for $X \subseteq N$, $|X| = m$.

The formulation of the problem above is a graph theoretical one. However the problem can also be formulated as an integer programming problem. The formulation is very similar to that of the Maximal Covering Location Problem $(MCLP)$ [223]. We define two sets of decision variables:

TABLE 17.1. List of paths and flow rates for the network in Figure 1

Path $p \in P$		f_p (flow rate per day)	Path $p \in P$		f_p (flow rate per day)
p_1	1-2-3-6	30	p_2	1-2	20
p_3	1-3-5	10	p_4	1-7-6	40
p_5	1-3	30	p_6	2-1	20
p_7	2-3-4	25	p_8	3-4	30
p_9	3-6	10	p_{10}	3-6-7	10
p_{11}	3-4-5	20	p_{12}	3-5	10
p_{13}	6-5-4	20	p_{14}	5-3	30
p_{15}	4-3	10	p_{16}	5-4-3-6	20
p_{17}	6-7-1	20	p_{18}	7-6-5	30
p_{19}	7-1-2	25	p_{20}	5-6-7-1	20
p_{21}	4-3-6	25	Total		455

$$X_j = \begin{cases} 1 & \text{if there is a facility located on node } j \quad j \in N \\ 0 & \text{otherwise} \end{cases}$$

and

$$Y_p = \begin{cases} 1 & \text{if at least one facility is located on path } p \quad p \in P \\ 0 & \text{otherwise} \end{cases}$$

The problem is

$$\max \sum_{p \in P} f_p Y_p$$

Subject to

$$\sum_{j=1}^{n} X_j = m \tag{17.1}$$

$$\sum_{j \in p} X_j \geq Y_p \quad p \in P \tag{17.2}$$

$$X_j = 0, 1 \quad Y_p = 0, 1 \tag{17.3}$$

Since Y_p is equal to 1 only if there is at least one facility located on path p, the flow rate f_p will be included exactly once (and double counting will be avoided) in the objective function when $Y_p = 1$. Constraints (17.2) guarantee that exactly m facilities are located. Constraints (17.3) make sure that if no facility is located on path p, Y_p is equal to zero and if there is at least one facility on path p since the objective function is a maximization of the sum of non negative terms, $Y_p = 1$.

For a branch and bound algorithm to solve the problem the reader can refer to [96]. Next we present a greedy heuristics to solve problem (P_1).

The Greedy Heuristic

The main idea of the greedy heuristic is to sequentially locate the facilities at nodes which intercept the remaining unintercepted flow in the network. At each step of the heuristic the node that intercepts the maximum remaining flow is selected as the next new location. We use P_j to denote the set of all paths that include node j.

The greedy heuristic for (P_1)

Step 0: $X^0 = \emptyset, N^0 = N$

Step 1: for $t = 1, \ldots, m$: $j_t = arg\ max_{j \in N^{t-1} - X^{t-1}} F(X^{t-1} \cup \{j\})$; $X^t = X^{t-1} \cup \{j_t\}$; $N^t = N^{t-1} - \{j_t\}$.

Step 2: Optimal solution is $X^* = X^m$ and optimal objective function is $F(X^*)$.

To illustrate the algorithm we refer to the example in Figure 1 for $m = 2$:
Step 1: $t = 1$ $j_1 = 3$ $(F(3) = 260)$ $X^1 = \{3\}$, $N^1 = \{1, 2, 4, 5, 6, 7\}$
$t = 2$ $j_2 = 1(F(1, 3) = 405)$ $X^2 = \{1, 3\}, N^2 = \{2, 4, 5, 6, 7\}$
Step 2: $X^* = \{1, 3\}$ $F(X^*) = 405$.
It can be verified that $X^* = \{1, 3\}$ is also the optimal solution.

Finally we note that extensive computational experience with the heuristic shows that, in most cases, it either provides the optimal solution, or a solution that is very close to the optimial one. We postpone the discussion on the worst case performance of the heuristic until Section 17.2.2.

Finding the required number of facilities

A related problem to (P_1) is the problem of finding the minimum number of service facilities that are needed to "capture an externally specified fraction of the flow." Let $1 - \alpha$ be the fraction of total flow required to be intercepted and $C_\alpha \equiv (1 - \alpha) \sum_{p \in P} f_p$. Since the nodal optimality holds also for our new problem, it can be formulated as:

$$min \sum_{j=1}^n X_j \qquad\qquad (P_2)$$
Subject to: $\sum_{j \in p} X_j \geq Y_p\ \forall p \in P$; $\sum_{p \in P} Y_p f_p \geq C_\alpha^*$; $X_j = 0, 1\ j \in N$, $Y_p = 0, 1\ p \in P$.

To find the optimal solution to the problem we utilize the greedy heuristic presented in the last section. The main idea is to avoid solving the exact problem (P_1) if possible. We define the relaxed problem RP_1 of (P_1) with noninteger constraints $0 \leq X_j \leq 1, 0 \leq Y_p \leq 1$. We call m_α^* the minimum number of facilities required to intercept C_α^*.

An algorithm to solve (P_2)

Step 0: $X^0 = \emptyset, N^0 = N, m_\alpha^* = 0, t = 0$
Step 1: $t = t + 1$: $j_t = arg\ max_{j \in N^{t-1} - X^{t-1}} F(X^{t-1} \cup \{j\})$; $X^t = X^{t-1} \cup \{j_t\}$; $N^t = N^{t-1} - \{j_t\}$.

If $F(X^t) \leq C_\alpha^*$ repeat step 1.

Step 2: Set $m_\alpha^* = t$ $m = m_\alpha^* - 1$ and solve RP$_1$

If optimal solution is $\leq C_\alpha^*$ then m_α^* is optimal

else solve (P_1)

If optimal solution of (P_1) is $> C_\alpha^*$ then $t = t - 1$, repeat step 2.

else m_α^* is optimal

end if

end if

As an example, suppose $C_\alpha^* = 410$ is specified for the network in Figure 1. Applying the algorithm, we get:

$t = 1, j_1 = 3$ $X^1 = \{3\}$ $N^1 = \{1, 2, 4, 5, 6, 7\}$, $F(X^1) = 260 < 410$

$t = 2, j_2 = 1$ $X^2 = \{1, 3\}$ $N^2 = \{2, 4, 5, 6, 7\}$, $F(X^2) = 405 < 410$

$t = 3, j_3 = 6$ $X^3 = \{1, 3, 6\}$ $N^3 = \{2, 4, 5, 7\}$, $F(X^3) = 410 > 405$

$m_\alpha^* = 3, m = 2$, optimal solution of RP$_1$ = $405 < 410$. The optimal solution is $m_\alpha^* = 3, X^3 = \{1, 3, 6\}$

Extensive computational experience show that over 75% of the time there is no need at all to solve problem (P_1) exactly. In all cases tested, the optimal number of facilities was either the same as or one less than that obtained with the greedy heuristic component of the algorithm.

17.2.2   Models That Allow Deviations From Pre-Planned Tours

In this section the assumption that customers do not deviate from their pre-planned trips is relaxed. Thus, a customer may deviate from his (her) pre-planned trip to travel to the facility and consume the service there. Therefore locations 'not far' from pre-planned paths may be also candidates for locating the facilities. Here, in contrast to the problem discussed earlier, interception and consumption of the service are equivalent.

We discuss three generalizations of problem (P_1) presented in [86] where deviations from pre-planned tours are considered. In the first one, called the "Delta Coverage Problem", flow along any path p is regarded as intercepted only if there is a facility at a distance of not more than a pre-specified number Δ from p (measured from the closest point on p to the facility). In the second generalization, which is called the "Maximal Market Size Problem", it is assumed that as the deviation distance (incurred when a customer deviates from a pre-planned tour) decreases, more and more flow will be intercepted. The problem is to maximize the expected number of customers who become actual users of the facilities.

In the third generalization, called the "Minimize Expected Inconvenience," it is assumed that all potential customers will deviate if necessary to a closest facility regardless of the actual deviation distance. The problem is to locate the facilities so as to minimize the total deviation distance per unit of time.

The Delta Coverage Problem

It is quite easy to show that the Delta Coverage Problem is a special case of problem (P_1). Let us define $I'(X, p)$ to be equal to 1 if there exists a node j on path p whose shortest distance to the closest facility in X is less than or equal to Δ. The constant Δ is a given number that is specified by the decision maker. In other words:

$$I'(X, p) = \begin{cases} 1 & \text{if there exists } j \in p \text{ such that } d(j, X) \leq \Delta \\ 0 & \text{otherwise} \end{cases}$$

where $d(j, X) = min_{x \in X}\{d(j, x)\}$.

Therefore, the problem which we call (P_3), is identical to (P_1), except for $I'(X, p)$ which replaces $I(X, p)$. Next we show how the Delta Coverage Problem can also be formulated as a linear integer programming problem. We define the set N' to include the set of nodes in N, and all points along the arcs in G that are exactly Δ units of distance away from a node.

As an example, consider again the network in Figure 1. Now the number next to each arc represents the length of this arc. If $\Delta = 2$ then $N' = \{$node 1, node 2, $\cdots$, node 7, (1,2,7), (1,2,2), (2,2,3), (3,2,2), (3,2,4), (3,2,5), (3,2,6), (4,2,5), (5,2,3), (5,2,4), (5,2,6), (6,2,5), (7,2,1)$\}$ (where (a, c, b) represents a point on arc (a, b) that is located at distance of c from node a). It is not difficult to show (see [86]) that an optimal set of locations to the problem exists in N'. Therefore the problem can be formulated identically to the integer programming version of (P_1), with the minor modification that X_j is now defined as a binary variable for any $j \in N'$ (instead for $j \in N$).

The Deviation Distance

The Deviation Distance can be defined as the additional distance incurred when a customer deviates from his (her) pre-planned trip. Suppose a customer travels on a pre-planned $p \in P$. For any path $p = \{1, 2, \cdots, l\}$ let node 1 designates the origin and node l designates the destination. We distinguish between two main cases:

1. All travel occur on shortest paths. In this case the deviation distance $D(p, X)$ is given by:

$$D(p, X) = \min_{x \in X} D(p, x) \equiv \min_{x \in X}\{d(1, x) + d(x, l) - d(1, l)\}.$$

Here we assume that a customer wishing to visit a facility will first take the shortest path to the facility, and then from the facility to the destination. The sum of these shortest distance minus the shortest distance from the origin to the destination is the deviation distance. As an example, refer to Figure 1. Let $p = \{2, 3, 6, 7\}$ and suppose there are 2 facilities at $X = \{1, 5\}$. Then $D(p, X) = min\{4 + 5 - 8; 8 + 4 - 8\} = 1$.

2. Travel in the network may not occur on shortest paths only but all nodes in p must be visited in a given sequence. The reason for maintaining the sequence could be for instance because of ordered deliveries or pickups at the nodes that must be visited. Here the deviation distance is given by:

$$D(p, X) = min_{j \in p, j \neq l}\{min_{x \in X}\{d(j, x) + d(x, j + 1) - l(j, j + 1)\}\},$$

where $l(j, j + 1)$ is the length of link $(j, j + 1)$. For example, refer to path $p = \{2, 1, 7\}$ and assume $X = \{3, 6\}$ in Figure 1. Then $D(p, X) = min\{3 + 2 - 4;\ 7 + 6 - 4;\ 2 + 5 - 5;\ 6 + 1 - 5\} = 1$.

The Maximal Market Size Problem

The main assumption of the Maximal Market Size Problem is that as the deviation distance get larger, less and less customers will visit (their) closest facility. We denote by $f_p g(D(p, X))$ the flow of customers along path p who visit the closest facility in X. The function $g(D(p, X))$ (with $g(0) \equiv 1$) represents the fraction of customers traveling on path p who deviate to the closest facility in X. The problem is:

$$max_{X \in G} \sum_{p \in P} f_p g(D(p, X)) \tag{P_4}$$

Similarly to problem (P_1), the result that an optimal set of locations exists in N also holds for problem (P_4). A formal proof of this result is included in [86]. The proof is based on the fact that if x is an interior point on arc (a, b), and $d(a, x) = \theta d(a, b)$, $\theta \in (0, 1)$, then $D(p, x)$ - the deviation distance from p to x, can be shown to be a piecewise linear function of θ. Since g is nondecreasing convex function, $g(D(p, x))$ is a convex function of θ as well. Therefore, the objective function in (P_4) is a convex function of θ and its maximum is obtained at $\theta = 0$ or $\theta = 1$.

Similar to problem (P_1) (and (P_3)), a greedy heuristic can be developed for problem (P_4). Essentially the heuristics are identical, except for the function $F(X)$ $X \subseteq N$ which is again a set function, but is now equal to $\sum_{p \in P} f_p g(D(p, X))$.

The Problem of Minimizing Expected Inconvenience

The main assumption here is that all customers will travel from their preplanned trips to a service facility which is closest in terms of the deviation distance. The problem (called (P_5)), which was first posed in [582], is to minimize the total (expected) deviation distance travelled per unit of time:

$$min_{X \subseteq G} \sum_{p \in P} f_p D(p, X) \tag{P_5}$$

Obviously, the objective function of (P_5) is identical to that of the m-median problem [509], and thus all the results available for the m-median problem can be applied to (P_5) as well.

An Example

FIGURE 2. A 4-node network

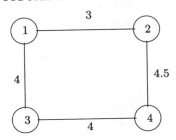

TABLE 17.2. Deviation Distances

$p \setminus x$	1	2	3	4	f_p
1-2-4	0	0	.5	0	100
3-1-2	0	0	0	1.5	150
4-3	7.5	7.5	0	0	75
4-2-1	0	0	.5	0	200
weighted average	562.5	562.5	150	225	

To illustrate problems $(P_3), (P_4)$ and (P_5) we refer to the very simple network in Figure 2. Suppose there are four paths with positive flow rates in the network: $P = \{1\text{-}2\text{-}4,\ 3\text{-}1\text{-}2,\ 4\text{-}3,\ 4\text{-}2\text{-}1\}$ with flow rates per unit of time given by $f_{1\text{-}2\text{-}4} = 100, f_{3\text{-}1\text{-}2} = 150, f_{4\text{-}3} = 75$ and $f_{4\text{-}2\text{-}1} = 200$. Suppose we locate only one facility, i.e. $m = 1$.

We start with the problem (P_3) for $\Delta = 0$ (which is essentially problem (P_1)). Since for nodes 1 and 2 the total flow intercepted is: $\sum_{p \in P} f_p I(1, p)$ $= 100 + 150 + 0 + 200 = 450$, and for nodes 3 and 4 the total flow intercepted is 225 and 375, respectively, either node 1 or 2 can be chosen as the optimal location. When $\Delta = 2$, the total flow $\sum_{p \in P} f_p I'(X, p) = 525$ is achieved by the points from the set $N' = \{(3, 2, 1), (3, 2, 4)\}$. Either one of them can be chosen as the optimal location (notice the absence of optimal nodal solutions).

For problems (P_4) and (P_5) we have to calculate deviation distances. We assume that travel allowed only along shortest paths. The following deviation distances are given in Table 17.2 (only case (i) distances are considered).

The objective function $\sum_{p \in P} f_p D(p, x)$ for $x = 1, 2, 3, 4$ is given in the bottom of Table 17.2, and node 3 is seen to be the optimal location for problem (P_5). Finally for problem (P_4) with the deviation function $g(D(p, x)) = e^{-bD(p, X)}$, the values of the objective function are $450 + 75e^{-7.5b}, 450 + 75e^{-7.5b}, 225 + 300e^{-.5b}, 375 + 150e^{-1.5b}$ for nodes 1, 2, 3, 4, respectively. Therefore, for sufficiency small b values ($b \leq \approx .49$) node 3 is optimal, otherwise node 1 or 2 is optimal.

Worst Case Analysis for the Greedy Heuristics

We showed that problems $(P_1), (P_3)$ and (P_4) can be formulated in the form $max_{X \subseteq \overline{N} = m} F(X)$, where for problems (P_1) and (P_4), $\overline{N} \equiv N$; for problem (P_3), $\overline{N} = N'$; for problems (P_1) and (P_3), $F(X) \equiv \sum_{p \in P} f_p I(p, X)$ (I' for (P_3)); and for problem (P_4), $F(X) = \sum_{p \in P} f_p g(D(p, X))$. We assume throughout our discussion in the current section that the first case of deviation distance holds, and thus $D(p, x)$ can be written as:

$$D(p, x) = min_{i,j \in p}[d(i, x) + d(x, j) - d(i, j)].$$

Based on the work of Nemhauser et. al. [888] it is sufficient to show that if $F(X)$ is submodular and nondecreasing then

$$\frac{Z^* - Z_G}{Z^*} \leq (1 - \frac{1}{m})^m \leq \frac{1}{e} \cong .37,$$

where Z^* is the optimal value of the problem at hand $((P_1), (P_3)$ or $(P_4))$, and Z_G is the result obtained by the corresponding greedy heuristic. Recall that a set function $F(X)$ is *submodular* if for any X^1 and X^2 $X^1, X^2 \subseteq \overline{N}$,

$$F(X^1 \cap X^2) + F(X^1 \cup X^2) \leq F(X^1) + F(X^2),$$

and *nondecreasing* if for any X^1 and X^2 $X' \subseteq X^2 \subseteq \overline{N}$, $F(X') \leq F(X^2)$. We can show that $F(X)$ is submodular and nondecreasing for (P_3) (and thus also for (P_1) when $\Delta = 0$), if $f_p \geq 0$ $\forall p \in P$. Let $N(x, \Delta)$ be the set of all nodes in N that are within radius Δ of x, i.e. $N(x, \Delta) = \{i \in N; d(x, i) \leq \Delta\}$. Let $N(X, \Delta) = \cup_{x \in X} N(x, \Delta)$ be the set of nodes that are "covered" by any subset X of potential facility locations. Obviously a path that includes at least one node in $N(X, \Delta)$ can be regarded as intercepted. First, if $X^1 \subseteq X^2$, $N(X^1, \Delta) \subseteq N(X^2, \Delta)$ and thus $F(X^1) \leq F(X^2)$ if $f_p \geq 0$, which implies that $F(X)$ is nondecreasing.

Let $P_X = \{p \in P; p \cap N(X, \Delta) = \emptyset\}$. Since $X^1 \subseteq X^2, N(X^1, \Delta) \subseteq N(X^2, \Delta)$ and thus $P_{X^2} \subseteq P_{X^1}$. To show submodularity, it is sufficient to show that if $k \notin X^2, F(X^2 \cup \{k\}) - F(X^2) \leq F(X^1 \cup \{k\}) - F(X^1)$. But $F(X^2 \cup \{k\}) - F(X^2) = \sum_{p \in P_{X^2}, \ p \cap N(k, \Delta) \neq \emptyset} f_p \leq \sum_{p \in P_{X^1}, \ p \cap N(k, \Delta) \neq \emptyset} f_p$ $= F(X^1 \cup \{k\}) - F(X^1)$.

For problem (P_4), since $D(p, X^1) \geq D(p, X^2)$ if $X^1 \subseteq X^2$ then $g(D(p, X^1)) \leq g(D(p, X^2))$ and thus $g(D(p, X))$ is nondecreasing, and consequently, since all $f_p \geq 0$, $F(X)$ is also nondecreasing. Now we will show that $g(D(p, X))$ is submodular. Consider $X^1 \subseteq X^2$ and $k \notin X^2$. Obviously, $D(p, X^2 \cup \{k\}) = min\{D(p, X^2), \alpha\}$, where $\alpha \equiv min_{i,j \in p}[d(i, k) + d(k, j) - d(i, j)]$. It is sufficient to show that:

$$g(min[D(p, X^2), \alpha]) - g(D(p, X^2)) \leq g(min[D(p, X^1), \alpha]) - g(D(p, X^1)).$$
$$(17.4)$$

We can distinguish three cases: (i) $\alpha \geq D(p, X^1) \geq D(p, X^2)$; (ii) $D(p, X^1) \geq \alpha \geq D(p, X^2)$; and (iii) $D(p, X^1) \geq d(p, X^1) \geq \alpha$.

In all three cases it is easy to show the condition (17.4) is satisfied. For example for case (i) condition (17.4) becomes $g(D(p, X^2)) - g(D(p, X^2)) \leq g(D(p, X^1)) - g(D(p, X^1))$ which is trivially satisfied. In the same way we can show that the condition holds for the other 2 cases. Therefore $g(D(p, X))$ is submodular, but since $f_p \geq 0$, it follows that $F(X)$ is also submodular as it is the sum of submodular set functions.

For problem (P_5) which is a variant of the p-median problem it is well known (see [887]) that the greedy heuristic solutions can be are arbitrarily far from the optimal solution.

Finally, we note that for the problem (P_1), it can be shown (see [86]) that the bound of $(1 - \frac{1}{m})^m$ is tight.

17.2.3   The Maximum Market Size Problem with Congestion

A major assumption of the Maximal Market Size problem (P_4) is that the time customers spend at the facility site is negligible, and thus $g(D(p, X))$ – the fraction of customers that travel to the facility – is a function of the deviation distance only. However this assumption becomes unrealistic when the service capacity of the facility is limited. Therefore, we will incorporate into the deviation distance the time that customers spend at the facility.

We assume that the number of customers travelling to a service facility is a Poisson distributed random variable, and the service time at the facility is a random variable with a general distribution. The system is therefore a $M/G/m$ queueing system with the additional difficulty that the general service time distributions at the facilities are not identical (for this queuing system a close form formula for the mean waiting time of customers in the system is not known). However when there is a single facility in the system i.e. the queuing system is $M/G/1$ a closed form expression of the mean waiting time of customers is readily available. We continue to assume that the flow from any path p to a facility is a non-increasing convex function of the (now modified) deviation distance. We further assume that we are given a finite set S of potential locations for the facilities (e.g. $S = N$). We will show how to optimally solve the problem for the single server case. Using the exact scheme to solve the problem for a single server, a location-allocation-type heuristic to solve the problem for a general number of facilities is developed in [85].

In this section we restrict our discussion to the single server case only.

The Single Server Problem

Suppose the facility is staffed with a single server who is capable of providing a general distributed service time with a known mean $\frac{1}{\mu}$ and a variance σ^2. The number of customers travelling per unit of time from any path p to the facility located at x is denoted by $S_p(x)$, and its mean is denoted by $\lambda_p(x)$. The new deviation distance denoted by $D'(p,x)$ incorporates into the orginal deviation distance $D(p,x)$ the mean waiting time of customers in the system $W(x)$: $D'(p,x) = D(p,x) + W(x)$. The total number of customers travelling to x from all paths in P, $S(x)$ is also Poisson distributed (since all f_p are non-negative) with a rate of $\lambda(x) = \sum_{p\in P}\lambda_p(x)$ per unit of time. Since the system is an $M/G/1$, it is known that (see [702]),

$$W(x) = \frac{\lambda(x)[\mu^2\sigma^2 - 1] + 2\mu}{2\mu(\mu - \lambda(x))}, \quad \lambda(x) < \mu.$$

Now we can formulate the problem of finding the location x that maximizes the total flow. The problem, called (P_6), is:

$\max_{x\in S} \lambda(x) = \sum_{p\in P} f_p g(D'(x,p)).$ $\hspace{2cm}(P_6)$

An interesting question is whether it is possible to indentify cases where the relative ranking of locations is independent of the service distribution parameters μ and σ^2. The answer to this question is positive [85]. It is shown there that if the service process is identical at each potential location, and if $\lambda(x)$ can be written as a product or sum (division or substraction) of two functions: $h_1(x)$ which depends only on the location x, and $h_2(W(x|\mu,\sigma^2)$ which depends on $W(x)$ (and thus also on μ and σ^2), then problem (P_6) is reduced to the problem of minimizing $h_1(x)$ over $x \in S$ (which is essentially problem (P_4)). Examples of the type of functions discussed above include the linear function: $\lambda(x) = \sum_{p\in P} f_p(1 - a(D(x,p) + W(x))$ (a is sufficiently small), or the negative exponential function $\lambda(x) = \sum_{p\in P} f_p e^{-b(D(x,p)+W(x))}$, $b > 0$. For the linear function: $h_1(x) = \sum_{p\in P} f_p[1 - a\,D(x,p)]$, $h_2(W(x|\mu,\sigma^2)) = a\sum_{p\in P} f_p W(x|\mu,\sigma^2)$, and $\lambda(x)$ is the sum of $h_1(x)$ and $h_2(W(x|\mu,\sigma^2)$. For the negative exponential function, $h_1(x) = \sum_{p\in P} f_p e^{-b(D(x,p))}$, $h_2(W(x|\mu,\sigma^2)) = e^{-b\,W(x|\mu,\sigma^2)}$ and $\lambda(x)$ is the product of the two functions.

Solving the Single Server Problem

It is not easy to evaluate the objective function $\lambda(x)$ for a given location $x \in S$ since $\lambda(x)$ appears on both sides of the equation:

$$\lambda(x) = \sum_{p\in P} f_p g\left(D(p,x) + \frac{\lambda(x)(\mu^2\sigma^2 - 1) + 2\mu}{2\mu(\mu - \lambda(x))}\right), \quad \lambda(x) < \mu.$$

Unfortunately it is impossible to isolate $\lambda(x)$ to only one side. Note, however, that the left hand side is a monotone increasing function of $\lambda(x)$, while the right hand side is a monotone decreasing function of $\lambda(x)$. Therefore, there is a unique intersection of the two functions. Moreover, this solution can be found by a simple binary search algorithm (see [85] for details).

Therefore, we can find the optimal solution of (P_6) by evaluating $\lambda(x)\ \forall x \in S$ and choosing the point x that yields the largest value. To illustrate, we refer again to the 4-node example discussed in Section 17.2.3, but now we consider $g(D(p,x)) = \frac{1}{D(p,x)+W(x)}$ and $S = N$. Suppose $\mu = 4$ and $\sigma^2 = 1$.

Applying binary search, we obtain $\lambda(1) = 3.936$, $\lambda(2) = 3.936$, $\lambda(3) = 3.938$, $\lambda(4) = 3.936$, and thus node 3 is the optimal location.

17.2.4   Allowing Multiple Interseptions

In this section we relax the assumption that each customer may only be intercepted by at most one facility. Now we assume that the total amount of flow intercepted along any path p is a function of the number of facilities located on that path. We denote by $q_p(n_p)$ the fraction of customers intercepted along p if there are n_p facilities on that path $(q_p(0) \equiv 0)$. We assume that $q_p(n_p)$ is a concave nondecreasing function that attains only a finited number of values. The assumption of concavity which means that the marginal contribution of additional facilities on p is decreasing seems realistic and consistent with the marketing literature [791]. Let b_p be the smallest integer such that $q_p(n_p) = 1$ for $n_p \geq b_p$.

Two problems are considered in [40]. In the first problem, (P_7), the objective is to locate at most m facilities in order to maximize the total amount of flow intercepted. In the second problem, (P_8), the objective is to find the minimum number of facilities needed to intercept the total amount of flow available on the network.

Again, as for the problem without double counting (where $b_p = 1 \; \forall p \in P$), it is easy to show that the search for optimal locations can be reduced to N. However, in contrast to the problems discussed until now it may be necessary to locate more than one facility at the same node in order to maintain optimality of nodal solutions. To illustrate this, we can refer to a simple path network composed of two nodes (1 and 2), with only path $1 - 2$ having non-zero flow.

Suppose $q_{1-2}(1) = \frac{1}{2}$, $q_{1-2}(2) = \frac{5}{6}$ and $q_{1-2}(i) = 1 \; i \geq 3$, i.e. $b_{1-2} = 3$. Then for both (P_7) and (P_8) with $m = 3$, if we cannot locate more than one facility at a node, the optimal solution is to locate, in addition to the two facilities at nodes 1 and 2, one facility at an inner point on arc $(1, 2)$.

Formulations and Analysis

To formulate the problem we require several more definitions. Let $\bar{x}(j)$ be the number of facilities located at node j given a location vector $\bar{x}$. Let a_{jp} be an indicator coefficient that assumes the value 1 when node j is included in p and 0 otherwise. Problems (P_7) and (P_8) can be now formulated as Integer Programs:

$$\max \sum_{p \in P} f_p q_p(n_p) \qquad (P_7)$$

Subject to: $n_p = \sum_{j \in N} a_{jp} \bar{x}(j) \;\; \forall p \in P$; $\sum_{j \in N} \bar{x}(j) \leq m$; $\bar{x}(j)$, n_p nonnegative integer variables $j \in N$ $p \in P$,

and

$$\min \sum_{j \in N} \bar{x}(j) \qquad (P_8)$$

Subject to: $\sum_{j \in N} a_{jp} \bar{x}(j) \geq b_p \;\; \forall p \in P$; $\bar{x}(j)$ nonnegative integer $j \in N$.

The first set of constraints of (P_7) counts the exact number of facilities located on path p. This number is required for calculating the fraction of customer flow intercepted in the nonlinear objective function of (P_7), and is required to be 'large' enough so that all flow is intercepted in (P_8).

Both problems (P_7) and (P_8) are NP-hard on general networks since the Node Cover Problem (see [913]) which is NP-hard can be easily reduced to either one of them.

The greedy heuristic that sequentially adds facilities that improve the objective function the most is found to be very efficient for solving (P_7). Moreover it is shown in [40] that the worst case error of $\frac{1}{e}$ also holds for (P_7) even though the function optimized in (P_7) is not a submodular set function (it is actually not a set function since the same location can appear several times in the location vector).

Problem (P_7) is NP-hard on a tree even without double counting (it is shown in [40] that the Node Cover Problem can be reduced to it). However problem (P_8) can be solved on a tree by a polynomial algorithm as shown below.

Solving Problem (P_8) on a tree

The main idea of the polynomial algorithm for (P_8) presented here is to delay the decision to locate facilities as late as possible during the implementation of the algorithm. Actually facilities will be located at a certain node only if a decision not to locate them at that node will result in an infeasible solution. Two definitions are required for the algorithm. A path $p \in P$ is called <u>active</u> if the number of facilites located on p is less than b_p. An active path is called <u>critical</u> with respect to node v if v is the only node of the 'current' tree that belongs to path p. Let $T(t)$ denote the tree in iteration t of the algorithm, $T(0) \equiv T$.

An algorithm to solve (P_8) on a tree

Step 0: $t = 0$

Step 1: choose any tip w of $T(t)$. If there exists on $T(t)$ a critical path in P (with respect to w) locate $\bar{x}^*(w) = \max\{b_p;\ p$ is critical with respect to $w\}$ facilities at node w. For any path $p \in P$ that includes node w, set $b_p = \max(0, b_p - \bar{x}^*(w))$. Delete all paths with $b_p = 0$ from P.

Step 2: if $P = \emptyset$ stop. Otherwise set $t = t + 1$. Let $T(t) = T(t-1) - [w, v]$ where $[w, v]$ is the link that connects w to node v (only one node v can exist) excluding node w. Go to Step 1.

As an example, refer to Figure 3 which depicts a tree with 7 nodes where $P = \{1\text{-}2\text{-}3\text{-}4,\ 2\text{-}5\text{-}7,\ 4\text{-}3\text{-}2\text{-}5\text{-}6,\ 7\text{-}5\text{-}6\}, b_{1-2-3-4} = 3, b_{2-5-7} = 2, b_{4-3-2-5-6} = 4, b_{7-5-6} = 1$. The algorithm proceeds as follows:

Step 0 : $t = 0$, Step 1 : node 1 is chosen. No critical path with respect to node 1 exists. Step 2 : $t = 1$ $T(1) = T(0) - [1, 2)$. Step 1 : node 4 is chosen. No critical path with respect to node 4 exists. Step 2 : $t = 2$ $T(2) = T(1) - [4, 3)$. Step 1 : node 3 is chosen. No critical path with

FIGURE 3. A 7-node tree.

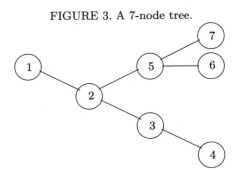

respect to node 3 exists. Step 2 : $t = 3$ $T(3) = T(2) - [3, 2)$. Step 1 : node 2 is chosen. Path 1-2-3-4 is critical with respect to node 2, $\bar{x}^*(2) = 3$, $b_{1-2-3-4} = 0$, $b_{2-5-7} = 0$, $b_{4-3-2-5-6} = 1$, $P = \{4\text{-}3\text{-}2\text{-}5\text{-}6, 7\text{-}5\text{-}6\}$. Step 2 : $t = 4$ $T(4) = T(3) - [2, 5)$. Step 1 : node 6 is chosen. No critical path with respect to node 6 exists. Step 2 : $t = 5$ $T(5) = T(4) - [6, 5)$. Step 1 : node 5 is chosen. Path $4 - 3 - 2 - 5 - 6$ is critical with respect to node 5, $\bar{x}^*(5) = 1$, $b_{4-3-2-5-6} = 0$, $b_{7-5-6} = 0$ $P = \emptyset$. Step 2 : stop. Therefore the optimal solution is to locate three facilities at node 2 and one facility at node 5.

17.2.5 Testing (P_1) on a Real-World Network

The early demonstrations of (P_1) used small test networks. The problem was first tested on a real urban network [586]. Real world transportation systems differ from contrived test problems in two important ways. First, urban transportation networks are much larger, and raise the question of our ability to solve the model effectively. Second, perhaps more importantly, urban transportation systems are not randomly structured; travel patterns tend to focus on central business districts, and movement tends to converge from local streets onto larger arterial roads and expressway systems. Such order could alter the operation and results of the problem substantially.

The problem was tested in the road network, with morning-peak traffic patterns, of Edmonton, Canada. The *City of Edmonton* provided estimates of intrazonal traffic, travel times and distances aggregated into 177 traffic zones. They provided full two-way network topology and link lengths thereby defining a system of 23350 non-zero flow pairs, 703 nodes and 2198 links; a large problem in terms of some of the solution procedures.

In the study [586] the efficiency and robustness of exact, greedy, and Teitz and Bart [1123] vertex substitution procedures were compared. They solved optimally for smaller problems ($p = 1, ..., 15$) with commercial software which relaxes the 0/1 condition, solves the problem as a continuous linear program, and then resolves non-binary solution variables through branch and abound. To do so *days* of mainframe CPU time were required. They solved larger problems ($p = 16, ..., 50$) with the vertex substitution heuristic (VSH), running the model many times from random starting solutions. Robust for small values of p, the VSH's performance deteriorated for larger values, bringing into question its robustness for problems beyond the reach of the optimizing procedure. Computation times were also high for the VSH.

They tested the greedy algorithm which locates facilities one by one, selecting the location at any stage which intercepts the most as yet unintercepted flow. The algorithm is very efficient, solving for $p = 1, ..., 50$ in 15 minutes on a 486/33. They compared the greedy algorithm's performance on the Edmonton data with the best solution (the optimal or best VSH solution) for $p = 1, ..., 50$; concluding that it is very robust. The greedy algorithm's performance also exceeds Berman et al.'s [86] worst case bound. They suggested that the algorithm's excellent performance is at least in part due to the structure of flows within the network. They hypothesized that in real world networks, with actual traffic flows bundled onto major arteries, the greedy heuristic's robustness will far exceed its theoretical limits.

The term *naive* was used in [586] to describe a strategy which completely ignores the problem of cannibalization and simply locates at the p intersections with the greatest flows. They compared the flows the best solution intercepts with those that the naive strategy expects to intercept and that actually would intercept.

Also illustrated was the spatial nature of cannibalization by mapping optimal and naive $p = 15$ solutions. When double counting is ignored, clusters of facilities are located in the central business district and on a major expressway, both major areas where early morning traffic is bundled.

The 15 optimally located facilities (double counting is not ignored) are much more dispersed; they are able, at only 15 of the network's 703 nodes, to intercept 68.9 percent of all flows.

In the study [586] the focus was on indentifying the flow cannibalization process, (intercepting flows more than once at the expense of flows elsewhere in the network not being intercepted at all) through the concept of flow intercepting *redundancy*. The naive strategy has a greater proclivity to intercept flows more than once, and through this redundancy, to leave more flows unserved than the greedy and optimal approaches. In their $p = 15$ solution, the naive model intercepts almost two percent of flows nine times, yet fails to intercept 53.8 percent of flows at all, whereas the greedy and optimal procedures intercept 45.8 percent of network flows only once, and none more than four times.

Hodgson's [583] observations that the greedy heuristic is very robust and that flow cannibalization is a serious problem are bolstered in this real world situation.

17.3   Probabilistic Flow Interception Problems

The data requirements for the model described in the previous section include the knowledge of the flow along all paths in the network that carry non-zero flow. This requirement can be very restrictive, since the number of such paths might be extremely large (generally, it is exponential in the number of nodes). Moreover, the path flow data is almost impossible to measure directly. Indeed, the data requirements make the deterministic formulation tractable only for the cases where the number of paths with non-zero flow is known, or can be assumed to be reasonably small (an important special case is when all flow occurs along the shortest paths connecting the origin-destination pairs).

In this section we re-formulate the Flow Interception Problem using only the data that can be "reasonably" measured through direct observation of the network. There are two such sources of data:

1. For each node i on the network we can measure the proportion b_i of all customers originating their travel from that node. We will use $\mathbf{b} = (b_1, b_2, \ldots, b_n)^T$ to represent the (column) vector of these quantities. We will refer to $\mathbf{b}$ as the "initial distribution vector" for the network.

2. For each arc (i, j) we can measure the proportion c_{ij} of customers at node i that use this arc to exit that node. Note that $\sum_{j \in N} c_{ij} = 1$ for every i (where c_{ii} represents the proportion of customers at i who complete their travel there). We will refer to quantities c_{ij} as the "turning probabilities" or transition probabilities.

The data requirements described above are much more realistic than the path-flow ones in Section 17.2. In particular, the initial distribution vector $\mathbf{b}$ can be obtained directly from the origin/destination (O/D) flow information, while the turning probabilities can be estimated by a number of techniques, including: direct traffic monitoring of all arc flows, a combination of traffic monitoring/approximation schemes, such as the FLAIR network [739], the traffic allocation procedures such as the STOCH algorithm (see Dial [282]). In addition, the memory storage requirements for this data are quite modest: only $|A|+|N|$ entries need to be stored.

Alternatively, the data given by $\mathbf{b}$ and (c_{ij} can be viewed as directly capturing the probabilistic aspect of the behavior of customers on the network.

The "original" Probabilistic Flow Interception problem (PFIP) was analyzed in Berman, Krass and Xu [89]. A "direct" formulation results in a highly non-linear Integer Programming (IP) problem, which is practically intractable. However, by applying the theory of constrained Markov Decision Processes (MDPs), it is possible to obtain a compact linear IP formulation to which standard procedures (e.g., CPLEX) can be applied.

Additional efficiency can be gained by constructing specialized Branch-and-Bound algorithms.

The basic PFIP model formulation assumes that: (1) all flow passing through a facility node is intercepted; (2) once a customer is intercepted, any subsequent visits to facility nodes are of zero value; (3) the total number of facilities to be located on the network is pre-specified, the fixed costs associated with opening these facilities can be ignored.

The first two of the above restrictions were removed in [90], while the last one was dealt with in [91]. It should be noted, however, that at the present time, no generalizations allowing for deviation of customer flows to nearby facilities, or for waiting lines to form in front of facilities are available - these should be the subject of future work.

Our development throughout this Section is roughly parallel to that of Section 17.2: we start by formulating the basic Probabilistic Flow Interception Problem in Section 17.3.1, the MDP-based approach and the linear IP formulation are discussed in Section 17.3.2. The latter section also outlines a "direct" proof of the correctness of the linear IP formulation, which is based on the fundamental properties of Markov Chains and thus bypasses the constrained MDPs. Section 17.3.3 describes the generalizations of the basic model to allow for the "imperfect" interception capability of some of the facilities (e.g., due to a poor location), the possibility of multiple interception of the same customer which is relevant in many marketing applications (each subsequent visit to the facilities is discounted - this model is related to the deterministic model described in Section 17.2.4), and the fixed costs associated with opening new facilities (the total number of facilities to be located now becomes one of the decision variables). Section 17.3.4 describes the Greedy Heuristic for the general model and presents worst-case bounds under some conditions. The results of the heuristic's performance in an extensive set of computational experiments are also discussed.

17.3.1   Formulation of the Basic Probabilistic Flow Interception Model

As in Section 17.2 above, we consider a transportation network $(N, \mathcal{A})$ where $N = \{1, 2, ..., n\}$ is the set of nodes, and $\mathcal{A}$ is the set of arcs. There are customers traveling on the network from a set of origin nodes to a set of destination nodes. We assume that the actual customer flows on the arcs are unknown, but there is a reliable knowledge about the probability (fraction of time) c_{ij} that a customer leaving node i will visit node j next, for all $i, j \in N$ such that $(i, j) \in A$ (with $c_{ij} = 0$ if $(i, j) \notin A$). It is also assumed that the initial distribution vector $\mathbf{b}$ is known.

Typically, a traveling customer enters the network via a certain node, and travels through a series of nodes until (s)he reaches a *destination* node. The nodes can thus be divided into two groups: *transient* nodes and destination nodes. A transient node is one from which a traveler will certainly visit another node within the network, while a destination node is one where a customer stops travelling (note that the origin nodes are transient). At a first glance, this classification seems incomplete, as it is possible that a node acts as a destination to some of the customers and as a transient node to others. Figure 4(a) gives such an example, where 60% of customers leaving node 1 travel to node 2, and the remaining 40% finish their travel at node 1. Node 1 is thus a 'mixed' node; it cannot be classified as either a 'pure' transient or a 'pure' destination node. Note however that this situation can be avoided by introducing an additional dummy node for each mixed node. In Figure 4(b), a dummy node 3 has been introduced for node 1. With this addition, node 3 is a (pure) destination node and node 1 becomes a (pure) transient node. Thus we can assume that each node can be classified as either a destination or a transient node.

FIGURE 4. Conversion of a 'mixed' node

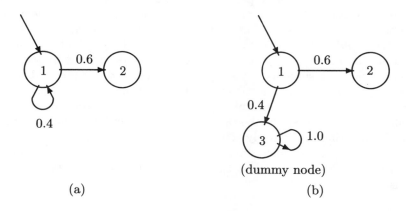

(a)   (b)

We consider the problem of locating a given number m of service facilities on a network so that the maximum expected amount of potential customer flow will be *intercepted* by these facilities (a customer is said to be intercepted if his (her) route passes through a node or arc where a facility is located).

While the facilities can be located anywhere on the network (at nodes, or along arcs), it is clear that an optimal set of locations exists at the nodes of the network (the reasoning here is exactly the same as in the deterministic case).

Given the probabilistic nature of the customer flows, it is convenient to introduce some of the terminology from the theory of Markov chains. We adjoin to the network an artificial "common sink" node 0, and set $b_0 = 0$, $c_{00} = 1$ and c_{i0} equal to the fraction of customers leaving the

FIGURE 5. Illustration of the Network Structure for the PFIP model.

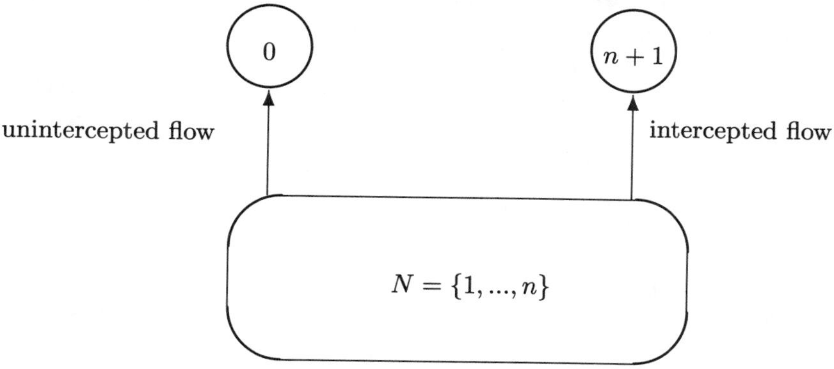

network (completing their travel) upon visiting node i. Thus, $c_{i0} = 1$ if i is a destination node, and $c_{i0} = 0$ if i is a transient node. We also add another artificial sink node $(n + 1)$, with $b_i = c_{i(n+1)} = 0$ for all $i \in N$ and $c_{(n+1)(n+1)} = 1$ (note that this node is initially not reachable from any other node on the network — it will be used later to model the intercepted flow). Thus, the customer flow data can be summarized by a $(n + 2) \times (n + 2)$-dimensional Markov matrix C^+ and an $(n + 2)$-dimensional vector $\mathbf{b}$ (we reserve the notation C to represent the $n \times n$ submatrix of C^+ corresponding to the non-artificial nodes). We make the following assumption:

Assumption 1 *All customers complete their travel in finite time. Equivalently, the only recurrent classes of C^+ are $\{0\}$ and $\{n + 1\}$.*

For the basic model considered in the current section, no "double-counting" is allowed, i.e., customers can be intercepted at most once during their sojourn through the network. Thus, once a customer flow is intercepted by a facility, the flow should be considered "removed" from the network. The following approach is adopted to avoid double-counting. If a facility is located at node i, all outgoing flow from i is immediately re-directed to the "interception" sink $(n + 1)$ by assigning $c_{i(n+1)} = 1$, $c_{ij} = 0$ for $j \neq i$, regardless of the original values of the c_{ij}'s. It is clear that by doing so, any customer flow that reaches node i will not reach any other node. The resulting model is pictorially presented on Figure 5.

The term "facility node" is used to denote a node with a facility, and "empty node" a node without a facility. The decision variables are

$$u_i = \begin{cases} 1 & \text{locate a facility at node } i \\ 0 & \text{otherwise} \end{cases} \quad , \quad i = 1, ..., n.$$

We will refer to the n-dimensional binary vector $\mathbf{u}$ as the "location vector". Let m be the total number of facilities to be located. A location vector is said to be feasible if $\sum_{i \in N} u_i = m$. Note that no facilities are allowed at the artificial nodes 0 and $(n+1)$.

We assume that the number of facilities m is less than that of destination nodes, because otherwise the problem becomes trivial (by locating a facility at every destination node all flow is intercepted). Every location vector $\mathbf{u}$ induces a new Markov transition matrix $C^+(\mathbf{u})$ with entries defined as follows:

$$
\begin{aligned}
c_{ij}(\mathbf{u}) &= c_{ij} && \text{if either } u_i = 0,\ i \in N,\ j \in N \cup \{0, n+1\} \text{ or } i \in \{0, n+1\} \\
c_{ij}(\mathbf{u}) &= 1 && \text{if } u_i = 1,\ i \in N,\ j = n+1 \\
c_{ij}(\mathbf{u}) &= 0 && \text{if } u_i = 1,\ i \in N,\ j \neq n+1.
\end{aligned}
$$

Let $f_{i(n+1)}(\mathbf{u})$ be the probability of ever reaching state $(n+1)$ in the Markov Chain $C^+(\mathbf{u})$ when the initial state is i. The total expected flow intercepted by all facilities under the initial distribution $\mathbf{b}$ is given by

$$
J(\mathbf{u}) = \sum_{i \in N} f_{i(n+1)}(\mathbf{u}) b_i.
$$

$J(\mathbf{u})$ can be re-written in terms of the original data by employing the standard Markov Chain techniques. Let $C(\mathbf{u})$ be the $n \times n$-dimensional submatrix of $C^+(\mathbf{u})$ corresponding to the nodes in N, and let $\mathbf{c_{n+1}}(\mathbf{u})$ be an n-dimensional column vector with entries $c_{i(n+1)}(\mathbf{u})$. Then,

$$
J(\mathbf{u}) = \mathbf{f}(\mathbf{u}) \mathbf{b}^T = \mathbf{b}^T [I - C(\mathbf{u})]^{-1} \mathbf{c_{n+1}}(\mathbf{u}),
$$

where the vector $\mathbf{f}(\mathbf{u})$ has components $f_{i(n+1)}(\mathbf{u})$, $i \in N$. Therefore we get the following Nonlinear Integer Program:

> max $\mathbf{b}^T [I - C(\mathbf{u})]^{-1} \mathbf{c_{n+1}}(\mathbf{u})$ (NLP)
>
> Subject to: $\sum_{i \in N} u_i = m$; $u_i \in \{0, 1\}$, $i \in N$.

Due to the presence of a matrix inversion in the objective, this problem is not easily amenable to the standard integer programming techniques. Furthermore, the matrix $C(\mathbf{u})$ is only defined for the integer values of $\mathbf{u}$; it is not at all clear how to define a "relaxed" version of (NLP) where fractional values of $\mathbf{u}$ would be allowed. This makes the development of a Branch-and-Bound-type procedure for (NLP) all but impossible. These considerations become our main motivation for exploring a different formulation of the problem within the linear integer programming framework, which will be discussed in Section 17.3.2.

Example 17.3.1 We illustrate the preceding discussion by the following example. Consider the 6-node network shown in Figure 6. Nodes 1, 2 and 3 are destination nodes and nodes 4, 5 and 6 are transient. The arrows represent the directions of customer flows and the numbers accompanying the arrows represent the transition probabilities. For instance, a customer, upon leaving node 4, has a transition probability of 0.3 to node 1 (and hence to be absorbed there), a transition probability of 0.2 to node 6, and

FIGURE 6. A 6-node network example. Numbers in parenthesis next to each node represent initial customer distribution probabilities; numbers next to each arc represent the corresponding transition probabilities.

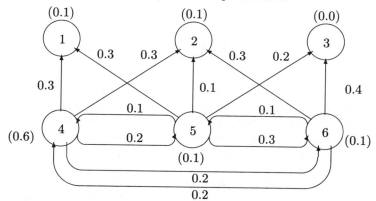

so on. The initial customer distribution is indicated by the figures in parentheses beside the nodes (e.g., node 2 has an initial customer distribution probability of 0.1). The corresponding $\mathbf{b}$ and C^+ are

$$\mathbf{b} = [0 \ 0.1 \ 0.1 \ 0 \ 0.6 \ 0.1 \ 0.1 \ 0]^T$$

and

$$C^+ = \begin{bmatrix} 1 & 0 & 0 & 0 & 0 & 0 & 0 & 0 \\ 1 & 0 & 0 & 0 & 0 & 0 & 0 & 0 \\ 1 & 0 & 0 & 0 & 0 & 0 & 0 & 0 \\ 1 & 0 & 0 & 0 & 0 & 0 & 0 & 0 \\ 0 & .3 & .3 & 0 & 0 & .2 & .2 & 0 \\ 0 & .3 & .1 & .2 & .1 & 0 & .3 & 0 \\ 0 & 0 & .3 & .4 & .2 & .1 & 0 & 0 \\ 0 & 0 & 0 & 0 & 0 & 0 & 0 & 1 \end{bmatrix}.$$

If there are $m = 2$ facilities, and they are placed at nodes 1 and 2 respectively (i.e., $\mathbf{u} = [1 \ 1 \ 0 \ 0 \ 0 \ 0 \ ]^T$), then

$$C(\mathbf{u}) = \begin{bmatrix} 0 & 0 & 0 & 0 & 0 & 0 \\ 0 & 0 & 0 & 0 & 0 & 0 \\ 0 & 0 & 0 & 0 & 0 & 0 \\ .3 & .3 & 0 & 0 & .2 & .2 \\ .3 & .1 & .2 & .1 & 0 & .3 \\ 0 & .3 & .4 & .2 & .1 & 0 \end{bmatrix},$$

and

$$\mathbf{c_{n+1}(u)}^T = [1 \ 1 \ 0 \ 0 \ 0 \ 0].$$

The corresponding interception probability vector $\mathbf{f}(\mathbf{u})$ can be computed to be $\mathbf{f}(\mathbf{u})^T = [1 \; 1 \; 0 \; .8348 \; .6429 \; .5313]$, with $J(\mathbf{u}) = \mathbf{f}(\mathbf{u})^T\mathbf{b} = .8183$, which happens to be the optimal solution, as can be verified by complete enumeration.

Other useful quantities to compute are

$$\mathbf{g}(\mathbf{u})^T = \mathbf{b}^T[I - C(\mathbf{u})]^{-1}. \tag{17.5}$$

For each $i \in N$, $g_i(\mathbf{u})$ is the expected number of visits to node i under the initial distribution $\mathbf{b}$. Note that facility nodes and destination nodes can be visited at most once. Thus, for a facility node, $g_i(\mathbf{u})$ represents the amount of flow intercepted by that facility. For a non-facility destination node, $g_i(\mathbf{u})$ represents the amount of unintercepted flow "escaping" through i. In the example above, $g_1(\mathbf{u}) = .3883, \quad g_2(\mathbf{u}) = .4300, \quad g_3(\mathbf{u}) = .1817$.

In general, the main difficulty in solving the PFIP is the "cannibalization" of flow between facilities: placing new facilities on the network can reduce the amount of flow intercepted at the existing facilities. One exception occurs when facilities are placed at the destination nodes (these facilities can not possibly affect the flows elsewhere on the network). This observation leads to the following bounds summarized in Theorem 1 below.

Theorem 17.3.1 *Let $J(i)$ represent the flow intercepted by a facility at node I when this is the only facility on the network. Let $\mathbf{u}$ be an arbitrary feasible location vector, $\mathbf{u}^*$ be the optimal location vector, and G be the set of destination nodes. Then*

1. *For any $i \in N$, $J(i) \geq g_i(\mathbf{u})$.*

2. *Let $\mathbf{u}$ be such that all facilities are located at the destination nodes. Then*
$$\sum_{\{i:u_i=1\}} J(i) = J(\mathbf{u}).$$

3. *Let $I \subset N$ be a subset of nodes with cardinality m. Then*
$$\max_{I \subset G, \; |I|=m} \sum_{i \in I} J(i) \leq J(\mathbf{u}^*) \leq \max_{|I|=m} \sum_{i \in I} J(i).$$

In our experience, the bounds provided in the last part of the Theorem are quite useful in the context of the Branch-and-Bound solution procedure.

17.3.2 Development of the Linear Integer Programming Formulation

The goal of this section is to re-formulate the Probabilistic Flow Interception Model defined in Section 17.3.1 above as a linear Integer Program.

The main technique employed to obtain this new formulation is the theory of constrained Total-reward Markov Decision Processes (TMDPs). This approach is briefly outlined below. Please refer to [89] and [90] for details.

We start with some basic introductions. Let Γ be a discrete-time Markov Decision Process with a finite state space $S = \{1, \ldots, s\}$, finite action spaces $A(i), i \in S$, immediate rewards $r_{ia}, i \in S, a \in A(i)$ (which are assumed to be non-negative), and the transition probabilities $p_{iaj}, \ i, j \in S, a \in A(i)$. We will use the notation $\pi = (\pi_1, \pi_2, \ldots)$ for a general (possibly randomized and history-dependent) policy in Γ (where π_t is the decision rule used at time t), and h for a deterministic (non-randomized, stationary) policy, with h_i denoting the action taken by h in state $i \in S$. Let X_t, A_t be the random variables representing the state and action at time t, respectively, and let P_π the probability measure induced by policy π. For a policy π the **expected total reward** is defined by

$$\psi_i(\pi) = \sum_{t=1}^{\infty} \sum_{j \in S} \sum_{a \in A(j)} P_\pi(X_t = j, A_t = a | X_1 = i) r_{ia}.$$

We will call Γ equipped with the expected total reward criterion the TMDP model, and policy π^* TMDP-optimal if

$$\psi_i(\pi^*) = \max_\pi \psi_i(\pi) \quad \text{for all } i \in S.$$

Given an instance of PFIP, we define the following TMDP Γ:

$S = N \cup \{0, n+1\}$
$A(i) = \{1, 2\}$ for $i \in N$
$A(0) = A(n+1) = \{1\}$
$r_{ia} = \begin{cases} 1 & \text{if } i \in N, a = 2 \\ 0 & \text{otherwise} \end{cases}$
$p_{iaj} = \begin{cases} c_{ij} & \text{if } i \in N \text{ and } a=1 \\ 1 & \text{if either } i \in N, j=n+1 \ \& \ a=2 \text{ or } i=j \in \{0, n+1\} \ \& \ a=1 \\ 0 & \text{if either } i \in N, j=0, \ \& \ a=2 \text{ or } i \in \{0, n+1\}, i \neq j, \ \& \ a=1. \end{cases}$

Intuitively, for the non-artificial states the first action corresponds to "treating" the state as an empty node, while the second action "treats" the state as a facility node. This suggests the following correspondence between the feasible location vectors in PFIP and certain deterministic policies in $\Gamma(1, \alpha)$: (1) for a given feasible location vector $\mathbf{u}$, let the deterministic policy $h(\mathbf{u})$ be defined by $h_i(\mathbf{u}) = 1$ if $u_i = 0$ and $h_i(\mathbf{u}) = 2$ if $u_i = 1, i = 1, \ldots, n$; (2) for a given deterministic policy h that takes action 2 in exactly m different states, let the location vector $\mathbf{u}(h)$ be defined by $u_i(h) = 0$ if $h_i = 1$ and $u_i(h) = 1$ if $h_i = 2$. This correspondence is formalized in the following result (see [90] for proof):

Theorem 17.3.2

1) *Let* **u** *be a feasible location vector. Then* $J(\mathbf{u}) = \sum_{i=1}^{n} \psi_i(h(\mathbf{u}))b_i$.

2) *Let* h *be a deterministic policy in* Γ *that takes action 2 in exactly* m *different states. Then* $\sum_{i=1}^{n} \psi_i(h)b_i = J(\mathbf{u}(h))$.

We define the following restricted (and finite) set of policies:

$D(m) = \{h$: a deterministic policy that takes action 2 in m different states$\}$.

The next result, which is a direct consequence of Theorem 17.3.2, shows that an optimal location vector in PFIP can be obtained by finding the best policy in the class $D(m)$.

Corollary 17.3.1 *Let* h^* *be such that*

$$\sum_{i=1}^{n} \psi_i(h^*)b_i = \max_{h \in D(m)} \sum_{i=1}^{n} \psi_i(h)b_i.$$

Then $\mathbf{u}(h^*)$ *is an optimal location vector in the PFIP.*

The preceding result shows that the PFIP is equivalent to a particular "constrained" TMDP (where the constraints are placed on the admissable policies). Usually, the term "constrained TMDP" is reserved for TMDPs with constraints on the expected total number of times each state- action pair is used (in our case, this would correspond to the expected number of times a given node is treated as a facility or an empty node). More precisely, for a policy π consider the following quantity:

$$x(\pi)_{ja} = \sum_{t=1}^{\infty} \sum_{i=1}^{n} P_\pi(X_t = j, A_t = a | X_t = i)b_i, \quad j \in \{1, \ldots, n\}, a \in A(j).$$

Then "constrained TMDP" refers to a model where additional (typically linear) constraints are placed on these quantities. Such models were studied by Kallenberg in [658], who developed a mathematical programming-based approach to finding an optimal policy. This approach formulates the constrained TMDP as a certain Mathematical Program (MP), and then reconstructs an optimal policy from an optimal solution to this MP. Below we will show how this approach results in a simplified Integer Programming formulation for the PFIP.

First note that for the MDP Γ, $x(\pi)_{ja} < \infty$ always holds by Assumption 1 of Section 17.3.1 above. From the reward structure of Γ, it also follows that

$$\sum_{i=1}^{n} \psi_i(\pi)b_i = \sum_{i=1}^{n} x(\pi)_{i2}$$

for every policy π. In particular, for any feasible location vector **u**, we have

$$J(\mathbf{u}) = \sum_{i=1}^{n} x(h(\mathbf{u}))_{i2}.$$

Note also that by the definition of $h(\mathbf{u})$,

$$x(h(\mathbf{u}))_{i2} = 0 \text{ whenever } u_i = 0$$

Intuitively, the quantities $\mathbf{x}(h(\mathbf{u}))$ are directly related to the quantities $\mathbf{g}(\mathbf{u})$ defined earlier. In order to apply Kallenberg's results, we need to "recast" our constraint $h \in D(m)$ in terms of quantities $x(\pi)$. The above observations suggest the following definition. Consider the following class of policies:

$$D'(m) = \{\pi | \exists \text{ a feasible location vector } \mathbf{u} \text{ s.t. } u_i = 0 \Rightarrow x(\pi)_{i2} = 0, \forall i.$$

It would certainly be convenient if $D(m)$ and $D'(m)$ were equivalent. However, while clearly

$$D(m) \subset D'(m),$$

the converse is not true (in particular, $D'(m)$ includes many non deterministic policies that can not be present in $D(m)$). However, it can be shown (see [90]) that an optimal policy to the constrained TMDP

$$max_{\pi \in D'(m)} \max_{h \in D(m)} \sum_{i=1}^{n} \psi_i(h)b_i$$

can always be found within the class $D(m)$. It follows that the TMDP with the policy set restricted to $D'(m)$ is equivalent to PFIP. We now apply Kallenberg's results to obtain the following linear Integer Program (**IP1**):

$$\begin{aligned}
\max \quad & \textstyle\sum_{i=1}^{n} x_{i2} \\
\text{s.t.} \quad & \textstyle\sum_{a=1}^{2} x_{ia} - \sum_{j=1}^{n} \sum_{a=1}^{2} p_{jai} x_{ja} = b_i \quad i = 1, \ldots, n \\
& u_i - x_{i2} \geq 0 \quad i = 1, \ldots, n \\
& \textstyle\sum_{i=1}^{n} u_i = m \\
& x_{ia} \geq 0, u_i = 0, 1 \quad i = 1, \ldots, n, \ a = 1, 2
\end{aligned}$$

The connection between PFIP and the (IP1) is formalized in the following result (see [90] for details of the proof).

Theorem 17.3.3 *Let* $(\mathbf{x}^*, \mathbf{u}^*)$ *be the optimal solution of the (IP1). Then* u^* *is the optimal facility location vector in the PFIP, and if* $u_i^* = 1$ *then* $x(h(\mathbf{u}))_{i2}$ *represents the total expected number of visits to a facility at node* i. *If the optimal solution of the (IP1) is unique with respect to* $\mathbf{x}$, *then* $\mathbf{x}^* = x(h(\mathbf{u}^*))$.

Remark 17.3.1 *Interpretation of Variables* The preceding result shows that not only does the (IP1) provide an optimal solution to the PFIP (through the integer variables u_i), but the continuous variables x_{ia} also contain valuable information. First note that in any optimal solution, either x_{i1} or x_{i2} must equal to 0 for every i (actually, exactly one of these variables must be 0 for every node that is accessible from the initial distribution). If $u_i = 1$, then i is a facility node. In that case $x_{i1} = 0$, and $x_{i2} = g_i(\mathbf{u})$ represents the *amount of flow intercepted by the facility at* i. If $u_i = 0$, then i is an empty node. In that case, $x_{i2} = 0$, and $x_{i1} = g_i(\mathbf{u}) > 0$. If i is a destination node, then x_{i1} represents the amount of unintercepted flow "escaping" through node i. If i is a transient node, x_{i1} represents the expected number of times node i is visited prior to interception at some facility node, or leaving the network through the common sink 0. Generally, transient nodes with high values of x_{i1} represent good candidates for constructing new facilities (however, such expansion might also reduce the flow intercepted by the current facilities).

While, technically, the above interpretations of the x variables in the (IP1) are only valid when the conditions of Theorem 3.3 are satisfied, i.e., the optimal solution of the (IP1) must be unique with respect to $\mathbf{x}$, we will show below how the IP formulation can be tightened to insure such uniqueness.

To summarize, we have shown that the optimal solution to the PFIP can be obtained directly from (IP1) - a mixed linear Integer Program with n binary and $2n$ continuous variables. This formulation can be attacked by any of the 'standard' integer programming techniques and seems much preferable to the original non-linear formulation of Section 17.3.1.

We illustrate the (IP1) with a small example.

Example 17.3.2 Refer again to the PFIP of Example 3.1 in Section 17.3.1 above. The (IP1) for this case is :

$$\max \quad x_{12} + x_{22} + x_{32} + x_{42} + x_{52} + x_{62}$$

Subject to

$$x_{11} + x_{12} - .3x_{41} - .3x_{51} = .1$$
$$x_{21} + x_{22} - .3x_{41} - .1x_{51} - .3x_{61} = .1$$
$$x_{31} + x_{32} - .2x_{51} - .4x_{61} = 0$$
$$x_{41} + x_{42} - .1x_{51} - .2x_{61} = .6$$
$$x_{51} + x_{52} - .2x_{41} - .1x_{61} = .1$$
$$x_{61} + x_{62} - .2x_{41} - .3x_{51} = .1$$
$$u_1 - x_{12} \geq 0$$
$$u_2 - x_{22} \geq 0$$
$$u_3 - x_{32} \geq 0$$
$$u_4 - x_{42} \geq 0$$
$$u_5 - x_{52} \geq 0$$
$$u_6 - x_{62} \geq 0$$
$$u_1 + u_2 + u_3 + u_4 + u_5 + u_6 = 2$$
$$u_1, u_2, u_3, u_4, u_5, u_6 = 0, 1; \quad x_{ia} \geq 0, \quad i = 1, \dots, 6; \quad a = 1, 2;$$

We used the LINDO software package to obtain the following optimal solution (specifying non-zero variables only):

$$u_1 = u_2 = 1$$
$$x_{12} = .3883, \quad x_{22} = .4300$$
$$x_{31} = .1817, \quad x_{41} = .6908, \quad x_{51} = .2701, \quad x_{61} = .3192,$$

with the objective function value of .8183. This agrees with the solution computed by exhaustive enumeration earlier (see Section 17.3.1 above), and the interpretation of variables provided in Remark 17.3.1.

The formulation (IP1) can be significantly "tightened" by observing that

1. Theorem 3.1 implies that the largest value x_{i2} variable can take in any solution is equal to the single-interception portability $J(i)$. Thus constraints $u_i - x_{i2} \geq 0$, $i \in N$ should be changed to $J(i)u_i - x_{i2} \geq 0$ $i \in N$ (the new constraints are much tighter; constants $J(i)$ can be computed in the pre-processing stage)

2. When u_i is set to 1, node i should always be "treated" as a facility node, i.e., the variable x_{i1} should be set to 0. Moreover, it can be shown that the largest value that x_{i1} achieves in any feasible solution is given by

$$G(i) = \mathbf{b}^T [I - C]^{-1} \tag{17.6}$$

(this is the expected number of visits to node I when there are no facilities on the network). This observation suggests the following new constraint:

$$G(i)(1 - u_i) - x_{i1} \geq 0$$

(refer to [91] for details).

The preceding discussion yields the following modified IP formulation for the PFIP, which we will refer to as **(IP2)**:

$$
\begin{aligned}
\max \quad & \sum_{i=1}^{n} x_{i2} \\
& \sum_{a=1}^{2} x_{ia} - \sum_{j=1}^{n} \sum_{a=1}^{2} p_{jai} x_{ja} = b_i \quad i = 1, \ldots, n \\
& J(i)u_i - x_{i2} \geq 0 \quad i = 1, \ldots, n \\
& G(i)(1 - u_i) - x_{i1} \geq 0 \quad i = 1, \ldots, n \\
& \sum_{i=1}^{n} u_i = m \\
& x_{ia} \geq 0, u_i = 0, 1 \quad i = 1, \ldots, n, \ a = 1, 2
\end{aligned}
\tag{17.7}
$$

Remark 17.3.2 In addition to being a much tighter formulation, (IP2) also has another important advantage over (IP1) - every feasible $\mathbf{u}$ vector induces a unique $\mathbf{x}$ vector in (IP2) (this is not the case in (IP1)). This implies that the interpretations of variables provided in Remark 17.3.1 are always valid for (IP2). In addition, the uniqueness of $\mathbf{x}$ with respect to $\mathbf{u}$ can be used to establish the connection between the PFIP and the (IP2) (i.e., Theorem 17.3.3) directly, bypassing the constrained MDP approach - see [91].

Computational experiments show that (IP2) instances with up to 100 nodes can be solved in reasonable time by using the CPLEX package on the DEC 3300 Alpha machine. A specialized branch-and-bound algorithm, utilizing some of the polyhedral structure of (IP2) significantly outperforms CPLEX and can handle instances with about 200 nodes. The details of the algorithm and of the computational experiments are presented in [92].

17.3.3 Generalized Probabilistic Flow Interception Models

One of the strengths of the general framework and the MDP-based approach presented in Sections 17.3.1 and 17.3.2 above is that it can be easily extended to models that are substantially more general than the basic PFIP model described earlier. In this section we present several such generalizations. While the direct formulations lead to non-linear IPs, equivalent linear IP formulations are readily obtained by employing the constrained MDP methodology described in the previous section.

Incorporating Fixed Costs and Different Visibilities

The basic PFIP model assumes that an "interception" occurs whenever a customer passes through a facility node, that is, a customer passing through a facility node always makes a conscious decision to either accept the service offered, or to decline it for the remainder of her sojourn through the network. However, there are a number of reasons why a customer needing service might, nevertheless, choose not to use the first facility encountered. This reasons might be due to poor visibility of the facility, inconvenient on/off ramp, etc. This case is especially relevant when the nodes in the network represent aggregate geographical locations (e.g., neighborhoods), rather than physical intersections. We will model this situation as follows: we assume that for each node i on the network we can measure an "obscureness index" β_i, such that if there is a facility at node i then a customer passing through this node has a $1 - \beta_i$ chance of being intercepted (as before, a facility may only intercept the customers that have not already been intercepted by the "upstream" facilities).

Another implicit assumption of PFIP is that the fixed costs are either negligible, or identical for all locations on the network. In reality, different locations have very different "desirability" characteristics, which are usually reflected in fixed costs (through rents, different costs for TV commercial slots, etc). Thus, we will assume that each node i has an associated fixed cost s_i, which includes both the set-up and the fixed operating costs. For a given facility location vector $\mathbf{u}$ and a facility node i, let $g_i(\mathbf{u})$ be the expected fraction of flow intercepted at i (this agrees with the notation defined in Section 3.1 earlier). We will assume that the expected revenue generated by the facility at i is equal to $rg_i(\mathbf{u})$, where r is the total revenue that could be earned by intercepting all customers in the system. The overall objective is to maximize the overall profit, which yields the following new objective function:

$$max_{\mathbf{u}} \sum_{i \in N} [rg_i(\mathbf{u}) - s_i u_i].$$

In order to model the "partial interception" of flows, the transition matrix $C^+(\mathbf{u})$ induced by a facility location vector $\mathbf{u}$ is redefined as follows:

$$c_{i,j}(\mathbf{u}) = c_{ij}, \quad \text{if } u_i = 0,\ i \in N,\ j \in N \cup \{0, n+1\} \text{ or } i \in \{0, n+1\}$$
$$c_{i,j}(\mathbf{u}) = \beta_i c_{ij}, \quad \text{if } u_i = 1,\ i \in N, j \in N \cup \{0\}$$
$$c_{i,n+1}(\mathbf{u}) = 1 - \beta_i, \quad \text{if } u_i = 1,\ i \in N$$

(intuitively, placing a facility at node i has the effect of diverting all of the incoming flow to node $(n+1)$ with probability $1 - \beta_i$; with probability β the flow is not affected by the presence of the facility).

One additional change resulting from the generalization of the basic model is that the constraint specifying that exactly m facilities should be placed is no longer reasonable - the number of facilities should be determined by the model itself. Thus, in principle, all $n-$dimensional binary vectors are feasible location vectors. However, in practice, budgetary and other constraints frequently limit the number of new facilities that may be considered. To reflect this, we will assume that a value m is now specified as an *upper bound* on the total number of new facilities that may be opened. We include the following constraint in our formulation:

$$\sum_{i \in N} u_i \leq m.$$

We will refer to the generalized mode described above as the **PFIP1**. The preceding discussion immediately leads to the following "direct" Non-linear Integer Programming Formulation for the PFIP1:

$$\max \quad r\mathbf{b}^T [I - C(\mathbf{u})]^{-1} \mathbf{c}_{n+1}(\mathbf{u}) - \sum_{i \in N} s_i u_i$$

Subject to

$$\sum_{i \in N} u_i \leq m$$
$$u_i = (0, 1), i \in N$$

(note that this formulation is undefined for fractional values of $\mathbf{u}$).

Using the techniques outlined in Section 17.3.2, we can re-formulate PFIP1 as the following Linear Integer Program **(IP3)**:

(MIP1) $\max \quad r \sum_{i \in N} (1 - \beta_i) x_{i2} - \sum_{i \in N} s_i u_i$

sub. to $x_{i1} + x_{i2} - \sum_{j \in N} c_{ji} x_{j1} - \sum_{j \in N} \beta_j c_{ji} x_{j2} = b_i, \quad i = 1, ..., n$

$$\sum_{i \in N} u_i \leq m$$

$$G(i)(1 - u_i) - x_{i1} \geq 0, \quad i = 1, ..., n$$

$$J'(i)u_i - x_{i2} \geq 0, \quad i = 1, ..., n$$

$$x_{i1}, x_{i2} \geq 0, u_i = (0,1), \quad i = 1, ..., n$$

The constants $G(i)$ for $i \in N$ have the same meanings as before (the expected total number of visits to i when no facilities are located. The constants $J'(i)$ are related to $J(i)$ in the previous model, but have a slightly different interpretation: for $i \in N$, $J'(i)$ represents the expected number of visits to i when it is the only facility node on the network. The formula for computing $J'(i)$, $i \in N$ is given by

$$J'(i) = (\mathbf{b}^T[I - (B_iC)]^{-1})_i,$$

where

$$B_i = diag[1, \ldots, 1, \beta_i, 1, \ldots, 1] \in R^{n \times n}.$$

Note also that variables x_{i2} represent the expected number of visits to i when it is a facility node. To obtain the amount of flow intercepted at i, the value of x_{i2} should be multiplied by $1 - \beta_i$. The profit margin at i is given by

$$r(1 - \beta_i)x_{i2} - s_i$$

(when i is a facility node).

Example 17.3.3 Suppose that for the network of Example 3.1, the following obscureness indices are defined:

$$\beta_1 = \beta_2 = \beta_3 = .2, \quad \beta_4 = \beta_5 = \beta_6 = .5$$

Furthermore, suppose that the fixed costs are given by

$$s_1 = s_2 = s_3 = 20, \quad s_4 = s_5 = s_6 = 10$$

and that $r = 100$. There is no limit on the number of facilities that may be opened (i.e., $m = 6$).

To solve this problem, we first compute the constants $G(i)$ and $J'(i)$ for $i = 1, \ldots, 6$ (the formulas for $J'(i)$ are given above, the corresponding formula for $G(i)$ can be found at the end of Section 3.2). This yields:

$G(1) = .3883, G(2) = .43, G(3) = .1817, G(4) = .6908, G(5) = .2701, G(6) = .3192$
$J'(1) = .3883, J'(2) = .4300, J'(3) = .1817, J'(4) = .6635, J(5) = .2608, J(6) = .3049.$

Using these parameters in (IP3) results in the following solution (obtained on LINDO, only non-zero variables are presented):

$$u_1 = 1, \ u_2 = 1, u_4 = 1$$
$$x_{12} = .2561, x_{22} = .2853, x_{42} = .6634$$
$$x_{31} = .12699, x_{51} = .1886, x_{61} = .2229,$$

with the optimal value of 26.4823. Thus, a total of three facilities should be located at nodes 1,2 and 4. The fraction of flow intercepted at each node is .2049, .2282, and .3317, respectively. It follows that the profit margins generated by the facilities are .488, 2.2824, and 23.173, respectively (i.e., the facility at node 1 is barely breaking even and would be a good candidate for closing).

The Multiple-Interception Case

A significant assumption in both, PFIP and PFIP1, is that a customer may only be intercepted *once* during their sojourn through the network. This assumes that once a customer has visited one of the facilities, the value of any subsequent visits to the facility set is zero. Clearly, this assumption is inappropriate in many marketing applications (where "facilities" are typically some advertising instruments and "visits" are exposures to them) - the value of the secondary visits may be quite high.

To reflect his situation we define a "multiple-interception" version of the PFIP1 in which customers are allowed to visit the facilities more than once. The value of each subsequent visit to the facility set is discounted by a constant discount factor $\alpha \in [0, 1]$. We will refer to this new model as **PFIP2**. As in PFIP1, both fixed and variable costs will be considered, with s_i representing the fixed costs of operating a facility at node i, and r representing the value of a single exposure (by *all* the customers) to the facility set. The model PFIP2 is obviously related to the (deterministic) "double-counting" model of Section 17.2.4. The deterministic model is somewhat more general because it allows the total amount of flow intercepted along a given path to be an arbitrary non-decreasing concave function of the number of facilities along the path, whereas PFIP2 restricts this function to α^k form (this restriction is natural, since in the probabilistic model the details of the "previous behaviour" of customers are not known).

The machinery needed to formally define the PFIP2 model is slightly different from that used earlier for PFIP1 (and PFIP). We proceed as follows. To the original node set N we adjoin a common sink node 0, and define the Markov Transition matrix C^+ as in Section 17.3.1 above (note that the "bookkeeping" sink $(n + 1)$ is not used in this model). Let P be the probability measure induced by C^+ and the initial distribution **b**. We define the following random variables:

$$T_k(\mathbf{u}), \quad k = 1, 2, \ldots - \text{the time of } k-\text{th visit to the facility set .}$$

We let $T_k(\mathbf{u}) = \infty$ if the $k-$th visit to the facility set never occurred. The formal definition of PFIP2 is given by:

$$\max \quad J(\mathbf{u}) = rE[\textstyle\sum_{k=1}^{\infty} \chi\{T_k(\mathbf{u}) < \infty\}\alpha^{k-1}v] - \textstyle\sum_{i \in N} s_i u_i$$

Subject to

$$\textstyle\sum_{i \in N} u_i \leq m$$
$$u_i \in \{0, 1\}, \quad i = 1, \ldots, n,$$

$$(17.8)$$

where $\chi\{\cdot\}$ is the indicator function, the expectation is taken with respect to P, and m represents the upper bound on the number of new facilities that may be opened. Assumption 1 in Section 17.3.1 implies that $\{0\}$ is the only recurrent class of C^+, and thus , for all k large enough, $T_k(\mathbf{u}) = \infty$ $P-a.s.$, and therefore the objective function in the model above is bounded.

Once again the direct formulation leads to a "nasty" non-linear IP, and we employ the constrained TMDP techniques of Section 17.3.2 to obtain the following linear IP formulation **(IP4)** (please refer to [90] for details):

(MIP1)   max $r \sum_{i \in N} x_{i2} - \sum_{i \in N} s_i u_i$

Subject to $x_{i1} + x_{i2} - \sum_{j \in N} c_{ji} x_{j1} - \alpha \sum_{j \in N} c_{ji} x_{j2} = b_i, \quad i = 1, ..., n$

$$\sum_{i \in N} u_i \leq m$$

$$G(i)(1 - u_i) - x_{i1} \geq 0, \quad i = 1, ..., n$$

$$J'(i)u_i - x_{i2} \geq 0, \quad i = 1, ..., n$$

$$x_{i1}, x_{i2} \geq 0, u_i = (0, 1), \quad i = 1, ..., n,$$

where constant $J'(i)$ and $G(i)$ were defined earlier.

The most striking feature of this formulation is its obvious similarity to the formulation (IP3) obtained earlier for PFIP1. Indeed, when all obscureness indices in PFIP1 are the same and equal to the discount factor in PFIP2 (i.e., $\alpha = \beta_i$, $i \in N$), the only difference between (IP4) and (IP3) is in the multiplicative factors $(1 - \beta_i)$ in the latter's objective. It follows that if we take the variable cost $r' = (1 - \alpha)r$ in PFIP2, the two models *are equivalent* (in a sense, the PFIP2 model is a special case of PFIP1). Intuitively, this implies that less-visible 'single-interception' facilities (i.e., the ones with high β values) would tend to be placed in similar locations as the 'multiple-interception' facilities, where the value of secondary visits is high (since $\alpha = \beta_i$ for equivalence).

An important special case of PFIP2 occurs when $\alpha = 1$ - i.e., each visit to the facility set has equal value. In this case the amount of flow intercepted by each facility is independent of the location of all other facilities, and the problem can be solved by a very simple greedy-type algorithm (see [90]).

Remark 17.3.3 Note that for both of the generalized models described in this section (PFIP1 and PFIP2), the set of allowable facility locations was implicitly restricted to nodes of the network. While such a restriction can be made without any loss of generality for the basic PFIP model (see section 17.3.1), this is not the case for generalized models. Indeed, if facilities can be located anywhere on the network, it is possible to construct examples where non-model locations are optimal [91]. Nevertheless, the restriction to nodal restriction is reasonable as long as *all* feasible facility locations are represented by the nodes.

17.3.4 Heuristic Approaches

In this section we define a heuristic-type algorithm for the PFIP1 model and present some analytical and computational results related to the algorithm's performance. Note that the equivalence of the PFIP1 and PFIP2 makes all results in this section immediately applicable to the latter model as well.

For any $A \subseteq N$, let $\mathbf{u}^A$ be a location vector such that $u_i^A = 1$ if $i \in A$ and $u_i^A = 0$ otherwise. Let $f(A)$ be the objective function value of PFIP1 corresponding to the location vector $\mathbf{u}^A$. The heuristic procedure is described below:

The Greedy-Type Heuristic Algorithm

Step 0: Set $k = 0$, $A_0 = \emptyset$.

Step 1: If $k = m$, Go to Step 4.

Step 2: Set $i_{k+1} = \arg\ \max_{j \in N - A_k} f(A_k \cup \{j\})$. Set $A_{k+1} = A_k \cup \{i_{k+1}\}$.

Step 3: IF $f(A_k) \geq f(A_{k+1})$, GO TO Step 4;
OTHERWISE, set k = k+1 and GO TO Step 1.

Step 4: (The Backward Search)
For $i = 1, \ldots, n$ do
IF $i \in A_k$ and $f(A_k) < f(A_k - \{i\})$ THEN
Set $A_{k-1} = A_k - \{i\}$, $k = k - 1$.

Step 5: Output the location vector $\mathbf{u}^{A_k}$, and the objective function value $f(A_k)$; The number of located facilities is k.

This procedure, in Step 2 attempts to expand the facility set by finding a (currently empty) location with the highest profit margin. If no location with positive profit margin is available, or the limit on the available number of facilities has been reached, the procedure goes to Step 4. Note that since placing new facilities might cannibalize some flow from the existing ones, as a result of the expansion of the facility set in Step 2 some of the previously placed facilities might now have negative profit margins. Thus, in Step 4 the facility set is checked, and the facilities with negative profit margins are discarded.

Note that Steps 1-3 describe the greedy heuristic procedure for the PFIP.

Remark 17.3.4 The "cleanup" in Step 4 could be performed after every iteration of Step 2, instead of just as a final step. This would result in improved accuracy, at the expense of longer running times: while this new procedure would still be finite (since the value of the objective function is required to (strictly) increase whenever the facility set is changed), the number of iterations could potentially be quite large - it is easy to construct examples where the same node enters and leaves the facility set more than once.

We would like to derive a worst-case bound for our Greedy-Type Algorithm by employing the submodularity-type arguments based on the results in Nemhauser *et al* [888]. This technique requires the objective function of the underlying optimization problem ($f(A)$ in our case) to be monotone and submodular. It can be shown that the submodularity property indeed holds for $f(A)$. However, in general, the monotony does not - a strict enlargement of a facility set might result in very small flows at some new facilities, leading to negative profit margins and a reduction in the objective function value (this is the reason for Step 4 above). This motivates the following (rather strong) assumption:

Assumption 2 *In any facility set A, every facility $i \in A$ generates non-negative profit margin.*

Note that under Assumption 2, the Heuristic Algorithm presented above is exactly equivalent to the Greedy Heuristic, since Step 4 can be skipped.

The following result establishes an easily-verifiable condition which is equivalent to Assumption 2 (see [91] for proof).

Proposition 17.3.1 *Let $B = diag(\beta_1, \ldots, \beta_n)$. Then Assumption 2 holds if and only if*

$$[\mathbf{b}^T (I - BC)^{-1}]_i \geq s_i/r \quad for \ all \ i \in N$$

The following worst-case bound for the Greedy-Type Algorithm can now be established.

Theorem 17.3.4 *Let f_{greedy} be the objective function value returned by the Greedy-Type Heuristic Algorithm, and let f_{opt} be the optimal objective function value. Suppose Assumption 2 holds. Then the worst-case bound on the relative error of the heuristic is given by*

$$\frac{f_{opt} - f_{greedy}}{f_{opt}} \leq (1 - \frac{1}{m})^m \leq e^{-1}.$$

The proof is based on submodularity-type arguments; it can be found in [91]. The worst-case bound established above is tight: for arbitrary n it is possible to construct an explicit example of a problem instance for which the bound is tight (see [91]). The result of Theorem 17.3.4 is very similar to the results obtained for the greedy heuristic in the case of deterministic models (see Section 17.2).

Extensive set of computational experiments of the Greedy-Type Algorithm was conducted in [91]. These experiments show that the performance of the algorithm for randomly generated problem instances is far better than the worst case bound: for different network sizes, the average relative error of the greedy solution compared to the optimal one ranges from 0% to 1%, with most cases having relative errors less that .01%. This results appear to be unaffected when the various problem parameters (e.g., the obscureness indices, the fixed and variable costs, the density of the network) are varied. Moreover, for randomly generated networks, the use of Step 4 in the Algorithm above is extremely rare (note however that in the test examples the upper bound on the number of facilities located was typically less than 10% of n, the use of Step 4 is probably much more common when the number of facilities is closer to n).

In summary, the Greedy-Type Algorithm appears to be an effective way to get near-optimal solutions for both the deterministic and the probabilistic versions of Flow Interception Problems.

17.4   Future Research

The study of Flow Interception Problems is very young - the field abounds with interesting open problems. Some of these have already been pointed out earlier. In this section we outline three directions for further research that we feel to be of particular importance.

The Relationship Between the Probabilistic and Deterministic Models While both the deterministic and probabilistic versions of the Flow Interception Problems are, in our opinion, quite important in their own right, the relationship between the two needs to be explored further. It is clear that the deterministic model employs a much richer (albeit probably unrealistic) set of initial data, and thus the two models are not, in general, equivalent (examples for which the two models yield different results are readily found [90]). One view is to regard the deterministic model as the fundamental "underlying" model for which the data is not available, and to regard the probabilistic model as an approximation to the deterministic one. This leads to the question of how large is the gap between two models' solutions. Some preliminary numerical experimentation indicates that the approximation provided by the probabilistic model is reasonably accurate, and is significantly better than the approximation obtained when all flows are assumed to occur along the shortests paths (see [90] for a description of the experiments and the results). However, this isuue clearly deserves further exploration.

Competitive Models Clearly, in most actual decisions regrading the placement of flow-intercepting facilities a major factor is the actions of the competitors. Thus, competitive location models would be particularly valuable in this context. It seems that the most useful framework for analyzing the competitive situation is the "leader-follower" games (the location decisions are unlikely to be taken simultaneously). For the case of two competitors, the followers' problem is essentially equivalent to the problems considered in this chapter (the leaders' actions essentially result in re-configuring some of the flows on the network; these new flows can be taken as input data by the follower). The leader's problem appears to be substantially more complicated (e.g., a particular intersection might not be sufficiently attractive to merit a facility in a non-competitive situation, however a competitor's facility later placed at the same location might prove disastrous). Both the deterministic and probabilistic versions of this problem appear to be very interesting.

Combined Models As pointed out in the introduction, most "classical" facility location models implicitly assume that *all* demand results from customers making special "planned" trips to the facility. The Flow Interception models essentially go to the opposite extreme, assuming that *all* demand comes from intercepting pre-existing customer flows (i.e., the customers never actually plan the trips to the facilities). For most real-life facilities (e.g., supermarkets) the actual situation lies somewhere in between these two extremes: part of the demand comes from pre- planned trips ("regular customers") and the rest comes from intercepting of flow-by traffic, with the importance of each component depending on the facility's type and location. Both deterministic and probabilistic versions of the "combined" models incorporating these two components of demand would be of great interest.

18

LOCATION-ROUTING PROBLEMS WITH UNCERTAINTY

Oded Berman
University of Toronto

Patrick Jaillet
The University of Texas at Austin
Laboratoire de Mathématiques Appliquées

David Simchi-Levi
Northwestern University

18.1   Introduction

In a wide variety of services, such as delivery, customer pick-up, repair and maintenance services, a service unit regularly visits several demand points during a given service tour. In location-routing problems that arise in such services, the objective is to find the home location of the service unit and design tours for servers so as to minimize total costs. Location-routing problems are considered to be very difficult since they require simultaneous solutions of location and routing problems. We will consider here models of location-routing problems for which the number and possibly the location of customers are not known exactly in advance but are described instead by a priori probability distributions.

Our objective in this chapter is to offer a broad overview that emphasizes the fundamental concepts and provides an up-to-date reference list for the numerous developments which have been published in this area during the last few years.

Two operational modes are possible in location-routing models. In the first, every potential instance of the customers is served by designing an optimal, or near-optimal, tour that starts at the service station visits all customers in that instance and then ends at the service center. We refer to this approach as the "re-optimization strategy" and to the location problem under this mode as the Traveling Salesman Location Problem (TSLP).

This approach suffers from one major disadvantage. Since the traveling salesman problem is NP-hard, one might have to solve every day an instance of a hard problem. With today computing power as well as recent development in exact algorithms for hard combinatorial problems (namely, cutting plane algorithms) this is clearly possible provided that resources are available. But, even if we assume that we have the resources available for re-optimizing every day, it might be the case that we either cannot re-optimize (due to lack of advance information, i.e., the information on who to serve on a given tour might not be available before the tour starts), or simply do not desire to reoptimize (for regularity of service). For all these reasons, we might need to use a different strategy, which we call an *a priori* strategy. Here, we wish to find an *a priori* tour through all the customers and then update it (possibly in real-time) in a simple way to accommodate each particular instance/variation. We refer to the problem of designing an efficient *a priori* tour as the Probabilistic Traveling Salesman Problem (PTSP) and to the location-routing problem under this mode as the Probabilistic Traveling Salesman Location Problem (PTSLP).

18.1.1   The Traveling Salesman Location Problem

A typical criterion for the traveling salesman location problem is the minisum objective. The minisum traveling salesman location problem is to locate the service unit so as to minimize the expected distance traveled (taking into consideration that on any given day a different service tour may be performed).

Traveling salesman location problems have received until recently little attention. Eilon, et al. [366] developed a heuristic for the problem on the plane with many customers that are uniformly distributed. Burness and White [153] suggested a heuristic that can handle only a small number of customers for the problem on the plane as it requires the solution of a large number of traveling salesman problems. Drezner, et al. [320] have a method for the problem on the plane with rectilinear distances. This method gives the optimal solution when the number of demand points is less than or equal to 3. Otherwise, the method is a heuristic that performs well only when few tours are required to pass through more than three customers. Wolf [1216] presented an exact, although inefficient, algorithm for the problem on a tree.

A basic assumption in all the studies discussed above is that the probabilities of the various tours to occur are given. During the last several years, the following version of the traveling salesman location problem has been analyzed. At each node of a network a service request might occur with a given probability. At a specific time of the day a service unit receives the list of service requests and immediately starts a traveling salesman tour that visits all the nodes in the list *at least once*.

The Traveling Salesman Location Problem: Let $G(V, A)$ be an undirected network with a set of nodes $V(|V| = n)$ and a set of links A. Calls for service are assumed to arrive solely from the nodes of the network and independently one from another. At any working day at most one call for service can arrive from any node i with a given probability p_i.

Starting at the beginning of each working day calls for service that arrive at the service center are registered in a service catalog. At a specific time t_0 the service unit receives the service catalog and unless it is empty it immediately leaves the center and starts a service tour that visits the first k calls that have arrived $k \leq b (b \leq n$ is the capacity bound on the number of calls that can be served by the unit). We assume, for simplicity, that the service unit spends zero time on servicing each call.

Let E be the collection of all the feasible lists ($|E| \leq 2^n$.) The probability of each list $S, p(S)$ when $b = n$ is given [100] by

$$p(S) = [\Pi_{i \in S} p_i][\Pi_{i \notin S}(1 - p_i)]. \tag{18.1}$$

Observe that (18.1) can be rewritten as $p(S) = K\Pi_{i \in S}(p_i/1 - p_i)$, where K is a constant (independent of S) $K = \Pi_{i \in V}(1 - p_i)$. For each finite set of points X on $G(V, A)$, let $L(X)$ be the length of the optimal traveling salesman tour that visits all the points in X at least once.

The Traveling Salesman Location Problem (TSLP) is to find a home location x^* for the service unit that minimizes the function

$$g(x) = \sum_{S \in E} p(S)L(S \cup \{x\}) \quad x \in G(V, A). \tag{18.2}$$

18.1.2 The Probabilistic Traveling Salesman Location Problem

Suppose $G = (V, A)$ is a complete graph on n nodes on which a routing problem is defined (for example the traveling salesman problem). Given a method $\mathcal{U}$ for updating an *a priori* solution f to the "full-scale" optimization problem on the original graph $G(V, A)$, $\mathcal{U}$ will then produce for problem instance S (there are 2^n instances), a feasible solution $t_f(S)$ with value ("cost") $L_f(S)$. (In the case of the TSP, $t_f(S)$ would be a tour through the subset S of nodes and $L_f(S)$ the length of that tour.) Then, given that we have already selected the updating method $\mathcal{U}$, the natural choice for the *a priori* solution f is to select f so as to minimize the expected cost

$$E[L_f] = \sum_{S \subseteq V} p(S)L_f(S). \tag{18.3}$$

This choice of a measure of effectiveness for the *a priori* solution f that we seek, namely the expected cost (18.3), gives a reasonable answer to our first question. But what properties should the updating method $\mathcal{U}$ have? The most desirable property of $\mathcal{U}$ would be for $L_f(S)$ to be "close" to the value of the optimal solution $L_{OPT}(S)$, for every instance S. A less

FIGURE 1. The PTSP methodology

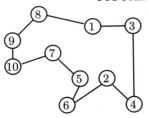 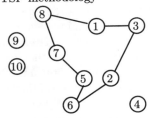

A priori tour through 10 points   The tour when points 4, 9,
and 10 need not be visited

restrictive and more global property is to require $E[L_f]$ to be "close" to
the expected cost $E[\Sigma]$, over all problem instances, of the re-optimization
strategy: $E[\Sigma] = \sum_{S \subseteq V} p(S) L_{OPT}(S)$. In addition, $\mathcal{U}$ must be able to update
efficiently the solution from one problem instance to the next.

In the following definitions of the updating methods $\mathcal{U}$, the choices of $\mathcal{U}$
may initially seem arbitrary. But these choices will turn out to be natural
ones. First, for every choice of $\mathcal{U}$ we are proposing, the updating of the
solution to a particular instance S can be done very easily. Moreover, these
updating methods are well suited for applications.

After this general discussion of the rationale behind the definitions which
follow, we describe informally the problems we are considering.

The Probabilistic Traveling Salesman Problem: The probabilistic
traveling salesman problem (PTSP) is perhaps the most fundamental stochastic routing problem that can be defined. It is essentially a traveling salesman problem (TSP), in which the number of points to be visited in each
problem instance is a random variable.

Consider a problem of routing through a set of n known points. We wish
to find *a priori* a tour through all n points. On any given instance of the
problem, the k points present ($0 \le k \le n$) will then be visited in the same
order as they appear in the *a priori* tour (see Figure 1 for an illustration).
The problem of finding such an *a priori* tour which is of minimum length
in the expected value sense is defined as the PTSP. The updating method
$\mathcal{U}$ for the PTSP is therefore to visit the points on every problem instance
in the same order as in the *a priori* tour, i.e. we simply skip those points
which are not present in that problem instance.

The expectation is computed over all possible instances of the problem,
i.e. over all subsets of the node set $V = \{1, 2, \dots n\}$. If we denote the
length of the a priori tour τ by L_τ (a random variable) and $L_\tau(S)$ as the
length of the tour required to visit $S \subseteq V$, then our problem is to find
an *a priori* tour through all n potential customers, which minimizes the
quantity $E[L_\tau] = \sum_{S \subseteq V} p(S) L_\tau(S)$.

The Probabilistic Traveling Salesman Location Problem: We are given a set of n nodes (customer locations) on a network. Each day a subset S of customers make a request for service with probability $p(S)$. By a specific time of each day, a service unit receives the list of calls for that day and starts a tour using the underlying network that visits all the customer locations in the list. The difficulty of having to compute an optimal tour for every instance can be overcome by using an *a priori* tour τ_p and then follow the PTSP updating method $\mathcal{U}$ described before, i.e., skip customer locations with no demand. The problem is then to find a point x in $G(V, A)$ and an *a priori* tour to minimize the expected distance traveled using the PTSP approach, i.e. to minimize $h(x, \tau) \triangleq \sum_{S \subseteq V} p(S) L_\tau(S \cup x)$. The problem of finding simultaneously an optimal location and an optimum *a priori* tour is called the probabilistic traveling salesman location problem (PTSLP).

18.1.3   Summary and Content of the Chapter

Before giving the detailed content of this chapter, let us emphasize that TSLPs and PTSLPs arise in the complex but very common contexts in which facility location, routing and, possibly, inventory-related decisions must be made simultaneously. Note the difference between these problems and the classical "median" (or "minisum") and "center" (or "minimax") problems in facility location theory. In the case of (P)TSLPs, once a facility is located, demands are visited through tours; therefore, the facility location problem must be "central" relative to the <u>ensemble</u> of the demand points, as ordered by the (yet unknown) tour through all of them. By contrast, in the classical problems the facility (or facilities) must be located by considering distances to <u>individual</u> demand points, thus making the problem more tractable.

In this chapter, we present a review of results on the TSLPs and PTSLPs. We start, in Section 18.2 with the TSLP and presents exact algorithms for special networks such as a tree [100] and a simple network (cactus) [101] and a heuristic for a general network [1080]. The heuristic can also be used for the problem in the plane with Euclidean or rectilinear distances [1079]. We then extend some of these results to the multiserver case [1077]. A generalization of the results in Section 18.2 to stochastic networks (where lengths of links are random variables) is given in [103].

In Section 18.3 we present a detailed treatment of the analysis of the probabilistic traveling salesman problem and extend the analysis to the probabilistic traveling salesman location problem. Most of the discussion here follows [110] as well as [632, 630, 631].

Finally, in Section 18.18.4, we show how all these results can be used in the context of design of distribution systems and for strategic planning.

FIGURE 2. Partition of $G(V, A)$ into $G(V_a, A_a)$ and $G(V_b, A_b)$

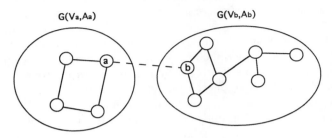

18.2   The Traveling Salesman Location Problem

A very useful result is that the search for the optimal solutions to our problem can be limited to the set of nodes V. The proof of this result which is included in [100] is similar to that of the 1-median problem [509, 533].

Theorem 1 *There exists at least one optimal solution to the traveling salesman location problem on a node of the network.*

Let us define a simple network to be a connected network where each link belongs to at most one cycle. Note that without any loss of generality (by adding for any node belonging to k cycles, $k > 1$, $k - 1$ new non-demand nodes that connect the $k - 1$ cycles and the original node by $k - 1$ new links of zero length) it can be assumed that in simple networks, each node also belongs to at most one cycle. Special cases of a simple network are trees and unicycles.

Next we discuss the minisum traveling salesman problem on these simple networks.

18.2.1   Simple Networks

Let us assume that $G(V, A)$ is a simple network but not a unicycle. Let a and b be two nodes that are connected by a link in G that does not belong to any cycle. Partition $G(V, A)$ into two mutually exclusive subnetworks $G(V_a, A_a)$ and $G(V_b, A_b)$ such that $a \in G(V_a, A_a)$ and $b \in G(V_b, A_b)$ (see Figure 2). Let E_C be the set of all lists that are composed of nodes that belong only to a subset $C \subseteq V$ (i.e., if $S \in E_C$ and $i \in S$ than $i \in C$) and let $E_{ab} = E - E_{V_a} \bigcup E_{V_b}$. Now we can present a key result for our analysis.

Theorem 2 *If $\sum_{S \in E_{V_b}} p(S) \geq \sum_{S \in E_{V_a}} p(S)$, then $x^* \in V_b$.*

The proof is given in [101].

The condition of Theorem 2 requires the knowledge of both sums $\sum_{S \in E_{V_a}} p(S)$ and $\sum_{S \in E_{V_b}} p(S)$. In the next lemma, we show that it is sufficient to calculate only one of the two :

Lemma 1 $\sum_{S \in E_{V_b}} p(S) \geq \sum_{S \in E_{V_a}} p(S)$ *IFF* $\sum_{S \in E_{V_a}} p(S) < \sqrt{K^2 + K} - K.$

The proof is given in [100].

Theorem 2 and Lemma 1 can be used to solve the problem on a tree.

Case of a tree: A direct consequence of Theorem 2 for a tree network is that the optimal location does not depend on the length of links of the tree. Moreover, as a result of Lemma 1, if we can calculate efficiently the sum of the probabilities in any given subtree, we can construct an efficient algorithm to solve our problem.

Suppose $G(V, A)$ is a tree network and let $G(V_i, A_i)$ be the subnetwork of $G(V, A)$ where node i is an end node (tip) of $G(V_i, A_i)$. Let $Y_i = \frac{1}{K} \sum_{S \in E_{V_i}} p(S)$. Suppose node j is connected by a link to node $i, j \notin V_i$. Let KY_j be the total probability of all the lists in $G(V_j, A_j) = G(V_i \bigcup j, A_i \bigcup (i, j))$, as depicted in Figure 3, then

$$Y_j = \frac{1}{K} \sum_{S \in E_{V_i}} p(S) + \frac{1}{K} \sum_{S \in E_{V_i}} p(S) \frac{p_j}{1 - p_j} + \frac{p_j}{1 - p_j} = Y_i + Y_i \frac{p_j}{1 - p_j} + \frac{p_j}{1 - p_j}$$

FIGURE 3. $G(V_i, A_i)$ and $G(V_j, A_j)$

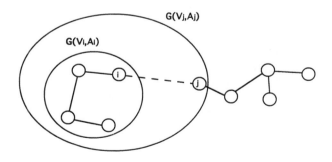

Now we can present the following algorithm to solve the problem on a tree in $O(n)$ time.

Algorithm 1 [100]:

Step 0: Define $Y_k = \frac{p_k}{1 - p_k}; k = 1, \cdots, n$

Step 1: If $|V| = 1$ Stop, the only node in V is optimal, otherwise go to Step 2.

Step 2: Choose any end node, call it i^*. If $KY_{i^*} \geq \sqrt{K^2 + K} - K$ go to Step 4, otherwise go to Step 3.

Step 3: Choose any node j directly connected by a link to node i^* and modify Y_j to be $Y_j = Y_j + Y_{i^*}Y_j + Y_{i^*}$, delete node i^* from V and link (i^*, j) from A and go to Step 2.

Step 4: Stop, node i^* is optimal.

Finally, when $p_i = p \; \forall i$, it can be shown [100] that Algorithm 1 is identical to Goldman's algorithm [489] to find the 1-median location on a tree. Moreover, in this case the optimal location is independent of the capacity bound b [100] (when $b = 1$ the problem is identical to the 1-median problem).

Case of a unicycle: Let us assume that $G(V, A)$ is a unicycle and that all the nodes are renumbered from 1 to n according to their appearance along the cycle. Suppose the total sum of all the lengths in $G(V, A)$ is $2Q$. We assume that for each node j there is a node j^* in $G(V, A)$ such that $d(j, j^*) = Q$ (if node j^* does not exist we can add such an artificial node with a demand weight, p_{j^*}, equal to zero).

Using the partition of $G(V, A)$ into the two subnetworks $G(V_t, A_t)$ and $G(V_{t^*}, A_{t^*})$ as depicted in Figure 4 it can be shown, following the same arguments as in Theorem 2, that

$$g(t) = g(t+1) + 2d(t, t+1) \left[\sum_{S \in E_{V_{t^*}}} p(S) - \sum_{S \in E_{V_t}} p(S). \right]$$

FIGURE 4. Partition of $G(V, A)$ into $G(V_t, A_t)$ and $G(V_{t^*}, A_{t^*})$

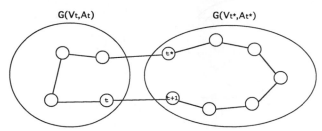

Let us show how to calculate $\sum_{S \in E_C} p(S)$, for every $C \subseteq V$. Suppose $C \subseteq V$ and E_C is the set of lists that includes only nodes of C, i.e., $E_C = \{S \in$

$E; j \in S$ implies $j \in C$}. By definition of $p(S)$

$$\sum_{S \in E_C} p(S) = K \sum_{S \in E_C} \prod_{j \in S} [\frac{p_j}{1 - p_j}].$$ (18.4)

But given that we focus only on lists in E_C

$$K_C \sum_{S \in E_C} \prod_{j \in S} [\frac{p_j}{1 - p_j}] = 1,$$ (18.5)

where

$$K_C = \prod_{i \in C} (1 - p_i),$$ (18.6)

$$\sum_{S \in E_C} p(S) = \frac{K}{\prod\limits_{i \in C} (1 - p_i)}.$$ (18.7)

Now, using (18.7)

$$\sum_{S \in E_{V_\ell}} p(S) = \frac{K}{\prod\limits_{i \in V_\ell} (1 - p_i)} \quad \ell = t, t^*.$$ (18.8)

In [101] an $O(n^2)$ algorithm to solve the problem on a unicycle is presented. It is shown there that a variant of such an algorithm can find the optimal location in $O(n)$ time.

Other simple networks: Let $G(V, A)$ be a simple network that contains at least two cycles. Now we show that $G(V, A)$ can be reduced to either a tree or a unicycle. A reduction to a tree is depicted in Figure 5 (a and b). As can be seen any cycle of $G(V, A)$ is replaced by a new node (double circle). A reduction of $G(V, A)$ to a unicycle is depicted in Figure 5 (c and d). Here, one of the original cycles (e.g., cycle 1 in Figure 5 a) becomes the unicycle where each node that has a degree greater than two in that cycle (nodes with double circles) represents the subnetwork rooted at that node.

The demand weight associated with the nodes that represent cycles or subnetworks (nodes with double circles) in either case of reduction can be calculated as follows. For cycle or subnetwork $G(V_k, A_k)$ from (18.7)

$$\sum_{S \in E_{V_k}} p(S) = \frac{K}{\prod\limits_{i \in V_k} (1 - p_i)}.$$ (18.9)

But if we call the node that replaces (represents) the cycle (subnetwork) node C_k, then for list $\{V_k\}$ in the reduced network we must have that $h_{C_k} = 1/[1 + \prod_{i \in V_k} (1 - p_i)]$.

To solve our problem on a simple network we can first apply Algorithm 1 to a tree (if $G(V, A)$ is not a tree we reduce $G(V, A)$ to one). If the optimal

FIGURE 5. A reduction to a tree and to a unicycle

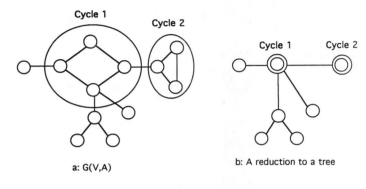

a: G(V,A)

b: A reduction to a tree

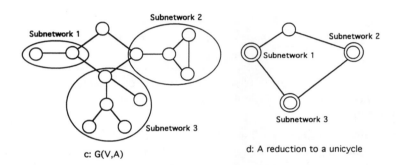

c: G(V,A)

d: A reduction to a unicycle

solution is an original node it is also the optimal solution to our problem. Otherwise, we reduce $G(V, A)$ to a unicycle using the cycle (node) obtained by Algorithm 1 as the unicycle and apply Algorithm 2 to find the optimal location.

18.2.2   A Heuristic to Solve the Problem on General Networks

There are two main contributors to the complexity of the problem on general networks:

1. Finding $L(S \cup \{x\}) \; \forall S \in E$

2. Calculation of 2^n probabilities of lists

We use an approximation $AP(S \cup \{x\})$ for $L(S \cup \{x\})$:

$$AP(S \cup \{x\}) = L(S) + 2d(x, S), \qquad (18.10)$$

where $d(x, S)$ is the distance between x and the closest node in S. In our heuristic we use the objective function

$$\min_{x \in V} \sum_{S \in E} p(S) AP(S \cup \{x\}), \tag{18.11}$$

instead of (18.2). But from (18.10) since $L(S)$ is not a function of x expression (18.11) is equivalent to

$$\min_{x \in V} \{f(x) = \sum_{S \in E} p(S) d(x, S)\}. \tag{18.12}$$

Next we show that (18.12) can be calculated efficiently. Let us consider any node x and let us rename the nodes in V, $j_1, j_2, \ldots, j_n$, such that $d(x, j_k) \leq d(x, j_\ell)$ for $\ell \geq k$ (since x is also a node, $j_1 \equiv x$). Let E_k be the set of all the lists where node j_k is the first node to be visited from x, i.e.,

$$E_k = \{S \in E; j_1, j_2, \cdots, j_{k-1} \notin S, j_k \in S\}, \quad k \geq 2.$$

Now $f(x)$ can be written as

$$f(x) = \sum_{k=2}^{n} d(x, j_k) \sum_{S \in E_k} p(S). \tag{18.13}$$

In [1080] it was shown that

$$\sum_{S \in E_k} p(S) = p_{j_k} \prod_{i=1}^{k-1} (1 - p_{j_i}). \tag{18.14}$$

In our heuristic we calculate (18.13) $\forall x \in V$ and thus obtain

$$\tilde{x} = \arg \min_{x \in V} f(x). \tag{18.15}$$

Observe that given the shortest distance matrix $D, \tilde{x}$ can be found in $O(n^2 \log n)$ time. The overall complexity of the heuristics can be further reduced to $O(n \log n)$, [107].

We now present a worst case analysis for our heuristic.

Theorem 3 $AP(S \cup \{x\}) \leq \frac{3}{2} L(S \cup \{x\}) \quad \forall S \in E, \quad \forall x \in V.$

The proof is given in [1080].

The bound obtained in Theorem 3 is tight as shown in the example depicted in Figure 6.

Finally we can present our worst case result

Lemma 2 $\dfrac{g(\tilde{x}) - g(x^*)}{g(x^*)} \leq \dfrac{1}{2}.$

FIGURE 6. Tightness of the bound

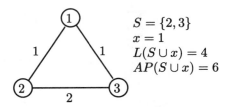

$$S = \{2, 3\}$$
$$x = 1$$
$$L(S \cup x) = 4$$
$$AP(S \cup x) = 6$$

The proof is given in [1080].

Notice that in our method we have not solved the problem of finding $L(X)$ the shortest TST that visits all the customers in the set X. We assume that $L(X)$ can be obtained for each day. Based on this assumption, we have found $\tilde{x}$. If there are subsets S with many vertices such that the traveling salesman problem cannot be solved exactly to obtain $L(S \cup \{x\})$ then one can use the following heuristic for $L(S \cup \{x\})$. $AP'(S \cup \{x\}) = 2d(x, S) + C(S)$, where $C(S)$ is the length of the traveling salesman tour through S according to the heuristic developed in [215]. Since $C(S) < \frac{3}{2}L(S)$ [215] and using Theorem 3 we get $AP'(S \cup \{x\}) < 2L(S \cup \{x\})$, hence the solution we have found will remain the same but the worst case relative error will increase by $\frac{1}{2}$.

The worst-case performance of the above heuristic can be improved, [107], to $(1 - p_{x^*})/2$, where x^* is the optimal location. Unfortunately, since the optimal location is unknown, p_{x^*} is unknown and therefore has to be replaced by a lower bound.

In the next section we show that our heuristic can be also used to solve the problem on the plane with either rectilinear or Euclidean distances.

18.2.3   The Traveling Salesman Location Problem in the Plane

Rectilinear distances: Now we focus on the problem in the plane with rectilinear distances. Let $W = \{w_1, w_2, \cdots w_m\}$ be the set of demand points (customers) where $w_i = (a_i, b_i)$ and let $x = (x_1, x_2)$ be a candidate location. The distance between any two demand points w_i and w_j is $|a_i - a_j| + |b_i - b_j|$.

In [1079], it is shown that the optimal solution can be obtained by solving the problem on its rectangular-grid network, $G_r(V, A)$. An example of a rectangular-grid network is depicted in Figure 7 (ignore (u, v) and x^* in the figure, notice that $W \subseteq V$). The major result is expressed in the following theorem.

Theorem 4 *There exists at least one optimal solution of our problem on* $G_r(V, A)$.

The proof of Theorem 4 is included in [1079]. Essentially it shows that

FIGURE 7. The problem and the grid-network presentation

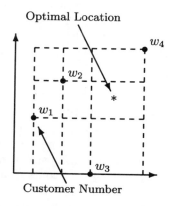

Optimal Location

Customer Number

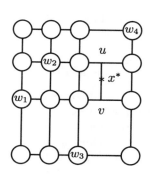

a: The Problem on the Plane
   with Four Customers

b: The Grid Network

if $x^* \notin G_r(V, A)$ then by adding a new link to $G_r(V, A)$ that contains x^* (see link (u, v) in Figure 6b), x^* must be optimal in the new grid-network, but by Theorem 1, either u or v must be optimal as well. Moreover due to Theorem 1, there must exists an optimal solution $x^* \in V$ to the problem. Therefore, our heuristic from the last section can find the optimal solution on $G_r(V, A)$ with the same worst case relative error.

Euclidean distances: The problem on the plane with Euclidean distances is very difficult, first of all, any point in the convex hall of W is a possible candidate to be optimal. Second, the quantity $\sum_{S \in E_k} p(S)$ required by our heuristic is a function of x (observe that now x is not necessarily nodal). In [1079] we show that the convex hall can be divided into convex polygons $A_1, A_2, \cdots, A_z$ such that on each $A_i, i = 1, \cdots, z$, the quantity $\sum_{S \in E_k} p(S)$ remains constant. Constructing A_i can be done in $O(|W|^4)$ time using any one of the algorithms [196, 363, 364].

Therefore minimizing $f(x)$ in (18.13) is equivalent to

$$\min_{i=1,\cdots,z} f(\tilde{x}_i), \qquad (18.16)$$

where (denoting $C_k = \sum_{S \in E_k} p(S)$),

$$f(\tilde{x}_i) = \min_{x \in A_i} \sum_{k=1}^{n} d(x, j_k) C_k. \qquad (18.17)$$

Notice that (18.17) is the classical Weber problem [724] which is the minisum location problem on the plane with Euclidean distances when there is no restriction that the solution is to be contained in A_i. But since the function minimized in (18.17) is convex, several algorithms (among others, Weiszfeld's [724] and Chandraseharan and Tamir's [189] ϵ−approximation) can find $\tilde{x}_i$ without the restriction that the solution must be in A_i. If the solution is not in A_i it is on the edges of A_i and then can be found by solving one-dimensional Weber problems. The worst case relative error is $\frac{1}{2}$ when using [1080] and $\frac{1}{2} + \epsilon$ when using [189].

18.2.4   The Multiserver Routing-Location Problem

Some of the previous results can be generalized for the multiserver case. Specifically, consider the problem of locating one service station in a given region. Assume that a single server can visit at most q customers on a single tour. Thus, if there are calls for service from a set of S customers, at least $\lceil |S|/q \rceil$ have to leave the center and perform a set of tours that visit each one of the customers that require service such that the server capacity is not exceeded and total length of routes is as small as possible. Our objective is to find a location for the service station so as to minimize expected distance traveled by all servers, over all possible lists.

The model is clearly related to vehicle routing problems in general and to the Capacitated Vehicle Routing Problem in particular. [1077] uses results from that literature to suggest a heuristic with a constant error guaranteed and whose relative error tends to zero as the number of customers increases. To present the result, let $g(x)$ be the minimal expected distance traveled by the servers, given that the service station is at point x. Let H be a heuristic algorithm for the capacitated routing problem, i.e., the problem that needs to be solved every day to generate efficient routes for the servers. Let $g^H(x)$ be expected distance traveled using this heuristic algorithm and when the service station is at point x. Let x^* be the optimal location for the service station and $\tilde{x}$ is a location associated with the heuristic H. The following result is proven in [1077] .

Theorem 5 *There exists a heuristic H and a location $\tilde{x}$ such that*
$$\frac{g^H(\tilde{x}) - g(x^*)}{g(x^*)} \leq 3 - \frac{2}{q}. \text{ Also, let the customers be uniformly distributed in}$$
a given area with $E[d(x)]$ being the expected distance from the depot, located at point x. Let $p_i = p$ for every customer. Then for every x, we have almost surely, $\lim_{n\to\infty} \frac{g(x)}{n} = \lim_{n\to\infty} \frac{g^H(x)}{n} = \frac{2p}{q} E[d(x)].$

Thus, the theorem says that for large number of customers, expected total distance traveled is dominated by the total radial cost between the depot and the customers. This insight will be used in the Section 18.4 to construct an efficient strategy for the design of distribution systems.

18.3   The Probabilistic Traveling Salesman Location Problem

18.3.1   The Probabilistic Traveling Salesman Problem

The expected length of a given PTSP tour τ: Let us consider a PTSP defined on a given complete graph $G = (V, A)$, with a cost $d : A \to R$. To facilitate the derivation of analytic results and without loss of generality, it will be again assumed that the vertices are mutually independent with regard to probabilities p_i, $1 \leq i \leq n$.

Consider now the tour $\tau = (1, 2, \ldots, n, 1)$ and let $E[L_\tau]$ be the expected length of τ. Note that in this case the tour length L_τ can take 2^n different values, the same as the number of instances involving "present" and "absent" vertices. For each such instance, we would require $O(n)$ additions to compute L_τ. Thus were we to use an enumeration approach, the computational effort to compute $E[L_\tau]$ would be $O(n2^n)$ for any given tour τ. Fortunately, a much more efficient approach exists [633, 634, 102].

Theorem 6

$$
\begin{aligned}
E[L_\tau] \;=\; & \sum_{i=1}^{n} \sum_{j=i+1}^{n} d(i,j) p_i p_j \prod_{k=i+1}^{j-1} (1 - p_k) \\
& + \sum_{j=1}^{n} \sum_{i=1}^{j-1} d(j,i) p_i p_j \prod_{k=j+1}^{n} (1 - p_k) \prod_{k=1}^{i-1} (1 - p_k). \quad (18.18)
\end{aligned}
$$

Sketch of the proof: This result follows directly from the following argument: the instances of τ for which $d(i,j)$ makes a contribution to $E[L_\tau]$ are those in which the vertices i and j are both present, while the vertices $i+1, \ldots, j-1$ are absent (and are thus skipped in traversing the tour). $\square$

Theorem 6 shows that $E[L_\tau]$ can be computed in $O(n^2)$ time under very general conditions. Jaillet [632, 630] gives a generalization of Theorem 6 for the case in which the independence assumption does not hold and also discusses a number of variations and extensions of the theorem.

Asymptotic comparison of re-optimization and a priori optimization: We turn next to the issue of characterizing and comparing the asymptotic behavior of the re-optimization and the *a priori* strategies for the PTSP, if the locations of the points are uniformly and independently distributed in the Euclidean plane. This comparison is important in order to assess the promise and potential usefulness of the *a priori* strategies. We will be quite informal; the interested reader can consult [628] for a detailed and rigorous treatment of these issues, as well as for generalizations.

Let $X^{(n)} = (X_1, \ldots, X_n)$ be n points uniformly and independently distributed in the unit square. Let L_{TSP}^n be the length of the TSP defined on $X^{(n)}$.

Let $E[\Sigma_{TSP}^n]$ be the expectation of the TSP solutions obtained under the re-optimization strategy defined on $X^{(n)}$.

Let $E[L_{PTSP}^n]$ be the expectation of the *a priori* strategy, i.e. the expected length of the optimal *a priori* solution to the PTSP defined on $X^{(n)}$.

It is well known [65] that we can characterize very sharply the solutions to the deterministic TSP.

Theorem 7 *With probability 1* $\lim\limits_{n\to\infty} \frac{L_{TSP}^n}{\sqrt{n}} = \beta_{TSP}$.

This almost sure convergence was later strengthened by Steele [1098] to include complete convergence, i.e.,

Theorem 8 $\forall \varepsilon > 0, \sum\limits_{n} P\left(\left|\frac{L_{TSP}^n}{\sqrt{n}} - \beta_{TSP}\right| > \varepsilon\right) < +\infty$.

We now characterize the expectation of the re-optimization strategy for the PTSP assuming that each of the n points is present with the same constant probability p, which is called the coverage probability. We remark that in the following theorem the expectation is taken over all the possible 2^n instances of the problem and the probability 1 statement refers to the random locations of the points.

Theorem 9 *With probability 1* $\lim\limits_{n\to\infty} \frac{E[\Sigma_{TSP}^n]}{\sqrt{n}} = \beta_{TSP}\sqrt{p}$.

Sketch of the proof: The intuitive idea in the proof is that the principal contribution to $E[\Sigma^n]$ comes from the sets S with $|S|$ close to np. The reason is that the number of points present is given by a Binomial random variable with parameters n, p and hence is almost surely asymptotically equivalent to np. In this range of $|S|$ we can apply Theorem 8 to obtain Theorem 9. Note that in [628], an explanation is given on why the complete convergence of Theorem 8 is crucial for this kind of argument. $\square$

Intuitively Theorem 9 means that solutions under the re-optimization strategy behave asymptotically similarly to those of the corresponding deterministic TSP but on np rather than n points. We next characterize asymptotically the *a priori* optimization strategy.

Theorem 10 *With probability 1* $\lim\limits_{n\to\infty} \frac{E[L_{PTSP}^n]}{\sqrt{n}} = \beta_{PTSP}(p)$.

Sketch of the proof: We first prove that the PTSP belongs to the class of subadditive Euclidean functionals whose asymptotic behavior has been characterized by [1099]. Their value is almost surely asymptotic to $c\sqrt{n}$, where c depends on the functional. For a detailed proof the reader is again referred to [628]. $\square$

Comparing Theorems 9 and 10 we can observe that the *a priori* and re-optimization strategies have similar asymptotic behaviors almost surely. Both theorems prove the existence of a constant but without determining the value of the constant analytically; in fact, for most similar asymptotic results, the respective limiting constants are unknown and only bounds or experimental estimations have been established (see [42, 629] for an important exception concerning the minimum spanning tree problem). In fact the current best known result on the relationship between $\beta_{PTSP}(p)$ and $\beta_{TSP}\sqrt{p}$ was obtained in [632] and is reproduced here: $\frac{5}{8}\sqrt{p} \leq \beta_{TSP}\sqrt{p} \leq \beta_{PTSP}(p) \leq \min(\beta_{TSP}, 0.9204\sqrt{p})$. On the other hand extensive experimental work by Johnson [642] suggests that $\beta_{TSP} \approx 0.71$. Note that it is tempting to conjecture that $\beta_{PTSP}(p) = \beta_{TSP}\sqrt{p}$, but no correct proof exists of this result [contrary to the erroneous claim in [110]]. Yet, the *a priori* strategy does seem to behave (asymptotically) equally well on average with the re-optimization strategy on Euclidean problems.

The complexity of a priori optimization: Having shown that *a priori* strategies are attractive compared with re-optimization strategies (at least for the PTSP) we now turn to the question of how difficult it is to find the optimal *a priori* solution from a computational complexity perspective.

We first introduce the decision version of a PTSP. Given a complete graph $G = (V, A)$, $|V| = n$, a cost $d : A \to R$, a vector $(p_1, \ldots, p_n)$ of the probabilities of presence of the vertices and a bound B, does there exist a PTSP tour f such that $E[L_f] \leq B$?

We can then characterize the complexity of the *a priori* strategy as follows:

Theorem 11 *The decision version of the PTSP is NP-complete.*

Sketch of the proof: We only need to show membership in NP, since the PTSP is a generalization of a well known NP-complete problem [445]. Membership in NP is seen to hold, since, given a solution f, we can compute $E[L_f]$ in $O(n^2)$ as we have shown in Theorem 6. □

Thus, although we can compute efficiently the expected length of any given a priori solution to a PTSP, it is still NP-hard to find an optimal *a priori* solution.

Theoretical approximations to optimal a priori solutions: In the previous section we found that it is still NP-hard to obtain optimal *a priori* solutions to the PTSP. In this section we address the question of approximating the optimal *a priori* solution with polynomial time heuristics, whose worst case behavior we can characterize.

The first natural question to address is how heuristic approaches to the deterministic problem perform when applied to the corresponding probabilistic problem. For example, what is the performance of the well-known Christofides heuristic for the TSP [740] if applied to the PTSP? In order to find useful bounds for the routing problems (PTSP) we assume below that the triangle inequality holds. We can then prove the following:

Theorem 12 *Let L_{TSP} be the length of the optimal solution to the deterministic TSP and let L_H be the length of a heuristic solution to the same* *problem. Let p be the coverage probability and $E[L_{PTSP}]$ the expected length of the optimum* a priori *solution to the corresponding PTSP. If the heuristic has the property that $\frac{L_H}{L_{TSP}} \le c$, then $\frac{E[L_H]}{E[L_{PTSP}]} \le \frac{c}{p}$.*

Sketch of the proof: Using the triangle inequality, we know that, for any tour f, $E[L_f] \le L_f$. In addition, we show that $E[L_{PTSP}] \ge pL_{PTSP}$. Combining these inequalities the result follows. $\Box$

Theorem 12 suggests that if the coverage probability is large then constant guarantee heuristics for the deterministic problem still behave well for the corresponding probabilistic problem. But if $p \to 0$ the bound is not informative and indeed one can find examples with $p \to 0, np \to \infty$ for which $\frac{E[L_{TSP}]}{E[L_{PTSP}]} \to \infty$, that is, even if $c = 1$, the optimal deterministic solution is an arbitrarily bad approximation to the optimal *a priori* solution. For a number of interesting examples see [632, 630]. As an indication of the rate at which the ratio $\frac{E[L_{TSP}]}{E[L_{PTSP}]}$ tends to infinity, Bertsimas [106] proves the following:

Theorem 13 *For the PTSP with triangle inequality $\frac{E[L_{TSP}]}{E[L_{PTSP}]} = O(\sqrt{n})$.*

We next investigate the existence of constant guarantee heuristics. We restrict our attention to Euclidean problems and examine the space filling curve heuristic, first proposed by [962] for the Euclidean TSP. The space filling curve heuristic can be described as follows:

1. Given the n coordinates (x_i, y_i) of the points in the plane compute the number $f(x_i, y_i)$ for each point. The function $f : R^2 \to R$ is called the Sierpinski curve. For details on the computation of $f(x, y)$ see [53].

2. Sort the numbers $f(x_i, y_i)$ and visit the corresponding initial points (x_i, y_i) in that order, producing a tour denoted SF.

The key property of the space filling curve heuristic that makes its analysis for the PTSP possible is the following: Consider an instance S of the problem. Suppose the space filling curve heuristic produces a tour $SF(S)$ if we run the heuristic on the instance S. Consider now the tour SF produced by the heuristic on the original instance of the problem, i.e. when all points are present. What is the tour that the PTSP strategy would produce in instance S if the *a priori* tour is SF?

The answer is precisely $SF(S)$, because sorting has the property of preserving the order in which the points in S will be visited by the space filling curve, which is exactly the property of the PTSP strategy as well. Based on this critical observation we can then analyze the spacefilling curve heuristic.

Theorem 14 *For the Euclidean PTSP the spacefilling curve heuristic produces a tour SF with the property $\frac{E[L_{SF}]}{E[L_{PTSP}]} \le \frac{E[L_{SF}]}{E[\Sigma_{TSP}]} = O(\log n)$.*

Sketch of the proof: In [962] it is proven that the length of the spacefilling curve heuristic satisfies: $\frac{L_{SF}}{L_{TSP}} = O(\log n)$. Consider an instance S of the problem. If the spacefilling curve heuristic is applied to the instance S, it will similarly produce a tour $SF(S)$ with length $\frac{L_{SF}(S)}{L_{TSP}(S)} = O(\log |S|) = O(\log n)$. But since $SF(S)$ is the tour produced by the PTSP strategy at instance S then

$$\frac{E[L_{SF}]}{E[\Sigma_{TSP}]} = \frac{\sum\limits_{S \subseteq V} p(S) L_{SF}(S)}{\sum\limits_{S \subseteq V} p(S) L_{TSP}(S)} \leq \frac{\sum\limits_{S \subseteq V} p(S) O(\log n) L_{TSP}(S)}{\sum\limits_{S \subseteq V} p(S) L_{TSP}(S)} = O(\log n).$$

$\square$

Note that this result does not depend on the probabilities of points being present. It holds even if there are dependencies on the presence of the points. Observe also that the spacefilling curve heuristic ignores the probabilistic nature of the problem but surprisingly produces a tour which is globally (in every instance) close to the optimal.

As a corollary to Theorem 14 we can compare the PTSP and the re-optimization strategies from a worst-case perspective. For the Euclidean PTSP, since $E[L_{PTSP}] \leq E[L_{SF}]$, $\frac{E[L_{PTSP}]}{E[\Sigma_{TSP}]} = O(\log n)$. Platzman and Bartholdi [962] conjecture that the spacefilling curve heuristic is a constant-guarantee heuristic. Unfortunately, Bertsimas and Grigni [109] showed this conjecture to be false, and the existence of a constant guarantee heuristic for the Euclidean PTSP remains open.

Heuristics and exact algorithms for the PSTP: In [632], several heuristics were proposed for the PTSP, and later were tested and extended by [640, 1024]. More recently more numerical works were conducted by [207, 73] who, among other things, studied various simulated annealing schemes.

In another set of numerical experiments, Bertsimas [106] has obtained solutions to Euclidean PTSPs by means of two different types of heuristics. The first of them is the spacefilling curve heuristic, while the second is based on seeking local optimality. For problems involving equal probabilities $p_i = p$, and not more than a few hundred nodes, some success was achieved with two separate iterative improvement algorithms based on the idea of local optimality.

Lin [760] used an iterative improvement algorithm for the TSP based on what he called the λ-opt local neighborhood. For a given tour τ consisting of n links between nodes, the neighborhood $S_\lambda(\tau)$ consists of those tours which differ from τ by no more than λ links. For $\lambda = 2$ this is the set of tours which can be obtained by reversing a section of τ; for $\lambda = 3$ it is the set of tours obtainable by removing a section of τ and inserting it, with or without a reversal, at another place in the tour. Both the 2-opt and 3-opt TSP algorithms were implemented, since when p is greater than about 0.5

the TSP solutions provide useful starting points for our more general PTSP routines.

Unlike the TSP case, the expected length $E[L_\tau]$ in the PTSP sense depends on all $(n^2 - n)/2$ independent elements of the distance matrix. We cannot, therefore, speak of some links leaving and others entering the tour; rather, it is only the weight given to each of the $d(i,j)$ by equation (18.18) which changes. We can still use Lin's λ-opt neighborhoods, but the computation of the changes in expected length becomes considerably more complicated. It takes $O(n^2)$ time to calculate the change in expected length from τ to an arbitrary tour in $S_2(\tau)$, so it would seem at first that testing for even 2-p-optimality (referred to heretofore as "2-p-opt") would take $O(n^4)$ time. We can, however, reduce this to $O(n^2)$ if we examine the tours in the proper sequence and maintain certain auxiliary arrays of information as the computation proceeds.

Another neighborhood tried by [106] consists of moving a single node to another point in the tour, rather than reversing an entire section. The corresponding neighborhood, which we call the 1-shift neighborhood, has roughly twice as many members as S_2, it is a subset of S_4, and yields much better results than S_2 in our experiments.

The best general approach described in this work seems to be to first use the spacefilling curve algorithm, followed by 3-opt if p is fairly large, and then finish by applying 1-shift. The threshold point below which 3-opt ceases to be helpful is uncertain and probably depends strongly on the specifics of the problem. For a detailed description of the numerical results the reader is referred to [106].

Finally, in terms of exact algorithms, very little has been done. In [632], various branch and bound schemes were proposed but not tested numerically. The only real exact algorithm development and testing has been done very recently by [735] who report solving optimally special cases of the PTSP with sizes of up to 50 nodes.

The a priori optimization models discussed above correspond to the case where expected value of TSP criterion is minimized. A priori routing-location problems with other criteria are analyzed in [41].

18.3.2 The Probabilistic Traveling Salesman Location Problem

Algorithms for the PTSLP: For this problem [102] also proved that locating the facility in one of the nodes is optimal. Thus, we only need to examine nodes as possible solution to the PTSLP.

No constant-guarantee heuristic is available for the PTSLP on general network. Berman and Simchi-Levi [102] proposed an $O(n^3)$ algorithm to find the optimal location if the a priori tour is given, which later was shown to be $O(n^2)$ by [106]. Also, [106, 107] has characterized the worst case performance of a Nearest Neighbor Location Heuristic and of a Spacefilling Curve Location Heuristic and showed that both provide a worst-case performance bound which is $O(logn)$, while being of complexity $O(n^2)$.

Also, the discerning reader may have already observed that any node i with $p_i = 1$ is an optimal location for the PTSLP (and as a matter of fact for the TSLP as well). If a node is to be visited on every instance of the problem anyway, we are assured that there will be no extra-distance penalty, if the facility is also located on that node! This observation, however, is true only if the distance matrix satisfies the triangular inequality.

Finally, it should be noted that only recently have some more formal attempts been made — by using large-scale mathematical programming formulations — to obtain optimal or heuristic solutions to stochastic problems, like the TSLP and the PTSLP, which combine locational and routing considerations [734, 735].

The PTSLP in the plane: For the PTSLP in the plane we easily have [110]:

$$\lim_{n\to\infty} \frac{E[L^n_{PTSLP}]}{\sqrt{n}} = \beta_{PTSP}(p). \qquad (18.19)$$

The limit in (18.19) is identical to that of Theorem 10, a result which is not surprising if one considers the fact that, in the limit, the location of the PTSLP facility is not important when the number of potential demands to be visited is arbitrarily large. Again, it is tempting to conjecture that the TSLP and PTSLP are equivalent, but this again depends on the proof that $\beta_{PTSP}(p) = \beta_{TSP}\sqrt{p}$, still an open problem.

18.3.3 The Multiserver Case

Some of the previous results can be generalized to more than one server. Let us first review generalizations of the PTSP.

The capacitated PTSP and capacitated m-PTSP: Specifically, we still consider demands which are probabilistic in nature and our problem is to determine *a priori* routes of minimal expected length for vehicles with finite capacity. The complications introduced by the finite capacity of the vehicles is a major point of interest. A first problem is to consider a single vehicle and design a giant *a priori* vehicle tour through all the demand points. While covering this tour the vehicle may run out of capacity and, in such an event, it will have to return to the depot — for instance, in order to deposit the load it has picked up at the points it has already visited. Thus, the expected tour length to be minimized must also include any additional distance traveled to and from the depot whenever the vehicle reaches its capacity. There is, of course, an alternative interpretation under which the very same problem can be viewed as a multi-vehicle probabilistic vehicle routing problem. This can be seen best if one sets p_i, the probability of visiting point i, equal to 1 for all i. Then the approach just described is identical to one of the two standard approaches to multi-vehicle deterministic vehicle routing problems (VRPs), namely "route first, cluster second". Under this interpretation, the returns of the vehicle to the depot result in multiple tours, so that we are dealing with multiple-VRP tours as solutions to the overall problem. However, in the general case when some of the p_i are strictly less than 1, some criterion or criteria must be used in order to break up the giant *a priori* tour into clusters of customers — with each cluster served by a different vehicle.

In order to consider all these aspects in a more specific way let us now define four generalized versions of the PTSP that can be classified as PVRPs.

The Capacitated Probabilistic Traveling Salesman Problem: Assume that each point x_i, requiring a visit with a probability p_i, independently of all others, has a unit demand, and that the salesman (vehicle) has a capacity q. We wish to find *a priori* a tour through all n points. On any given instance, the subset of points present will then be visited in the *same order* as they appear in the *a priori* tour. Moreover if the demand on the route exceeds the capacity of the vehicle, the salesman has to go back to the depot before continuing on his route. The problem of finding such a tour of minimum expected total length (the expected length of the tour in the PTSP sense plus the expected extra distance due to overloading) is defined as a Capacitated PTSP. This problem is the "Probabilistic Vehicle Routing Problem under Updating Method $\mathcal{U}_b$" as described in [634] (see also Figure 8).

The m-Probabilistic Traveling Salesmen Problem (m-PTSP): Consider the problem of routing through a set of n points starting from and ending at a depot. On any given instance of the problem, only a random subset of points (each unit-demand point x_i being present with a probability p_i, independently of the others) has to be visited. We wish to find, *a priori*, m subtours, each starting from and ending at the depot, such that each point is included in exactly one tour, and of minimum total expected length under the skipping strategy (method $\mathcal{U}_b$). For this problem we have m vehicles with no capacity limits.

The Capacitated m-Probabilistic Traveling Salesmen Problem: This capacitated vehicle problem is a natural combination of the Capacitated PTSP and of the m-PTSP.

The General Probabilistic Vehicle Routing Problem: This is the same as the Capacitated m-PTSP except that the demand of each customer x_i is no longer modeled by a Bernoulli random variable with parameter p_i, but rather by a more general random variable.

We now summarize our main results on the m-PTSP, Capacitated-PTSP, and Capacitated m-PTSP in the plane. For details, the reader is referred to [633, 634, 106, 631, 108].

FIGURE 8. The two updating methods

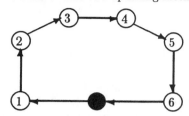

(i) A priori route through 6 customers (each with a
demand of 0 or 1 unit) by a vehicle of capacity 2

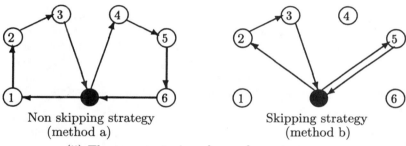

Non skipping strategy   Skipping strategy
(method a)   (method b)

(ii) The two strategies when only customers
2, 3, and 5 have a positive demand

As mentioned above, several "good" heuristics have been proposed for the
PTSP but none of them has been so far amenable to a theoretical analysis
which would prove that the heuristic provides a "constant guarantee" (in
the worst-case sense).

For the analysis of the other problems, we will nevertheless assume that
it is possible to obtain optimal or near-optimal solutions to the PTSP. Any
progress on the PTSP will automatically translate into progress on the
more complicated probabilistic vehicle routing problems.

The m-PTSP: In contrast to the relationship between the TSP and the
m-TSP, the m-PTSP cannot be transformed into an equivalent PTSP. Nevertheless the heuristic which consists of assigning the first $m-1$ vehicles to
the $m-1$ closest points to the depot, and the last vehicle to the remaining
points can be easily analyzed mathematically, [631].

The Capacitated PTSP: If we denote by $H^c(x^{(n)})$ the value of the
heuristic, say $\mathcal{H}^c$, which consists of simply using the optimal PTSP tour
for this problem, we have a heuristic with a worst-case ratio of $q+1$ ([631]),
which is not very good especially when q is big. Let us analyze an improve-

ment of this algorithm, introduced in [106] for a simpler version of this problem. This heuristic, based on ideas contained in [508], is the following:

Cyclic Heuristic:

Step 1: Find an optimal PTSP tour $t_1 = (x_0, x_{\sigma^*(1)}, x_{\sigma^*(2)}, \ldots, x_{\sigma^*(n)}, x_0)$ and consider the tours

$t_k^* = (x_0, x_{\sigma^*(k)}, x_{\sigma^*(k+1)}, \ldots, x_{\sigma^*(n)}, x_{\sigma^*(1)}, \ldots, x_{\sigma^*(k-1)}, x_0)$ for $k \in [2..n]$.

Step 2: Compute the objective values $E_{t_k}^c$ for $k \in [1..n]$.

Step 3: Take the tour with the minimum objective value.

We then have a heuristic with a worst-case ratio of $2(1 + np/q)\overline{d}$ ([631]). When n goes to infinity, we then have a constant-guarantee (independent of q) heuristic.

Moreover, with the help of the Cyclic Heuristic, one can get the asymptotic behavior of the Capacitated PTSP. Indeed, one can show [631] that the Cyclic Heuristic is asymptotically optimal with probability one.

The Capacitated m-PTSP: Let us consider the heuristic, here called the m-Cyclic Heuristic, which consists of assigning the first $m - 1$ vehicles to the $m - 1$ closest points to the depot, and using the Cyclic Heuristic on the remaining $n - m + 1$ points. Then, letting $n_1 = n - m + 1$, we have, as before, a heuristic with a worst-case ratio of $2 + \frac{1}{n_1}(\frac{q}{p} - 1)$. One can also show the asymptotic optimality of the m-Cyclic Heuristic [631].

More general probabilistic vehicle routing problems: We have discussed here cases of unit demand at each demand point. It is in fact possible to consider the case of unequal demands in much the same way as in [508] (for example [108]). Following [633], one can also consider more general probability distributions for the demand at each point, such as binomial distributions, and re-derive most of the previous results. Finally in our multi-vehicle models, we did not consider the importance of balanced routes. If this is an issue, modifications of several heuristics proposed in [508] could be analyzed successfully for a probabilistic environment, as well.

The multiserver PTSLP: The extensions of the PTSLP to the multiserver case can now be done without major difficulties. As in Section 18.2.4, assume that a single server can visit at most q customers on a single tour. Thus, if there are calls for service from a set of S customers, at least $\lceil |S|/q \rceil$ have to leave the center and perform a set of tours that visit each one of the customers that require service such that the server capacity is not exceeded and total length of routes is as small as possible. Our objective is to find a location for the service station so as to minimize expected distance traveled by all servers (using the *a priori* strategy). For each possible node location, one can use the results of the previous section in order to design the various *a priori* tours for the different servers.

18.4   Applications to Systems Design and Strategic Planning

The results described in the previous sections enable us to develop analytical models to assist the design and control of distribution systems.

As an example, consider a company that delivers consumer goods to a number of stores located in an area of size A. The company has decided to open a number of warehouses in the region and has carried out a market survey to estimate the number of potential customers, denoted by n, and the probability distribution of their demands. At this preliminary stage of the analysis the company assumes that all potential customers have the same probability distribution with $\overline{w}$ being the expected customer's demand.

Based on the information available from the survey, the company wants to determine the number and locations of service stations; how to allocate customers to depots and what should be the routing strategies so as to minimize total system cost. This cost includes a cost associated with the average distance traveled by all vehicles plus a fixed set-up cost, denoted by c, for each established depot.

The insight obtained from the analysis of the capacitated vehicle routing problem as well as the previous location-routing problems can be used to propose a three-stage hierarchical approach in which decisions about the number of centers and their locations (first stage), customers allocations (second stage) and routing strategies (third stage) are combined to reduce total system cost.

As we have seen, the total radial cost between the depot and the customers dominates the objective function. This cost is, of course, related to the cost of the K-median problem. Thus, the K-median problem provides an insight to our model. For instance, when the demand of a customer can be split over several vehicles, the total distance traveled in a distribution system with K centers optimally located in the area is asymptotically $\frac{2\overline{w}n}{q}\beta\sqrt{\frac{A}{K}}$, where $\beta = 0.377196\dots$ and q is the vehicles' capacity. This is true as long as $1/K = o(n)$ and $K = O(n/\log n)$. Furthermore, this asymptotic value is achieved by placing centers in a regular hexagonal pattern throughout the area and each service center serves all the customers inside its hexagon.

It follows that, for large enough n, the best number of service stations is the minimizer of the following function (recall that c is a fixed set-up cost for establishing a depot) $TC(K) = \frac{2\overline{w}n}{q}\beta\sqrt{\frac{A}{K}} + cK$.

Let $\delta = (\beta\frac{\overline{w}n}{qc}\sqrt{A})^{2/3}$, then the best number of stations is $\lceil\delta\rceil$ or $\lfloor\delta\rfloor$ whichever yields the best $TC(K)$. Furthermore, the analysis also shows where to locate the centers and how to allocate customers to each center, thus providing answers to the strategic and tactical problems.

What should be the routing strategy used on a daily basis? Note that in this model, we assume that every working day the servers have exact information on the customers that need service and their actual demands. Hence, every working day, each center faces an instance of the single depot capacitated vehicle routing problem, for which efficient heuristics exist.

We can now summarize the hierarchical design: choose the number of stations as the one that minimizes $TC(K)$, locate the facilities at the center of hexagonal patterns (strategic decision), each customer will be served by its closest service station (tactical decision) and use a special heuristic, called OP heuristic, on a daily basis to find efficient routes for servers (operational control).

The main difficulty in using a hierarchical approach to design is to evaluate the overall performance of the system. In theory, the approach provides a design in which the total system cost approaches the minimal total cost, as the number of customers tends to infinity. In practice, however, the rate of convergence of the hierarchical design cost to the optimal cost may be quite slow. To estimate this rate of convergence [1078] describes a series of numerical experiments for problems of **moderate** size. For instances, for three distribution centers, located according to the hierarchical design, the relative error between the hierarchical design cost and a lower bound on the optimal solution value decreases from 35% to 16% as the number of customers increases from 100 to 500.

18.5   A Different Class of Stochastic Facility Location Problems

To conclude this chapter, we note that an extensive literature also exists on a different set of stochastic facility location problems which, however do not involve any routing considerations. These problems are characterized by demands which appear at random times and random locations on a network and are served by stationary facilities or traveling vehicles that can serve only one demand at a time. Coupled with service times whose duration is a random variable, these conditions generate queueing phenomena. Thus, in this set of problems "optimal facility (or vehicle) locations" are typically defined to be those that minimize the expected total time that elapses between the occurrence of a demand and the completion of service to that demand. Note that this total time includes any time spent waiting for a facility or vehicle to become available. Comprehensive reviews of this class of facility location problems are provided in [892, 56].

19

LOCATION, ROUTING AND THE ENVIRONMENT

Brian Boffey
University of Liverpool

John Karkazis
University of the Aegean

19.1   Introduction

Increasing interest in the environment has been witnessed and concern for its well-being has been expressed over the last two decades or so. This is largely due to a better understanding of the serious and long lasting effects of human polluting activities on local and global ecosystems. Accidents related to the nuclear and petrochemical industries, whether experienced or merely feared, have contributed much to this end. This chapter will concentrate on airborne pollutants and their importance when locating obnoxious facilities (often involving producing/processing/storing hazardous materials) and the routing of vehicles carrying hazardous materials (called obnoxious vehicles by [57]). For a more detailed and complete account of work in this area the interested reader may refer to the survey articles [385, 1148, 117] and Chapter 20 in this book. The essence of the problems is that, in addition to cost minimisation, benefit maximisation etc, it is required to minimise some detrimental interaction 'with a neighbourhood' of the facility and/or route used. In principle this interaction may be in either direction but we shall only be interested in the effect the facility or vehicle has on the environment as happens, for example, when acid rain results from pollutants emitted from the stack of a coal-fired power station or when chlorine gas escapes after an accident from a vehicle carrying the gas. Problems involving obnoxious facilities and/or vehicles will be termed (for want of a better term) *hazard* problems as opposed to *ordinary* problems.

Hazard problems deal with situations which are fundamentally multi-objective. Also the environmental interaction is often complex involving a variety of costs, with risk distributed asymmetrically - the benefits and disbenefits of a location/routing decision often being felt by different groups - and with a variety of actors (manufacturers, carriers, legislators, man-in-the-street etc).

Research in this area, though still relatively sparse, is increasing rapidly.

Much is restricted to (sometimes unrealistically) simplified models, which is natural in the early stages of development of a complex area. In addition to the need to continue work with simplified models (to gain additional theoretical understanding) there is also a need for increased attention to be paid to more complex and realistic models with problems being solved by heuristic and interactive methods, and use being made, where appropriate, of computer graphics. Also, at least some of the more complex problems can be reduced to simpler ones if k-best solution methods are used rather than just a method to find a single 'best' solution [117, 118, 115].

Generally, the effects of pollution may be expected to decrease with increasing distance from the pollution source. That is, if the spread of pollution is symmetric, then the effect at point r of a facility at m is expected to be proportional to $1/d(m,r)$ where $d(m,r)$ is the distance between r and m (or some other monotonic decreasing function of $d(m,r)$), but see figure 4. However, the assumption of symmetry is not always sound. For example, airborne pollution might be serious downwind of an accident yet mild upwind, and wind direction varies from time to time. The noise footprint near an airport, while not symmetrical, is determined by the runway orientation. Waterborne pollution will depend on the drainage pattern.

In this chapter the focus of attention will be on extending the basic location/routing models to take account of wind effects. For this it is important to understand the process involved and section 19.2 contains a description of the mechanism of the spread of airborne pollutants by means of wind power. There seems relatively little work done on incorporating wind effects, exceptions being [677, 668, 920, 1221]. While pollution load is calculated with regard to the prevailing distribution of wind directions and strengths, no account is taken of topographical features.

Section 19.3 gives a short description of the characteristics and methods of multiobjective problems that are of interest in this chapter. The next section looks at locating a facility which emits low grade pollutants and section 19.5 deals with the routing of obnoxious vehicles. Finally, in section 19.6 some possible suggestions for further research are indicated.

19.2   The Mechanism of Airborne Pollution Spread

Analytical mathematical formulations for airborne pollution spread using a Gaussian model [931] are based on the following assumptions:

1. pollutants are chemically relatively inert during dispersion (this being the case for sulphur dioxide and nitrogen monoxide for example);

2. the ground over which the pollutants are spreading is relatively flat (more specifically, it is assumed that slopes are not in excess of 5

degrees);

3. meteorological conditions are constant during dispersion.

The Gaussian plume as developed by [919, 1146] is the most widely used mathematical model for airborne spread of pollutants. A detailed discussion of the assumptions on which the Gaussian plume model is based is given in [1105].

Before presenting specific formulae we first discuss the general shape and other characteristics of the plume [1055]. The main characteristics are as follows.

1. The elevation of effluent gases (called the *equivalent stack height*) is, in general, considerably higher than the *actual* stack height. This elevation, which sometimes can reach an altitude of 2.5 times the actual stack height, is caused by the speed at which gases leave the chimney mouth (*cf* figure 1). For a conventional coal-fired power plant the gases leave the stack with a high speed caused by their high temperature. Thus an elevation of 200 - 300 metres may be attained at which height the wind speed is likely to be much higher than at ground level.

2. Because of the relatively high wind speeds at the equivalent stack height the pollutants are often carried in the plume for great distances along the direction of the wind. For example, for the meteorological conditions that are typical of the Thessaloniki region of northern Greece and an actual stack height of 25 metres the pollutant's ground level concentration reaches a maximum at a distance of 3.0 - 3.5 kilometres from the stack, whereas for a stack height of 50 metres the pollution concentration reaches a maximum at a distance of 15 - 20 kilometres with pollutant concentration remaining at relatively high levels for more than 100 kilometres from the polluting unit [667].

3. The plume exhibits an expansion with time along the direction of the prevailing wind in the following sense. At any instant the plume leaving the stack occupies only a very narrow angle. However, the plume tends to be sinuous so that even though wind direction is 'constant' the direction of the plume leaving the stack mouth will vary. Thus, averaged over a period of time the angle of the plume will increase. Expansion of the time averaged plume is most rapid in the first few minutes thereafter continuing at a reduced rate. For example, the angle of the plume after 5 minutes may reach 10 - 15 degrees but after a further 60 minutes may be no more than 20 - 30 degrees. For practical purposes pollutant concentration may be taken to be zero outside an angle of 25 degrees (or less). In summary, the active plume is very narrow and pollutant concentration decreases

FIGURE 1. Because gases often leave the stack at fairly high speeds, the equivalent point of emission is displaced from the stack mouth

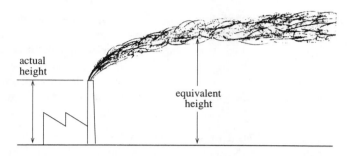

rapidly (in fact 'exponentially' in the Gaussian plume model) in a transverse direction away from the axis of the wind.

4. An important characteristic of the plume for computational purposes is that for a downwind distance between 100 metres and 120 kilometres the transverse pollution concentration exhibits a single maximum with monotonic decrease away from the maximum [1146, 1105]. That is, the ground level concentration in the transverse direction has a unimodal or 'bell-like' shape (*cf* figure 2).

19.2.1 *Mathematical formulae for pollution dispersion*

For the Gaussian plume model the mathematical expression for ground level pollutant concentration is

$$f(x,y) = \frac{Qe^{(-0.5(y^2/q^2(x))-0.5(h^2/v^2(x)))}}{\pi u q(x) v(x)}$$

where f is the pollutant concentration, Q is the pollution emission rate from the stack, u is the mean wind speed, $q(x)$ and $v(x)$ are functions whose explicit form is given below, h is the equivalent stack height and $\pi = 3.14159....$ Note that f is symmetric in y as might be expected.

The greatest concentration of pollutants occurs on the x-axis in the downwind direction. For the particular case of Thessaloniki's meteorological conditions and $h = 300$ metres the largest concentration (maximum of f) is a distance of 118 kilometres downwind from the stack (*cf* figure 2)!

19.2.2 *The diffusion parameters*

Several methods have been suggested for the evaluation of the diffusion parameters based on a vast amount of experimental data. Of these, the

FIGURE 2. Profile of pollutant concentration (a) along the axis of the plume, and (b) transverse to the plume axis

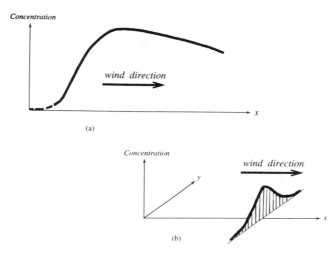

(a)

(b)

method of [919] as modified by [1146] is today the most widely used. The 'parameters', q and v, are functions of both the meteorological conditions (mainly atmospheric stability) and the distance from the stack, and values given by [919] correspond to ground profiles of 'average' roughness and an equivalent stack height not exceeding a few hundred metres. The atmospheric turbulence and the speed of the winds are taken into account when the atmospheric stability type is determined. Six atmospheric stability types, namely A, B, ..., F, have been defined which depend on the wind speed at an altitude of 10 m above ground level and the intensity of solar radiation. Types A, B and C characterize unstable atmospheres, type D corresponds to a neutral atmosphere and E and F correspond to stable and very stable atmospheres respectively.

The determination of the form of functions q and v depends upon the atmospheric stability type. The diffusion parameters in the form of 'logarithmically scaled' curves approximated by the following functions are [919, 1146]: $q(x) = ax^b$ for the horizontal diffusion parameter and $v(x) = a_0 + a_1 x^{1/2} + a_2 x^{1/3} + a_3 x^{1/4}$ for the vertical parameter.

For the meteorological conditions prevailing in the Thessaloniki region the diffusion function coefficients are $a = 0.0804$, $b = 0.8929$, $a_0 = 8.3913$, $a_1 = -1.069$, $a_2 = 19.445$, $a_3 = -27.3357$. Note that the range of validity for the above approximations is $x \in [100m, 120000m]$.

19.2.3 Practical application of the model

The analytical expressions for the dispersion model have been derived on the assumption that the ground is horizontal, but the model performs satisfactorily for more realistic ground profiles. Firstly, the model is applicable in the case of flat terrain with natural vegetation, orchards, forests or houses of more or less uniform height. These elements are considered as the *roughness* profile of the ground and are easily accounted for by appropriate increases in the diffusion parameters [1105].

Furthermore, the model is applicable in cases where objects (such as hills or high buildings) are present, provided modifications are made to the diffusion parameters [474] to account for the formation of recirculation zones on the downwind side of the object. However, such applications of the model are tentative since the recirculation zones are too complicated to be treated with the simple analytical expressions suggested by Gifford.

In the case of streamlined obstacles such as smooth hills with low gradients (less than 1/5 upwind and less than 1/7 downwind) no recirculation occurs and the model is applicable.

Even though the equations of the model were derived using the assumption of constant meteorological conditions the model is also applicable during *slow* meteorological changes in which the system assumes a succession of equilibrium states. Thus, if meteorological statistics (wind speed and direction, and the atmospheric type) are available, the model can be applied for the estimation of time averaged pollutant concentrations resulting in a more realistic estimation of the pollution load.

Wind data is normally recorded for 8 directions: N, NE, E, SE, S, SW, W, NW. Thus, for example, the frequency F_N associated with direction N corresponds to winds observed in a 45 degree sector centred on due North, and similarly for other frequencies. This means that the data is effectively in 'histogram form' (*cf* figure 3), and needs to be smoothed' in some appropriate way prior to input into the Gaussian plume model.

Finally, to illustrate that pollution load is not simply a decreasing function of distance, figure 4 shows the situation for the idealised case in which wind speed is always the same and wind direction is uniformly spread over all possible directions.

19.3 Relevant Features of Bicriterion Problems

In this section only those aspects of immediate relevance to the topic on hand will be discussed; for a more general reference the reader is referred to [234]. A hazard problem may be expressed in general terms as

P1:   minimise $\{o(x), h(x)\}$
       subject to: $x \in O \cap H = \Omega$.

FIGURE 3. Wind frequency data in histogram form (for Thessaloniki) together with a smoothed wind profile

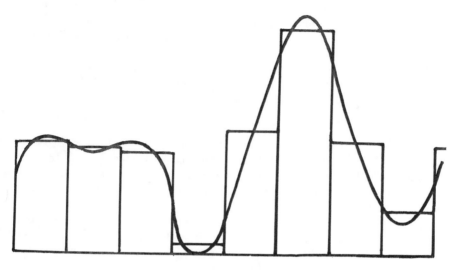

FIGURE 4. Pollution load corresponding to constant wind speed and uniform distribution of wind directions (figure 2a)

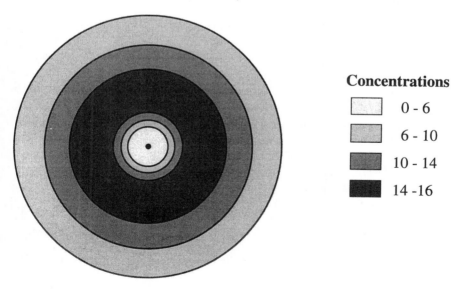

Concentrations

0 - 6

6 - 10

10 - 14

14 -16

The solution, x, is typically a 0-1 vector (with $x_i = 1$ or 0 according as site i is occupied or route i used) but may be a vector with real- or integer-valued components, or be a continuous set etc. O is a set containing those solutions which are *ordinary-feasible* (that is, satisfy ordinary constraints such as 'all customers must be assigned to a facility'). H is the set of *hazard-feasible* solutions (that is, satisfy hazard constraints such as 'no facility may be within a distance s of a populated area'). For simplicity, we have assumed that there is just one 'ordinary objective' $o(x)$ and one 'hazard objective' $h(x)$ but this restriction could easily be removed. In general, P1 will not possess a solution x^* in the sense that

$$o(x^*) = \min \{o(x) \mid x \in W\} \quad \text{and} \quad h(x^*) = \min \{h(x) \mid x \in W\}.$$

However, if x and y are such that $o(x) \leq o(y)$ and $h(x) < h(y)$, or $o(x) < o(y)$ and $h(x) \leq h(y)$, then x is clearly 'better than' y. In such cases x is said to *dominate* y and y is said to be *inferior* to x. The set of all non-inferior solutions (or NIS for short) may be regarded as effectively being the solution of P1.

To proceed further the decision-maker may be involved, with the final choice of solution being on the basis of his or her (implicit) tradeoff function between the two objectives and knowledge of non-modelled features. Of course the analyst can still usefully contribute at this stage (perhaps by helping to elicit the decision maker's preferences as implied by the tradeoff function). Indeed, if the tradeoff may be assumed to be linear, and this assumption is frequently made, then calculation of the NIS may be achieved by the *weighting method*. This requires the solution of a parametrised set of single-objective problems

P1(w):   minimise $(1 - w)o(x) + wh(x)$
     subject to: $x \in O \cap H = \Omega$

for all $0 \leq w \leq 1$. Whatever the approach, it is usually the case that, in reality, parts of the NIS may quickly be ruled out by a decision-maker and attention focussed on the parts of the NIS that are of interest. [252] suggest involving the decision-maker interactively in the problem-solving process to steer effort towards getting a good approximation of the 'interesting' part of the NIS. That is, extra constraints such as $o(x) \leq o_0$, $h(x) \leq h_0$ may be added. This will correspond to restricting w to some range $0 \leq w_L \leq w \leq w_U \leq 1$. They also go on to discuss ways that 'interesting' *non extreme point* solutions in the NIS (which are not generated by the weighting method) may be investigated.

Another approach that is potentially of use in connection with hazard problems is the *NISE* method [234]. Very briefly this operates as follows:

1. Find solutions which optimise the two objectives (*cf* points A and B in figure 5).

FIGURE 5. First steps in an application of the NISE method

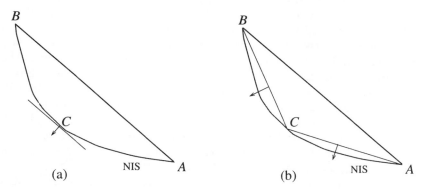

(a)   NIS   A   (b)   NIS   A

2. Solve the problem $P(w_{AB})$, where w_{AB} corresponds to optimization in a direction orthogonal to AB, thereby obtaining a solution C (*cf* figure 5).

3. Continue this procedure by solving $P(w)$ for $w = w_{AC}$ and $w = w_{BC}$ corresponding to directions orthogonal to AC and BC, and so on until the NIS is approximated to the desired accuracy.

19.4   Location of Obnoxious Facilities

Relatively much of the work in this area has been concerned with the *maximin* problem which requires the location of a single facility in the plane so as to maximise the minimum distance to some *protected* set or region (eg. towns, nature reserve, ...). The papers [260, 331, 829, 830, 823, 675, 661, 32] all fall into this general category. A maximin problem in which a facility is to be located on a network has also been considered [221].

For 'low grade' pollutants, such as those from a coal-fired power station, it is the cumulative effects that are important and so it is required that the sum of negative effects be minimised. [549, 547] implicitly consider a *minisum* problem in the plane in which it is required to locate a facility so as to minimise the *sum* of the pollution effects from the polluting facility. (Note that the polluting effects of a facility might not simply be proportional to distance from the facility, as would be the case for airborne pollutants spread by winds whose frequency distribution is asymmetrical about the facility. Also, as indicated in section 19.2, a 'near' facility may in fact be 10's of kilometres away.) A solution technique termed 'Big-Square Small-Square' (or BSSS for short) was introduced by [549] (see also [677, 668]).

Minisum problems in which a facility is to be located on a network have also been considered [221].

The BSSS method can be used to locate a single facility that produces airborne pollutants. To see why this is so consider a rectangular area, A, astride the wind axis and upwind from a particular (point) site P, and with edges parallel and orthogonal to the wind direction. The form of the Gaussian plume results, for site P, in the minimum value of the pollution load corresponding to the facility being at one of the corners of A [668]. When several sites are being considered the sum of the individual minimal values provides a lower bound for the pollution load S experienced by *all* sites P_i. (Note that for different sites P_i the minima may occur at different corners of A. Also, S may possess several local minima as the location of the facility varies over A, and the minimum of S need not lie at a corner of A.) In general terms BSSS may be described as follows. It starts with a rectangle containing the whole region of interest. The centre of the rectangle normally provides an initial feasible location for the facility and hence an upper bound (incumbent value) on the optimal solution value. (If the region is very irregular then a feasible solution might not be obtained so soon.) Rectangle A is now divided into 4 equal rectangles each with a corner at the centre of A. The procedure continues with at each iteration:

- selection of a rectangle;

- evaluation of the solution corresponding to the centre of the rectangle and if necessary the update of the incumbent solution and incumbent value;

- division of the rectangle into four similar subrectangles with the lower bound on solution value in each computed;

- elimination of subrectangles not containing any feasible solution and those whose minimum value exceeds the incumbent value.

A more detailed description is contained in [677, 668]. The major inconvenience with this application lies in the condition that the rectangle must have sides parallel to the wind direction. This makes calculations involving wind directions not along the 8 principle directions N, NE, ..., W, NW less efficient. However, calculations along non principle directions are certainly necessary because of the relatively small angle of the plumes.

If the cross section of the plume in any direction were to give a unimodal pollution concentration then the minimum of pollution load over any rectangle will be at a corner point and the necessity to consider rectangles oriented along the wind direction would be removed. In fact this condition, which we will refer to as the *unimodality condition*, is very nearly met, and bearing in mind other, more severe, approximations made in modelling the unimodality condition may reasonably be assumed to hold.

Another aspect of location, not recognised above, is that facilities will normally be located at or near to a road and so the fact that location must take place on a network may be incorporated from the outset. Roads can be represented (to the accuracy required) by a collection of straight line segments. Now the unimodality condition implies that the minimum pollution concentration will occur at one end of each line segment. This in turn means that a (simplified) BSSS algorithm may be applied to find the best location(s) (or a set of good potential locations) on a network at which to locate a polluting facility [671].

19.5   Routing of Obnoxious Vehicles

Traditionally, vehicles have been routed to minimise cost (time etc) to the carrier subject, perhaps, to constraints on delivery times. However, when hazardous materials are being carried it is relevant also to try to minimise the expected risk to those living alongside or near to the route used (and perhaps to natural features such as nature parks). The simplest way of treating this is to impose restrictions on which arcs may not be used; that is, the problem is a shortest path problem on a subnetwork [6]. We now look at alternative ways of dealing with risk. At this point it is convenient to define some terms.

$\pi_i = a_1 a_2 ... a_t$ is the path comprising arcs $a_1, a_2, ..., a_t$,
d_a = 'expected' damage due to an accident on arc a,
p_a = probability of an accident on arc a if the vehicle *starts to travel* on a,
$h_a = p_a d_a$,
$h(\pi) = h_{a_1} + h_{a_2} + h_{a_3} + ... + h_{a_t}$.

The damage d_a may be expected to be some function f of distance from the site of the accident. Unless f is a simple function, it will be necessary to resort to a numerical approximation procedure. If f decays very rapidly, as might be expected for example from the effects of an explosion, then d_a becomes just 'tons past people' [996]. For situations in which airborne pollutants are emitted the situation is more complicated. Wind effects may be incorporated by using Monte Carlo methods [672] or a curve may be fitted, explicitly of implicitly, to the data and integrations performed numerically. The expected risk $e(\pi)$ associated with travel on path π is given by $e(\pi) = h_{a_1} + (1 - p_{a_1})h_{a_2} + ... + ((1 - p_{a_1})...(1 - p_{a_{t-1}}))h_{a_t}$ and the basic risk minimisation problem is to mininimise $e(\pi)$ over π in the set of *loopless* paths from O to D.

Of course, the probabilities p_a are (hopefully) very small in most situations and the minimisation of $e(\pi)$ can be replaced by minimisation of $h(\pi)$ over all paths from O to D the 'loopless' condition now being redundant.

If it is desired also to minimise the physical length of π then a bicriterion shortest path problem results. [210] applied the weighting method to such a problem arising in connection with spent nuclear fuel.

Typically, one component in the routing of hazardous materials is minimising route length. [Note that 'length' is here used in a general sense and could mean a combination of risk and physical distance.] If there is another criterion, C say, such as equity, then a potentially useful strategy in many situations is to generate the 1-st, 2-nd, 3-rd, ... shortest paths from origin to destination, testing each path encountered against the criterion C, and terminating when an acceptable solution has been found or when a specified number of paths have been generated. At termination the 'best' solution found is adopted. This general approach has been adopted by [835] in relation to the transport of nuclear spent fuel. Generally, such strategies are potentially useful in other hazard problems involving the routing of obnoxious vehicles as there are very efficient ways of generating the k shortest loopless paths [932].

It is desirable that the adopted solution should also appear to be fair, or equitable. [498] use the criterion that the risks felt by any two zones should not differ by too much. There seems a danger here that this might lead to a 'levelling up' of risk and it is not difficult to devise an example in which no zone has a reduction in risk but some have an increase. Hence, there is a need to exercise caution as to which zones should have their risks compared. [1229] try to ensure equity by putting constraints on the volume of hazardous material carrying vehicles that are permitted to use arcs associated with high risk. This in turn requires that a way be found for prescribing these capacities in an equitable way. Another alternative is to minimise the maximum risk felt by any zone; this can be achieved by adapting appropriately the model of [498]. A possibility that might have some merit is to minimise a linear combination of the total risk felt by all zones and the maximum risk felt. When several routes are being used, [761] consider spreading risk 'fairly over time' as well as 'over space'. [245] looked at a routing - scheduling problem in the presence of curfews though with a different objective in mind. Finally, [334] discuss a maximin routing problem in the plane in which a route with fixed end points keeps as far as possible from given (undesirable) points. Currently much interest is being shown in incorporating equity into location models and there remains scope for more work in this area.

19.5.1   Combined Location and Routing

There is little work on combined location - routing models [1230, 996, 766, 768, 117] and even less that incorporates wind effects. A notable exception to this is [1221]. This paper is concerned *inter alia* with technology choice, and waste disposal based on photolysis is assessed. This leads to a solution in which relatively many units are located *near to the point of production* thus resulting in an equitable solution in that the benefits and disbenefits of waste disposal are associated with the same area. Combined location -

routing problems may be expected often to encounter economies of scale tradeoffs with large units of a polluting facility being likely to result in greater economic efficiency but result in greater exposure through extra transportation being involved.

19.6   Future Directions

Hazardous materials are involved in many industrial processes whether as a raw material, product or waste. Publicity surrounding accidents involving such materials has led to heightened awareness of the associated danger, and increasing attention from operations researchers and others. Some trends and directions for further research are emerging.

Models are generally becoming more detailed and it is desirable for the trend towards more realism to be continued. In this chapter we have discussed one direction in which realism may be improved and that is through the incorporation of wind effects. In general, however, a very detailed and complicated model does not of itself imply that it is the best model or even a good one. What is required is a careful assessment, *in conjunction with specialists in the relevant area*, to decide which factors are significant and which are not. A good model will then be one which includes the former but not the latter. When such a model has been devised for a particular situation it is relevant to consider data requirements and solution methods.

A model may well need to be simplified in the light of data availability but development of the initial model is of value in its own right since it may lead to increased understanding. The second potential difficulty, namely that a model may be too complicated to solve exactly by existing techniques, is not necessarily a valid reason for abandoning a good model. Instead, a modern heuristic method (such as simulated annealing, tabu search or a genetic algorithm) may usefully be employed. There seems to be considerable scope for the use of heuristic methods in locational analysis generally, and perhaps of parallel processing also in order to accommodate computation requirements.

For obnoxious facility location, and for this aspect of a larger problem, interactive computer graphics appear to provide a powerful tool [123, 124, 473] particularly when a maximin criterion is being used. For location and routing hazard problems interactive graphics might be useful also. Another fruitful approach is that based on the use of k-best methods (and particularly k-shortest path methods, where applicable, in view of the effective algorithms available). With this and improvement heuristics (local search) generally, a range of (hopefully) high quality solutions will be on hand for the decision maker to make his or her final choice. This provision of a range of solutions is an important feature and may well outweigh not having a (guaranteed) optimal solution (which is only optimal for the model problem anyway!).

For combined location - routing the potential sites at which to locate the facility(ies) form a discrete set. Because of the risk associated with the facilities the objective function will tend to 'peak' at points on arcs between population centres. Thus, it is reasonable for generation of the candidate sites, to consider all points on the network and not merely at nodes. But more than this, off-network points for facilities could be considered as this would tend to avoid risk to population along roads and traffic on them. (Of course an extra cost for connecting the site to the network must be taken into account.) There is also the general question of mixed-metric models which deserves further consideration.

Another feature that has hardly been touched on in this chapter is that of *equity*, a topic which is currently receiving considerable attention. A discussion of equity in the context of hazardous materials logistics is provided in Chapter 20 in this book. Here we merely mention that equity may be *environtal* in that it concerns spreading the effect of a location decision fairly over the affected 'environment' including the population contained therein, or *economic* in that it treats the polluting units in an even handed way.

Finally, since the transportation of hazardous materials poses a risk, there is a pressure towards locating origins and destinations close together, or even at the same sites. This might suggest increased risk around the combined origin - destinations, but [1221] argue that this is a matter of technological development and that locational analysis should provide the stimulus for better use of technology.

20
HAZARDOUS MATERIALS LOGISTICS

Erhan Erkut
University of Alberta

Vedat Verter
University of Alberta

20.1 Introduction

As a result of the lifestyle choices industrial societies seem to have made, a very large number of hazardous materials (hazmats) are produced for consumption and as a by-product of industrial activity in developed countries. Hazmats, or dangerous goods, include explosives, gases, flammable liquids and solids, oxidizing substances, poisonous and infectious substances, radioactive materials, corrosive substances, and hazardous wastes. Well-known examples of facilities that deal with hazmats are petroleum refineries, chemical processing plants, paper mills, and nuclear power plants. Further, transportation of dangerous goods, such as soda from a caustic soda supplier to a refinery, chlorine from a chlorine producer to a chemical manufacturing plant, or hazardous waste from industrial facilities to an incinerator has become common practice. Hazmats, however, can be extremely harmful to the environment and to human health, since exposure to their toxic chemical ingredients may lead to injury or death of plants, animals, and humans.

To give an idea of the magnitude of the problem, we use two figures: 1) It is estimated that four billion tons of hazmats are transported annually across the U.S. [944]; 2) The world-wide generation of hazardous wastes is estimated to be about 300-400 million tons per year [1153]. These figures suggest that hazmats constitute an ever-increasing potential danger for society and the environment unless they receive special treatment during their transportation, utilization, and disposal, to make them harmless or less dangerous. Fortunately, the hazmats management problem has attracted the attention of many governments who use legislation and regulations, as well as taxes and economic incentives, to secure public and environmental safety. For example, the Hazardous Materials Transportation Uniform Safety Act of 1990 requires the U.S. Department of Transportation to determine standards for designating the routes to be used for hazmat shipments. The Canadian Environmental Protection Act of 1988 requires a thorough review of all the chemicals currently in use in Canada in order to determine the extent and nature of the risk associated with each substance. This is an effort to design and enforce the necessary control mechanisms for toxic chemicals. In addition to these acts by individual governments, the United Nations recognizes the severity of the problem and expends efforts for minimizing risks associated with hazmats. For example, a U.N. initiative, The Basel Convention [1153] adopted in 1989 by 105 countries, states the international consensus to minimize the generation of hazardous wastes via improvements in production technology, recycling and reuse. The Basel Convention also regulates the transboundary movements of hazardous wastes to secure their disposal under environmentally sound conditions.

Despite the continuous effort to mitigate the adverse effects of hazmats, accidents do happen during their use, loading/unloading, transport and disposal. Although these accidents are quite rare events, they can lead to very undesirable consequences. For example, in 1976 the town of Seveso, Italy, was showered with dioxins and furans due to an explosion at a chemical plant which killed thousands of birds and small animals, and injured 190 people. In Niagara Falls, New York, over one thousand households were evacuated when leachate was discovered at the nearby Love Canal hazardous waste disposal site in 1978. A year later, a train carrying toxic chemicals was derailed in Mississauga, Ontario and chlorine leaking from damaged tank cars forced the evacuation of 200,000 people. In 1982, 2, 700 fatalities were reported due to a gasoline truck explosion in a tunnel in Afghanistan. An accidental discharge of methyl isocyanate from a chemical plant in Bhopal, India killed 4,000 people in 1984. In addition to its devastating consequences in the vicinity of the nuclear reactor, the 1986 Chernobyl accident also caused long term problems, even in the neighboring countries, mainly related to the contamination of vegetable and animal produce.

Decisions involving hazmats are very difficult, not only because of the potential for spectacular accidents, but also due to public sensitivity, which is primarily driven by the perception of associated risks. The perceived risks associated with transportation of hazmats, and hazardous facilities, are much higher than the risk assessments provided by experts. For example, according to the U.S. Nuclear Regulatory Commission, the estimated frequency of events involving a given number of fatalities is several orders of magnitude smaller for nuclear power plant accidents than for airplane crashes. However, based on revealed preferences, it is fair to assume that the public perceives air travel risks as acceptable, whereas nuclear power plant risks are perceived as unacceptable. Furthermore, public reaction to hazmat-related accidents is much stronger than accidents that do not involve hazmats. For example, the average annual number of deaths in hazmats accidents is reported to be about 10 compared to the 35,000 fatalities in "ordinary" traffic accidents in the U.S. Nevertheless, almost every single hazmat accident makes it to the front pages of newspapers, and results in a call for some sort of action (tighter transport regulations, banning of hazmats, etc.). As a consequence of the public perceptions of hazmat risks, hazmat logistics decisions are more difficult to make and to implement than other logistics decisions.

Another contributing factor to public sensitivity may be the inequity in the distribution of risks associated with hazardous facilities and hazmat movements. Although the benefits of hazardous facilities, such as nuclear power plants and hazardous waste treatment and disposal facilities, accrue across the nation, it is usually the nearby population that is exposed to the burdens. For example, people living close to a hazardous facility might suffer from emissions from the facility, and experience a loss in property values. Thus, local residents are very quick in raising opposition to the location of hazardous facilities in their region, or to the transportation of hazmats via nearby roads. Many governments are reluctant to force these issues in the face of very vocal protests. For example, the provincial government in Ontario, Canada has been attempting to site an integrated hazardous waste management facility for the last ten years, and is still unsuccessful to this day.

The factors briefly discussed above complicate the decision-making process in the area of hazmat logistics. However, they also provide a challenge and an opportunity for quantitative analysis to make a contribution to the solution of these difficult problems. From a problem-solving perspective, the field of hazmats logistics is concerned with the configurational decisions for potentially hazardous facilities, such as their location, size and technology content, and the routing decisions regarding the transportation of hazmats. Due to their very nature, these problems involve achievement of a multiplicity of objectives, the most prominent ones being minimization of cost, minimization of risk, and maximization of equity in the spatial distribution of risk. Furthermore, there exists a broad range of stakeholders in the management of hazmats in various levels of government, in industry, and the public. Note that each party might have a different set of priorities and a different viewpoint. For example, although federal governments develop country-wide policies on hazardous waste treatment and disposal, it is usually the responsibility of local governments to establish treatment and disposal facilities within their jurisdiction. Concerning the transportation of dangerous goods within a city, although the transportation department is responsible for designating the allowable routes, a carrier company would in general try to identify the most efficient route between the origin and destination for each shipment.

Analytical approaches for hazmat logistics are of significant value in generating and evaluating alternative solutions to the problems that arise in the management of hazmats. Without quantitative models, governments would choose one of the few solutions suggested by their technical staff, and the public would be left with a difficult choice of accepting or rejecting this solution. We conjecture that this is one of the primary reasons for public opposition to the strategic decisions regarding the management of hazmats. Implementation of quantitative methods in assessment of the risks, and in solving the hazmat logistics problems, would certainly facilitate our understanding of the tradeoffs among the prevailing objectives, and may also enhance communication between stakeholders.

Hazmat logistics is currently a very active area of research: *Transportation Research Record* published two special issues on hazmat transportation in 1988 and 1989, and *Transportation Science* devoted an issue to hazmat logistics in 1991. There is a special section on hazmat transportation in the March/April 1993 issue of the *Journal of Transportation Engineering*. The June 1994 issue of the *European Journal of Operational Research* contains a special section on emerging trends in risk management with several papers pertaining to hazmat logistics. A special issue of *INFOR* on this topic will be published in 1995, and *Location Science* is also planning a special issue. In this chapter, we provide an overview of the literature that deals with analytical approaches for hazmat logistics. Most of the papers we discuss are in the areas of risk assessment, operations research, and decision analysis. We refer the reader to [1164] for a more extensive bibliography on this topic.

The major concerns in strategic management of hazmats are the risks associated with them, the spatial distribution of these risks, and the costs incurred during the management process. Thus, in Section 20.2 we review the methodology for assessment of the risks imposed on the public by potentially hazardous activities. The prevailing approaches for modeling equity in the spatial distribution of risk are presented in Section 20.3. Section 20.4 covers the cost aspects of hazmat management. The structural decisions in designing hazmat management systems can be analyzed in two broad categories: the configurational decisions regarding hazardous facilities, and transportation logistics. We devote Sections 20.5 and 20.6 to the discussion of analytical approaches for facility design and transportation planning respectively. The current trend towards the development of integrated hazmats logistics models, and the previous studies that would serve as building blocks in this endeavor are presented in Section 20.7. In the final section, we suggest directions for future research in decision-making models, and for implementation of decision-support-systems, in strategic management of hazmats.

20.2   Risk Assessment

In the context of this chapter, risk is a measure of the possible undesirable consequences of a release of hazmats during their use, storage, transport, or disposal. A release of hazmats can lead to a variety of incidents, for example, a spill, a fire or an explosion in the case of flammable liquids, or a toxic cloud or plume in the case of pressure-liquefied gasses. As indicated by the examples in the previous section, the undesirable consequences of these incidents include fatalities, injuries, damages to property, losses in property values, and environmental damages. The release event can be caused by an accident, such as a core melt in a nuclear reactor or derailment of a railcar carrying hazmats. However, releases can also occur without an accident, such as leakages from a hazmats container, or toxic emissions from a hazardous waste incinerator. To differentiate between these two categories of risk, we call the former accident risk, whereas the latter constitutes exposure risk.

In designing facilities that deal with hazmats and designating routes for hazmat transport, it is crucial to be able to compare the risks of alternative decisions. This requires quantification of the risks associated with the various transportation, storage, and disposal options for hazmats. Several risk assessment methodologies are available for this task.

The most common measure of the societal risk associated with a potentially hazardous activity is its expected undesirable consequence. That is, risk is usually defined as the probability of a release event during the activity, multiplied by the consequence of that event. This definition is appropriate only if a single release event is possible, such as a single shipment of hazmats between an origin and destination pair. In the case of multiple shipments, or the operation of a hazardous facility, the expected total consequence of all possible incidents needs to be computed. Thus, the major components of a risk assessment study are estimation of (i) the incident probabilities, (ii) the consequences of each incident, and (iii) the volume of the activity. The probability of an incident needs to be estimated per shipment, per unit distance, per unit time period, or per unit amount. Hazmats accidents are very rare events, which makes these estimation tasks nontrivial. The volume of the activity can be represented as the number of shipments to be made, the total distance to be traveled, the total time the hazardous facility would be operated, or the amount of hazardous waste to be processed.

Despite the fact that a variety of undesirable consequences are possible, most of the risk assessment literature focuses on fatalities due to the release of hazmats. Although, this approach simplifies the risk assessment process, its end result might be far from representing the absolute risk of a potentially hazardous activity. (An exception to this is the early work of [1037] suggesting the assessment of dollar figures to fatalities and injuries.) Even the estimation of the number of fatalities as a result of an incident is, however, quite difficult since for most hazmats the direct impact on human life is not well known. Fortunately, for many decisions in the strategic management of hazmats, a comparison of the relative risks of alternatives is needed rather than a quantification of the absolute risk of each alternative decision. It is however, crucial to be consistent in data collection and estimation in order to produce reliable assessments of the relative risks.

In a review of risk assessment methodologies for hazmat transportation, Abkowitz and Cheng [4] identified statistical inference as the most commonly used procedure for estimating risk. This technique presumes that sufficient historical data exist to determine the frequency and consequences of the release incidents, and that past observations can adequately be used to infer future expectations. A number of studies have appeared in the literature identifying potential sources of information for hazmats risk assessment. In a series of papers List et al. [765], List and Abkowitz [764], and List and Abkowitz [763] reported on the major statistical sources in the U.S. for gathering data on the movement of dangerous substances over the network of highways, railroads and waterways. They focused on issues of completeness, consistency, and compatibility both between and within the source databases commonly available. Stough and Hoffman [1103] focused on estimating the frequency of hazmat types on the highways of Indiana. Their results indicated that volume of the transportation activity was higher during day time, and that flammable liquids and corrosives were the most frequent types of hazardous truck loads. Harwood et al. [565] presented analyses of data from several databases in the U.S. to document the types of accidents and incidents that occur when transporting hazmats by truck. They identified traffic accidents as a major cause of severe hazmat incidents, and attempted to estimate the probability of a release given an accident. Using 1982 data, Glickman [480] provided estimates of release accident rates in the U.S. in transporting hazmats by rail and by truck. He concluded that there can be no general answer to the question of which mode is safer. Harwood et al. [564] used a weighted average of 1985-87 data from California, Illinois, and Michigan in estimating truck accident and release rates. Their detailed study incorporates factors such as roadway type, accident type, and demographic characteristics of the surrounding environment (i.e. urban vs. rural). We prefer not to quote any benchmark estimates of release accident rates here due to the significant differences between the figures published in the papers referenced above. Furthermore, there exists considerable criticism regarding the statistical reliability of the source databases due to the under-reporting of release events, and thus the quality as well as quantity of available data.

Note that the statistical risk assessment studies that appeared in the academic literature are mostly confined to the transportation of hazmats. Although the societal risks of several potentially hazardous facilities have also been estimated, usually as part of an environmental risk assessment, these studies unfortunately remain as consulting reports, and, hence, are not easily accessible to the academic community.

An alternative way to estimate the frequency and consequences of hazmat release incidents is the use of logical diagrams such as fault trees and event trees. Fault trees show how a top event can occur in relation to the occurrence of a set of basic events. Thus, construction of a fault tree starts by identifying the top event of interest and then determining unions and intersections of lower level events that would lead to the top event. This procedure is repeated for each event at the lower level, and is iterated until the basic events are obtained. Fault trees enable us to determine the probability of the top event (e.g. containment tank failure) on the basis of the probabilities of the basic events for which sufficient historical data exist or expert judgments are reliable. Event trees however, identify and quantify the possible consequences of the top event. Construction of an event tree is an iterative process. At each iteration, the outcomes of an event are identified and the conditional probability of each subsequent event is assessed given the preceding event. This facilitates a thorough analysis of the consequences of a top event (e.g. a traffic accident involving a hazmats truck). Note that the probability estimation process constitutes a top-down approach in event trees, in comparison to the bottom-up approach adopted in fault trees. Boykin et al. [122] presented an application of fault and event trees in the analysis of the risks associated with technology alternatives in a chemical storage facility. Applications of these techniques in hazmat transportation risk assessment are reported in [20]. Although fault and event tree analyses present alternatives to statistical risk assessment, there are difficulties associated with these methods as well. For example, estimation of probabilities of certain events in a fault tree may be no easier than estimating the probability of the event at the top of the tree. Perhaps more importantly, it is impossible to enumerate all possible ways in which an accident could happen. Such problems with these techniques are discussed in [1201] and [986].

A method, closely related to event tree analysis, makes use of Bayes' theorem, which facilitates estimation of the probability of an event as a product of a sequence of conditional probabilities and the probability of an initial event. For example, the probability of a gasoline explosion during its transportation can be estimated by multiplying the probability of a traffic accident, the probability of a gasoline release given the accident, and the probability of an explosion given the release. This technique is especially useful, when the historical records are not sufficient for a direct assessment of the probability of interest. Further, it is also possible to incorporate subjective estimates in the Bayesian approach, as pointed out by Abkowitz and Cheng [4] Note that calculation of an event probability is equivalent to tracing the corresponding branch of an event tree. Glickman [479] presented an application of this method in the assessment of the risks of highway transportation of flammable liquids in bulk. In a related study, Chow et al. [213] used Bayesian methods to estimate the chance of severe nuclear accidents and their risk. Apsimon and Wilson [35], however, suggested the use of numerical models in the assessment of the possible consequences of a nuclear accident.

Once the risk imposed on each individual due to hazmats is estimated using the techniques discussed above, a societal risk can be calculated as an aggregation of the individual risks. This societal risk, in turn, can be used as input for analytical models in hazmat logistics decisions. (Although in the remainder of this section we concentrate on the societal risk, the levels of individual risks are also relevant in decision-making. In a recent paper, Saccomanno and Shortreed [1033] emphasized the dichotomy between societal and individual risks associated with hazmat transport.) In developing the concept of societal risk, we focus on the shipment of material m through path p as an example. Similar ideas, however, apply to the use, storage, and disposal of hazmats. We define R_{ipm} to be the societal risk at population center i due to the potentially hazardous activity p involving material m. Assuming that each individual in a population center will incur the same risk, the societal risk can be expressed as a product of the individual risk and the population size.

$$R_{ipm} = \text{IR}_{ipm}\text{POP}_i \qquad (20.1)$$

where,

IR_{ipm} = Individual risk at population center i due to the shipment of material m through path p, and

POP_i = Population at population center i.

For ease of exposition (and without loss of generality) we will assume that only a single type of accident, release, incident, and consequence are possible during the potentially hazardous activity. That is, if m is a toxic gas such as chlorine, then a traffic accident involving the chlorine truck could lead to the release of a certain percentage of the contents into the air which would cause a toxic plume and hence fatalities. (Clearly, more than one type of accident, release, incident, and consequence can occur, and all possibilities need to be incorporated in the expected risk calculation.) Thus, individual risk could be defined as the probability of death for a person at population center i due to the potentially hazardous activity. In general, a path p can be decomposed into a set S_p of road segments (indexed by s), where the road characteristics, such as accident rate and population density nearby, are quite uniform within each segment. Thus,

$$\text{IR}_{ipm} = \sum_s P_s(A)P_{sm}(R|A)P_m(I|R)P_{ism}(D|I) \qquad (20.2)$$

where,

$P_s(A)$ = probability of an accident on road segment s per shipment,

$P_{sm}(R|A)$ = probability of the release of material m given an accident on road segment s,

$P_m(I|R)$ = probability of the incident given the release of material m, and

$P_{ism}(D|I)$ = probability of death for an individual at population center i, due to the material m incident on road segment s.

Note that the individual risk in (20.2) represents the probability of death per hazmat shipment. The prevailing methods for the estimation of accident, release and incident probabilities in (20.2) were presented in the preceding paragraphs. Estimation of $P_{ism}(D|I)$, however, constitutes one of the major weaknesses of the state of the art in hazmat risk assessment. There seems to be no consensus on how to extrapolate the results of toxicological experiments performed on animals to infer human fatalities. In the case of toxic gases, the probability of death for an individual is usually expressed as a function of the toxic plume concentration at that location which is a highly random event by itself. The concentration level depends on the distance between the individual and the accident location as well as the direction of wind and atmospheric stability at the time of the accident. Karkazis and Boffey [672], and Patel and Horowitz [920], presented models that incorporate the effect of winds in determining the diffusion of gases, over wide areas, from possible spills during transport. The effect of winds in the assessment of facility risks are taken into account in the studies by Karkazis and Papadimitriou [677] and Karkazis et al. [668]. The use of a Gaussian type dispersion model in the determination of the toxic plume concentration levels is the common feature of the above models, whereas Alp [20] suggested that a probit function would be more appropriate in estimating the possible effects of hazmats on an individual.

The study reported in [1034] is a good example of a comprehensive study that uses all of the methods discussed above. The authors of this paper provided an analysis of the risks associated with truck and rail transportation of LPG and liquefied chlorine. Statistical inference was used in the estimation of accident probabilities, whereas construction of fault trees was necessary in assessing spill probabilities, mainly due to the lack of data. The authors determined a hazard area around the spill associated with each possible undesirable consequence by the use of damage propagation models. The size and rate of spill of the hazmat are the primary determinants of the damage range. The authors also provided a comparison of the potential hazard areas suggested in the literature for a variety of materials, and pointed out the lack of consensus in these estimates. The results of their study suggested that rail constitutes a safer mode of transportation for LPG, whereas the risks associated with truck transportation are less in the case of chlorine.

It is common to skip some steps in the risk assessment process, and hence to use approximations in derivation of the risk parameters of a hazmat logistics model. One alternative is a worst-case approach that ignores the probability of a release incident, and assumes that all inhabitants of the hazard area will face the same undesirable consequences of exposure to hazmats. Batta and Chiu [57] and Revelle et al. [996] adopted this approach in modeling the risks of hazmats transportation assuming $P_{ism}(D|I) = 1$ within a threshold distance from the road segment and zero otherwise. List and Mirchandani [766] incorporated the probability of a release incident as well as $P_{ism}(D|I)$ in their model, where the latter varies with the inverse-square of the distance from the incident. An alternative approach amounts to ignoring the consequences of an incident and concentrating on minimizing accident probability in designing hazmat management systems. This approximation has been adopted by Saccomanno and Chan [1032] and Abkowitz et al. [6] in forming one of the objectives in their multi-objective hazmat transportation models.

Not only do such approximations to the modeling of risk simplify the construction of analytical models, but also valid arguments exist to justify their use. At one end of the risk modeling spectrum, if we were to use the threshold-distance model to find origin-destination routes for hazmat shipments, we would be minimizing the number of people exposed to transport risks. Proponents of this approach claim that, with its emphasis on population instead of accident probabilities, this measure is suitable for modeling the exposure risk as perceived by the people living near potentially hazardous activities. Exposure minimization may, in turn, result in the minimization of public opposition. At the other end of the spectrum, we could minimize the accident probability, ignoring the population data. In defense of this approach, one may suggest that each hazmat incident receives a considerable amount of negative publicity independent of its consequences, and hence the likelihood of an incident should be minimized to mitigate public opposition. We caution here that the risk proxy used (traditional vs. exposure-only vs. probability-only) may influence the solution of a mathematical model. We found it quite easy to construct simple examples where different proxies of risk resulted in the selection of different routes.

The traditional expected consequence representation of risk in (20.1) is deemed inappropriate for hazmat logistics by many authors since this representation assumes a risk-neutral public. Most people, however, would judge a low-probability-high-consequence event as more undesirable than a high-probability-low-consequence event even if the expected consequences of the two events are equal. This risk-averse attitude is not reflected by the traditional representation of risk, although it is prevalent in the decisions involving hazmats. One way to incorporate risk aversion in hazmat logistics models is to account for the variance in associated consequences. Wijeratne et al. [1204] and Ahmad [12] treated the expected consequence and the variance of the consequence as two quasi-independent evaluation criteria. This allowed them to address the low-probability-high-consequence nature of the release incidents in a multi-objective framework.

Although the addition of variance information to the expected consequence models results in a more accurate representation of risk, some information about the nature of risk is still lost in the process. A complete representation of risk would require the use of a risk profile, which is a cumulative distribution function of the random consequences of a potentially hazardous activity. It gives the probability that consequences will exceed a given level. Most popular risk profiles in the literature are cumulative F-N curves, where N denotes the number of fatalities, and F denotes the frequency of events with greater than N fatalities. Saccomanno and Shortreed [1033] presented risk profiles of the societal risks for chlorine transport by rail and truck. This technique was also adopted by Glickman and Rosenfield [482] in analyzing the risks of catastrophic hazmats releases in train derailments. Although the use of risk profiles enriches the presentation of risk assessment results concerning a few decision alternatives, it is not clear how one might use risk profiles in optimization models.

There are other ways of overcoming the deficiency of the risk-neutral expected consequence definition, which are more amenable for inclusion in optimization models. For example, Abkowitz et al. [6] suggested modeling perceived risk by using a risk preference parameter q in the following form:

$$PR_{ipm} = IR_{ipm}(POP_i)^q \qquad (20.3)$$

A risk-averse attitude regarding the consequences of potentially hazardous activities can be represented by using $q > 1$ in (20.3). The above form can also be considered as a surrogate disutility function for perceived societal risk. Clearly, for a given value of q, the perceived risk of an action can be quantified and used as data in an optimization model. Note that, (20.3) reduces to the traditional definition of risk when $q = 1$, which is sometimes referred to as the "technical risk". Although (20.3) provides us with a pragmatic way of accounting for risk aversion, it is perhaps preferable to model risk aversion explicitly by deriving risk disutility functions empirically. In their recent work, McCord and Leu [812] presented the use of convex disutility functions within a decision-theoretic framework for modeling risk aversion for a single hazmat shipment.

The traditional representation of risk is also criticized for not being appropriate in the case of multiple hazmat shipments between an origin and a destination. The expected consequence definition assumes that a selected path can be taken as many times as necessary regardless of the number of accidents on that path. However, the shipments are likely to be suspended to allow for a re-evaluation of the routing policy when a catastrophic accident occurs during the transportation of an extremely hazardous substance. Sivakumar et al. [1084] and Sivakumar et al. [1083] proposed the use of conditional risk (i.e. expected consequence given the occurrence of the first accident), to deal with this type of routing problems. Focusing on the same weakness of the traditional definition of risk, Jin [641] studied alternative risk measures such as the expected total consequence, given that shipments will continue until a threshold number of accidents occur or a fixed number of shipments are completed. She viewed each shipment as a probabilistic experiment and observed that the number of accidents in a finite number of shipments is Binomially distributed. (In fact, when the number of shipments is quite large, the probability distribution of the number of accidents can be approximated by a Poisson distribution, since the accident probabilities are very small. Representation of the hazardous materials accidents via Poisson models has been suggested in the early works of Kalelkar and Brooks [657] and Robbins [1010].)

A final objection to the use of the traditional expected consequence risk model is raised on the grounds that it ignores the safety measures taken by the communities around a potentially hazardous activity to mitigate the undesirable consequences of release incidents. For example, risks imposed on a community due to hazmat shipments on nearby roads can be reduced by establishment of an effective emergency response system. Pijawka et al. [944] and Scanlon and Cantilli [1039] devised community-preparedness indices, in an attempt to assess the absolute risks of hazmats transportation.

20.3   Equity

Assessment of the amount of risk imposed on the public by the activities that involve hazmats is a primary step towards the development of analytical models for hazmat management. A comprehensive evaluation of the public risks associated with a potentially dangerous activity, however, should also address the issue of equity. Equity implies the distribution of impacts resulting from a policy decision in a fair manner among the members of the public. In the context of hazmat logistics, equity corresponds to fairness in the spatial distribution of risk. Burdening a small segment of the society with the risks associated with hazmats while a much larger group of people reaps the benefits from the activities that involve these materials, conflicts with the principle of equity. Morell [866] suggested that the perceptions of inequity may be one of the primary reasons for public opposition to the location of potentially hazardous facilities. Hence, a more-equitable decision may not only be superior to a less-equitable decision from a philosophical standpoint, but also from a very practical standpoint if the more-equitable decision has a better chance of being approved by the public.

Before moving on to mathematical programming models that incorporated equity as one of their criteria, we first discuss the measurement of equity. Most of the pioneering work on the notions of equity appeared in the economics literature. The management science literature in this area is relatively sparse. One exception is the decision analysis literature, where a number of attempts have been made to provide a decision-theoretic framework for modeling equity. These studies focus on fatalities as the major undesirable consequence of activities posing public risk. Thus, as a result of the potentially hazardous activity, each individual will either be a fatality or not. A possible consequence can be represented by an N-vector e.g.

$(0, 1, \ldots, 0)$ denoting that individual 1 is alive, individual 2 is a fatality, etc. Given the set of possible consequences, each hazardous activity can be thought of as a lottery over this set. We can express the individual risk vector as follows:

$$IR = \{IR_1, IR_2, \ldots, IR_N\} \tag{20.4}$$

where, IR_i denotes the marginal probability of death for individual i. (Note that it is possible to perceive person i as a representative of the population center i.) Within this probabilistic context, there are two ways of assessing the equity of a decision: ex-ante and ex-post. Ex-ante equity refers to the fairness of the distribution of individuals' marginal probabilities of death; ex-post equity refers to that of the individuals' fates in the actual outcome of the decision.

Assuming that each individual at risk is valued identically, we can concentrate on the number of fatalities rather than on who they are in analyzing the ex-post equity of possible consequences of a decision. Keeney [683] showed that ex-post risk equity implies a risk prone utility function over the number of fatalities, which is in conflict with the preference for catastrophe avoidance. This result can be motivated using a simple comparison of two alternatives: Alternative A, where each of two individuals has a 50% chance of becoming a fatality (independent of the probability of death of the other person), and Alternative B, where it is determined a priori which one of the two individuals will die. Using an ex-post risk equity measure, Alternative A is judged to be more equitable than Alternative B. Note that, however, under Alternative A there is a 0.25 probability that both individuals will die (i.e. a catastrophe). In a subsequent paper, Keeney [683] also provided the specific forms of the von Neumann-Morgenstern utility functions that are suitable for modeling ex-post risk equity. Sarin [1037], and Keeney and Winkler [690] devised similar measures for ex-post risk equity of a decision. One such measure provided in Sarin [1036] is $\sum_y ay(N - y)P(y)$ where y denotes the number of fatalities and $a < 0$ is a scaling constant. Note that, according to this measure, the most equitable consequences are the ones with $y = 0$ (no one dies), or $y = N$ (everyone dies), whereas $y = \frac{N}{2}$ (half the population dies) constitutes the least equitable consequence.

We now turn to ex-ante risk equity. Let $IR' = \{IR_1, \ldots, IR_i + \epsilon, \ldots, IR_j - \epsilon, \ldots, IR_N\}$. An alternative that results in IR is more ex-ante-equitable than an alternative with IR' if the absolute difference between IR_i and IR_j is less than the absolute difference between $IR_i + \epsilon$ and $IR_j - \epsilon$. This is called the principle of transfers which implies that the most equitable case is where the marginal probability of death is the same for each individual. Two commonly used measures for ex-ante equity are $\sum_i |IR_i - \underline{IR}|$ and $\sum_i (IR_i - \underline{IR})^2$, where $\underline{IR}$ denotes the average probability of death.

Broome [149] showed that the von Neumann-Morgenstern utility functions are inappropriate for modeling preferences for ex-ante risk equity. Using an axiomatic approach, Fishburn [404] identified maximal subsets of mutually compatible equity axioms from a set of six axioms for a population of two individuals. This study suggests that, although von Neuman-Morgenstern utility functions are inappropriate, ordinal value functions can be developed for ex-ante risk equity. One such ordinal value function is provided by Sarin [1037]. Later, Fishburn and Straffin [405] extended the earlier axiomatic approach of Fisburn [404] to $N > 2$.

Keeney and Winkler [690] suggested a multiattribute utility model that combines risk, ex-ante risk equity, and ex-post risk equity. Hammerton et al. [529], and Keller and Sarin [691] reported empirical studies on equity and public risk. For an overview of the approaches to modeling equity in public risk, we refer the reader to Sarin [1036].

In an attempt to quantify inequality in locational decisions, Erkut [383] focused on the distances between a single public service facility and its users. Location of a facility implies a set of distances

$$d = \{d_1, d_2, \ldots, d_N\} \tag{20.5}$$

where d_i denotes the distance between population center i and the facility. Erkut [383] proposed the proximity of d to a perfectly equitable ideal solution as a measure of equality. He generated a family of inequality measures by varying the definition of the ideal solution, and the distance metric used in quantifying the proximity of d to the ideal solution. He also showed that some popular inequality measures such as the maximum distance, the range of distances, and the mean absolute deviation of distances are not scale invariant, and do not satisfy the principle of transfers, which is a widely-accepted equity axiom. Two measures that are suggested as appropriate means for quantifying inequality are $\dfrac{\sqrt{\sum_i (d_i - \underline{d})^2}}{N\underline{d}}$ and $\dfrac{\sum_i \sum_j |d_i - d_j|}{N^2 \underline{d}}$ where $\underline{d}$ is the average distance associated with the set d. The former measure is the coefficient of variation and the latter is equal to twice the Gini coefficient, perhaps the most popular inequality measure in economics. (The use of the Gini coefficient in analytical models for public service delivery problems has also been suggested by Mandell [797].) Both of these measures are improved as one moves further away from the set of users. Although this property may be in conflict with efficiency-oriented criteria for locating service facilities, it may be suitable for locating hazardous facilities, where remote locations are usually preferable.

In another paper that directly addresses equality in facility location, Berman and Kaplan [88] provided a transfer payment scheme to achieve equitable distribution of the benefits associated with a desirable facility. Service recipients would pay taxes or receive side payments depending on how close they are to the facility. They showed that the distance minimizing facility location models also maximize the total user benefits when perfect benefit equity is assured by a constraint. In the event that transfer payments cannot be implemented, however, they argued that the mean absolute deviation of travel distances would be an appropriate measure for inequality. Although this paper deals with desirable facilities, it may be possible to extend its results to hazardous facilities.

We now move on to alternative ways that have been used in the literature for addressing the equity issue in hazmat logistics models. A pragmatic way to impose equity in a mathematical programming model for hazmat logistics is to use upper bounds on the undesirable consequences imposed on the individuals in the system. For example, in their hazmat transport model, Zografos and Davis [1229] used a set of constraints on the amount of flow on links of the transport network. A closely-related technique imposes an upper bound on the range of undesirable consequences. As an example, Gopalan et al. [498] presented a model where the difference in total risk between all pairs of population centers remains below a prespecified level. (In a follow-up paper, Lindner-Dutton et al. [761] extended the Gopalan et al. [498] model to facilitate equitable sequencing of hazmat shipments.) Another popular technique used to address equity is minimizing the maximum undesirable consequence for any individual. This alternative is used by Erkut and Neuman [387], and List and Mirchandani [766], in multicriteria models.

In philosophically-parallel papers that suggested a new paradigm for hazmat logistics, Ratick and White [983] suggested the simultaneous siting of multiple facilities rather than a large central facility, and Gopalan et al. [498] suggested the use of multiple routes for hazmat shipments between a given origin-destination pair instead of a single path. These (similar) proposals are based on the fundamental argument that the concentration of hazardous activity in one location, or on one route, violates the principle of equity in the spatial distribution of risk. Given the required capacity for the potentially hazardous activity, Ratick and White [983] (and later White and Ratick [1198]), presented analytical approaches for locating a set of facilities to maximize equity which may help mitigate public opposition. Similarly, Gopalan et al. [498] developed an algorithm to spread the transport risks equitably on the transport network.

Although the strategy of distributing potentially hazardous activities to several locations and routes may improve risk equity, it may also increase the total societal risk . Furthermore, the reduction in transport costs due to the spatial distribution of hazardous facilities may be offset by the loss of scale economies in the construction and operation costs in smaller facilities. Thus, equity should be considered in connection with total risk and system cost, and multiobjective models should be used to analyze the tradeoffs between total risk, equity, and total cost in designing hazmat management systems.

20.4   Cost Aspects

From the viewpoint of the public, the major concern regarding hazmat logistics decisions seems to be the possible undesirable consequences. However, economic consequences of logistics decisions are clearly very important to businesses that deal with hazmats. In fact, risk and equity are societal concerns and it is unrealistic to expect industry to consider them in their logistics decisions unless they are required by regulations, or unless there are some tangible benefits (such as an improved company image). Given the importance of cost in these problems, estimation of the costs associated with hazmat logistics decisions is crucial for industry. Evaluation of the cost implications of policy decisions regarding hazmats is also important for government agencies. For example, the routes designated for dangerous goods shipments, and the technologies identified for treatment of hazardous wastes must be economically feasible. Otherwise, enforcement of the rules and regulations can be a problem, and illegal practices in dealing with hazmats, such as the export of hazardous wastes to developing countries, may increase.

Industrial companies that use hazmats in their production process need to construct appropriate storage facilities for these materials, and adopt safe manufacturing technologies. Many industrial activities also generate hazardous wastes as a by-product. The amount of hazardous waste generated, however, can be reduced by the use of waste-friendly process technologies. Thus, the hazmat logistics decisions in industrial companies involve tradeoffs between the costs and the (public and environmental) risks of the process technology and storage design alternatives. However, there are other facilities that deal with significantly larger quantities of toxic substances, which are potentially more hazardous than industrial manufacturing plants. For example, hazardous waste treatment and disposal facilities, nuclear reactors, and high-level nuclear waste repository sites. For these facilities, a more comprehensive approach, that incorporates the decisions regarding facility location, size, and technology content as well as the routes to be used for hazmat shipments to and from the facilities, is needed. Clearly, effectiveness of the logistics policies depends on the reliability of the cost estimates as well as the risk assessments associated with alternative configurational and routing decisions.

Cost estimation in the context of hazmat management can turn out to be quite involved. For example, establishment of a hazardous waste treatment and disposal plant usually requires the assessment of site acquisition and construction costs at several candidate locations. These costs partly depend on geological properties of the candidate sites, which may make cost estimation a costly activity with considerable uncertainty. Further, there are alternative technologies for processing a given type of hazardous waste, e.g. organic wastes can be incinerated, or disposed of in a hazardous waste landfill, or processed via solar detoxification. Hence, cost estimates are needed for a number of technologies. The estimation process is also complicated by the existence of economies of scale in technology acquisition costs (for example, quadrupling the capacity of a biological waste incinerator only doubles the investment cost), and economies of scope in integrated hazardous waste treatment facilities (for example, the construction cost of an integrated facility treating three types of waste would be less than the sum of the construction costs of three dedicated facilities). The cost estimation process should incorporate all possible technology alternatives, and all relevant economies, so that an optimum configuration for the facility can be found.

Another complicating factor in the estimation of costs for hazmat logistics decision is the uncertainty in the cost of the hazard mitigation and compensation measures which must be negotiated with the host community for locating a hazardous facility. Although for intelligent decisions all costs should be known with some accuracy at the planning stage, the mitigation and compensation costs are not finalized until shortly before the location decision is finalized. This may limit the use of optimization models in locating hazardous facilities. Similarly, for transport decisions, estimation of the cost of an accident involves considerable uncertainty. The accident cost includes the cost of cleanup, property losses, payments to the survivors of fatality victims, and the opportunity cost of the congestion caused by the accident. Wright [1217] pointed out that the average cost per accident, based on 70 dangerous goods accidents in Canada, was determined to be $600,000, although the cost has been as high as $10M in the case of a propane tank accident. Clearly these accident costs vary according to the routes taken. Whether or not to explicitly include these accident costs in logistics models depends on the objective of the analysis. When modeling the system at a strategic level, these costs can be seen as costs to society, and hence should be included in the model. However, from the viewpoint of a waste shipper, these accident costs may or may not be relevant, depending on the insurance structure and who pays for the damages arising from the accident.

20.5   Planning Potentially Hazardous Facilities

In this section we will focus on analytical approaches for planning potentially hazardous facilities, such as, nuclear power plants, nuclear waste repositories, and hazardous waste treatment and disposal facilities. Unfortunately, process technology choice and storage design decisions for the management of hazmats in industrial plants has not received much attention in the academic literature. Although the potentially hazardous facilities are undesirable from a public and environmental safety viewpoint, their services, e.g. production of gasoline, or safe disposal of the nuclear and other hazardous wastes, are certainly required components of our contemporary lifestyle. Thus, a first step in designing such a facility is the identification of the geographical region it will serve. This facilitates estimation of the types and volumes of the hazmats to be dealt with. Given the service area and the type of service to be provided, facility planning involves the decisions regarding the number of facilities to be established, their location, the technologies to be used at each facility, their size as well as the allocation of facilities to demand zones. The designed system must have sufficient capacity to serve the region, and the service should be provided with minimum possible adverse impacts and cost.

The above definition implies the multicriteria nature of the facility planning problem pertaining to potentially hazardous facilities. Minimization of total cost and risk, and maximization of equity are the primary objectives of the design problem. However, in many cases a more detailed analysis may be necessary, focusing on more specific objectives such as the undesirable impacts on environmental aesthetics, endangered species, or archeological sites. Analytical approaches for planning potentially hazardous facilities range from multiobjective mathematical programming models to multicriteria decision analysis techniques. The former is especially useful for problems with many alternative solutions and only a few objectives such as the planning of a set of hazardous waste treatment centers, whereas the latter is more suitable for problems with many objectives and only a few alternative solutions, such as the location of a high-level nuclear waste repository. Siting of potentially hazardous facilities has received much more attention in the academic literature than the other configurational decisions. This can partially be explained by the public's sensitivity over the locations of these facilities, and hence the controversies that need to be resolved in finalizing the siting decisions. Clearly, location of a facility determines its proximity to the population centers, and therefore is a crucial factor in determining its risk. However, toxicity of the materials used in the facility, as well as the capacity and technology configurations of the facility, are also significant determinants of facility risk. The dependencies among configurational decisions suggest the development of integrated models that facilitate global optimization.

In practice, the planning of potentially hazardous facilities is a two-stage process. The first stage constitutes a prescreening of the alternative solutions in an attempt to identify the candidates for the detailed comparative analysis to be performed in a second stage. In their recent review of the normative approaches for siting hazardous facilities, Kleindorfer and Kunreuther [701] emphasized that operations research methodologies should be coordinated with a broad process view of siting. We suggest that this applies to all configurational decisions. In the past, the common approach for the establishment of hazardous facilities was a decide-announce-defend approach. This is a process where the developer of the facility finalizes the plans regarding its configuration and then assumes a defensive position against the criticism from other stakeholders. Due to the failure of this approach in practice, a legislated siting approach became popular to facilitate construction of the most controversial plants. This involves the setting of certain design standards by a government agency prior to the planning process. A well-known example of this approach is the Nuclear Waste Policy Act of 1982 issued by the U.S. Congress that charged the Department of Energy with constructing and operating the first high-level nuclear waste repository in the U.S.. The most effective approach in mitigating the public opposition to hazardous facilities, however, seems to be voluntary siting. This approach incorporates all the stakeholders in the facility design process. The aim is to make the facility attractive rather than undesirable for the host community, by the use of compensation packages, benefit sharing, and shared management, as well as by convincing the public that the necessary hazard mitigation mechanisms will be incorporated. This approach proved to be instrumental in both the construction and the expansion of the Alberta Special Waste Treatment Center at Swan Hills, Canada. The models presented in Kleindorfer and Kunreuther [701] and Richardson [1006] explicitly incorporate the risk compensation issues relevant to siting of hazardous facilities.

Analytical approaches for the evaluation of alternative solutions to the facility design problem provide a quantitative framework for the negotiations between stakeholders. Despite the multiobjective nature of the problem, earlier attempts were confined to the development of single objective models. Although such models enhanced our understanding of the various dimensions of the design problem, they are unable to incorporate the numerous tradeoffs associated with the establishment of hazardous facilities. For example, in one of the earlier papers, Dutton et al. [349] proposed a model which minimizes the total cost of the facilities so as to identify the optimal locations of nuclear power plants in the Pacific Northwest region of the U.S. (However, they did point out that the final locations of the facilities will depend on political realities and community acceptance.) Many authors, who worked on single-objective models, modeled the location of undesirable facilities on the basis of their distance to the surrounding population centers. One way to protect the people from adverse impacts of the facilities is to assure that the facilities are not closer to the population centers than a threshold distance. This can be implemented in analytical models by the use of lower bounds on distances between the facility and the population centers. An alternative (and more common) technique is to maximize some function of the distances between the facilities and the population centers. Erkut and Neuman [385] provided an extensive review of the location models with distance maximization objectives. The two prevailing approaches are maximization of the sum of distances and maximization of the minimum distance. Unfortunately, these are very simplistic ways of modeling the undesirability of hazardous facilities. Presumably, the perceived risk of a facility, and the associated public opposition, decrease with distance. Using this concept, a pragmatic approach for siting potentially hazardous facilities may be the incorporation of public opposition to the facility in the analytical model as an objective to be minimized. Thus, the determination of the exact form of the opposition function (for a given facility) is a relevant research topic. However, Rahman et al. [981] is the only published paper we are aware of that develops an empirical distance-based opposition function for an undesirable facility.

In a paper, which is still limited to a single objective, but goes beyond considering distances between the facility and population centers, Karkazis and Papadimitriou [677] focused on minimization of the total atmospheric pollution caused by the emissions from hazardous facilities. The determinants of the pollutant burden on a population center are its distance to the facility and the direction of winds. This study is extended in Karkazis et al. [668] and applied to the Salonica district of Greece.

As discussed in Erkut and Neuman [385], single-objective models are usually inadequate in modeling hazardous facility location problems, which are characterized by multiple stakeholders and multiple objectives. Cohon et al. [235] are among the first who presented a multiobjective mathematical model for planning potentially hazardous facilities. Their aim was to select sites, types and sizes of the nuclear power plants to be established in the six-state region of the eastern U.S. The population impact of the power plants were considered as well as their construction and operating costs. The authors showed that remote sites can be found for locating these facilities in the prespecified region with little or no cost penalty. Before moving on to the discussion of other papers with multiple objectives, we present a generic multiobjective model for planning potentially hazardous facilities in order to expose the structure of the facility design problem. The notation provided in the preceding sections will be used, and hence index i denotes population centers. We will assume that the facilities provide a service that involves a single type of hazmat, and therefore index m will be suppressed. Potentially hazardous activity p however, will be represented by two indices, j and k. Index j denotes the alternative facility locations and index k denotes the alternative facility configurations, i.e. facility size and technology content. The generic model is as follows:

$$\min \quad \{g_1(x, y), g_2(y), -g_3(y)\} \tag{20.6}$$

$$g_1(x, y) = \sum_j \sum_k F_{jk} y_{jk} + \sum_i \sum_j f_{ij} x_{ij} \tag{20.7}$$

$$g_2(y) = \sum_i \sum_j \sum_k R_{ijk} y_{jk} \tag{20.8}$$

$$g_3(y) = f_e \left(\sum_j \sum_k R_{1jk} y_{jk}, \ldots, \sum_j \sum_k R_{Njk} y_{jk} \right) \tag{20.9}$$

subject to

$$\sum_j x_{ij} = D_i, \quad \forall i \tag{20.10}$$

$$\sum_i x_{ij} \leq \sum_k a_k y_{jk}, \quad \forall j \tag{20.11}$$

$$\sum_k y_{jk} \leq 1, \quad \forall j \tag{20.12}$$

$$x_{ij} \geq 0, \ y_{jk} \in \{0,1\} \quad \forall \ i,j,k \tag{20.13}$$

where

F_{jk} = annualized construction and operation cost of configuration k at site j,

f_{ij} = unit transport cost between population center i and facility j,

R_{ijk} = annual risk imposed on population center i due to the establishment of a facility with configuration k at site j,

$f_e(.)$ = a measure of equity in the spatial distribution of risk,

D_i = annual demand for service at population center i,

a_k = annual capacity of configuration k.

The decision variables are:

x_{ij} = amount of service provided to population center i from a facility located at j,

y_{jk} = 1 if configuration k is established at site j, 0 otherwise.

Note that the population centers that are exposed to the risks of the hazardous facilities also receive service from them. The constraint set (20.10) ensures that the service demand at each population center is satisfied. The set (20.11) imposes the capacity limit associated with the selected technology at each facility, and allows the population centers to receive service only from open facilities. The set (20.12) assures that at most one facility configuration can be selected at each site. The first objective (20.7) is minimization of total system cost which is the sum of investment, operation and transportation costs. Transportation costs are associated with hazardous waste shipments from the population centers in the case of a hazardous waste treatment center, whereas they would represent the transmission costs in the case of a power plant. The second objective (20.8) constitutes minimization of total system risk, and the third objective (20.9) involves maximization of a measure of equity defined over the set of individual risks. The model presumes that facility risk is a function of the configurational decisions, and that the operational decisions (i.e. the amount of hazmat actually dealt with in the facility) will not have a significant contribution to risk. Nevertheless, the above model constitutes a framework to present the multiobjective approaches that have appeared in the literature for hazardous facility planning. (Note that, although it is possible to incorporate the scale economies in the investment and operation costs with the above model, evaluation of the economies of scope in designing hazardous facilities would require the development of a multicommodity model.)

We now describe several papers that use a model similar to the generic model introduced above. Ratick and White [983] presented a multiobjective model for locating undesirable facilities and selecting their size. Their cost objective does not include transportation costs. Public opposition to the facilities is modeled as an increasing function of facility size, and is minimized as their second objective. The third objective is maximization of the minimum equity-outcome index, which is related to the number of facilities that pose a risk to each population center. The model was solved by treating the first and third objectives as constraints. White and Ratick [1198] translated the public opposition and equity objectives of the Ratick-White model into dollar figures which represent the compensation required to mitigate the opposition. This allowed them to solve the model as a single objective mathematical program. Erkut and Neuman [387] expanded on the Ratick-White model, where the disutility of an individual at population center i due to a facility at site j is formulated as a nonlinearly increasing function of facility size and a nonlinearly decreasing function of the distance between i and j. Their objectives were minimization of total system cost, total disutility from the facilities, and the maximum individual disutility in the system. They suggested a branch-and-bound algorithm to generate the set of all efficient solutions to the problem.

In a paper closely-related to [1198, 387], Wyman and Kuby [1219] described a multiobjective model that explicitly considers the technology selection decisions in hazardous waste processing centers. Based on this study, Wyman and Kuby [1220] suggested that solar detoxification, a new technology alternative for managing toxic waste, is superior to the conventional methods, e.g. incineration, in terms of all three objectives, risk, cost, and equity. Using a bicriterion model, Rahman and Kuby [980] examined the tradeoffs between minimizing cost and public opposition within the context of locating solid waste transfer stations. The unique feature of their model is the opposition function derived empirically from the opinion survey data presented in Rahman et al. [981]. Our survey of this literature reveals that more work is needed on the development of multicommodity facility design models with an emphasis on technology selection and sizing decisions.

Multicriteria decision analysis approaches have been more popular than multiobjective mathematical programming problems, in designing potentially hazardous facilities, especially nuclear power plants, and nuclear waste repositories. Keeney and Nair [688] presented one of the first studies implementing a multiattribute utility approach for selecting nuclear power plant sites. They used the following six major criteria in evaluating nine alternative sites in the Pacific Northwest region of the U.S.: (i) health and security of the population in the surrounding region, (ii) the loss of salmonids in streams absorbing the heat from the power station, (iii) other biological effects on the surrounding region, (iv) socioeconomic impact of the installation, (v) aesthetic impact of the power lines, and (vi) investment and operating costs of the power station. Assessment of the decision maker's preferences led to the use of a multiplicative multiattribute utility function in aggregating the individual utility functions, each associated with one of the above attributes. Roy and Bouyssou [1027] demonstrated an application of the ELECTRE III method to solve the siting problem reported in Keeney and Nair [688]. This allowed the authors to compare the results of their outranking approach with that of Keeney-Nair's multiattribute utility model. Multiattribute utility approach presumes the existence of a complete and coherent preference structure of the decision maker which constitutes the basis for ranking the alternatives. In contrast, the outranking model does not claim to deal with such a preference system to be described, but aims at construction of the relationship the decision maker would like to have with reality. Due to the lack of a complete preference structure, an outranking model may only provide a partial ordering of the alternatives. Although their results were similar to those of Keeney and Nair, Roy and Bouyssou [1027] emphasized the differences caused by the conflicting assumptions of the two approaches. For example, outranking models rule out the possibility that a major disadvantage in one attribute could be compensated for by a number of minor advantages in other attributes, which is clearly allowed for in the multiattribute utility model. Another case study on nuclear power plants that uses an additive multiattribute utility function in siting the facilities and their water sources for cooling purposes is reported in Kirkwood [698].

A specific hazardous facility location problem that received considerable attention in the decision analysis literature is the location of a high-level nuclear waste repository in the U.S. Spent nuclear fuel removed from a power reactor produces heat as part of the radioactive decay process. Therefore, it is necessary to keep the fuel rods in water pools at the power plant for a cool-down period. However, a permanent storage site for the spent nuclear fuel rods is necessary. In the U.S. construction of a permanent geological repository at the Yucca Mountain in Nevada is in progress, and according to the latest announcement from the Department of Energy it will not be operational until 2010. Multiattribute utility approach has been suggested as a suitable decision aid for a variety of problems at several stages of the planning process of this facility. Merkhofer and Keeney [834] used an additive multiattribute utility model in a comparative analysis of the five candidate sites in order to select three sites for characterization. (Site characterization is a detailed data gathering process to determine the suitability of the site for a repository, and involves the construction of exploratory shafts for underground testing which was estimated to cost $1,000 million per site.) They used fourteen attributes regarding the construction and operation period of the repository, and two attributes regarding the postclosure impacts in order to evaluate the alternative sites. Although the Department of Energy announced a different ranking of the five sites than proposed by Merkhofer and Keeney [834], their study clarified some uncertainties regarding repository performance. Furthermore, Keeney [685] showed that characterization of all three sites was a waste of approximately $1,500 million and suggested a sequential characterization procedure.

Recently, Keeney and von Winterfeldt [689] suggested a reconsideration of the U.S. national policy for the construction of an underground repository at Yucca Mountain. They claimed that there exist superior alternative strategies, such as temporary above-ground storage at a centralized facility or next to nuclear power plants until the year 2100. The objectives used to evaluate the alternative strategies include concerns for health and safety, environmental and socioeconomic impacts, direct and indirect costs, equity concerns, federal government's responsibility to manage nuclear waste, and implications of theft and misuse of nuclear waste. Concerning the operation of the Yucca Mountain repository, Rautman et al. [984] suggested an analytical approach based on the classical transportation model of linear programming for scheduling the burial of the spent fuel rods. Their model incorporates the fuel type, burnup level and the time of discharge from a nuclear reactor in an attempt to optimize the allocation of the limited space at the repository for disposal of the current nuclear waste stocks in the U.S. In related studies on the location of nuclear waste repositories in Europe, Briggs et al. [134] presented the use of the outranking method PROMETHEE for planning the disposal of medium and high-level nuclear waste. Based on the same case study, Delhaye et al. [273] demonstrated an application of the ORESTE method. These authors suggested that the outranking methods are suitable for designing potentially hazardous facilities since their information requirements are usually less than that of other multicriteria decision analysis techniques.

20.6   Hazardous Materials Transport Planning

In this section, we will review the state-of-the-art in hazmat transportation logistics. The strategic decisions pertaining to the transportation activity are transport mode selection, transport vehicle selection and routing. The distinguishing feature of hazmat transportation is its possible adverse impacts on people and the environment. Therefore, the associated risks need to be taken into account in all transport logistics decisions, as well as the costs. Mode selection involves identification of the most appropriate mode of transport between an origin-destination pair for a given hazmat.

Given the transport mode (i.e. rail or road), vehicle selection deals with determination of the best vehicle design to carry the hazmat. Note that mode selection, in effect, determines the transport network available for the transportation activity.

Routing decisions involve selection of the paths between origin-destination pairs to be used in transporting the hazmats. The routing problem can be perceived at two different levels. At the level of a hazmat carrier, the problem is to select the route(s) between a given origin-destination pair for a given hazmat, transport mode, and vehicle type. We can identify this as the single commodity, single origin-destination, local route planning problem. Hazmat carriers need to solve a local route planning problem for each shipment order they receive, and depending on the amount of material to be shipped, this may involve planning multiple shipments between the origin and the destination. At the level of a government agency, however, all the hazmats that need to be transported should be taken into account in designating the dangerous goods roads in a region. This is clearly a multicommodity, multiple origin-destination problem which we identify as global route planning. This strategic logistics problem needs to be solved when there are significant changes in the types and amounts of the hazmats to be carried or significant changes in the characteristics of the transport network.

Mode and vehicle selection problems are usually dealt with in the context of risk assessment studies. Complete enumeration of alternatives is a suitable solution methodology for these problems since the number of alternative transport modes or vehicle designs is usually very small. Many authors, including Glickman [480], Saccomanno et al. [1034], and Purdy [978], compared highway transport with rail transport for the shipment of a specific substance between two points. Their results show that the preferred mode of transport depends on the nature of the material shipped, the characteristics of the containers used during transportation (e.g. size and reliability), the topology of the transport network for each mode, as well as the spatial distribution of population around the network links. Thus, there is no transport mode which is safer for all possible instances of the hazmat transportation problem. Moving on to vehicle selection, Glickman [479] compared the risks of transporting flammable liquid chemicals through New York City under two alternative tank truck designs, specified by the N.Y.C. Fire Department and by the U.S. Department of Transportation. Expected number of fatalities for each vehicle design is estimated under four scenarios involving a typical route or a most hazardous route (more city streets and bridges), and average accident rates (dry pavement and unrestricted visibility) or worst case accident rates. Results of the risk assessment were subjected to an uncertainty analysis to incorporate the randomness in the associated risks.

The local route planning problem has attracted the attention of many researchers and has been studied extensively. Thus, we will devote the remainder of this section to the analytical approaches for this problem. It is interesting to note here that very little attention has been paid to the global routing problem. However, this may be the only way to optimize the performance of a hazmat transport system in terms of its total cost, total risk, and equity in the spatial distribution of risk. Allowing individual shippers to choose their routes may result in a poor overall system performance by overloading certain segments of the transport network. This may be undesirable due to increased risk caused by traffic congestion, or because of an inequitable distribution of risk. Therefore, regulators need to search for ways of making the transport system safer and/or more equitable. Global routing models would facilitate the evaluation of alternative strategic decisions, such as prohibiting the use of certain links for dangerous goods transport or construction of new roads so as to improve the transport network. These models, however, can be very difficult in terms of analytical tractability. Some of the papers we discuss in the section on integrated models deal with the global routing problem.

Paralleling our discussion of facility planning literature in the previous section, we present a generic model for the local route planning problem. Let $G = (V, S)$ be a connected, undirected and planar transportation network with node set V and link set S. It is necessary to partition the actual transport links into road segments each having uniform accident and consequence probabilities. This is required by many of the prevailing risk assessment techniques as discussed in Section 20.2. Thus, links of G represent road segments. Note that not all of the population centers (indexed by i) need be connected to the transport network of interest, but they may be put at risk due to their proximity to the transport links. Nodes of G represent population centers, potentially hazardous facilities, road intersections, and endpoints of road segments. Two nodes are of special interest within the context of local route planning, namely the origin O and the destination D of the shipment. The trip of the vehicle from O to D will be stopped if an accident occurs during the transport activity. Consequently, an accident can occur at road segment s only if there were no accidents in the preceding segments traversed by the vehicle. Thus, the probability of an accident at road segment s is,

$$P_s^*(A) = (1 - P_1(A))(1 - P_2(A)) \ldots (1 - P_{s-1}(A))P_s(A) \qquad (20.14)$$

where the vehicle is traveling along the path $1, 2, \ldots, s - 1, s$, and $P_s(A)$ is the accident probability on road segment s, given that the vehicle is traversing segment s. As discussed in Section 20.2, one has to examine the historical records in determining the past frequency of accidents on road segment s as an estimate of $P_s(A)$. Nevertheless, given that the accident probabilities are usually on the order of 10^{-6} per vehicle-mile, we will assume that

$$P_s^*(A) \approx P_s(A) \qquad (20.15)$$

The expression in (20.15) implies that the occurrence of accidents along a chosen path are independent random events. In effect, the risk associated with each road segment is assumed to be independent of the accident probabilities of other road segments. This greatly simplifies model formulation since the risk associated with each path can now be represented as an additive measure across road segments.

As in the previous section, we will denote the potentially hazardous activity p with two indices, u and v. Index u denotes the starting node of road segment s, and index v denotes its end node. We again will suppress the index m in the notation since local route planning is a single commodity problem by definition. Since we do not view equity as a relevant objective for this version of the routing problem the following generic model has two objectives.

$$\min\{h_1(z),\ h_2(z)\} \tag{20.16}$$

$$h_1(z) = \sum_u \sum_v f_{uv} z_{uv} \tag{20.17}$$

$$h_2(z) = \sum_i \sum_u \sum_v R_{iuv} z_{uv} \tag{20.18}$$

$$\tag{20.19}$$

subject to

$$\sum_v z_{uv} - \sum_v z_{vu} = \begin{cases} 1 & \text{if } u = 0 \\ -1 & \text{if } u = D, \quad \forall u \\ 0 & \text{otherwise} \end{cases} \tag{20.20}$$

$$z_{uv} \in \{0, 1\}, \quad \forall u, v \tag{20.21}$$

where

f_{uv} = cost of transporting one load of shipment across road segment (u, v).

R_{iuv} = risk imposed on population center i per shipment on road segment (u, v).

The decision variables are

z_{uv} = 1 if road segment (u, v) is used for the trip, 0 otherwise.

The above model (20.16)-(20.21) constitutes a bicriteria shortest path problem formulation where the first objective (20.17) represents the minimization of total transport cost, and the second objective (20.18) represents the minimization of total transport risk. Constraint set (20.20) ensures that the feasible solutions correspond to paths from O to D. (Note that there is no constraint to prevent looping, since no efficient solution to the above problem could contain loops, given assumption (20.15). If one wishes to add equity considerations to the above model, then it may be necessary to explicitly state a set of constraints to prevent looping.) The generic model constitutes a framework for presenting the prevailing analytical approaches for the local route planning problem.

Although the generic model presented above is a multiobjective model, many authors studied single objective versions of the local routing problem. In the only example we know of, that considers cost alone, Yu and Judd [1225] provided a cost effectiveness analysis of transportation strategies for ranking the five candidate sites for a nuclear waste repository in the U.S. Clearly, the transport cost figures provided by such an analysis are relevant in the siting of the repository. However, transport risks are probably just as important. In fact, many authors suggest that minimization of risk is the highest priority objective in the transportation of dangerous goods. This is perhaps an explanation of why the majority of the local routing models appearing in the literature are single objective formulations focusing on risk minimization. Next we provide a brief review of this literature.

In one of the earlier papers, Glickman [481] showed that rerouting of railroad shipments of hazmats can significantly reduce population exposure, and could lead to about a 50% reduction in the expected casualties. He also noted that extensive track upgrading can be even more effective in mitigating accidental releases of hazmats. Batta and Chiu [57] analyzed a local routing problem where the population is continuously distributed along transport links as well as being concentrated at the nodes of the network. Their basic model aims at finding a path that minimizes a weighted sum of the lengths over which the vehicle is within a threshold distance from the populated zones. That is, all the people living within a prespecified bandwidth around the path used by the vehicle are assumed to be exposed to the same adverse impacts of hazmats transportation. The authors then incorporated the probability of an accidental release in the basic model. Thus, the objective of the first model is to minimize total population exposure, whereas that of the second model is to minimize expected consequence. In both models, they showed that assuming (20.15) allows for a shortest path problem formulation with modified link lengths.

More recently, Karkazis and Boffey [672] presented an analytical approach where the total pollutant burden at each population center is calculated via a dispersion model which takes into account the characteristics of the hazmat being shipped, the distance between the accident location and the population center, as well as the meteorological conditions at the time of the accidental release. A series of accidents are simulated along each arc of the transport network by randomly generating the accident site and the associated wind direction. This enables the authors to estimate an expected damage parameter for each road segment which, in turn, serves as an input to their mathematical model to find a path between the origin-destination pair that minimizes the expected total pollutant burden. In this paper, Karkazis and Boffey did not assume (20.15) to simplify the model which required them to develop a branch-and-bound type solution strategy. In a related study, Boffey and Karkazis [118] analyzed the relationship between a linear model that results from assumption (20.15) and a nonlinear model that results when (20.15) is not assumed.

We now mention two single-objective papers that do not fit our generic local transport model. Drezner and Wesolowsky [334] presented a model for designing a hazmat route that maximizes the minimum weighted distance from the population centers to the route. This can be viewed as a network design problem with an equity objective. As discussed in the risk assessment section, Sivakumar's [1082] work on both the single shipment and the multiple shipment versions of the local routing problem is focused on minimization of expected risk at the occurrence of the first accident. The suggested column generation heuristics take advantage of the inherent fractional programming structure of the model, as well as the assumption in (20.15).

Before moving on to papers that pose multiobjective models, we point out a relatively well-known single-objective network problem that has potential for being useful in hazmat transport modeling and practice, namely the k-shortest path problem. In practice, hazmat carriers need to know a set of alternative paths between an origin-destination pair so that they can change their routing strategy when the best route is congested or temporarily closed. This suggests the relevance of k-shortest path algorithms for solving the local route planning problem. Furthermore, the k-shortest path problem also arises as a subproblem in some hazmats transport models with a variety of risk minimization objectives as demonstrated by the work of Jin [641]. When one is solving a bicriteria shortest path problem, like our generic local transport model, a k-shortest path algorithm can provide a quick way of generating a subset of the efficient frontier. Thus, this familiar network problem can be useful in the solution of multiobjective models, as well as in the generation of alternative solutions for the single-objective models. A computational comparison of four k-shortest path algorithms within the context of highway transportation of nuclear spent fuel rods is provided by Miaou and Chin [835]. Although the k-shortest path algorithm can be useful to solve multiobjective hazmat routing problems, more direct approaches are available. For example, Turnquist [1147], Wijeratne [1203], Ahmad [12] and List and Turnquist [767] used the lexicographic multiobjective shortest path algorithm of Martins [808] to solve different hazmat routing problems with multiple objectives. We refer the reader to Current and Min [249] for a review of multiobjective shortest path algorithms.

In a seminal paper, Saccomanno and Chan [1032] pointed out the relevance of multiple objectives to the hazmat transport problem. They adopted a shortest path algorithm to identify the best route under each of the following three strategies: (i) minimize truck operating costs, (ii) minimize accident likelihood, and (iii) minimize risk exposure which is defined as the product of accident probability and consequent damages. Their analysis incorporated stochastic factors such as visibility and pavement conditions (dry or wet), and deterministic factors such as road type, in the estimation of accident probabilities. Concerning the consequent damages, the costs of risk exposure to the individuals directly involved in the accident and the individuals indirectly effected by the accident are estimated as well as the possible property damages. Saccomanno and Chan concluded that the minimization of accident likelihood is an inferior strategy since the risk reduction benefits usually do not compensate for the increase in transport costs. Thus, they suggested the use of the risk minimization strategy.

One of the first studies that explicitly adopted a multiobjective approach for routing hazmat shipments was provided by Robbins [1010]. His objectives were to minimize the total length of the route, and to minimize the population exposure due to the routing decision. The statistical analysis on 105 origin-destination pairs selected from the U.S. Interstate highway system showed that route length weakly explains population exposure, and that the reduction in the exposure risk provided by the minimum population route compared to the minimum length route is statistically significant. The multiobjective analysis demonstrated that there are routes which can reduce the size of the endangered population but do not increase the length of route to the same extent as does the minimum population route. Chin and Cheng [210] provided a similar bicriteria routing model for nuclear spent fuel transportation in the U.S. Using 1980 population data, they showed that the minimum population route is sensitive to the threshold distance used for modeling the risk exposure during transport. They also demonstrated the nature of the tradeoffs between route length and population exposure.

Abkowitz and Cheng [5] used Saccomanno and Chan's [1032] estimates for the costs of possible adverse impacts of hazmat accidents in their multiobjective model. The model minimizes a weighted sum of the transportation and risk exposure costs. Abkowitz and Cheng paid particular attention to estimation of the indirect effects of accidental hazmat releases. They examined the possible impacts of toluene (a flammable liquid) and benzyl chloride (a corrosive substance) dispersions through air, when an accident occurs during a shipment of five tons of the material. Zografos and Davis [1229] suggested a preemptive goal programming approach for the local route planning problem. Their objectives were to minimize (i) total population at risk, (ii) risk of special population categories, (iii) property damages, and (iv) travel time as a surrogate for the transport cost. More recently, McCord and Leu [812] presented a multiattribute utility approach for local routing. They demonstrated the assumptions necessary for reducing the cost-risk minimization problem to a shortest path problem. They argued that the parameters of the multiattribute utility function could not be estimated with accuracy. Nevertheless, in the numerical example they provided, the model is able to generate a small set of noninferior paths even without exact preference information.

A closely related problem to routing of hazmats shipments is the scheduling of these shipments, which was first studied by Cox [244]. Cox and Turnquist [245] presented a model for finding the departure time for any given route to minimize the delay en route when there are curfews on hazmats transportation imposed by government agencies. Turnquist [1147] also addressed the uncertainties associated with the parameters of the Cox-Turnquist model. Wijeratne [1203] developed a stochastic multiobjective shortest path algorithm for the routing and scheduling decisions for hazmats shipments.

All of the prevalent approaches for hazmat transportation presume the availability of reliable estimates of accidental release probabilities. A unique approach is due to Klein [700] who suggested the use of fuzzy sets for the incorporation of imprecise information in the logistics models.

A current trend in hazmat routing is the integration of analytical approaches with the databases containing population, road and toxic substances data, and geographical information systems (GIS). This significantly enhances the data input and solution output stages in the decision process. Further, graphical representation of the selected routes on maps through the use of GIS also facilitates interactive decision making. Weigkricht and Fedra [1184] reported three such applications in Europe. Patel and Horowitz [920] made use of a GIS in the analysis of the results of their routing model that incorporates the diffusion of gases over wide areas from accidental releases during transport. Abkowitz et al. [6] developed a software (HazTrans) which incorporates the following criteria (and their weighted sums) in aiding the local routing decisions: (i) shipment distance, (ii) travel time, (iii) accident probability, (iv) population exposure, and (v) expected consequence. Recently, Glickman [478] presented another decision support system (PC-HazRoute) for solving the local route problem with the following criteria (and their weighted sums) among others: (i) shortest path, (ii) minimum population exposure path, (iii) minimum accident probability path, (iv) minimum societal risk path, (v) minimum user-defined risk path. Such systems can be useful for analyzing the cost-risk tradeoffs in road transport of dangerous goods, as well as for identifying desirable routes with respect to several criteria.

20.7   Integrated Models

The facility design and transportation logistics decisions are interrelated within the context of hazmat management systems. In particular, the location of a potentially hazardous facility determines either the origin or the destination of the hazmat shipments, and therefore interacts with the routing decisions. The total risk of a hazmat management system is the sum of the risks due to the facilities and the transportation activity. Further, the spatial distribution of risk is also determined by the location, size, and technology content of the facilities as well as the type and amount of the hazmats carried across the transport network. Thus, analytical approaches for the management of hazmats need to incorporate all the relevant strategic decisions in order to optimize the overall system performance. The recent reviews by List et al. [768] and Boffey and Karkazis [117] reveal the sparsity of the literature on integrated models for hazmat logistics. Only a few analytical models have been developed that integrate the location and routing decisions so as to minimize total cost and total risk, and maximize equity in managing hazmats. This section is devoted to the analytical approaches that would serve as building blocks in the development of integrated models for the strategic management of hazmats.

To motivate the need for integrated models on a concrete example, we briefly focus on hazardous waste. An effective way to model a hazardous waste management system is to view it as a network of waste generator sites, transfer stations, treatment centers, and disposal sites. This enables one to facilitate the strategic management decisions by solving the associated network design problem. The generator sites produce a variety of hazardous wastes as a by-product of their industrial activity. These wastes should be routed either to a treatment center with compatible technology, or to a disposal site. All wastes however, should move through the network until they are completely detoxified, or they reach an ultimate disposal site. Jennings and Scholar [639] presented the use of simulation methodology in evaluating alternative strategies to improve the hazardous waste management system. Schwartz et al. [1054] provided a model to estimate the adverse impacts of the accidents during hazardous waste transportation and the accidents at the treatment facilities. An illustrative application for siting incinerators in the Los Angeles area indicated that although the transport risks were similar, the facility risks were much higher at the urban sites compared to that of the rural sites. Mahoney [792] analyzed the trade-offs between the on-site and off-site hazardous waste management methods. He proposed a mixed integer programming model that minimizes the total system cost which is composed of fixed and variable costs for production process modifications at generator sites, treatment, storage, transport, and disposal. Each generator has to satisfy the demand for a finished product while hazardous waste is created as a by-product. The model is applied to the electroplating industry in New England where the management options for each generator included alternative product rinsing systems, alternative wastewater treatment systems, sludge drying, alternative types of on-site storage and transport, temporary storage and treatment at off-site facilities, and eventual landfill disposal. The case study identified a collaborative action by waste generators to establish an off-site treatment center as the best strategy.

To the extent of our knowledge, the earliest integrated approach for siting potentially hazardous facilities and routing hazmats shipments is due to Shobrys [1076]. The proposed multiobjective model is used in the analysis of temporary storage locations for spent nuclear fuel in the eastern U.S. Recently, Revelle et al. [996] published a paper based on this study. Their model simultaneously locates the storage facilities, assigns nuclear reactors to these facilities, and selects paths for the spent fuel shipments. A shipment route for each generator and storage site pair is chosen so as to minimize the total cost and the total risk of transportation. In fact, a weighted sum of the two criteria is used in a shortest path formulation, and the weights assigned to the criteria determine which member of the set of nondominated paths between a reactor-storage site pair will be used for the shipments. The selected routes are then used as input to a p-median formulation for siting the storage facilities. The risks associated with storage sites, however, are not incorporated in their model.

In contrast, Zografos and Samara [1230] suggested a goal programming model for the combined location-routing problem which incorporates both facility and transport risks. Their objectives were to minimize (i) the total distance between the population centers and hazardous waste disposal sites, (ii) the total transport risk, and (iii) the total travel time. The model also allows for the achievement of a certain level of risk equity by constraining the amount of hazardous waste flow on the transport links. Another comprehensive location and routing model is proposed by List and Mirchandani [766] which considered cost, risk, and equity in a multiobjective framework. At each population center, there exist risks due to shipments crossing nearby roads as well as potentially hazardous activities at nearby facilities. The measure of equity is the maximum risk imposed on any individual in the system. A distinguishing feature of the List-Mirchandani model is the incorporation of accident probabilities and alternative treatment technologies for hazardous wastes. Although the proposed model is very general, the authors made certain simplifications in applying it to a real-world case, and devising a solution procedure.

All of the above models assume that only a finite set of alternative sites would be available for potentially hazardous facilities. A recent paper that assumes an infinite feasible region is by Stowers and Palekar [1104], where the authors considered the integration of routing decisions with the location of a single undesirable facility on a network. Their risk minimization model shares many of the structural assumptions of Batta and Chiu [57] in formulating risk exposure. They exploited the geometric properties of the exposure function in analyzing several special cases of the problem.

The two final topics we discuss in this section are the location of emergency response teams and inspection stations. Due to the dangerous nature of hazmats, specialized emergency response teams are required for containing and cleaning up spills resulting from transport accidents. The response time of specialized teams is extremely important in mitigating the adverse impacts of accidental releases, and it is a function of the distance between their base station and the accident location. Hence, the determination of the number of emergency response stations needed in a region, and their locations, is a strategic decision that is relevant to the design of a hazmats management system. In a related paper, Psaraftis et al. [976] provided an analytical model for optimal response to oil spills. Their model locates appropriate levels and types of cleanup capability in a region and allocates the response teams to the zones with high oil spill potential. An illustrative application of the model in covering the high-risk harbors and off-shore zones of New England is also presented. Saccomanno and Allen [1031] presented a set covering formulation for locating specialized teams on a road network to respond to hazmats accidents at the nodes of the road network. The response team location problem has a similar structure to the well studied set covering problem of location analysis. However, the objective here is to cover the entire transport network, rather than to cover a set of discrete points as in set covering models. List [762] presented a multi-objective model for analyzing tradeoffs among response time, risk, risk equity and cost in locating emergency response teams. More recently, List and Turnquist [767] presented a GIS-based analysis of the routing and response team siting alternatives for high-level radioactive waste shipments in the U.S.

The location of inspection stations for inspecting hazmats shipments on the road is another important problem relevant to designing hazmats management systems. One preventable cause of accidental hazmats releases is the violation of loading rules and other regulations regarding truck and driver safety. The responsibility of an inspection team is to detect such violations and discontinue the trip of such a vehicle. The location of the inspection stations becomes a critical issue when there is heavy hazmat traffic on the transport network. A typical objective is to minimize uninspected vehicle travel. Recently, Mirchandani et al. [856] viewed this problem as a capacitated facility location problem and suggested an extension to their model to explicitly consider the point on a trip at which a vehicle is inspected. Although emergency response team siting and inspection station siting problems are important components of hazmat management, there seems to be very little reported research on these topics.

20.8   Conclusions and Suggestions

In this chapter, we presented a review of the existing analytical approaches for strategic management of hazmats. After discussing the building blocks of analytical models in this area, namely risk assessment, modeling of equity, and cost aspects, we focused on facility planning models, transport models, and integrated models. In this final section, we will reflect on the state-of-the-art, and point out certain directions for future research. In our judgment, more work is needed in all phases of research in this area to produce credible models. Below, we first discuss potential improvements in the building blocks, then we move on to our suggestions regarding models, and we finish with directions for implementation. We focus on a separate idea in each paragraph, which is summarized by the paragraph heading.

20.8.1   About the definition of risk

The risk modeling spectrum, discussed in the risk assessment section, suggests that there can be many ways of combining incident probabilities and population data for selecting solutions to hazmat logistics problems. Within this spectrum, in addition to the traditional definition of risk, we only encountered models at the two extremes (population-only and probability-only) in the literature. It is clearly possible to view the risk-minimization problem as a bicriteria problem of minimizing incident probability and minimizing exposed population, and then finding all efficient solutions. In the context of hazmat routing, the bicriteria approach may result in a number of risk-minimizing efficient paths, instead of a single "optimal" risk path that would result from using one specific risk formula. Although the single optimal risk path may be more desirable from an implementation perspective, producing one optimal risk path, in the absence of a consensus on the modeling of risk, may amount to little more than oversimplification. Unless we have good insights into the concept of risk in the minds of the decision-makers, the safest thing to do may be to report all efficient solutions to the population-probability minimization problem.

20.8.2   More about the definition of risk

A valid criticism to the expected consequence definition of risk is that it ignores variances. For example, a route that goes through two small towns may be judged to be just as risky as a second route that goes through one large town, if one considers only expected consequences. However, the variance of the consequences is larger for the second route which may make it less desirable than the first one. In Section 20.2, we discussed several techniques suggested in the literature to overcome the deficiency of the traditional definition of risk in capturing the low-probability high-consequence nature of hazmat incidents. Another way to overcome this deficiency is the development of mean-variance models that express route impedances as expected consequences plus the variance multiplied by some risk aversion constant. By doing a parametric analysis on this constant, one may be able to generate a set of efficient solutions from a risk-averse perspective. If one is particularly concerned with catastrophe-avoidance, however, a pragmatic alternative would be finding routes where the maximum consequence is below some threshold.

20.8.3   On the scope of risk assessment

There is a vast literature presenting techniques for assessment of the risks associated with potentially hazardous activities. The common feature of these studies, however, is their focus on fatalities as the major undesirable consequence of hazmat incidents. There are only a few papers which make an attempt to incorporate also the possibility of injuries and property damages in risk assessment. Furthermore, environmental risks are usually ignored in the suggested assessment methods. Possible damages to specific components of the environment (e.g. habitats, ecosystems) are included only in multiattribute utility models. Hazmat logistics decisions should take into account all the risks to humans and environment for broader acceptance by the public. This requires development of risk assessment techniques that incorporate all possible undesirable consequences of activities involving hazmats.

20.8.4   On perceived risks

Quantifying the risks associated with alternative hazmat management strategies is clearly very important for decision-making in this area. Technical risk assessments, however, have limited value in mitigating public opposition to potentially hazardous activities. This is partly because public opposition is a function of perceived risks, as opposed to technical risks provided by experts, and perceived risks of hazardous substances can be much higher than their technical risks. Unfortunately, very little effort has been spent on quantifying perceived risks. Given the disagreement between probability estimates in the risk assessment literature, and considering that we will (hopefully) never have sufficient accident data to improve on our estimates, it may be a wise strategy to turn to quantifying perceived risks. We believe that more work is needed to improve our understanding of the way perceived risks vary as a function of the hazardous substance, the distance to a hazardous activity, and the volume of the activity. A better understanding of perceived risks would allow us to more accurately estimate mitigation and compensation costs, as well as public opposition.

20.8.5   About inclusion of risk equity in models

It is suggested in the academic literature that equity in the spatial distribution of risk is a critical concern in designing hazmat management strategies acceptable to the public. On the other hand, risk equity is a non-issue to the hazmat industry, and, thus, current practice violates risk equity principles. An analysis of the dynamics of public opposition to hazardous activities may be helpful in determining the significance of risk equity from the public's perspective. It is highly unlikely, however, that the hazmat industry will ever be concerned with risk equity, unless they are required to do so. Risk equity concepts are more likely to be accepted and applied in hazardous facility design problems where governments play substantial roles.

20.8.6   About the modeling of risk equity

There is no agreement among proponents of equity on the proper way of modeling equity in the spatial distribution of risk. In fact, all of the measures used in the mathematical programming models are rather simplistic, and they violate a basic axiom in equity measurement, namely the principle of transfers. We suggest that more care should be given to the modeling of equity in mathematical programming models.

20.8.7   About multicommodity models

Most of the mathematical models suggested for strategic management of hazmats deal with only one commodity. It is true that some of the logistics problems encountered in practice can be decomposed into a set of single commodity problems. For example, a hazmat carrier would treat each shipment order as a separate logistics problem. However, there are other logistics problems where the concerns for overall risk and equity constitute a coupling factor for the variety of materials involved in the hazardous activity. The task of a government agency to designate dangerous goods roads in its jurisdiction constitutes a good example for this family of logistics problems. In this case, multicommodity representations are needed to increase model realism.

20.8.8   On the scope of facility design models

Current focus of the analytical models for planning potentially hazardous facilities is on the location of plants for processing or storage of a single commodity. This approach is clearly insufficient for two reasons: (i) it reduces the strategic design problem to a location problem by ignoring other dimensions of configuration of a facility, such as size and technology content; (ii) it overlooks the fact that many of the available technologies for treatment and disposal of hazardous wastes have the flexibility to process a variety of materials, and hence there exist opportunities for economies of scope. Analytical models are capable of assisting real-world decisions to the extent that they incorporate the multiplicity of strategic decision components in the configuration of a hazardous facility. We suggest developing models that include technology and capacity selection, as well as location selection, for hazardous facilities. Clearly, such models would be of multicommodity nature.

20.8.9   Shortest paths vs. network flows

An overwhelming majority of the hazmat routing literature focuses on the problem of transporting a single commodity between a given origin and destination. Although the approaches may differ in the way risk is incorporated in the routing model, almost all of the models in the literature reduce to a shortest path formulation. However, the network flow model, of which the shortest path is a special case, seems quite suitable for more extensive studies of hazmat transport problems. The network flow model allows for multiple origins and destinations, as well as edge costs and capacities. It is also possible to capture the multiple commodity nature (such as the aggregation of risks) of the strategic hazmat transport problem using a multicommodity network flow model. Unfortunately, no one has yet used the multicommodity network flow model as a basis to solve the general multicommodity, multiple origin-destination, hazmat routing problem which we call global routing. We believe that such analytical models for global route planning may have a good chance of being accepted and used in strategic decision-making for hazmat management.

20.8.10   Interaction between research disciplines

Research in hazmat logistics is clearly a multidisciplinary venture. It is futile to limit oneself to one discipline in such a venture. However, we have observed a lack of interaction between different disciplines in this area. For example, although considerable research has appeared in the decision analysis literature on risk equity, researchers working on mathematical programming models seem to have ignored this literature. Another example is the lack of objective functions in mathematical programming models that are based on multiattribute utility theory. In almost all papers that use a multiobjective mathematical program, the solution algorithm either converts all- but-one of the objectives into constraints, or uses arbitrary weights for adding the objectives. As another example, operations researchers have preferred to view risk as a network edge attribute, without paying much attention to how these numbers are calculated. In return, the risk assessment literature has been somewhat slow in picking up on certain operations research tools, such as bicriteria shortest path algorithms and probabilistic analysis.

20.8.11   On applications potential

Although we advocate simultaneous optimization of the facility configuration and routing decisions in integrated models, it is clear, from an implementation perspective that establishment of new facilities is much more difficult than designating (or choosing) the routes to be used. In particular, the location of a hazardous facility is primarily a political problem, which does not lend itself easily to analysis by mathematical models. However, we believe that public debate on hazardous facility location can be enhanced significantly through the development of analytical models which provide a rational framework for discussion and negotiation. Although we view the role of analytical models as a secondary one in facilities design, the opportunity to apply analytical models to routing problems is substantial. The more difficult versions of the routing problems (and the more interesting ones from an academic perspective) are those that incorporate multiple objectives. However, currently there is no pressing reason for private hazmat carriers to use multiobjective models. In fact, in many instances carriers do not even consider the risks in routing. This may change, however, were a government agency, such as the Environmental Protection Agency in the U.S., to require consideration of public risks in hazmat transport through the use of analytical decision-support-systems. Certain actions by insurance companies and banks may also provide economic incentives for the carriers to consider transport risks.

20.8.12   On implementation

Integration of optimization models with a geographical information system (GIS) is a necessary condition for successful implementation. Without a GIS, data collection and input is a tedious process, and results are little more than unwieldy lists of numbers. Perhaps because of these reasons, most papers that propose a new model either do not implement the suggested model on a realistic problem (and choose to use a made-up numerical example), or use a realistic problem that appeared in another paper. There are very few papers in this area that utilize a GIS. In fact, combination of optimization models with GIS is a relatively young area of research. We suggest that hazmat logistics research be in the forefront of this endeavor. In fact, the use of a GIS may be the only practical way to include environmental risks in hazmat logistics decision-support-systems. (For example, one can find routes that avoid a set of sensitive aquatic ecosystems quite easily using a GIS.) Another necessary condition for successful implementation is the inclusion of a multiobjective decision-making component in a decision-support-system. Tradeoffs between objectives must be highlighted for the selection of a preferred solution from among many nondominated solutions to a problem. Interactive multicriteria analysis seems promising for this purpose.

In closing, we would like to point out that the general area of applying analytical models to hazmat logistics is relatively young. In fact, almost all of the papers we reference here are published after 1980, and more than half are published after 1990. Considering this, we think that the accomplishments are quite substantial, although we have been critical about some aspects of this literature. Given the amount of attention hazmat logistics problems received in the recent past, and continue to receive currently, from academics, practitioners, governments and the public in general, we think the future of this research area is quite bright. However, to make a real contribution to the solution of problems, researchers must pay attention to the realities of the problems. The importance of collaboration with researchers in other disciplines, and with practitioners and government agencies dealing with hazmats, cannot be overemphasized.

REFERENCES

[1] Aashtiani, H.Z. (1979) *the Multi-Modal Traffic Assignment Problem*, Unpublished Ph.D. Dissertation, Operations Research Center, MIT, Cambridge, MA.

[2] Aashtiani, H.Z. and T.L. Magnanti (1981) "Equilibria on a Congested Transportation Network," *SIAM Journal on Algebriac and Discrete Methods*, 2, 213-226.

[3] Abdel-Malek, L.L. (1985) "Optimum Positioning of a Moving Service Facility," *Computers and Operations Research*, 12, 437-444.

[4] Abkowitz, M. and P. Cheng (1989) "Hazardous Materials Transport Risk Estimation Under Conditions of Limited Data Availability," *Transportation Research Record*, 1245, 14-22.

[5] Abkowitz, M. and P. Cheng (1988) "Developing a Risk-Cost Framework For Routing Truck Movements of Hazardous Materials," *Accident Analysis Prevention*, 20, 39-51.

[6] Abkowitz, M., M. Lepofsky and P. Cheng (1992) "Selecting Criteria For Designating Hazardous Materials Highway Routes," *Transportation Research Record*, 1333, 30-35.

[7] Abramowitz, M. and I.A. Stegun (1972) *Handbook Of Mathematical Functions*, National Bureau of Standards, 9th Printing.

[8] Achabal, D., W.L. Gorr and V. Mahajan (1982) "MULTILOC: a Multiple Store Location Decision Model," *Journal of Retailing*, 58, 5-25.

[9] Adam, E.E. Jr. and P.M. Swamidass (1989) "Assessing Operations Management from a Strategic Perspective," *Journal of Management*, 15, 181-203.

[10] Affholder, J.G. (1991) "Les Centres De La France ... Et D'ailleurs: Aspects Techniques Et Mediatiques," *Moyenne, Milieu, Centre, Histoires Et Usages*, J. Feldman, G. Lagneau and B. Matalon, Editors, De L'Ecole Des Hautes Etudes En Sciences Sociales, Paris.

[11] Ahituv, N. and O. Berman (1988) *Operations Management of Distributed Service Networks*, Plenum Press, NY.

[12] Ahmad, A. (1993) "A Procedure for Deriving a Small Portfolio of Non-Dominated Routes from a Stochastic Multi-Objective Network," Ph.D. Dissertation, Rensselaer Polytechnic Institute.

[13] Aikens, C.H. (1985) "Facility Location Models for Distribution Planning," *European Journal of Operational Research* 22, 263-279.

[14] Akinc, U. (1985) "Multi-Activity Facility Design and Location Problems," *Management Science*, 31, 275-283.

[15] Akinc, U. and B. M. Khumawala (1977) "An Efficient Branch and Bound Algorithm for the Capacitated Warehouse Location Problem," *Management Science*, 23, 585-594.

[16] Al Khayyal, F.A. and J.E. Falk (1983) "Jointly Constrained Biconvex Programming," *Mathematics of Operations Research*, 8, 273-286.

[17] Al Khayyal, F.A. (1990) "Jointly Constrained Bilinear Programs and Related Problems: an Overview," *Computers and Mathematics with Applications*, 19, 53-62.

[18] Allison, P.D. (1978) "Measures of Inequality," *American Sociological Review* 43, 865-880.

[19] Alonso, W. (1967) "A Reformulation of Classical Location Theory and Its Relation to Rent Theory," *Papers of the Regional Science Association*, 19, 23-44.

[20] Alp, E. (1995) "Risk-Based Transportation Planning Practice: Overall Methodology and a Case Example," *INFOR*, to appear.

[21] Aly, A.A., D.C. Kay and D.W. Jr. Litwhiler (1979) "Location Dominance on Spherical Surfaces," *Operations Research*, 27, 972-981.

[22] Aly, A.A. and A.S. Marucheck (1982) "Generalized Weber Problem with Rectangular Regions," *Journal of the Operational Research Society*, 33, 983-989.

[23] Aly, A.A. and B. Rrahli (1990) "Analysis of a Bicriteria Location Model," *Naval Research Logistics*, 37, 937-944.

[24] Ambrose, P.G. (1968) "An Analysis of Intra-Urban Shopping Behavior," *Town Planning Review*, 38, 327-334.

[25] Anderson, L.R. and R.A. Fontenot (1992) "Optimal Positioning of Service Units Along a Coordinate Line," *Transportation Science*, 26, 346-351.

[26] Andreatta, G. and F.M. Mason (1985) "Properties of the K-Centra on a Tree," *Transportation Science*, 11, 243-252.

[27] Andreatta, G. and F.M. Mason (1986) "Review of Recent Results About the K-Centrum of a Tree," *Annals of Operations Research*, 6, 129-136.

[28] Aneja, Y.P., R. Chandrasekaran and K.P.K. Nair (1988) "A Note on the M-Center Problem with Rectilinear Distances," *European Journal Of Operational Research*, 35, 118-123.

[29] Aneja, Y.P. and M. Parlar (1994) "Algorithms for Weber Facility Location in the Presence of Forbidden Regions and/or Barriers to Travel," *Transportation Science*, 28, 70-76.

[30] Anonymous (1993) "Development Brief: a Better Prescription," *The Economist*, London.

[31] Aoyagi, M. and A. Okabe (1993) "Spatial Competition of Firms in a Bounded Two-Dimensional Market," *Regional Science and Urban Economics*, 23, 259-289 .

[32] Appa, G.M. and I. Giannikos (1994) "Is Linear Programming Necessary for Single Facility Location with Maximin of Rectilinear Distance?," *Journal of the Operational Research Society*, 45, 97-107.

[33] Applebaum, W. (1966) "Methods for Determining Store Trade Areas, Market Penetration and Potential Sales," *Journal of Marketing Research*, 3: 127-141.

[34] Applebaum, W. (1968) *Guide to Store Location Research with Emphasis on Supermarkets*, Reading, MA, Addison-Wesley.

[35] Apsimon, H. M. and J. Wilson (1991) "The Application of Numerical Models to Assess Dispersion and Deposition in the Event of a Nuclear Accident," *Journal of Forecasting*, 10, 91-103.

[36] Arnott, R. (1986) *Location Theory*, Editor, Harwood Academic, Chur.

[37] Atkinson, M.D., J.-R. Sack, N. Santoro and T. Strothotte (1986) "Min-Max Heaps and Generalized Priority Queues," *Communications of the ACM*, 29, 996-1000.

[38] Atkinson, D.S. and P.M. Vaidya (1992) "A Scaling Technique for Finding the Weighted Analytical Center of a Polytope," *Mathematical Programming*, 57, 163-192.

[39] Aurenhammer, F. (1991) "Voronoi Diagrams - a Survey of a Fundamental Geometric Data Structure," *ACM Computing Surveys*, 23, 345-405.

[40] Averbakh, I. and O. Berman "Locating Flow Capturing Units on a Network with Multi-Counting and Diminishing Returns to Scale," *E.J.O.R.*, to appear.

[41] Averbakh, I., O. Berman and D. Simchi-Levi (1994) "Probabilistic a Priori Routing-Location Problems," *Naval Reserach Logistics*, 41, 973-980.

[42] Avram, J. and D. Bertsimas (1992) "The Minimum Spanning Tree Constant in Geometrical Probability and Under the Independent Model; an Unified Approach," *Annals of Applied Probability*, 2, 113-130.

[43] Aykin, T. (1988) "On the Location of Hub Facilities," *Transportation Science*, 22, 155-157.

[44] Aykin, T. and A.J.G. Babu (1987) "Constrained Large-Region Multifacility Location Problems," *Journal of the Operational Research Society*, 38, 241-252.

[45] Aykin, T. and G.F. Brown (1992) "Interacting New Facilities and Location-Allocation Problems," *Transportation Science*, 26, 212-222.

[46] Bajaj, C. (1988) "The Algebraic Degree of Geometric Optimisation Problems," *Discrete Computational Geometry*, 3, 177-191.

[47] Bacon, R. (1984) *Consumer Spatial Behavior*, New York, Oxford University Press.

[48] Balakrishnan, P.V. and J.E. Storbeck (1990) "McThresh: Modeling Maximum Coverage with Threshold Constraints," Working Paper No. 90-46, College of Business, Ohio State University, Columbus, Ohio.

[49] Balinski, M.L., and P. Wolfe (1963) "On Benders Decomposition and a Plant Location Problem," Mathematica Working Paper, ARO-27, Princeton, NJ.

[50] Balinski, M.L. (1965) "Integer Programming: Methods, Uses, Computation," *Management Science*, 12, 253-313.

[51] Ball, M. and F. Lin (1993) "A Reliability Model Applied to Emergency Service Vehicle Location," *Operations Research*, 41, 18-36.

[52] Bannock, G., R.E. Baxter and E. Davis (1987) *Dictionary of Economics*, Penguin, London, UK.

[53] Bartholdi, J. and L. Platzman (1982) "An $O(n \log N)$ Planar Traveling Salesman Heuristic Based on Space Filling Curves," *Operations Research Letters*, 1, 121-125.

[54] Batta, R. (1988) "Single Server Queueing - Location Models with Rejection," *Transportation Science*, 22, 209-216.

[55] Batta, R. (1989) "A Queueing - Location Model with Expected Service Time-Dependent Queueing Disciplines," *European Journal of Operational Research*, 192-205.

[56] Batta, R., O. Berman, S. Chiu, R. Larson, and A. Odoni (1990) "Discrete Location Theory," *Discrete Location Theory*, P. Mirchandani and R. Francis, Editors, John Wiley and Sons, NY.

[57] Batta, R. and S.S. Chiu (1988) "Optimal Obnoxious Paths on a Network: Transportation of Hazardous Materials," *Operations Research*, 36, 84-92.

[58] Batta, R., J. Dolan, and N. Krishnamurthy (1989) "The Maximal Expected Covering Location Problem: Revisited," *Transportation Science*, 23, 277-287.

[59] Batta, R., A. Ghose and U. Palekar (1989) "Locating Facilities on the Manhattan Metric with Arbitrarily Shaped Barriers and Convex Forbidden Regions," *Transportation Science*, 23, 26-36.

[60] Batta, R., R. Larson and A. Odoni (1988) "A Single - Server Priority Queueing - Location Model," *Networks*, 87-103.

[61] Batta, R. and L.A. Leifer (1988) "On the Accuracy of Demand Point Solutions to the Planar, Manhattan Metric, P-Median Problem, with and Without Barriers to Travel," *Computers and Operations Research*, 15, 253-262.

[62] Batta, R. and U.S. Palekar (1988) "Mixed Planar/Network Facility Location Problems," *Computers and Operations Research*, 15, 61-67.

[63] Baumol, W.J. and P. Wolfe (1958), "A Warehouse-Location Problem," *Operations Research*, 6, 252-263.

[64] Bazaraa, M.S. and C.M. Shetty (1979) *Nonlinear Programming - Theory and Algorithms*, John Wiley & Sons, NY.

[65] Beardwood J., J. Halton, and J. Hammersley (1959) "The Shortest Path Through Many Points," *Proc. Camb. Phil. Soc.*, 55, 299-327.

[66] Beasley, J.E. (1988) "An Algorithm for Solving Large Capacitated Warehouse Location Problems," *European Journal of Operational Research* 33, 314-325.

[67] Beaumont, J.R. (1980) "Spatial Interaction Models and the Location Allocation Problem," *Journal of Regional Science*, 20, 37-50.

[68] Beaumont, J.R. (1987) "Location Allocation Models and Central Place Theory," *Spatial Analysis and Location Allocation Models*, A. Ghosh and G. Rushton, Editors, NY, Van Nostrand Reinhold.

[69] Beauzamy, B. and B. Maurey (1977) "Points Minimaux Et Ensembles Optimaux Dans Les Espaces De Banach," *Journal of Functional Analysis*, 24, 107-139.

[70] Beckmann, M.J. (1976) "Spatial Price Policies Revisited," *Bell Journal of Economics*, 7, 619-629.

[71] Beckman, M.J., C.B. McGuire, and C.B. Winsten (1956) *Studies In the Economics of Transportation*, Yale University Press, New Haven, CT.

[72] Beckmann, M.J. and J.-F. Thisse (1986) "The Location of Production Activities," P. Nijkamp (ed.), *Handbook of Regional and Urban Economics*, Vol. I, North-Holland, Amsterdam, 21-95.

[73] Bellalouna, M. (1993) *Problèmes D'Optimization Combinatoires Probabilistes*, Ph.D. Thesis, Centre De Mathématiques Appliquées, ENPC, France.

[74] Bellman, R. (1965) "An Application of Dynamic Programming to Location Allocation Problems," *SIAM Review*, 7, 126-128.

[75] Benders, J.F. (1962) "Partitioning Procedures for Solving Mixed-Variables Programming Problems," *Numeriche Mathematik*, 4, 238-252.

[76] Benedict, J. (1983) "Three Hierarchical Objective Models Which Incorporate the Concept of Excess Coverage for Locating EMS Vehicles Or Hospitals," M.Sc. Thesis, Northwestern University.

[77] Bennett, C.D. and A. Mirakhor (1974) "Optimal Facility Location with Respect to Several Regions," *Journal of Regional Science*, 14, 131-136.

[78] Benkherouf (1993) "Single Facility Location with Rectilinear Distance and Rotated Axes," *Arabian Journal of Science and Engineering*, 18, 343-349.

[79] Berens, W. (1988) "The Suitability of the Weighted ℓ_p-Norm in Estimating Actual Road Distances," *European Journal of Operational Research*, 34, 39-43.

[80] Berens, W. and F.J. Koerling (1985) "Estimating Road Distances by Mathematical Functions," *European Journal of Operational Research*, 21, 54-56.

[81] Berens, W. and F.J. Koerling (1988) "On Estimating Road Distances by Mathematical Functions - a Rejoinder," *European Journal of Operational Research*, 36, 254-255.

[82] Berlin, G. (1972) *Facility Location and Vehicle Allocation for Provision of an Emergency Service*, Ph.D. Dissertation, the Johns Hopkins University, Baltimore, MD.

[83] Berman, O. (1990) "Mean-Variance Location Problems," *Transportation Science*, 24, 287-293.

[84] Berman, O. (1994) "The P-Maximal Cover - P Partial Center Problem on Networks," *European Journal of Operational Research*, 72, 432-442.

[85] Berman, O. "The Maximizing Market Size Discretionary Facility Location Problem with Congestion," Revised for *Socio-Economic Planning Sciences*.

[86] Berman, O., D. Bertsimas and R.C. Larson "Locating Discretionary Service Facilities II: Maximizing Market Size, Minimizing Inconvenience," Revised for *Management Science*.

[87] Berman, O. and E.H. Kaplan (1987) "Facility Location and Capacity Planning with Delay Dependent Demand," *International Journal of Production Research*, 25, 1773-1780.

[88] Berman, O. and E.H. Kaplan (1990) "Equity Maximizing Facility Location Schemes," *Transportation Science*, 24, 137-144.

[89] Berman, O., D. Krass and C.W. Xu "Locating Discretionary Service Facilities Based on Probabilistic Customer Flows," *Transportation Science*, to appear.

[90] Berman, O., D. Krass and C.W. Xu (1994) "Generalized Discretionary Service Facility Location Models with Probabilistic Customer Flows," Working Paper, University of Toronto.

[91] Berman, O., D. Krass and C.W. Xu (1994) "Locating Flow-Intercepting Facilities: New Approaches and Results," Working Paper, University of Toronto.

[92] Berman, O., D. Krass and C.W. Xu (1994) "Valid Inequalities, Facets, and Computational Results for the Discretionary Service Facility Location Problem," Working Paper, University of Toronto.

[93] Berman, O. and R.C. Larson (1982) "The Median Problem with Congestion," *Computers and Operations Research*, 4, 119-126.

[94] Berman, O. and R.C. Larson (1985) "Optimal 2-Facility Network Districting in the Presence of Queuing," *Transportation Science*, 19, 261-277.

[95] Berman, O., R.C. Larson and S.S. Chiu (1985) "Optimal Server Location on a Network Operating as an M/G/1 Queue," *Operations Research*, 33, 746-771.

[96] Berman, O., R.C. Larson and N. Fouska (1992) "Optimal Location of Discretionary Facilities," *Transportation Science*, 26, 201-211.

[97] Berman, O., R. Larson and C. Parkan (1987) "The Stochastic Queue p-Median Problem," *Transportation Science*, 21, 207-216.

[98] Berman, O. and R.R. Mandowsky (1986) "Location - Allocation on Congested Networks," *European Journal of Operational Research*, 26, 238-250.

[99] Berman, O. and C. Parkan (1981) "A Facility Location Problem with Distance Dependent Demand," *Decision Sciences*, 12, 623-632.

[100] Berman, O. and D. Simchi-Levi (1986) "Minisum Location of a Traveling Salesman," *Networks*, 16, 239-254.

[101] Berman, O. and D. Simchi-Levi (1988) "Minisum Location of Traveling Salesman on Simple Networks," *European Journal of Operations Research*, 36, 241-250.

[102] Berman, O. and D. Simchi-Levi (1988) "Finding the Optimal a Priori Tour and Location of a Traveling Salesman with Non-Homogeneous Customers," *Transportation Science*, 22, 148-154.

[103] Berman, O. and D. Simchi-Levi (1989) "The Salesman Location Problem on Stochastic Networks," *Transportation Science*, 23, 54-57.

[104] Berman, O. and D. Simchi-Levi (1990) "Conditional Location Problems on Networks," *Transportation Science*, 24, 77-78.

[105] Berman, O. and E.K. Yang (1991) "Medi-Center Location Problems," *Journal of the Operational Research Society*, 42, 313-322.

[106] Bertsimas, D. (1988) *Probabilistic Combinatorial Optimization Problems*, Technical Report 194, Operations Research Center, MIT.

[107] Bertsimas, D.J. (1989) "Traveling Salesman Facility Location Problems," *Transportation Science*, 23, 184-191.

[108] Bertsimas, D. (1992) "A Vehicle Routing Problem with Stochastic Demand," *Operations Research*, 40, 574-585.

[109] Bertsimas, D. and M. Grigni (1989) "Worst-Case Examples for the Space Filling Curve Heuristic for the Euclidean Traveling Salesman Problem," *Operations Research Letters*, 5, 241-244.

[110] Bertsimas, D., P. Jaillet and A. Odoni (1990) "A Priori Optimization," *Operations Research*, 38, 1019-1033.

[111] Bhadury, J., R. Chandrasekharan and V. Padmanabhan "Competitive Location and Entry Deterrence: a Variant of Hotelling's Duopoly Model," Unpublished Working Paper, University of New Brunswick.

[112] Bianchi, C. and R. Church (1988) "A Hybrid FLEET Model for Emergency Medical Service System Design," *Social Sciences in Medicine*, 163-171.

[113] Bihlmayer, W. and W. Forst (1989) "Approximation by Circles," *Zeitschrift Für Operations Research*, 33, 315-339.

[114] Bodily, S.E. (1978) "Police Sector Design Incorporating Preferences of Interest Groups for Equality and Efficiency," *Management Science*, 24, 1301-1313.

[115] Boffey T.B. (1995) "Multiobjective Routing Problems," *TOP (Spanish Journal of Operations Research and Statistics*, to appear.

[116] Boffey, T.B. and J. Karkazis (1984) "p-Medians and Multi-Medians," *Journal of the Operational Research Society*, 35, 57-64.

[117] Boffey, B. and J. Karkazis (1993) "Models and Methods for Location And Routing Decisions Relating to Hazardous Materials," *Studies in Locational Analysis*, 5, 149-166.

[118] Boffey, B. and J. Karkazis (1995) "Linear Versus Nonlinear Models For Hazardous Materials Routing," to appear in *INFOR*.

[119] Bongartz, I., P.H. Calamai and A.R. Conn (1994) "A Projection Method for ℓ_p Norm Location-Allocation Problems," *Mathematical Programming*, 66, 283-312.

[120] Bouliane, J. and G. Laporte (1992) "Locating Postal Relay Boxes Using a Set Covering Algorithm," *American Journal of Mathematical and Management Sciences*, 12, 65-74.

[121] Boyer, M., J.J. Laffont, P. Mahenc and M. Moreaux (1994) "Location Distortions Under Incomplete Information," *Regional Science And Urban Economics*, 24, 409-440.

[122] Boykin, R.F., R.A. Freeman and R.R. Levary (1984) "Risk Assessment In a Chemical Storage Facility," *Management Science*, 30(4), 512-517.

[123] Brady, S.D. and R.E. Rosenthal (1980) "Interactive Computer Graphical Solutions of Constrained Minimax Location Problems," *AIIE Transactions*, 12, 241-248.

[124] Brady S.D., R. E. Rosenthal and D. Young (1983) "Interactive Graphical Minimax Location of Multiple Facilities With General Constraints," it IIE Transactions, 15, 242-253.

[125] Braid, R.M. (1991) "The Locations of Congestible Facilities in Adjacent Jurisdictions," *Regional Science and Urban Economics*, 21, 617-626.

[126] Brandeau, M.L. (1992) "Characterization of the Stochastic Median Queue Trajectory in a Plane with Generalized Distances," *Operations Research*, 40, 331-341.

[127] Brandeau, M.L. and S.S. Chiu (1989) "An Overview of Representative Problems in Location Research," *Management Science*, 35, 645-674.

[128] Brandeau, M.L. and S.S. Chiu (1990) "Trajectory Analysis of the Stochastic Queue Median in a Plane with Rectilinear Distances," *Transportation Science*, 24, 230-243.

[129] Brandeau, M.L. and S.S. Chiu (1992) "A Center Location Problem With Congestion," *Annals of Operations Research*, 40, 17-32.

[130] Brandeau, M.L. and S.S. Chiu (1993) "Sequential Location and Allocation: Worst Case Performance and Statistical Estimation," *Location Science*, 1, 289-298.

[131] Brandeau, M.L. and S.S. Chiu (1994) "Facility Location in a User-Optimizing Environment with Market Externalities: Analysis of Customer Equilibria and Optimal Public Facility Locations," *Location Science*, 2, 129-147.

[132] Brandeau, M.L. and S.S. Chiu (1994) "Location of Competing Private Facilities in a User-Optimizing Environment with Market Externalities," *Transportation Science*, 28, 125-140.

[133] Brans, J.P. (1985) "European Summer Institute on Location, First Meeting Proceedings," *European Journal of Operational Research*, Editor, 20, 299-396.

[134] Briggs, T., P.L. Kunsch and B. Mareschal (1990) "Nuclear Waste Management: an Application of the Multicriteria PROMETHEE Methods," *European Journal of Operational Research*, 44, 1-10.

[135] Brimberg, J. (1989) "Properties of Distance Functions and Minisum Location Models," Ph.D. Thesis, McMaster University, Hamilton, Ontario, Canada.

[136] Brimberg, J., P.D. Dowling and R.F. Love (1994) "The Weighted One-Two Norm Distance Model: Empirical Validation and Confidence Interval Estimation," *Location Science*, 2, 91-100.

[137] Brimberg, J. and R.F. Love (1991) "Estimating Travel Distances by the Weighted ℓ_p Norm," *Naval Research Logistics*, 38, 241-259.

[138] Brimberg, J. and R.F. Love (1992) "A New Distance Function for Modeling Travel Distances in a Transportation Network," *Transportation Science*, 26, 129-137.

[139] Brimberg, J. and R.F. Love (1992) "Local Convergence in a Generalized Fermat-Weber Problem," *Annals of Operations Research*, 40, 33-66.

[140] Brimberg, J. and R.F. Love (1993) "Directional Bias of the ℓ_p-Norm," *European Journal of Operational Research*, 67, 287-294.

[141] Brimberg, J. and R.F. Love (1993) "General Considerations on the Use of the Weighted ℓ_p-Norm as an Empirical Distance Measure," *Transportation Science*, 27, 341-349.

[142] Brimberg, J. and R.F. Love (1993) "Global Convergence of a Generalized Iterative Procedure for the Minisum Location Problem with ℓ_p Distances," *Operations Research*, 41, 1153-1163.

[143] Brimberg, J. and R.F. Love (1994) "A Location Problem with Economies of Scale," *Studies in Locational Analysis*, 7, 9-19.

[144] Brimberg, J., P.D. Dowling and R.F. Love (1995) "Estimating the Parameters of the Weighted ℓ_p Norm by Linear Regression," Submitted for Publication.

[145] Brimberg, J., R.F. Love, and J.H. Walker (1994) "The Effect of Axis Rotation on Distance Estimation," *European Journal of Operational Research*, 68, to appear.

[146] Brimberg, J. and A. Mehrez (1994) "Multi-Facility Location Using a Maximin Criterion and Rectangular Distances," *Location Science*, 2, 11-19.

[147] Brimberg, J. and G.O. Wesolowsky (1992) "Probabilistic ℓ_p Distances in Location Models," *Annals of Operations Research*, 40, 67-75.

[148] Brooke, A., D. Kendrick and A. Meeraus (1988) *GAMS: a User's Guide*, the Scientific Press, South San Francisco, CA.

[149] Broome, J. (1982) "Equity in Risk Bearing," *Operations Research*, 30(2), 412-414.

[150] Brown, G.G., G.W. Graves and M.D. Honczarenko (1987) "Design and Operation of a Multicommodity Production/Distribution System Using Primal Goal Decomposition," *Management Science*, 33, 1469-1480.

[151] Brown, G.G. and M.P. Olson (1995) "Dynamic Factorization in Large-Scale Optimization," *Mathematical Programming*, to appear.

[152] Buhl, H.U. (1988) "Axiomatic Considerations in Multi-Objective Location Theory," *European Journal of Operational Research*, 37, 363-367.

[153] Burness, R.C. and J.A. White (1976) "The Traveling Salesman Location Problem," *Transportation Science*, 10, 348-360.

[154] Burwell, T., J. Jarvis and M. McKnew (1993) "Modeling Co-located Servers and Dispatch Ties in the Hypercube Model," *Computers and Operations Research*, 113-119.

[155] Cabot, A.V., R.L. Francis and M.A. Stary (1970) "A Network Flow Solution to a Rectilinear Distance Facility Location Problem," *AIIE Transactions*, 2, 132-141.

[156] Cadwallader, M. (1975) "Behavioral Model of Consumer Spatial Decision Making," *Economic Geography*, 51, 339-349.

[157] Calamai, P.H. and A.R. Conn (1987) "A Projected Newton Method for ℓ_p Norm Location Problems," *Mathematical Programming*, 38, 75-109.

[158] Campbell, J.F. (1990) "Designing Logistics Systems by Analyzing Transportation, Inventory and Terminal Cost Tradeoffs," *Journal of Business Logistics*, 11, 159-179.

[159] Campbell, J.F. (1990) "Freight Consolidation and Routing with Transportation Economies of Scale," *Transportation Research*, 24B, 345-361.

[160] Campbell, J.F. (1990) "Locating Transportation Terminals to Serve an Expanding Demand," *Transportation Research*, 24B, 173-192.

[161] Campbell, J.F. (1992) "Location-allocation for Distribution to a Uniform Demand with Transshipments," *Naval Research Logistics*, 39, 5, 635-649.

[162] Campbell, J.F. (1993) "One-to-many Distribution with Transshipments: an Analytic Model," *Transportation Science*, 27, 4, 330-340.

[163] Campbell, J.F. (1993) "Continuous and Discrete Demand Hub Location Problems," *Transportation Research*, B, 27B, 6, 473-482.

[164] Canovas, L. and B. Pelegrin (1992) "Improving Some Heuristic Algorithms for the Rectangular p-cover Problem," *Proceedings of the VI Meeting of the EURO Working Group on Locational Analysis*, J.A.Moreno-Perez, Editor, Universidad De La Laguna, Tenerife, Spain, 23-31.

[165] Carbone, R. and A. Mehrez (1980) "The Single Facility Minimax Distance Problem Under Stochastic Location of Demand," *Management Science*, 26, 113-115.

[166] Carrizosa, E. (1992) "Problemas De Localizacion Multiobjetivo (Multiobjective Location Problems - Spanish)," Ph.D. Thesis, Facultad De Matematicas, Universidad De Sevilla, Spain.

[167] Carrizosa, E., E. Conde, F.R. Fernandez and J. Puerto (1993) "Efficiency in Euclidean Constrained Location Problems," *Operations Research Letters*, 14, 291-295.

[168] Carrizosa, E.J., E. Conde, M. Munoz and J. Puerto (1993) "Condorcet Points with the Manhattan Distance," Working Paper, Departamento De Estadistica E Investigacion Operational, Facultad De Matematicas, Universidad De Sevilla, 41012 Sevilla, Spain.

[169] Carrizosa, E.J., E. Conde, M. Munoz and J. Puerto (1993) "Simpson Points in Planar Problems with Locational Constraints. The Polyhedral-gauge Case," Working Paper, Departamento De Estadistica E Investigacion Operational, Facultad De Matematicas, Universidad De Sevilla, 41012 Sevilla, Spain.

[170] Carrizosa, E.J., E. Conde, M. Munoz and J. Puerto (1993) "Simpson Points in Planar Problems with Locational Constraints. the Round-norm Case," Working Paper, Departamento De Estadistica E Investigacion Operational, Facultad De Matematicas, Universidad De Sevilla, 41012 Sevilla, Spain.

[171] Carrizosa, E.J., E. Conde, M. Munoz and J. Puerto (1994) "The Generalized Weber Problem with Expected Distances," *RAIRO*.

[172] Carrizosa, E.J., E. Conde, M. Munoz and J. Puerto (1994) "Planar Point-objective Location Problems with Nonconvex Constraints: a Geometrical Construction," *Journal of Global Optimisation*, to appear.

[173] Carrizosa, E. and F.R. Fernandez (1990) "El Conjunto Eficiente En Problemas De Localizacion Con Normas Mixtas(ℓ_p) - (The Efficient Set In Location Problems with Mixed Norms) - Spanish," *Trabajos De Investigacion Operativa (Spain)*, 6, 61-69.

[174] Carrizosa, E. and F.R. Fernandez (1993) "A Polygonal Upper Bound for the Efficient Set for Single-location Problems with Mixed Norms," *Top (Spanish Journal of Operations Research and Statistics)*, 1, 107-116.

[175] Carrizosa, E., F.R. Fernandez and J. Puerto (1990) "Determination of a Pseudoefficient Set for Single-location Problems with Mixed Polyhedral Norms," *Proceedings of the Fifth Meeting of the EURO Working Group on Locational Analysis*, F. Orban and J.P. Rasson, Editors, FUNDP, Namur, Belgium, 27-39.

[176] Carrizosa, E., F.R. Fernandez and J. Puerto (1990) "An Axiomatic Approach to Location Criteria," *Proceedings of the Fifth Meeting of the EURO Working Group on Locational Analysis*, F. Orban and J.P. Rasson, Editors, FUNDP, Namur, Belgium, 40-53.

[177] Carrizosa, E.J., M. Munoz and J. Puerto (1994) "On the Set of Optimal Points to the Weighted Maxmin Problem," *Studies in Locational Analysis*, 7, 21-33.

[178] Carrizosa, E. and F. Plastria (1992) "Polynomial Algorithms for Parametric Minquantile and Maxcovering Planar Location Problems with Locational Constraints," Report BEIF/47, Vrije Universiteit Brussel, Belgium.

[179] Carrizosa, E. and F. Plastria (1993) "A Characterization of Efficient Points in Constrained Location Problems with Regional Demand," Working Paper, BEIF/53, Vrije Universiteit Brussel, Brussels, Belgium.

[180] Carrisoza, E. and J. Puerto (1995) "A Discretizing Algorithm for Location Problems," *European Journal of Operational Research*, 80, 166-174.

[181] Carter, G.M., J.M. Chaiken and E. Ignall (1982) "Response Areas for Two Emergency Units," *Operations Research*, 571-594.

[182] Casini, E. and P.L. Papini (1986) "Un Algoritmo Per Determinare Nel Piano Un Certo Cerchio Massimo (an Algorithm to Determine a Certain Maximal Circle in the Plane) -italian, Cifarelli, Ed. *Scritti in Onore Di F.Brambilla*, Bocconi Comunicazione, Milano, Italy.

[183] Casini, E. and P.L. Papini (1992) "Different Metrics and Location Problems," S.P. Singh, Editor, *Approximation Theory, Spline Functions and Applications*, Kluwer, Netherlands, 243-253.

[184] Chakerian, G.D. and M.A. Ghandehari (1985) "The Fermat Problem in Minkowski Spaces," *Geometriae Dedicata*, 17, 227-238.

[185] Chalmet, L.G., R.L. Francis and A. Kolen (1981) "Finding Efficient Solutions for Rectilinear Distance Location Problems Efficiently," *European Journal of Operational Research*, 6, 117-124.

[186] Chan, A.W. and D.W. Hearn (1977) "A Rectilinear Distance Round-trip Location Problem," *Transportation Science*, 11, 107-123.

[187] Chandrasekaran, R. and A. Daughety (1981) "Location on Tree Networks: P-Centre and N-Dispersion Problems," *Mathematics of Operations Research*, 6, 50-57.

[188] Chandrasekaran, R. and M.J.A.P. Pacca (1980) "Weighted Min-max Location Problems: Polynomially Bounded Algorithms," *OPSEARCH*, 17, 172-180.

[189] Chandrasekaran, R. and A. Tamir (1989) "Open Questions Concerning Weiszfeld's Algorithm for the Fermat-Weber Location Problem," *Mathematical Programming*, 44, 293-295.

[190] Chandrasekaran, R. and A. Tamir (1990) "Algebraic Optimization: the Fermat-Weber Location Problem," *Mathematical Programming*, 46, 219-224.

[191] Chapman, S.C. and J.A. White (1974) "Probabilistic Formulations of Emergency Service Facilities Location Problems," ORSA/TIMS Conference, San Juan, Puerto Rico.

[192] Charalambous, C. (1982) "Extension of the Elzinga-Hearn Algorithm to the Weighted Case," *Operations Research*, 30, 591-594.

[193] Charalambous, C. (1985) "Acceleration of the HAP Approach for the Multifacility Location Problem," *Naval Research Logistics Quarterly*, 32, 373-389.

[194] Chatelon, J.A., D.W. Hearn and T.J. Lowe (1982) "A Subgradient Algorithm for Certain Minimax and Minisum Problems - the Constrained Case," *SIAM Journal on Control and Optimization*, 20, 455-469.

[195] Chaudhry, S.S., S.T. McCormick and I.D. Moon (1986) "Locating Independent Facilities with Maximum Weight: Greedy Heuristics," *Omega International Journal of Management Science* 14, 383-389.

[196] Chazelle, B., J. Guibas and P. Lee (1985) "The Power of Geometric Duality," *BIT*, 25, 76-90.

[197] Chen, P.C., P. Hansen and B. Jaumard (1991) "On-line and Off-line Vertex Enumeration by Adjacency Lists," *Operations Research Letters*, 10, 403-409.

[198] Chen, P.C., P. Hansen, B. Jaumard and H. Tuy (1992) "Weber's Problem with Attraction and Repulsion," *Journal of Regional Science*, 32, 467-486.

[199] Chen, P.C., P. Hansen, B. Jaumard and H. Tuy "Solution of the Mutisource Weber and Conditional Weber Problems by D.C. Programming," *Operations Research*, to appear.

[200] Chen, R. (1983) "Solution of Minisum and Minimax Location-allocation Problems with Euclidean Distances," *Naval Research Logistics Quarterly*, 30, 449-459.

[201] Chen, R. (1984) "Location Problems with Costs Being Sums of Powers of Euclidean Distances," *Computers and Operations Research*, 11, 285-294.

[202] Chen, R. (1984) "Solution of Location Problems with Radial Cost Functions," *Computers and Mathematics with Applications*, 10, 87-94.

[203] Chen, R. (1988) "Conditional Minisum and Minimax Location-allocation Problems in Euclidean Space," *Transportation Science*, 22, 157-160.

[204] Chen, R. (1991) "An Improved Method for the Solution of the Problem of Location on an Inclined Plane," *Revue D'Automatique,d'Informatique Et De Recherche Operationnelle-Oper.Res.*, 25, 45-53.

[205] Chen, R. and G.Y. Handler (1993) "The Conditional P-center Problem in the Plane," *Naval Research Logistics*, 40, 117-127.

[206] Chen, R. and G.Y. Handler (1987) "A Relaxation Method for the Solution of the Minimax Location-allocation Problem in Euclidean Space," *Naval Research Logistics Quarterly*, 34, 775-788.

[207] Chervi, P. (1990) *a Computational Approach to Probabilistic Vehicle Routing Problems*, Master's Thesis, Operations Research Center, MIT.

[208] Cheung, T.Y. (1980) "Multifacility Location Problem with Rectilinear Distance by the Minimum Cut Approach," *ACM Transactions on Mathematical Software*, 6, 387-390.

[209] Chhajed, D., R. Francis and T. Lowe (1993) "Contributions of Operations Research to Location Analysis," *Location Science*, 1, 4, 263-287.

[210] Chin, S. and P. Cheng (1989) "Bicriterion Routing Scheme for Nuclear Spent Fuel Transportation," *Transportation Research Record*, 1245, 60-64.

[211] Chiu, S.S. and R.C. Larson (1985) "Locating an n-Server Facility in a Stochastic Environment," *Computers and Operations Research*, 12, 509-516.

[212] Choi, I.C. and S.S. Chaudhry (1993) "The P-Median Problem with Maximum Distance Constraints: a Direct Approach," *Location Science*, 1, 235-243.

[213] Chow, T.C., R.M. Oliver and G.A. Vignaux (1990) "A Bayesian Escalation Model to Predict Nuclear Accidents and Risk," *Operations Research*, 38, 2, 265-277.

[214] Chrissis, J., R. Daves and D. Miller (1982) "The Dynamic Set Covering Problem," *Applied Mathematical Modelling*, 6, 1-6.

[215] Christofides, N. (1976) *Worst-case Analysis of a New Heuristic for the Traveling Salesman Problem*, Technical Report, GSIA, Carnegie-Mellon University.

[216] Christofides, N. and P. Viola (1971) "The Optimum Location of Multi-centers on a Graph," *Operational Research Quarterly*, 145.

[217] Christaller, W. (1933) *Central Places in Southern Germany*, Translated in 1966, Prentice-Hall, NJ.

[218] Chrystal, G. (1885) "On the Problem to Construct the Minimum Circle Enclosing n Given Points in the Plane," *Proceedings of the Edinburgh Mathematical Society*, 3, 30-33.

[219] Chung, C.H. (1986) "Recent Applications of the Maximal Covering Location Planning (M.C.L.P.) Model," *Journal of the Operational Research Society*, 37, 735-746.

[220] Church, R.L. and D.J. Eaton (1987) "Hierarchical Location Analysis Using Covering Objectives," in *Spatial Analysis and Location-Allocation Models*, A. Ghosh and G. Rushton, Editors, 163-185.

[221] Church, R.L. and R.S. Garfinkel (1978) "Locating an Obnoxious Facility on a Network," *Transportation Science*, 12, 107-118.

[222] Church, R.L. and M.E. Meadows (1979) "Location Modeling Utilizing Maximum Service Distance Criteria," *Geographical Analysis*, 11, 358-373.

[223] Church, R.L. and C. ReVelle (1974) "The Maximal Covering Loction Problem," *Papers of the Regional Science Association*, 32, 101-118.

[224] Clark, W.A.V. and G. Rushton (1970) "Models of Intra-Urban Consumer Behavior and Their Implications for Central Place Theory," *Economic Geography*, 46, 486-97.

[225] Cockayne, E.J. (1967) "On the Steiner Problem," *Canadian Mathematical Bulletin*, 10, 431-450.

[226] Coehlo, J.D. and A.G. Wilson (1976) "The Optimum Location and Size of Shopping Centers," *Regional Studies*, 10, 413-21.

[227] Cohen, M.A., M. Fisher and R. Jaikumar (1989) "International Manufacturing and Distribution Networks: a Normative Model Framework," *Managing International Manufacturing*, K. Ferdows, Editor, Elsevier Science Publishers, 67-93.

[228] Cohen, M.A. and H.L. Lee (1985) "Manufacturing Strategy: Concepts and Methods," *the Management of Productivity and Technology in Manufacturing*, P.R. Kleindorfer, Editor, Plenum Press, 153-188.

[229] Cohen, M.A. and H.L. Lee (1989) "Resource Deployment Analysis of Global Manufacturing and Distribution Networks," *Journal of Manufacturing and Operations Management*, 2, 81-104.

[230] Cohen, M.A., H.L. Lee and S. Moon (1987) "An Integrated Model for Manufacturing and Distribution System Design," Technical Report, Wharton School, University of Pennsylvania.

[231] Cohen, M.A. and S. Moon (1990) "Impact of Production Scale Economies, Manufacturing Complexity and Transportation Costs on Supply Chain Facility Networks," *Journal of Manufacturing and Operations Management*, 3, 269-292.

[232] Cohen, M.A. and S. Moon (1991) "An Integrated Plant Loading Model with Economies of Scale and Scope," *European Journal of Operational Research*, 50, 266-279.

[233] Cohen, S. and W. Applebaum (1960) "Evaluating Store Sites and Determining Store Rents," *Economic Geography*, 36, 1-35.

[234] Cohon, J.L. (1978) *Multiobjective Programming and Planning*, New York: Academic Press.

[235] Cohon, J.L., C. ReVelle, J. Current, T. Eagles, R. Eberhart and R. Church (1980) "Application of a Multiobjective Facility Location Model to Power Plant Siting in a Six-state Region of the US," *Computers & Operations Research*, 7, 107-123.

[236] Converse, P.D. (1949) "New Laws of Retail Gravitation," *Journal of Marketing*, 14, 379-384.

[237] Cooper, L. (1963) "Location-allocation Problems," *Operations Research*, 11, 331-343.

[238] Cooper, L. (1964) "Heuristic Methods for Location-allocation Problems," *SIAM Review*, 6, 37-53.

[239] Cooper, L. (1967) "Solutions of Generalized Locational Equilibrium Models," *Journal of Regional Science*, 7, 1-18.

[240] Cooper, L. (1968) "An Extension of the Generalized Weber Problem," *Journal of Regional Science*, 8, 181-197.

[241] Cooper, L. (1973) "n-Dimensional Location Models: an Application to Cluster Analysis," *Journal of Regional Science*, 13, 41-54.

[242] Cornuejols, G., M.L. Fisher and G.L. Nemhauser (1977) "Location of Bank Accounts to Optimize the Float," *Management Science*, 23, 8, 789-810.

[243] Cornuéjols, G., G.L. Nemhauser and L.A. Wolsey (1990) "The Uncapacitated Facility Location Problem," In: P.B. Mirchandani and R.L. Francis, Editors, *Discrete Location Theory*, John Wiley & Sons, NY, 119-171.

[244] Cox, R.G. (1984) "Routing and Scheduling of Hazardous Materials Shipments: Algorithmic Approaches to Managing Spent Nuclear Fuel Transport," Ph.D. Dissertation, Cornell University, Ithaca, NY.

[245] Cox, R.G. and M.A. Turnquist (1986) "Scheduling Truck Shipments of Hazardous Materials in the Presence of Curfews," *Transportation Research Record*, 1063, 21-26.

[246] Craig, C.S., A. Ghosh and S. McLafferty (1984) "Models of Retail Location Process: a Review," *Journal of Retailing*, 60, 5-36.

[247] Crama, Y., P. Hansen and B. Jaumard (1991) "Complexity of Product Positioning and Ball Intersection Problems," *Mathematics of Operations Research*, to appear.

[248] Current, J.R. (1988) "Special Issue on Location Analysis," *Environment and Planning B*, 15, 127-236.

[249] Current, J.R. and H. Min (1986) "Multiobjective Design of Transportation Networks: Taxonomy and Annotation," *European Journal of Operational Research*, 26, 187-201.

[250] Current, J.R., H. Min and D. Schilling (1990) "Multiobjective Analysis of Facility Location Decisions," *European Journal of Operational Research*, 49, 295-307.

[251] Current, J.R. and S. Ratick (1992) "Special Issue: Facility Location Modeling," *Papers in Regional Science*, 71, 3.

[252] Current J.R., C. S. ReVelle and J. L. Cohon (1990) "An Interactive Approach to Identify the Best Compromise Solution for Two Objective Shortest Path Problems," *Computers and Operations Research*, 17, 187-198.

[253] Current, J.R. and D.A. Schilling (1990) "Special Issue on Location Analysis and Modeling," *Geographical Analysis*, 22, 1.

[254] Current, J.R. and D.A. Schilling (1991) "Special Issue on Location Analysis," *Information Systems and Operations Research*, 29, 130-183.

[255] Curry, B. and L. Moutinho (1991) "Expert Systems for Site Location Decisions," *Logistics Information Management*, 4, 19-27.

[256] Daganzo, C.F. (1987) "The Break-bulk Role of Terminals in Many-to-many Logistic Networks," *Operations Research*, 35, 4, 542-555.

[257] Dahlquist, G. and Å. Björck (1974) *Numerical Methods*, Translated by N. Anderson, Prentice Hall, Englewood Cliffs, New York.

[258] Dakin, R.J. (1965) "A Tree Search Algorithm for Mixed Integer Programming Problems," *Computer Journal*, 8, 250-255.

[259] Dammico, G., U. Mocci and G. Pesamosca (1994) "Facility Location in Hierarchical Networks," *Studies in Locational Analysis*, 6, 19-30.

[260] Dasarathy, B. and L.J. White (1980) "A Maximin Location Problem," *Operations Research*, 28, 1385-1401.

[261] Dasgupta, P., A. Sen and D. Starrett (1973) "Notes on the Measurement of Inequality," *Journal of Economic Theory*, 6, 180-187.

[262] Daskin, M.S. (1983) "A Maximum Expected Covering Location Model: Formulation, Properties and Heuristic Solution," *Transportation Science*, 17, 48-70.

[263] Daskin, M.S. (1995) "Network and Discrete Location: Models, Algorithms, and Applications," John Wiley & Sons, New York.

[264] Daskin, M.S., K. Hogan and C. ReVelle (1988) "Integration of Multiple, Excess, Backup, and Expected Covering Models," *Environment and Planning, B: Planning and Design*, 15-35.

[265] Daskin, M.S. and E.H. Stern (1981) "A Hierarchical Objective Set Covering Model for Emergency Medical Service Vehicle Deployment," *Transportation Science*, 15, 137-152.

[266] Davies, R. (1973) "Evaluation of Retail Store Attributes and Sales Performance," *European Journal of Marketing* 7, 89-102.

[267] Davies, R.L. and D.S. Rogers (1984) *Store Location and Store Assessment Research*, New York.

[268] Davis, P.J. and P. Rabinowitz (1984) *Methods of Numerical Integration,* 2nd Edition, Academic Press, Boston.

[269] Dax, A. (1986) "An Efficient Algorithm for Solving the Rectilinear Multifacility Location Problem," *IMA Journal of Numerical Analysis,* 6, 343-355.

[270] Dax, A. (1986) "A Note on Optimality Conditions for the Euclidean Multifacility Location Problem," *Mathematical Programming,* 36, 72-80.

[271] Day, D.L., J.U. Farley and J. Wind (1990) "New Perspectives on Strategy Research: a View from the Management Sciences," *Management Science,* 36, 1142.

[272] Dearing, P.M. and R.L. Francis (1974) "A Network Flow Solution to a Multifacility Minimax Location Problem Involving Rectilinear Distances," *Transportation Science,* 8, 126-141.

[273] Delhaye, C., J. Teghem and P. Kunsch (1991) "Application of the ORESTE Method to a Nuclear Waste Management Problem," *International Journal of Production Economics,* 24, 1, 2, 29-39.

[274] Demange, G. (1982) "A Limit Theorem on the Minmax Set," *Journal of Mathematical Economics,* 9, 145-164.

[275] Demange, G. (1983) "Spatial Models of Collective Choice," In: J.-F. Thisse and H.G. Zoller, *Locational Analysis of Public Facilities,* North-Holland, Amsterdam.

[276] De Meyer, A. (1992) "An Empirical Investigation of Manufacturing Strategies in European Industry," *Manufacturing Strategy - Process and Content,* C.A. Voss, Editor, 221-238, Chapman & Hall.

[277] De Meyer, A. and K. Ferdows (1985) "Integration of Information Systems in Manufacturing," *International Journal of Operations and Production Management,* 5, 5-12.

[278] Demjanov, V.F. (1968) "Algorithms for Some Minimax Problems," *Journal of Computer and Systems Science,* 2, 352-380.

[279] Dewan, S. and H. Mendelson (1990) "User Delay Costs and Internal Pricing for a Service Facility," *Management Science,* 36, 1502-1517.

[280] Dhar, U.R. and J.R. Rao (1982) "Domain Approximation Method for Solving Multifacility Location Problems on a Sphere," *Journal of the Operational Research Society,* 33, 639-645.

[281] Dhar, U.R. and J.R. Rao (1982) "An Efficient Algorithm for Solving Area-constrained Location Problems on a Sphere," *OPSEARCH,* 19, 23-32.

[282] Dial, R.B. (1971) "A Probabilistic Multipath Traffic Assignment Model Which Obviates Path Enumeration," *Transportation Research,* 5, 83-111.

[283] Dobson, G. and U. Karmarkar (1987) "Competitive Location on a Network," *Operations Research,* 35, 565-574.

[284] Domschke, W. and A. Drexl (1985) "Location and Layout Planning: an International Bibliography," *Lecture Notes in Economics and Mathematical Systems,* Springer, Berlin, 238.

[285] Dornsland, K.A. (1993) "Simulated Annealing," in C.R. Reeves (editor), *Modern Heuristic Techniques for Combinatorial Problems,* London: Blackwell, 20-69.

[286] Dowling, P.D. and R.F. Love (1990) "Floor Layouts Using a Multifacility Location Model," *Naval Research Logistics,* 37, 945-952.

[287] Downs, R.M. (1970) "The Cognitive Structure of an Urban Shopping Center," *Environment and Behavior,* 2, 13-39.

[288] Drezner, T. (1994) "Locating a Single New Facility Among Existing Un-equally Attractive Facilities," *Journal of Regional Science*, 34, 237-252.

[289] Drezner, T. (1994) "Optimal Continuous Location of a Retail Facility, Facility Attractiveness, and Market Share: an Interactive Model," *Journal of Retailing*, 70, 49-64.

[290] Drezner, T. and Z. Drezner "Competitive Facilities: Market Share and Location with Random Utility," *Journal of Regional Science*, to appear.

[291] Drezner, T. and Z. Drezner "Location of Multiple Retail Facilities in a Continuous Space," Submitted for Publication.

[292] Drezner, T. and Z. Drezner "Replacing Discrete Demand with Continu-ous Demand: the Impact on Optimal Facility Location," Submitted for Publication.

[293] Drezner, Z. (1979) "Bounds on the Optimal Location to the Weber Prob-lem Under Conditions of Uncertainty," *Journal of the Operational Re-search Society*, 30, 923-931.

[294] Drezner, Z. (1981) "On a Modified One-center Model, *Management Sci-ence*, 27, 848-851.

[295] Drezner, Z. (1981) "On Location Dominance on Spherical Surfaces," *Op-erations Research*, 29, 1218-1219.

[296] Drezner, Z. (1982) "Competitive Location Strategies for Two Facilities," *Regional Science and Urban Economics*, 12, 485-493.

[297] Drezner, Z. (1983) "Constrained Location Problems in the Plane and on a Sphere," *IIE Transactions*, 15, 300-304.

[298] Drezner, Z. (1984) "The P-centre Problem - Heuristic and Optimal Algo-rithms," *Journal of the Operational Research Society*, 35, 741-748.

[299] Drezner, Z. (1984) "The Planar Two Center and Two Median Problems," *Transportation Science*, 18, 351-361.

[300] Drezner, Z. (1985) "Sensitivity Analysis of the Optimal Location of a Facility," *Naval Research Logistics Quarterly*, 32, 209-224.

[301] Drezner, Z. (1985) "A Solution to the Weber Location Problem on the Sphere," *Journal of the Operational Research Society*, 36, 333-334.

[302] Drezner, Z. (1985) "O(nlogn) Algorithm for the Rectilinear Round-trip Location Problem," *Transportation Science*, 19, 91-93.

[303] Drezner, Z. (1986) "Location of Regional Facilities," *Naval Research Lo-gistics Quarterly*, 32, 209-224.

[304] Drezner, Z. (1987) "On the Rectangular P-center Problem," *Naval Re-search Logistics*, 34, 229-234.

[305] Drezner, Z. (1987) "Heuristic Solution Methods for Two Location Prob-lems With Unreliable Facilities," *Journal of the Operational Research So-ciety*, 38, 509-514.

[306] Drezner, Z. (1988) "Maximizing the Minimum Sight Angle of a Set of Shapes," *IIE Transactions*, 20, 194-200.

[307] Drezner, Z. (1989) "Stochastic Analysis of the Weber Problem on the Sphere," *Journal of the Operational Research Society*, 40, 1137-1144.

[308] Drezner, Z. (1989) "Conditional P-center Problems," *Transportation Sci-ence*, 23, 51-53.

[309] Drezner, Z. (1991) "The Weighted Minimax Location Problem with Set-up Costs and Extensions," *Revue D'Automatique D'Informatique Et De Recherche Operationelle - Oper. Res.*, 25, 55-64.

[310] Drezner, Z. (1992) "A Note on the Weber Location Problem," *Annals of Operations Research*, 40, 153-161.

[311] Drezner, Z. (1992) "Locational Decisions," *Annals of Operations Research*, 40, J.C. Baltzer AG, Editor, Scientific Publishing Company, Basel, Switzerland.

[312] Drezner, Z. (1992) "On the Computation of the Multivariate Normal Integral," *ACM Transactions on Mathematical Software*, 18, 470-480.

[313] Drezner, Z. "Converting an Area to Discrete Points," Submitted for Publication.

[314] Drezner, Z. and B. Gavish (1985) "ϵ-approximations for Multidimensional Weighted Location Problems," *Operations Research*, 33, 772-783.

[315] Drezner, Z. and A.J. Goldman (1991) "On the Set of Optimal Points to the Weber Problem," *Transportation Science*, 25, 3-8.

[316] Drezner, Z., A. Mehrez and G.O. Wesolovsky (1991) "The Facility Location Problem with Limited Distances," *Transportation Science*, 25, 183-187.

[317] Drezner, Z., S. Schaible and D. Simchi-Levi (1990) "Queueing-location Problems on the Plane," *Naval Research Logistics*, 37, 929-935.

[318] Drezner, Z. and S. Shelah (1987) "On the Complexity of the Elzinga-Hearn Algorithm for the 1-center Problem," *Mathematics of Operations Research*, 12, 255-261.

[319] Drezner, Z. and D. Simchi-Levi (1992) "Asymptotic Behavior of the Weber Location Problem on the Plane," *Annals of Operations Research*, 40, 163-172.

[320] Drezner, Z., G. Steiner and G.O. Wesolowsky (1985) "One-facility Location with Rectilinear Tour Distances," *Naval Research Logistics Quarterly*, 32, 391-405.

[321] Drezner, Z., J.-F. Thisse and G.O. Wesolowsky (1986) "The Minimax-min Location Problem," *Journal of Regional Science*, 26, 87-101.

[322] Drezner, Z. and G.O. Wesolowsky (1978) "A Trajectory Method for the Optimization of the Multifacility Location Problem with ℓ_p Distances," *Management Science*, 24, 1507-1514.

[323] Drezner, Z. and G.O. Wesolowsky (1978) "A New Method for the Multi-facility Minimax Location Problem," *Journal of the Operational Research Society*, 29, 1095-1101.

[324] Drezner, Z. and G.O. Wesolowsky (1978) "Facility Location on a Sphere," *Journal of the Operational Research Society*, 29, 997-1004.

[325] Drezner, Z. and G.O. Wesolowsky (1980) "Single Facility ℓ_p Distance Minimax Location," *SIAM Journal on Algebraic and Discrete Methods*, 1, 315-321.

[326] Drezner, Z. and G.O. Wesolowsky (1980) "A Maximin Location Problem with Maximum Distance Constraints," *AIIE Transactions*, 12, 249-252.

[327] Drezner, Z. and G.O. Wesolowsky (1980) "Optimal Location of a Demand Facility Relative to Area Demand," *Naval Research Logistics Quarterly*, 27, 199-206.

[328] Drezner, Z. and G.O. Wesolowsky (1981) "Optimum Location Probabilities in the ℓ_p Distance Weber Problem," *Transportation Science*, 15, 85-97.

[329] Drezner, Z. and G.O. Wesolowsky (1982) "A Trajectory Approach to the Round-trip Location Problem," *Transportation Science*, 16, 56-66.

[330] Drezner, Z. and G.O. Wesolowsky (1983) "Minimax and Maximin Facility Location on a Sphere," *Naval Research Logistics Quarterly*, 30, 305-312.

[331] Drezner, Z. and G.O. Wesolowsky (1983) "Location of an Obnoxious Facility with Rectangular Distances," *Journal of Regional Science*, 23, 241-248.

[332] Drezner, Z. and G.O. Wesolowsky (1985) "Location of Multiple Obnoxious Facilities," *Transportation Science*, 19, 193-202.

[333] Drezner, Z. and G.O. Wesolowsky (1989) "The Asymmetric Distance Location Problem," *Transportation Science*, 23, 201-207.

[334] Drezner, Z. and G.O. Wesolowsky (1989) "Location of an Obnoxious Route," *Journal of the Operational Research Society*, 40, 1011-1018.

[335] Drezner, Z. and G.O. Wesolowsky (1990) "The Weber Problem on the Plane with Some Negative Weights," *INFOR*, 29, 87-99.

[336] Drezner, Z. and G.O. Wesolowsky (1991) "Facility Location When Demand Is Time Dependent," *Naval Research Logistics*, 38, 763-777.

[337] Drezner, Z. and G.O. Wesolowsky (1994) "Finding the Circle Or Rectangle Containing the Minimum Weight of Points," *Location Science*, 2, 83-90.

[338] Drezner, Z. and G.O. Wesolowsky (1995) "Obnoxious Facility Location in the Interior of a Planar Network," *Journal of Regional Science*, to appear.

[339] Drezner, Z. and G.O. Wesolowsky (1995) "Location on a One-Way Rectilinear Grid," *Journal of the Operational Research Society*, to appear.

[340] Drezner, Z., G.O. Wesolowsky, and T. Drezner "On the Logit Approach to Competitive Facility Location," Submitted for Publication.

[341] Du, D.Z. and Y. Zhang (1992) "On Better Heuristics for Steiner Minimum Trees," *Mathematical Programming*, 57, 193-202.

[342] Durier, R. (1989) "Continuous Location Theory Under Majority Rule," *Mathematics of Operations Research*, 14, 258-274.

[343] Durier, R. (1990) "On Pareto Optima, the Fermat-Weber Problem and Polyhedral Gauges," *Mathematical Programming*, 47, 65-79.

[344] Durier, R. (1994) "The Fermat-Weber Problem and Inner Product Spaces," *Journal of Approximation Theory*, 78, 161-173.

[345] Durier, R. and C. Michelot (1985) "Geometrical Properties of the Fermat-Weber Problem," *European Journal of Operational Research*, 20, 332-343.

[346] Durier, R. and C. Michelot (1986) "Sets of Efficient Points in a Normed Space," *Journal of Mathematical Analysis and Applications*, 117, 506-528.

[347] Durier, R. and C. Michelot (1994) "On the Set of Optimal Points to the Weber Problem: Further Results," *Transportation Science*, 28, 141-149.

[348] Dutta, B.K. and W.R. King (1980) "Metagame Analysis of Competitive Strategy," *Strategic Management Journal*, 1, 4-13.

[349] Dutton, R., G. Hinman and C.B. Millham (1974) "The Optimal Location Of Nuclear Power Facilities in the Pacific Northwest," *Operations Research*, 22, 478-487.

[350] Dwyer and Evans (1981) "A Branch and Bound Algorithm for the List Selection Problem in Direct Mail Advertising," *Management Science*, 29, 658-667.

[351] Dyer, M.E. (1983) the Complexity of Vertex Enumeration Methods, *Mathematics of Operations Research*, 8, 381-402.

[352] Dyer, M.E. (1984) "Linear Time Algorithms for Two- and Three-variable Linear Programs," *SIAM Journal on Computing*, 13, 31-45.

[353] Dyer, M.E. (1986) "On the Multidimensional Search Technique and Its Application to the Euclidean One-Centre Problem," *SIAM Journal on Computing*, 15, 725-738.

[354] Dyer, M.E. and A.M. Frieze (1985) "A Simple Heuristic for the p-centre Problem," *Operations Research Letters*, 3, 285-288.

[355] Eagle, T.C. (1984) "Parameter Stability in Disaggregate Retail Choice Models: Experimental Evidence," *Journal of Retailing*, 60, 1, 101-23.

[356] Eastin, R.V. (1975) "Entropy Maximization and Inferred Ideal Weights in Public Facility Location," *Environment and Planning A*, 7, 191-198.

[357] Eaton, D., R. Church, V. Bennet and B. Namon (1981) "On Deployment of Health Resources in Rural Colombia," *TIMS Studies in the Management Sciences*, 331-359.

[358] Eaton, D., M. Hector, V. Sanchez, R. Lantigua and J. Morgan (1986) "Determining Ambulance Deployment in Santo Domingo, Dominican Republic," *Journal of the Operational Research Society*, 113.

[359] Eaton, C. and R. Lipsey (1975) "The Principle of Minimum Differentiation Reconsidered: Some New Developments in the Theory of Spatial Competition," *Review of Economic Studies*, 42, 27-50.

[360] Eaton, B. and R. Lipsey (1979) "The Theory of Market Preemption: the Persistence of Excess Capacity and Monopoly in Growing Spatial Markets," *Economica*, 46, 149-158.

[361] Eaton, B.C. and R.G. Lipsey (1980) "The Block Metric and the Law of Markets," *Journal of Urban Economics*, 7, 337-347.

[362] Edelsbrunner, H. (1987) *Algorithms in Combinatorial Geometry*, Springer Verlag, Berlin.

[363] Edelsbrummer H. and L. Guibas (1986) "Topologically Sweeping an Arrangement," *Proc. of the 18th Ann. ACM Symp. on Theory of Computing*, 389-407.

[364] Edelsbrummer H., J. O'Rourke and R. Seidel (1993) "Constructing Arrangements of Lines and Hyperplane with Applications," *SIAM Journal on Computing*, to appear.

[365] Efroymson, M.A. and T.L. Ray (1966) "A Branch-and-bound Algorithm for Plant Location," *Operations Research*, 14, 361-368.

[366] Eilon, S., C.D.T. Watson-Gandy and N. Christofides (1971) *Distribution Management: Mathematical Modelling and Practical Analysis*, Griffin, London.

[367] Eiselt, H.A. (1992) "Location Modeling in Practice," *American Journal of Mathematical and Management Sciences*, 12, 3-18.

[368] Eiselt, H.A. (1992) "Hotelling's Duopoly on a Tree," *Annals of Operations Research*, 40, 195-207.

[369] Eiselt, H.A. and G. Charlesworth (1986) "A Note on p-centre Problems in the Plane," *Transportation Science*, 20, 130-133.

[370] Eiselt, H.A. and M. Gendreau (1991) "An Optimal Algorithm for Weighted Minimax Flow Centers on Trees," *Transportation Science*, 25, 314-316.

[371] Eiselt, H.A. and G. Laporte (1989) "Competitive Spatial Models," *European Journal of Operational Research*, 39, 231-242.

[372] Eiselt, H.A. and G. Laporte (1991) "Locational Equilibrium of Two Facilities on a Tree," *Recherche Operationnelle/Operations Research*, 25, 5-18.

[373] Eiselt, H.A. and G. Laporte (1993) "The Existence of Equilibria in the 3-facility Hotelling Model in a Tree," *Transportation Science*, 27, 39-43.

[374] Eiselt, H.A., G. Laporte and J.-F. Thisse (1993) "Competitive Location Models: a Framework and Bibliography," *Transportation Science*, 27, 44-54.

[375] Eiselt, H.A. and G. Pederzoli (1984) "A Location Problem in Graphs," *New Zealand Operational Research*, 12, 49-53.

[376] Eiselt, H. and G. Pederzoli (1986) "Voronoi Diagrams and Their Uses - a Survey, Part I: Theory, Part Ii: Applications," *Proceedings of the Administrative Sciences Association of Canada*, 7, 98-115.

[377] Elson, D.G. (1972) "Site Location Via Mixed-integer Programming," *Operational Research Quarterly*, 23, 31-43.

[378] Elzinga, D.J. and D.W. Hearn (1972) "The Minimum Covering Sphere Problem," *Management Science*, 19, 96-104.

[379] Elzinga, D.J. and D.W. Hearn (1972) "Geometrical Solutions for Some Minimax Location Problems," *Transportation Science*, 6, 379-394.

[380] Elzinga, D.J. and D.W. Hearn (1973) "A Note on a Minimax Location Problem," *Transportation Science*, 7, 100-103.

[381] Elzinga, D.J., D.W. Hearn and W.D. Randolph (1976) "Minimax Multifacility Location with Euclidean Distances," *Transportation Science*, 10, 321-336.

[382] Erkut, E. (1990) "The Discrete p-Dispersion Problem," *European Journal of Operational Research*, 46, 48-60.

[383] Erkut, E. (1993) "Inequality Measures for Location Problems," *Location Science*, 1, 199-217.

[384] Erkut, E., T. Baptie and B. Von Hohenbalken (1990) "The Discrete p-Maxian Location Problem," *Computers & Operations Research*, 17, 51-61.

[385] Erkut, E. and S. Neuman (1989) "Analytical Models for Locating Undesirable Facilities," *European Journal of Operational Research*, 40, 275-291.

[386] Erkut, E. and S. Neuman (1991) "Comparison of Four Models for Dispersing Facilities," *INFOR*, 29, 68-85.

[387] Erkut, E. and S. Neuman (1992) "A Multiobjective Model for Locating Undesirable Facilities," *Annals of Operations Research*, 40, 209-227.

[388] Erkut, E. and T.S. Öncü (1991) "A Parametric 1-maximin Location Problem," *Journal of the Operational Research Society*, 42, 49-55.

[389] Erlenkotter, D. (1977) "Facility Location with Price Sensitive Demands: Private, Public, and Quasi Public," *Management Science*, 24, 378-386.

[390] Erlenkotter, D. (1978) "A Dual-based Procedure for Uncapacitated Facility Location," *Operations Research*, 26, 992-1009.

[391] Erlenkotter, D. (1989) "The General Optimal Market Area Model," *Annals of Operations Research*, 18, 45-70.

[392] Eyster, J.W. and J.A. White (1973) "Some Properties of the Squared Euclidean Distance Location Problem," *AIIE Transactions*, 5, 275-280.

[393] Eyster, J.W., J.A. White and W.W. Wierwille (1973) "On Solving Multifacility Location Problems Using a Hyperboloid Approximation Procedure," *AIIE Transactions*, 5, 1-6.

[394] Ezra, N., G.Y. Handler and R. Chen (1994) "Solving Infinite p-center Problems in Euclidean Space Using an Interactive Graphical Method," *Location Science*, 2, 101-109.

[395] Falk, J.E. and R.M. Soland (1969) "An Algorithm for Separable Nonconvex Programming Problems," *Management Science*, 15, 550-569.

[396] Fawcett, S.E. (1992) "Strategic Logistics in Co-ordinated Global Manufacturing Success." *International Journal of Production Research*, 30, 1081-1099.

[397] Fejer, L. (1922) "Uber Die Lage Der Nulstellen Von Polynomen, Die Aus Minimumforderungen Gewisser Art Entspringen," *Mathematische Annalen*, 85, 41-48.

[398] Ferdows, K. Editor (1989) *Managing International Manufacturing*, North-Holland.

[399] Ferdows, K. (1989) "Mapping International Factory Networks," *Managing International Manufacturing*, K. Ferdows, Editor, North-Holland, 3-21.

[400] Ferdows, K. and J. Miller, Et.al, (1986) "Evolving Global Manufacturing Strategies: Projections Into the 1990's," *International Journal of Operations and Production Management*, 16, 6-16.

[401] Fiacco, A.V. and G.P. McCormick (1968) *Nonlinear Programming : Sequential Unconstrained Minimization Techniques*, John Wiley and Sons, New York and Toronto.

[402] Fine, C.H. and R.M. Freund (1990) "Optimal Investment in Product-flexible Manufacturing Capacity," *Management Science*, 36, 449-466.

[403] Fine, C.H. and A.C. Hax (1985) "Manufacturing Strategy: a Methodology and an Illustration," *Interfaces*, 15, 28-46.

[404] Fishburn, P.C. (1984) "Equity Axioms for Public Risks," *Operations Research*, 32, 901-908.

[405] Fishburn, P.C. and P.D. Straffin (1989) "Equity Considerations in Public Risks Evaluation," *Operations Research*, 37, 229-239.

[406] Flaherty, T. (1986) "Coordinating International Manufacturing and Technology," *Competition in Global Industries*, M.E. Porter, Editor, Harvard Business School Press, 83-109.

[407] Flaherty, T. (1989) "International Sourcing: Beyond Catalog Shopping and Franchising," *Managing International Manufacturing*, K. Ferdows, Editor, North-Holland, 95-124.

[408] Fliege, J. (1994) "Some New Coincidence Conditions in Minisum Multifacility Location Problems with Mixed Gauges," *Studies in Locational Analysis*, 7, 49-60.

[409] Florian, M and S. Nguyen (1974) "A Method for Computing Network Equilibrium with Elastic Demand," *Transportation Science*, 8, 321-332.

[410] Floudas, C.A. and V. Visweswaran (1993) "A Primal-relaxed Dual Global Optimization Approach," *Journal of Optimization Theory and Its Applications*, 78c, 187.

[411] Folland, S.T. (1983) "Predicting Hospital Market Shares," *Inquiry*, 20, 34-44.

[412] Fortune, S. (1987) "A Sweepline Algorithm for Voronoi Diagrams," *Algorithmica*, 2, 153-174.

[413] Fotheringham, A.S. (1983) "A New Set of Spatial-Interaction Models: the Theory of Competing Destinations," *Environment and Planning A*, 15, 15-36.

[414] Fotheringham, A.S. (1988) "Market Share Analysis Techniques: a Review and Illustration of Current US Practice," *Store Choice, Store Location and Market Analysis*, N. Wrigley, Editor, London, Routledge.

[415] Fotheringham, A.S. and M.E. O'Kelly (1989) *Spatial Interaction Models: Formulations and Applications*, Kluwer.

[416] Fotheringham, A.S., and M. Webber (1980) "Spatial Structure and the Parameters of Spatial Interaction Models," *Geographical Analysis*, 12, 33-46.

[417] Foulds, L.R. and H.W. Hamacher (1993) "Optimal Bin Location and Sequencing in Printed Circuit Board Assembly," *European Journal of Operational Research*, 66, 279-290.

[418] Fourer, R., D.M. Gay and B.W. Kernighan (1993) "AMPL a Modelling Language for Mathematical Programming," *the Scientific Press*, South San Francisco.

[419] Francis, R.L. (1967) "Some Aspects of a Minimax Location Problem," *Operations Research*, 15, 1163-1169.

[420] Francis, R.L. and A.V. Cabot (1972) "Properties of a Multifacility Location Problem Involving Euclidean Distances," *Naval Research Logistics Quarterly*, 19, 335-353.

[421] Francis, R.L., H.W. Hamacher, C.Y. Lee and S. Yeralan (1994) "Finding Placement Sequences and Bin Locations for Cartesian Robots," *IIE Transactions*, 26, 47-59.

[422] Francis, R.L. Ans T.J. Lowe (1992) "On Worst-case Aggregation Analysis for Network Location Problems," *Annals of Operations Research*, 40, 229-246.

[423] Francis, R.L., T.J. Lowe and M.B. Rayco (1993) "Row-column Aggregation for Rectilinear Distance P-median Problems," *Research Report*, Department of Industrial & Systems Engineering, University of Florida, Gainesville.

[424] Francis, R.L., L.F. McGinnis Jr. and J.A. White (1992) *Facility Layout and Location: an Analytical Approach*, 2nd Edition, International Series in Industrial and Systems Engineering, Prentice Hall, Englewood Cliffs, NJ.

[425] Francis, R.L. and J.A. White (1974) *Facility Layout and Location: an Analytical Approach*, Prentice Hall, Englewood Cliffs, NJ.

[426] Frenk, J.B.G., J. Gromicho, F. Plastria and S. Zhang (1994) "A Deep Cut Ellipsoid Algorithm and Quasiconvex Programming," in S. Komlosi (Ed.), *Proceedings IVth Workshop on Generalized Convexity, PECS, Hungary*, Springer, Lect. Notes Econ. Math. Systems, 405, 153-170.

[427] Frenk, J.B.G., J. Gromicho and S. Zhang (1994) "General Models in Min-max Continuous Location: Theory and Solution Techniques," *Journal Of Optimisation Theory and Applications*, to appear.

[428] Frenk, J.B.G., J. Gromicho and S. Zhang (1994) "General Models in Min-max Planar Location: Checking Optimality Conditions," *Journal of Optimisation Theory and Applications*, to appear.

[429] Frenk, J.B.G. and M.J. Kleijn (1994) "On Miehle's Algorithm and the Perturbed ℓ_p-distance Multifacility Location Problem," *Studies in Locational Analysis*, 7, 61-75.

[430] Frenk, J.B.G., M. Labbe, R.J. Visscher and S.Z. Zhang (1989) "The Stochastic Queue Location Problem in the Plane," Report 8948/A, Econometric Institute, Erasmus University Rotterdam, the Netherlands.

[431] Frenk, J.B.G., M. Labbe and S.Z. Zhang (1989) "The Stochastic K-priority Queue Location Problem," Report 8951/A, Econometric Institute, Erasmus University Rotterdam, the Netherlands.

[432] Frenk, J.B.G., M.T. Melo and S. Zhang (1994) "The Weiszfeld Method in Single Facility Location," *InvestigaÇao Operacional (Portugal)*, 14, 35-59.

[433] Frenk, J.B.G., M.T. Melo and S. Zhang (1994) "A Weiszfeld Method for a Generalized ℓ_p Distance Minisum Location Model in Continuous Space," *Location Science*, 2, 111-127.

[434] Frick, H. (1985) "On the Solution of the Facility Location Problem When the Dominance Method Fails," *Zeitschrift Für Operations Research*, 29, B101-B106.

[435] Friedrich, C.J. (1929) *Alfred Weber's Theory of Location of Industries*, University of Chicago Press, Chicago.

[436] Friesz, T.L., R.L. Tobin, H.J. Cho and N.J. Mehta (1990) "Sensitivity Analysis Based Heuristic Algorithms for Mathematical Programs with Variational Inequality Constraints," *Mathematical Programming*, 48, 265-284.

[437] Friesz, T., L. Tobin and T. Miller (1989) "Existence Theory for Spatially Competitive Network Facility Location Models," *Annals of Operations Research*, 18, 267-276.

[438] Friez, T., T. Miller and R. Tobin (1988) "Competitive Network Facility Location Models: a Survey," *Papers of the Regional Science Association*, 65, 47-57.

[439] Furlong, W.H. and G.A. Slotsve (1983) "Will That Be Pickup Or Delivery? an Alternative Spatial Pricing Strategy," *Bell Journal of Economics*, 14, 271-274.

[440] Gabszewicz, J.J. and J.-F. Thisse (1986) "Spatial Competition and the Location of Firms," *Fundamentals of Pure and Applied Economics*, 5, 1-71.

[441] Gabszewicz, J. and J.-F. Thisse (1992) "Location," *Handbook of Game Theory with Economic Applications*, R. Aumann and S. Hart, Editors, Elsevier Science Publishers, Amsterdam, 281-304.

[442] Gao, L. and E.P. Robinson (1992) "A Dual-based Optimization Procedure for the Two-echelon Uncapacitated Facility Location Problem," *Naval Research Logistics*, 39, 191-212.

[443] Garey, M.R., R.L. Grahm and D.S. Johnson (1977) "The Complexity of Computing Steiner Minimal Trees," *SIAM Journal on Applied Mathematics*, 32, 835-859.

[444] Garey, M.R. and D.S. Johnson (1977) "The Rectilinear Steiner Tree Problem Is NP-complete," *SIAM Journal on Applied Mathematics*, 32, 826-834.

[445] Garey, M. and D. Johnson (1979) *Computers and Intractibility: a Guide to the Theory of NP-completeness*, Freeman, San Francisco.

[446] Gautschi, D.A. (1981) "Specification of Patronage Models of Retail Center Choice," *Journal of Marketing Research*, 18, 162-174.

[447] Gelders, L.F., L.M. Pintelon and L.N. Van Wassenhove (1987) "A Location-allocation Problem in a Large Belgian Brewery," *European Journal of Operational Research*, 28, 196-206.

[448] Geoffrion, A.M. (1968) "Proper Efficiency and the Theory of Vector Optimisation," *Journal of Mathematical Analysis and Applications*, 22, 618-630.

[449] Geoffrion, A.M. (1972) "Generalized Benders Decomposition," *Journal of Optimization Theory and Applications*, 10, 237-260.

[450] Geoffrion, A.M. (1975) "A Guide to Computer-assisted Methods for Distribution Systems Planning," *Sloan Management Review*, 16, 17-41.

[451] Geoffrion, A.M. (1976) "The Purpose of Mathematical Programming Is Insight, Not Numbers," *Interfaces*, 7, 81-92.

[452] Geoffrion, A.M. (1977) "Objective Function Approximations in Mathematical Programming," *Mathematical Programming*, 13, 23-37.

[453] Geoffrion, A.M. (1979) "Making Better Use of Optimization Capability in Distribution System Planning," *AIIE Transactions*, 11, 96-106.

[454] Geoffrion, A.M. and G.W. Graves (1974) "Multicommodity Distribution System Design by Benders Decomposition," *Management Science*, 20, 822-844.

[455] Geoffrion, A.M., G.W. Graves and S.J. Lee (1982) "A Management Support System for Distribution Planning," *INFOR*, 20, 287-314.

[456] Geoffrion, A.M. and R.E. Marsten (1972) "Integer Programming Algorithms - a Framework and State-of-the-Art Survey," *Management Science*, 18, 465-491.

[457] Geoffrion, A.M. and R.F. Powers (1980) "Facility Location Analysis Is Just the Beginning (if You Do It Right)," *Interfaces*, 10, 22-30.

[458] Geoffrion, A.M. and R.F. Powers (1995) "20 Years of Strategic Distribution System Design: an Evolutionary Perspective," *Interfaces*, to appear.

[459] Geoffrion, A.M. and T.J. Van Roy (1979) "Caution: Common Sense Planning Methods Can Be Hazardous to Your Corporate Health," *Sloan Management Review*, 20, 31-42.

[460] Ghosh, A. (1984) "Parameter Nonstationarity in Retail Choice Models," *Journal of Business Research*, 12, 425-36.

[461] Ghosh, A. and B. Buchanan (1988) "Multiple Entries in a Duopoly: a First Entry Paradox," *Geographical Analysis*, 20, 111-121.

[462] Ghosh, A. and C.S. Craig (1983) "Formulating Retail Location Strategy in a Changing Environment," *Journal of Marketing*, 47, 56-68.

[463] Ghosh, A. and C.S. Craig (1984) "A Location Allocation Model for Facility Planning in a Competitive Environment," *Geographical Analysis*, 16, 39-51.

[464] Ghosh, A. and C.S. Craig (1986) "An Approach to Determining Optimal Locations for New Services," *Journal of Marketing Research*, 23, 354-362.

[465] Ghosh, A. and C.S. Craig (1991) "FRANSYS: a Franchise Location Model," *Journal of Retailing*, 67, 212-234.

[466] Ghosh, A. and F. Harche (1993) "Location-Allocation Models in the Private Sector: Progress, Problems, and Prospects," *Location Science*, 1, 81-106.

[467] Ghosh, A. and S. McLafferty (1982) "Locating Stores in Uncertain Environments: a Scenario Planning Approach," *Journal of Retailing*, 58, 5-22.

[468] Ghosh, A. and S. McLafferty (1984) "A Model of Consumer Propensity for Multipurpose Shopping," *Geographical Nalysis*, 16, 244-249.

[469] Ghosh, A., and S. McLafferty (1987) *Location Strategies for Retail and Service Firms*, Lexington, Mass., Lexington Books.

[470] Ghosh, A. and G. Rushton (1987) *Spatial Analysis and Location-Allocation Models*, Van Nostrand Reinhold Company, NY.

[471] Ghosh, A. and G. Rushton (1987) "Progress in Location Allocation Modeling," *Spatial Analysis and Location Allocation Models*, A. Ghosh and G. Rushton, Editors, NY, Van Nostrand Reinhold.

[472] Ghosh, A. and V. Tibrewala (1992) "Optimal Timing and Location in Competitive Markets," *Geographical Analysis*, 24, 317-334.

[473] Giannakis Y. (1993) PhD Thesis, London School of Economics and Political Science.

[474] Gifford F.A.(1961) "Diffusion Around Buildings," *Nuclear Safety*, 47, 364-369.

[475] Gilbert, R. (1986) "Preemptive Competition," *New Developments in the Analysis of Market Structure*, J. Stiglitz and F. Mathewson, Editors, 3, 90-125, MIT Press, Cambridge, MA.

[476] Gilmore, P.C. and R.E. Gomory (1961) "A Linear Programming Approach to the Cutting Stock Problem," *Operations Research* 9, 849-859.

[477] Gleason, J. (1975) "A Set Covering Approach to Bus Stop Location," *Omega*, 605-608.

[478] Glickman, T.S. (1994) "The Cost-Risk Tradeoffs Associated with Rerouting Interstate Highway Shipments of Hazardous Materials To Minimize Risk," *Resources for the Future*, Washington DC, 24.

[479] Glickman, T.S. (1991) "An Expeditious Risk Assessment of the Highway Transportation of Flammable Liquids in Bulk," *Transportation Science*, 25, 115-123.

[480] Glickman, T.S. (1988) "Benchmark Estimates of Release Accident Rates in Hazardous Materials Transportation of Rail and Truck," *Transportation Research Record*, 1193, 22-28.

[481] Glickman, T.S. (1983) "Rerouting Railroad Shipments of Hazardous Materials to Avoid Populated Areas," *Accident Analysis Prevention*, 15, 329-335.

[482] Glickman, T.S. and D.B. Rosenfield (1984) "Risks of Catastrophic Derailments Involving the Release of Hazardous Materials," *Management Science*, 30, 503-511.

[483] Glover, F., G. Jones, D. Karney, D. Klingman and J. Mote (1979) "An Integrated Production, Distribution, and Inventory Planning System," *Interfaces*, 9, 21-35.

[484] Goetze, B., E.N. Gordejev, W. Nehrlich and D.W. Kochetkov (1990) "Maximal Empty Boxes in Higher Dimensional Spaces," *Journal of Information Processing and Cybernetics (form. Elektron.Inf.ver.Kyb.)*, 26, 537-545.

[485] Goldberg, J. and L. Paz (1991) "Locating Emergency Vehicle Bases When Service Time Depends on Call Location," *Transportation Science*, 264-280.

[486] Golden, B.L. and H.A. Eiselt (1992) *Special Issue of the American Journal of Mathematical and Management Sciences*, Editors, 12, 1-94.

[487] Golden, B. and E. Wasil (1987) "Computerized Vehicle Routing in the Soft Drink Industry," *Operations Research*, 35, 6-17.

[488] Goldman, A.J. (1969) "Optimal Locations for Centers in a Network," *Transportation Science*, 3, 352-360.

[489] Goldman, A.J. (1971) "Optimal Center Location in Simple Networks," *Transportation Science*, 5, 212-221.

[490] Goldman, A.J. (1972) "Minimax Location of a Facility in a Network," *Transportation Science*, 6, 407-418.

[491] Goldman, A.J. (1975) "Concepts of Optimal Locations for Partially Noxious Facilities," ORSA/TIMS National Meeting, Chicago, IL, *Bulletin of the Operations Research Society of America*, 23, 1, B-85.

[492] Goldman, A.J. and P.M. Dearing (1975) "Concepts of Optimal Location for Partially Noxious Facilities," *ORSA Bulletin*, 23, 1, B-31.

[493] Goldman, A.J. and P.R. Meyers (1965) "A Domination Theorem for Optimal Locations," Presented At the ORSA/TIMS Meeting, Houston, TX, *ORSA Bulletin*, 13, 2, B-147.

[494] Golledge, R.S., W.A.V. Clark and G. Rushton (1966) "The Implications of the Consumer Behavior of a Dispersed Farm Population in Iowa," *Economic Geography*, 42, 265-72.

[495] Golledge, R.G. and H. Timmermans (1988) *Behavioural Modelling in Geography and Planning*, Editors, Croom Helm, London.

[496] Goodchild, M.F. (1984) "ILACS: a Location Allocation Model for Retail Site Selection," *Journal of Retailing*, 60, 84-100.

[497] Goodchild, M.F. and V.T. Noronha (1987) "Location Allocation and Impulsive Shopping: the Case of Gasoline Retailing," in *Spatial Analysis and Location Allocation Models*, A. Ghosh and G. Rushton, Editors, NY, Van Nostrand Reinhold.

[498] Gopalan, R., K.S. Kolluri, R. Batta and M.H. Karwan (1990) "Modeling Equity of Risk in the Transportation of Hazardous Materials," *Operations Research*, 38, 961-973.

[499] Graitson, D. (1987) "Spatial Competition a La Hotelling: a Selective Survey," *Journal of Industrial Economics*, 31, 13-25.

[500] Graves, S.C. (1981) "A Review of Production Scheduling," *Operations Research*, 29, 646-675.

[501] Greenhut, M.L. (1956) *Plant Location in Theory and Practice*, University of North-Carolina Press, Chapel Hill.

[502] Greenhut, M.L., G. Norman and C.S. Hung (1987) *the Economics of Imperfect Competition*, Cambridge University Press, Cambridge.

[503] Grossman, T.A. (1993) *Models for Optimal Facility Utilization Decisions*, Unpublished Ph.D. Dissertation, Department of Industrial Engineering and Engineering Management, Stanford University, Stanford, CA.

[504] Grossman, T.A. and M.L. Brandeau (1994) "Optimal Investment of A Facility Improvement Budget in a User-Optimizing Environment With Market Externalities," Working Paper, University of Calgary, Faculty of Management, Calgary, Alberta, Canada.

[505] Grossman, T.A. and M.L. Brandeau (1994) "Optimal Investment of A Facility Improvement Budget in the Presence of Market Externalities," Working Paper, University of Calgary, Faculty of Management, Calgary, Alberta, Canada.

[506] Grossman, T.A. and M.L. Brandeau (1994) "Inducing Self-Optimizing Facility Users to Behave in a Socially Optimal Manner," Working Paper, University of Calgary, Faculty of Management, Calgary, Alberta, Canada.

[507] Gunawardane, G. (1982) "Dynamic Versions of Set Coverings Type Public Facility Location Problems," *European Journal of Operational Research*, 10, 190-195.

[508] Haimovich, M., A. Rinnooy Kan and L. Stougie (1988) "Analysis of Heuristics for Vehicle Routing Problems," B. Golden and A. Assad, Editors, *Vehicle Routing: Methods and Studies*, North Holland, Amsterdam.

[509] Hakimi, S.L. (1964) "Optimal Location of Switching Centers and the Absolute Centers and Medians of a Graph," *Operations Research*, 12, 450-459.

[510] Hakimi, S.L. (1965) "Optimum Distribution of Switching Centers in a Communication Network and Some Related Graph Theoretic Problems," *Operations Research*, 13, 462-475.

[511] Hakimi, S.L. (1983) "On Locating New Facilities in a Competitive Environment," *European Journal of Operational Research*, 12, 29-35.

[512] Hakimi, S.L. (1986) "p-median Theorems for Competitive Location," *Annals of Operations Research*, 5, 79-88.

[513] Hakimi, S.L. (1990) "Locations with Spatial Interactions: Competitive Locations and Games," *Discrete Location Theory*, R.L. Francis and P.B. Mirchandani, Editors, Wiley-Interscience, New York, NY, 439-478.

[514] Hakimi, S.L., M. Labbé and E. Schmeichel (1992) "The Voronoi Partition of a Network and Its Implications in Location Theory," *ORSA Journal on Computing*, 4, 412-417.

[515] Hakimi, S.L., E.F. Schmeichel and S.G. Pierce (1978) "On p-centers in Networks," *Transportation Science*, 12, 1-15.

[516] Hakimi, S.L. and S.N. Maheshwari (1972) "Optimum Locations of Centers in Networks," *Operations Research*, 20, 967-973.

[517] Halfin, S. (1974) "On Finding the Absolute and Vertex Centers of a Tree with Distances," *Transportation Science*, 8, 75-77.

[518] Hall, R.W. (1986) "Discrete Models/Continuous Models," *OMEGA, The International Journal of Management Science*, 14, 213-220.

[519] Hall, R.W. (1988) "Median, Mean and Optimum as Facility Locations," *Journal of Regional Science*, 28, 65-82.

[520] Halpern, J. (1976) "The Location of a Center-Median Convex Combination on an Undirected Tree," *Journal of Regional Science*, 16, 237-245.

[521] Halpern, J. (1978) "Finding Minimal Center - Median Convex Combination Cent-Dian of a Graph," *Management Science*, 24, 535-544.

[522] Halpern, J. (1980) "Duality in the Cent-Dian of a Graph," *Operations Research*, 28, 722-735.

[523] Halpern, J. (1981) "ISOLDE I Proceedings," *European Journal of Operational Research*, Editor, 6, 93-231.

[524] Halpern, J. and O. Maimon (1983) "Accord and Conflict Among Several Objectives in Locational Decisions on Tree Networks," *Locational Analysis of Public Facilities*, J.-F. Thisse and H.G. Zoller, Editors, North Holland Publ. Co., Amsterdam, the Netherlands, 301-314.

[525] Hamacher, H.W. (1994) *Mathematische Methoden Der Standortplanung in Der Ebene*, Research Monograph, Zentrum Für Techno- Und Wirtschaftsmathematik, University of Kaiserslautern, Germany.

[526] Hamacher, H.W. and S. Nickel (1992) "Restricted Planar Location Problems and Applications," Preprint 227, Fachbereich Mathematik, Universität Kaiserslautern, Germany.

[527] Hamacher, H.W. and S. Nickel (1993) "Multicriterial Planar Location Problems," Preprint 243, Fachbereich Mathematik, Universität Kaiserslautern, Germany.

[528] Hamacher, H.W. and S. Nickel (1994) "Combinatorial Algorithms for Some 1-facility Median Problems in the Plane," *European Journal of Operational Research*, 79, 340-351.

[529] Hammerton, M., M.W. Jones-Lee and V. Abbott (1982) "Equity and Public Risk: Some Empirical Results," *Operations Research*, 30, 203-207.

[530] Hampel, F.R., E.M. Ronchetti, P.J. Rousseeuw and W.A. Stahel (1986) *Robust Statistics - the Approach Based on Influence Functions*, Wiley, NY.

[531] Handler, G.Y. (1985) "Medi-Centers of a Tree," *Transportation Science*, 19, 246-260.

[532] Handler, G.Y. (1973) "Minimax Location of a Facility in an Undirected Tree Graph," *Transportation Science*, 7, 287-293.

[533] Handler, G.Y. and P.B. Mirchandani (1979) *Location on Networks: Theory and Algorithms*, the MIT Press, Cambridge, MA.

[534] Hanjoul, P., H. Beguin and J.C. Thill (1986) "Shapes of Market Areas Under Euclidean Distance," *Cahiers De Géographie De Besançon*, 28, 107-110.

[535] Hanjoul, P., H. Beguin and J.C. Thill (1989) "Advances in the Theory of Market Areas," *Geographical Analysis*, 21, 185-196.

[536] Hanjoul, P., P. Hansen, D. Peeters and J.-F. Thisse (1990) "Uncapacitated Plant Location Under Alternative Spatial Price Policies," *Management Science*, 36, 41-57.

[537] Hanjoul, P. and J.C. Thill (1987) "Elements of Planar Analysis in Spatial Competition," *Regional Science and Urban Economics*, 17, 423-439.

[538] Hansen, M.M. and C.B. Weinberg (1979) "Retail Market Share in a Competitive Environment," *Journal of Retailing*, 55, 37-46.

[539] Hansen, P. and B. Jaumard (1994) "Lipschitz Optimization," *Handbook of Global Optimization*, R. Horst and P. Pardalos, Editors, Kluwer, Dordrecht.

[540] Hansen, P., B. Jaumard and S. Krau (1994) "Weber's Problem on the Sphere," Cahiers Du GERAD, G-94-27, Montréal, Canada.

[541] Hansen, P., B. Jaumard and S. Krau "A Column Generation Algorithm for the Multisource Weber Problem," in Preparation.

[542] Hansen, P., B. Jaumard and S.H. Lu (1991) "An Analytical Approach to Global Optimization," *Mathematical Programming*, 52, 227-254.

[543] Hansen, P., B. Jaumard and S.H. Lu (1992) "Global Optimization of Univariate Lipschitz Functions: I. Survey and Properties," *Mathematical Programming*, 55, 251-272.

[544] Hansen, P., B. Jaumard and S.H. Lu (1992) "Global Optimization of Univariate Lipschitz Functions: II. New Algorithms and Computational Comparison," *Mathematical Programming*, 55, 273-292.

[545] Hansen, P., M. Labbé, D. Peeters and J.-F. Thisse (1987) "Facility Location Analysis," *Fundamentals of Pure and Applied Economics*, 22, 1-70.

[546] Hansen, P., M. Labbe, D. Peeters, J.-F. Thisse and J.V. Henderson (1987) "Systems of Cities and Facility Location," Chur, Harwood Academic Publisher.

[547] Hansen, P., D. Peeters, D. Richard and J.-F. Thisse (1985) "The Minisum and Minimax Location Problems Revisited," *Operations Research*, 33, 1251-1265.

[548] Hansen, P., D. Peeters and J.-F. Thisse (1981) "Constrained Location and the Weber-Rawls Problem," *Annals of Operations Research*, 11, 147-166.

[549] Hansen, P., D. Peeters and J.-F. Thisse (1981) "On the Location of an Obnoxious Facility," *Sistemi Urbani*, 3, 299-317.

[550] Hansen, P., D. Peeters and J.-F. Thisse (1982) "An Algorithm for a Constrained Weber Problem," *Management Science*, 28, 1285-1295.

[551] Hansen, P., D. Peeters and J.-F. Thisse (1983) "Public Facility Location Models: a Selective Survey," *Locational Analysis of Public Facilities*, J.-F. Thisse and H.G. Zoller, Editots, NY, North Holland.

[552] Hansen, P., D. Peeters, and J.-F. Thisse (1992) "Facility Location Under Zone Pricing," *Cahiers Du GERAD*, G-92-29, Montréal, Canada.

[553] Hansen, P., D. Peeters and J.-F. Thisse "The Profit Maximizing Weber Problem," *Cahiers Du GERAD*, Forthcoming.

[554] Hansen, P., J. Perreur and J.-F. Thisse (1980) "Location Theory, Dominance and Convexity: Some Further Results," *Operations Research*, 28, 1241-1250.

[555] Hansen, P. and J.-F. Thisse (1977) "Multiplant Location for Profit Maximization," *Environment and Planning*, A9, 63-73.

[556] Hansen, P. and J.-F. Thisse (1981) "The Generalized Weber-Rawls Problem," *Operations Research Hamburg* July 1981, J.P. Brans, Editor, North-Holland, Amsterdam.

[557] Hansen, P. and J.-F. Thisse (1983) "Recent Advances in Continuous Location Theory," *Sistemi Urbani*, 1, 33-54.

[558] Hansen, P., J.-F. Thisse and P. Hanjoul (1981) "Simple Plant Location Under Uniform Delivered Pricing," *European Journal of Operations Research*, 6, 94-103.

[559] Hansen, P., J.-F. Thisse and R. Wendell (1986) "Equivalence of Solutions to Network Location Problems," *Mathematics of Operations Research*, 11, 672-678.

[560] Harker, P. (1984) "A Variational Inequality Approach for the Determination of Oligopolistic Market Equilibrium," *Mathematical Programming*, 30, 105-111.

[561] Harris, B., A. Farhi and J. Dufour (1972) *Aspects of a Problem in Clustering*, University of Pennsylvania Press, Philadelphia, PA.

[562] Harris C.D. and E.L. Ullman (1945) "The Nature of Cities," *the Annals of the American Academy of Political and Social Science*, 242, 7-17.

[563] Harsanyi, J.C. (1975) "Can the Maximin Principle Serve as a Basis for Morality? a Critique of John Rawls's Theory," *American Political Science Review*, 69, 594-606.

[564] Harwood, D.W., J.G. Viner and E.R. Russell (1993) "Procedure for Developing Truck Accident and Release Rates for Hazardous Material Routing," *Journal of Transportation Engineering*, 119, 189-199.

[565] Harwood, D.W., E.R. Russell and J.G. Viner (1990) "Characteristics Of Accidents and Incidents in Highway Transportation of Hazardous Materials," *Transportation Research Record*, 1245, 23-33.

[566] Hax, A.C. and D. Candea (1984) *Production and Inventory Management*, Prentice-Hall, NY.

[567] Hay, D. (1976) "Sequential Entry and Entry-deterring Strategies," *Oxford Economic Papers*, 28, 240-257.

[568] Hayes, R.H. and S.C. Wheelwright (1979) "Link Manufacturing Process and Product Life Cycles," *Harvard Business Review*, 133-140.

[569] Hayes, R.H. and S.C. Wheelwright (1979) "The Dynamics of Process-product Life Cycles," *Harvard Business Review*, Mar-Apr, 127-136.

[570] Haynes, K. and A.S. Fotheringham (1988) *Gravity and Spatial Interaction Models*, Beverly Hills, Sage.

[571] Hearn, D.W. and J. Vijay (1982) "Efficient Algorithms for the (weighted) Minimum Circle Problem," *Operations Research*, 30, 777-795.

[572] Hestenes, M.R. (1969) "Multiplier and Gradient Methods," *Journal of Optimization Theory and Applications*, 4, 303-320.

[573] Hill, T.J. (1989) *Manufacturing Strategy: Text and Cases*, Irwin, IL.

[574] Hillsman, E.L. (1984) "The *p*-median Structure as a Unified Linear Model for Location-allocation Analysis," *Environment and Planning A*, 16, 305-318.

[575] Hitt, M., R. Hoskisson and R. Ireland (1991) "Global Diversification: Interactive Effects W. Product Diversification on Innovation and Performance," *Technical Report*, Texas A&M University.

[576] Hodder, J.E. and M.C. Dincer (1986) "A Multifactor Model for International Plant Location and Financing Under Uncertainty," *Computers & Operations Research*, 13, 601-609.

[577] Hodder, J.E. and J.V. Jucker (1982) "Plant Location Modeling for the Multinational Firm," *1982 AIB Asia-Pacific Conference*, 248-258.

[578] Hodder, J.E. and J.V. Jucker (1985) "International Plant Location Under Price and Exchange Rate Uncertainty," *Production Economics - Trends and Issues*, R. Grubbstrom and Hinterhuber, Editors, Elsevier, 225-230.

[579] Hodgart (1978) "Optimizing Access to Public Services: a Review of Problems, Models, and Methods of Locating Central Facilities," *Progress in Human Geography*, 2, 17-48.

[580] Hodgson, M.J. (1978) "Toward More Realistic Allocation in Location Allocation Models: an Interaction Approach," *Environment and Planning A*, 10, 1273-85.

[581] Hodgson, J.M. (1981) "A Location-Allocation Model Maximizing Consumers' Welfare," *Regional Studies*, 15, 493-506.

[582] Hodgson, M.J. (1981) "The Location of Public Facilities Intermediate to the Journey to Work," *European Journal of Operational Research*, 6, 199-204.

[583] Hodgson, M.J. (1990) "A Flow-Capturing Location-Allocation Model," *Geographical Analysis*, 22, 270-279.

[584] Hodgson, M.J. and S. Neuman (1993) "A GIS Approach to Eliminating Source C Aggregation Error in p-median Models," *Location Science*, 1, 155-170.

[585] Hodgson, M.J., K.E. Rosing and F. Shmulevitz (1993) "A Review of Location-allocation Applications Literature," *Isolde VI Survey Papers, Studies in Locational Analysis*, 5, 3-29.

[586] Hodgson, M.J., K.E. Rosing and A.L.G. Stornier "Applying the Flow Capturing Location-Allocation Model to Realistic Network Traffic Problems: the Case of Edmonton, Canada," *European Journal of Operational Research*, to appear.

[587] Hodgson, M.J., R.T. Wong and J. Honsaker (1987) "The p-centroid Problem on an Inclined Plane," *Operations Research*, 35, 221-233.

[588] Hofmann, J.E. (1969) "Uber Die Geometrische Behandlung Einer Fermatschen Extremwert-Aufgabe Durch Italiener Des 17. Jahrhunderts (On the Geometric Treatment of an Extremal Problem Fermat's by Italians of the 17th Century) - in German," *Sudhoffs Archive*, 53, 86-99.

[589] Hogan, K. and C. ReVelle (1986) "Concepts and Applications of Backup Coverage," *Management Science*, 34, 1434-1444.

[590] Holzman, R. (1990) "An Axiomatic Approach to Location on Networks," *Mathematics of Operations Research*, 15, 553-563.

[591] Hoover, E. (1936) "Spatial Price Discrimination," *Review of Economic Studies*, 4, 182-191.

[592] Hoover, E.M. (1948) *the Location of Economic Activity*, McGraw-Hill, NY.

[593] Hopmans, A. (1986) "A Spatial Interaction Model for Branch Bank Accounts," *European Journal of Operational Research*, 27, 242-250.

[594] Horst, R. (1976) "An Algorithm for Nonconvex Programming Problems," *Mathematical Programming*, 10, 312-321.

[595] Horst, R. and P. Pardalos (1994) *Handbook of Global Optimization*, Editors, Dordrecht, Kluwer.

[596] Horst, R. and H. Tuy (1990) *Global Optimisation, Deterministic Approaches*, Springer Verlag, Berlin, Germany.

[597] Horst, R. and H. Tuy (1993) *Global Optimization: Deterministic Approaches*, 2nd Edition, NY, Springer.

[598] Horst, R., J. De Vries and N.V. Thoai (1988) "On Finding New Vertices and Redundant Constraints in Cutting-Plane Algorithms for Global Optimization," *Operations Research Letters*, 7, 85-90.

[599] Hotelling, H. (1929) "Stability in Competition," *Economic Journal*, 39, 41-57.

[600] Hua, Lo-Keng and Others (1962) "Application of Mathematical Methods to Wheat Harvesting," *Chinese Mathematics*, 2, 77-91.

[601] Hubbard, R. (1978) "A Review of Selected Factors Conditioning Consumer Travel Behavior," *Journal of Consumer Research*, 5, 7-21.

[602] Huchzermeier, A. (1991) *Global Manufacturing Strategy Planning Under Exchange Rate Uncertainty*, Ph.D. Thesis, Decision Sciences, University of Pennsylvania.

[603] Huff, D.L. (1964) "Defining and Estimating a Trade Area," *Journal of Marketing*, 28, 34-38.

[604] Huff, D.L. (1966) "A Programmed Solution for Approximating an Optimum Retail Location," *Land Economics*, 42, 293-303.

[605] Huriot, J.M. and J. Perreur (1973) "Modeles De Localisation Et Distance Rectilineaire," *Revue D'Economie Politique*, 83, 640-662.

[606] Hurter, A.P. Jr. and J.S. Martinich (1989) *Facility Location and the Theory of Production*, Kluwer Academic Publishers,Boston, Dordrecht, London.

[607] Hurter, A.P., M.K. Schaeffer and R.E. Wendell (1975) "Solution of Constrained Location Problems," *Management Science*, 22, 51-56.

[608] Hwang, F.K. (1991) "A Primer of the Euclidean Steiner Problem," *Annals of Operations Research*, 33.

[609] Hwang, F.K., D.S. Richards and P. Winter (1992) "The Steiner Tree Problem," *Annals of Discrete Mathematics*, North-Holland, Amsterdam, 53.

[610] Hwang, F.K. and Y.C. Yao (1990) "Comments on Bern's Probabilistic Results on Rectilinear Steiner Trees," *Algorithmica*, 5, 591-598.

[611] Ichimori, T. and T. Nishida (1985) "Note on a Rectilinear Distance Round-trip Location Problem," *Transportation Science*, 19, 84-91.

[612] Idrissi, H.F., O. Lefebvre and C. Michelot (1988) "A Primal Dual Algorithm for a Constrained Fermat-Weber Problem Involving Mixed Gauges," *Revue D'Automatique D'Informatique Et De Recherche Operationelle - Oper. Res.*, 22, 313-330.

[613] Idrissi, H.F., O. Lefebvre and C. Michelot (1989) "Duality for Constrained Multifacility Location Problems with Mixed Norms and Applications," *Annals of Operations Research*, 18, 71-92.

[614] Idrissi, H.F., O. Lefebvre and C. Michelot (1989) "Applications and Numerical Convergence of the Partial Inverse Method," *Optimisation - Fifth French-German Conference*, Castel Novel 1988, Lect. Not. Math. 1405, Springer Verlag, Berlin.

[615] Idrissi, H.F., O. Lefebvre and C. Michelot (1991) "Solving Constrained Multifacility Minimax Location Problems," Working Paper, Centre De Recherches De Mathematiques Statistiques Et Economie Mathematique, Universite De Paris 1, Pantheon-Sorbonne, Paris, France.

[616] Idrissi, H.F., P. Loridan and C. Michelot (1988) "Approximation of Solutions for Location Problems," *Journal of Optimization Theory and Applications*, 56, 127-143.

[617] Illgen, A. (1979) "Das Verhalten Von Abstiegsverfahren an Einer Singularitat Des Gradientes," *Mathematik, Operationsforschung Und Statistik*, Serie Optimisation, 10, 39-55.

[618] Imai, H., D.T. Lee and C.D. Yang (1992) "1-segment Center Problems," *ORSA Journal on Computing*, 4, 426-434.

[619] Imai, T. (1989) "Numerically Robust Algorithm for Constructing Voronoi Diagrams for Line Segments," Master's Thesis, Department of Mathematical Engineering and Information Physics, Faculty of Engineering, University of Tokyo [in Japanese].

[620] Imai, T. (1994) "Manual of SEGVOR: a FORTRAN Program for Constructing the Voronoi Diagram of Line Segments," Department of Mathematical Engineering and Information Physics, University of Tokyo.

[621] Ingene, C.A. and A. Ghosh (1990) "Consumer and Producer Behavior in a Multipurpose Shopping Environment," *Geographical Analysis*, 22, 70-93.

[622] Insurance Services Office (1974) *Grading Schedule for Municipal Fire Protection*, Insurance Services Office, New York.

[623] Iri ,M., K. Kubota and K. Murota (1991) "Geometrical/geographical Optimization and Fast Automatic Differentiation," *Yugoslav Journal of Operations Research*, 1, 121-134.

[624] Iri, M., K. Murota and T. Ohya (1984) "A Fast Voronoi-diagram Algorithm with Applications to Geographical Optimization Problems, in P. Throft-Christensen," (ed.) *Lecture Notes in Control and Information Sciences, Vol.59, System Modelling and Optimization*, Proceedings of the 11th IFIP Conference, Copenhagen, Springer-Verlag, Berlin, 273-288.

[625] Isard, W. (1956) *Location and Space Economy*, MIT Press, Cambridge, Mass.

[626] Ittig, P.T. (1985) "Capacity Planning in Service Operations : The Case of Hospital Outpatient Facilities," *Socio-Economic Planning Science*, 19, 425-429.

[627] Jacobsen, S.K. and O.G.B. Madsen (1983) ISOLDE II Proceedings, *European Journal of Operational Research*, Editors, 12, 15-111, 217-313.

[628] Jaillet, P. (1993) Analysis of Combinatorial Optimization Problems in Euclidean Spaces, *Mathematics of Operations Research*, 18.

[629] Jaillet, P. (1993) Cube Versus Torus Models for Combinatorial Optimization Problems and the Euclidean Minimum Spanning Tree Constant, *Annals of Applied Probability*, to appear.

[630] Jaillet, P. (1988) "A Priori Solution of a Traveling Salesman Problem in Which a Random Subset of the Customers Are Visited," *Operations Research*, 36, 929-936.

[631] Jaillet, P. (1991) "Probabilistic Routing Problems in the Plane," *Operational Research 90*, H. Bradley, Editor, Pergamon Press, London.

[632] Jaillet, P. (1985) *the Probabilistic Traveling Salesman Problems*, Technical Report 185, Operations Research Center, MIT.

[633] Jaillet, P. (1987) "Stochastic Routing Problems," *Stochastics in Combinatorial Optimization*, G. Andreatta, F. Mason and P. Serafini, Editors, World Scientific, Singapore.

[634] Jaillet, P. and A. Odoni (1988) "Probabilistic Vehicle Routing Problems," *Vehicle Routing: Methods and Studies*, B. Golden and A. Assad, Editors, North Holland, Amsterdam.

[635] Jain, A.K. and V. Mahajan (1979) "Evaluating the Competitive Environment in Retailing Using Multiplicative Competitive Interactive Models," *Research in Marketing*, Sheth and J. Greenwich, Editors, CT, JAI Press.

[636] Jarvis, J. (1975) "Optimization in Stochastic Systems with Distinguishable Servers," Technical Report No. 19-75, Operations Research Centre, MIT.

[637] Jarvis, J. (1985) "Approximating the Equilibrium Behavior of Multi-server Loss Systems," *Management Science*, 32, 235-239.

[638] Jaumard, B., H. Ribault and T. Herrmann (1994) "An On-line Cone Intersection Algorithm for Global Optimization of Mutlivariate Lipschitz Functions," *Working Paper*, 94-27, Institute of Informatics, University of Fribourg, Switzerland.

[639] Jennings, A. and R. Sholar (1984) "Hazardous Waste Disposal Network Analysis," *Journal of Environmental Engineering ASCE*, 110, 325-342.

[640] Jezequel, A. (1985) *Probabilistic Vehicle Routing Problems*, Master's Thesis, Operations Research Center, MIT.

[641] Jin, H. (1993) "Hazardous Materials Routing: a Probabilistic Perspective," Ph.D. Dissertation, Department of Industrial Engineering, State University of New York At Buffalo, Buffalo, NY.

[642] Johnson, D. (1993) *Private Communication*.

[643] Johnson, N.L. and S. Kotz (1972) *Distributions in Statistics: Continuous Multivariate Distributions*, John Wiley & Sons, NY.

[644] Jordan, C. (1869) "Sur Les Assamblages Des Lignes" *Zeitschrift Für Die Reine Und Angewandte Mathematik*, 70, 185-190.

[645] Jorjani, S., C.H. Scott and T.R. Jefferson (1993) "Quality Locations for The Constrained Minimax Location Model," *International Journal of Systems Science*, 24, 1009-1016.

[646] Juel, H. (1981) "Bounds in the Generalized Weber Problem Under Conditions of Uncertainty," *Operations Research*, 29, 1219-1227.

[647] Juel, H. (1983) "Coincident Optima for the Two-facility Weber Problems," *Transportation Science*, 17, 110-113.

[648] Juel, H. and R.F. Love (1976) "An Efficient Computational Procedure for Solving the Multifacility Rectilinear Facility Location Problem," *Operational Research Quarterly*, 27, 697-703.

[649] Juel, H. and R.F. Love (1980) "Sufficient Conditions for Optimal Facility Locations to Coincide," *Transportation Science*, 14, 125-129.

[650] Juel, H. and R.F. Love (1981) "Fixed Point Optimality Criteria for the Weber Problem with Arbitrary Norms," *Journal of the Operational Research Society*, 32, 891-897.

[651] Juel, H. and R.F. Love (1981) "On the Dual of the Linearly Constrained Multifacility Location Problem with Arbitrary Norms," *Transportation Science*, 15, 329-337.

[652] Juel, H. and R.F. Love (1985) "The Facility Location Problem for Hyperrectilinear Distances," *IIE Transactions*, 17, 94-98.

[653] Juel, H. and R.F. Love (1992) "Dual of a Generalized Minimax Location Problem," Locational Decisions, *Annals of Operations Research*, Z. Drezner, Editor, 40, 261-264.

[654] Jung, K.H. and T.M. Cavalier (1988) "Locating a Traveling M/G/1 Server Using Euclidean Distances to a Finite Set of Demand Points," W.P., Pennsylvania State University, University Park, PA.

[655] Kaiser, M.J. and T.L. Morin (1993) "Algorithms for Computing Centroids," *Computers and Operations Research*, 20, 151-165.

[656] Kaiser, M.J. and T.L. Morin (1993) "Center Points, Equilibrium Positions, and the Obnoxious Location Probem," *Journal of Regional Science*, 33, 237-249.

[657] Kalelkar, A.S. and R.E. Brooks (1978) "Use of Multidimensional Utility Functions in Hazardous Shipment Decisions," *Accident Analysis Prevention*, 10, 251-265.

[658] Kallenberg, L.C.M. (1983) *Linear Programming and Finite Markovian Control Problems*, Math. Centre Tracks 148, Holland.

[659] Kariv, O. and S.L. Hakimi (1979) "An Algorithmic Approach to Network Location Problems, Part I: the P-Centers," *SIAM Journal of Applied Mathematics*, 37, 513-538.

[660] Kariv, O. and S.L. Hakimi (1979) "An Algorithmic Approach to Network Location Problems, Part II: the P-median," *SIAM Journal of Applied Mathematics*, 37, 539-560.

[661] Karkazis, J. (1988) "The General Unweighted Problem of Locating Obnoxious Facilities on the Plane," *Belgian Journal of Operations Research Statistics and Computer Science*, 28, 43-49.

[662] Karkazis, J. (1989) "Facilities Location in a Competitive Environment: a Promethee Based Multiple Criteria Analysis," *European Journal of Operational Research*, 42, 294-304.

[663] Karkazis, J. (1990) "The Discrete N-wind Approach for Solving the Problem of Locating Facilities Causing Airborne Pollution," *Proceedings of the Fifth Meeting of the EURO Working Group on Locational Analysis*, F. Orban and J.P. Rasson, Editors, FUNDP, Namur, Belgium, 75-85.

[664] Karkazis, J. (1991) "The Problem of Locating Facilities Causing Airbone Pollution," *OR Spektrum*, 13, 159-166.

[665] Karkazis, J. (1992) "Routing of Protected Materials/commodities on the Plane: Determining Routes in an Unfriendly Environment," *Proceedings of the VI Meeting of the EURO Working Group on Locational Analysis*, J.A. Moreno-Perez, Editor, Universidad De La Laguna, Tenerife, Spain, 105-112.

[666] Karkazis, J. (1994) "Location of a Facility in a Competitive Environment: a Planar-based Approach," *Studies in Locational Analysis*, 7, 77-90.

[667] Karkazis J., T.B. Boffey and D. Andrioti (1994) "Determining Location and Stack Height of Polluting Units in an Industrial Area," *Studies in Regional & Urban Planning*, 3, 185-198. Also submitted to *Journal of the Operational Research Society*.

[668] Karkazis, J., B. Boffey and N. Malevris (1992) "Locations of Facilities Producing Airborne Pollution," *Journal of Operational Research Society*, 43, 313-320.

[669] Karkazis J. and T.B. Boffey (1993) *Studies in Locational Analysis: ISOLDE VI Proceedings*, Editors, Issue 4, University of the Agean.

[670] Karkazis, J. and T.B. Boffey (1993) *Studies in Locational Analysis: ISOLDE VI Survey Papers*, Editors, Issue 5, Planning and Regional Development Studies Association.

[671] Karkazis, J. and B. Boffey (1994) "Modelling Pollution Spread," *Studies in Locational Analysis*, 7, 91-104.

[672] Karkazis, J. and B. Boffey (1994) "Optimal Location of Routes for Vehicles Transporting Hazardous Materials," *European Journal of Operational Research*, to appear.

[673] Karkazis, J., B. Boffey and N. Malevris (1993) "Location of Facilities Producing Airborne Pollution," *Journal of the Operational Research Society*, 43, 313-320.

[674] Karkazis, J. and P. Karagiorgis (1986) "A Method to Locate the Maximum Circle(s) Inscribed in a Polygon," *Belgian Journal of Operations Research Statistics and Computer Science*, 26, 3-36.

[675] Karkazis, J. and P. Karagiorgis (1987) "The General Problem of Locating Obnoxious Facilities in the Plane," *Proceedings of the 11th IFORS Conference*, Buenos Aires, Brazil.

[676] Karkazis, J. and C. Papadimitriou (1992) "A Branch-and-bound Algorithm for the Location of Facilities Causing Atmospheric Pollution," *European Journal of Operational Research*, 58, 363-373.

[677] Karkazis, J. and C. Papadimitrou (1992) "Optimal Location of Facilities Causing Atmospheric Pollution," *European Journal of Operational Research*, 58, 363-373.

[678] Katz, I.N. and L. Cooper (1980) "Optimal Location of a Sphere," *Computers and Mathematics with Applications*, 6, 175-196.

[679] Katz, I.N. and L. Cooper (1981) "Facility Location in the Presence of Forbidden Regions, 1: Formulation and the Case of Euclidean Distance

with One Forbidden Circle," *European Journal of Operational Research*, 6, 166-173.

[680] Kaufman, L., Van Den Eede and P. Hansen (1977) "A Plant and Warehouse Location Problem," *Operational Research Quarterly*, 28, 547-554.

[681] Kaufman, L. and F. Plastria (1988) "The Weber Problem with Supply Surplus," *Belgian Journal of Operations Research Statistics and Computer Science*, 28, 15-31.

[682] Kaufmann, P. and V.K. Rangan (1990) "A Model for Managing System Conflict During Franchise Expansion," *Journal of Retailing*, 66, 155-73.

[683] Keeney, R.L. (1980) "Equity and Public Risk," *Operations Research*, 28, 527-534.

[684] Keeney, R.L. (1980) "Utility Functions for Equity and Public Risk," *Management Science*, 26, 345-352.

[685] Keeney, R.L. (1987) "An Analysis of the Portfolio of Sites to Characterize for Selecting a Nuclear Repository," *Risk Analysis*, 7, 195-218.

[686] Keeney, R.L. (1992) *Value-Focussed Thinking: a Path to Creative Decisionmaking*, Harvard University Press, Cambridge, MA.

[687] Keeney, R.L. (1994) "Using Values in Operations Research," *Operations Research*, 42, 793-813.

[688] Keeney, R.L. and K. Nair (1977) "Selecting Nuclear Power Plant Sites In the Pacific Northwest Using Decision Analysis," in Conflicting Objectives in Decisions, D.E. Bell, R.L. Keeney and H. Raiffa, Editors, John Wiley & Sons, New York., and Also in Siting Energy Facilities, R. L. Keeney, Academic Press, NY, 49-79.

[689] Keeney, R.L. and D. Von Winterfeldt (1994) "Managing Nuclear Waste From Power Plants," *Risk Analysis*, 14, 107-130.

[690] Keeney, R.L. and R.L. Winkler (1985) "Evaluating Decision Strategies for Equity of Public Risks," *Operations Research*, 33, 955-970.

[691] Keller, L.R. and R.K. Sarin (1988) "Equity in Social Risk: Some Empirical Observations," *Risk Analysis*, 8, 135-146.

[692] Kendrick, D.A. and A.J. Stoutjesdijk (1978) *the Planning of Industrial Investment Programs*, World Bank.

[693] Khumawala, B.M. (1972)" an Efficient Branch and Bound Algorithm for the Warehouse Location Problem," *Management Science*, 18, B, 718-731.

[694] Khumawala, B.M. (1973) "An Efficient Algorithm for the P-median Problem with Maximum-Distance Constraint," *Geographical Analysis*, 5, 309-321.

[695] Kincaid, R.K. (1992) "Good Solutions to Discrete Noxious Location Problems Via Metaheuristics," *Annals of Operations Research*, 40, 265-281.

[696] Kincaid, R.K. and R.T. Berger (1994) "The Maxminsum Problem on Trees," *Location Science*, 2, 1-9.

[697] Kirkpatrick, S., C.D. Gelatt and M.P. Vecchi (1983) "Optimization by Simulated Annealing," *Science*, 220, 671-680.

[698] Kirkwood, C.W. (1982) "A Case History of Nuclear Power Plant Site Selection," *Journal of the Operational Research Society*, 33, 353-363.

[699] Klein, R. (1989) "Concrete and Abstract Voronoi Diagrams," *Lecture Notes in Computer Science*, 400, Springer, Berlin.

[700] Klein, C.M. (1991) "A Model for the Transportation of Hazardous Waste," *Decision Sciences*, 22, 1091-1108.

[701] Kleindorfer, P.R. and H.C. Kunreuther (1994) "Siting of Hazardous Facilities," *Operations Research and the Public Sector*, S.M. Pollock, M.H. Rothkopf and A. Barnett, Editors, North-Holland.

[702] Klienrock, L. (1975) *Queuing System I: Theory*, John Wiley and Sons, NY, 5.

[703] Klincewicz, J.G. and H. Luss (1987) "A Dual-based Algorithm for Multiproduct Uncapacitated Facility Location," *Transportation Science*, 21, 198-206.

[704] Klincewicz, J.G., H. Luss and E. Rosenberg (1986) "Optimal and Heuristic Algorithms for Multiproduct Uncapacitated Facility Location" *European Journal of Operational Research*, 26, 251-258.

[705] Ko, M.T., R.C.T. Lee and J.S. Chang (1990) "Rectilinear M-center Problem," *Naval Research Logistics*, 37, 419-427.

[706] Kogut, B. (1985) "Designing Global Strategies: Comparative and Competitive Value-added Chains," *Sloan Management Review*, Summer, 15-28.

[707] Kogut, B. (1985) "Designing Global Strategies: Profiting from Operational Flexibility," *Sloan Management Review*, Fall, 27-38.

[708] Kohlberg, E. (1983) "Equilibrium Store Locations When Consumers Minimize Travel Time Plus Waiting Time," *Economic Letters*, 11, 211-216.

[709] Kolesar, P. (1980) "Testing for Vision Loss in Glaucoma Suspects," *Management Science*, 439-450.

[710] Kolstar, C.D. (1985) "A Review of the Literature on Bilevel Mathematical Programming," Report No. LA-10284-MS, Los Alamos National Laboratory, Los Alamos, New Mexico.

[711] Korneenko, N.M. and H. Martini (1993) "Hyperplane Approximation and Related Topics," *New Trends in Discrete and Computational Geometry*, J. Pach, Editor, Springer, Berlin, 135-161.

[712] Koshizuka, T. (1991) "Efficiencies of a Sequential Algorithm and an Intuitive Selection Method in a Planar Location Problem," *Operational Research '90, Proceedings of the Twelfth International Conference on Operational Research*, H.E. Bradley, Editor, Athens 1990, Pergamon, Oxford.

[713] Koshizuka, T. and O. Kurita (1991) "Approximate Formulas of Average Distances Associated with Regions and Their Applications to Location Problems," *Mathematical Programming*, 52, 99-123.

[714] Kotha, S. and D. Orne (1989) " Generic Manufacturing Strategies: a Conceptual Synthesis," *Strategic Management Journal*, 10, 211-231.

[715] Kramer, G.H. (1977) "A Dynamical Model of Political Equilibrium," *Journal of Economic Theory*, 16, 310-334.

[716] Krarup, J. and P.M. Pruzan (1980) "The Impact of Distance on Location Problems," *European Journal of Operational Research*, 4, 256-269.

[717] Krarup, J. and P.M. Pruzan (1983) "The Simple Plant Location Problem: Survey and Synthesis," *European Journal of Operational Research*, 12, 36-81.

[718] Kubota, K. (1991) "PADRE2, a FORTRAN Precompiler Yielding Error Estimates and Second Derivatives," *Proceedings of SIAM Workshop on "Automatic Differentiation of Algorithms – Theory and Application"*.

[719] Kuby, M.J. (1987) "Programming Models for Facility Dispersion: the P-Dispersion and Maxisum Dispersion Models," *Geographical Analysis*, 19, 315-329.

[720] Kucera, L., K. Mehlhorn, B. Preis and E. Schwarzenecker (1992) "Exact Algorithms for a Geometric Packing Problem," Working Paper, Max-Planck Institut Fur Informatik Im Stadtwald, D-6600 Saarbrucken, Germany.

[721] Kuehn, A.A. and M.J. Hamburger (1963) "A Heuristic Program for Locating Warehouses," *Management Science*, 9, 643-666.

[722] Kuenne, R.E. and R.M. Soland (1972) "Exact and Approximate Solutions to the Multisource Weber Problem," *Mathematical Programming*, 3, 193-209.

[723] Kuhn, H.W. (1967) "On a Pair of Dual Nonlinear Programs," *Methods of Nonlinear Programming*, J. Abadie, Editor, North Holland, Amsterdam, Netherlands.

[724] Kuhn, H.W. (1973) "A Note on Fermat's Problem," *Mathematical Programming*, 4, 98-107.

[725] Kuhn, H.W. and R.E. Kuenne (1962) "An Efficient Algorithm for the Numerical Solution of the Generalized Weber Problem in Spatial Economics," *Journal of Regional Science*, 4, 21-33.

[726] Kumar, S. (1993) *Location Decisions for Private Facilities in The Presence of Elastic Demand and Externalities*, Unpublished Ph.D. Dissertation, Department of Engineering-Economic Systems, Stanford University, Stanford, CA.

[727] Kumar, S. and S.S. Chiu (1994) "A Single Facility Location Model with an Embedded Demand Equilibrium," Working Paper, Department of Engineering-Economic Systems, Stanford University, Stanford, CA.

[728] Kumar, S. and S.S. Chiu (1994) "Optimal Competitive Positioning in a User-Choice Environment with Congestion Elastic Demand," Working Paper, Department of Engineering-Economic Systems, Stanford University, Stanford, CA.

[729] Kumar, S. and S.S. Chiu (1994) "Location-Allocation Decision in Franchising Non-Essential Services," Working Paper, Department of Engineering-Economic Systems, Stanford University, Stanford, CA.

[730] Kyparisis, J. (1986) "Uniqueness and Differentiability of Solutions of Parametric Nonlinear Complementarity Problems," *Mathematical Programming*, 36, 105-113.

[731] Labbé, M. and S.L. Hakimi (1991) "Market and Locational Equilibrium for Two Competitors," *Operations Research*, 39, 749-756.

[732] Labbé, M., D. Peeters and J.-F. Thisse (1994) "Location on Networks," *Handbook of Operations Research and Management Science: Networks*, M. Ball, T. Magnanti, C. Monma, and G. Nemhauser, Editors, North Holland, Amsterdam, Forthcoming.

[733] Lakshmanan, T.R. and W.A. Hansen (1965) "A Retail Market Potential Model," *Journal of American Institute of Planners*, 31, 134-43.

[734] Laporte, G. (1988) "Location-Routing Problems," *Vehicle Routing: Methods and Studies*, B.L. Golden and A.A. Assad, Editors, 163-198, North Holland, Amsterdam.

[735] Laporte, G., F. Louveaux and H. Mercure (1989) "Models and Exact Solutions for a Class of Stochastic Location-routing Problems," *European Journal of Operations Research*, 39, 71-78.

[736] Laporte, G., F.V. Louveaux and L. Van Hamm (1994) "Exact Solution to a Location Problem with Stochastic Demands," *Transportation Science*, 28, 95-103.

[737] Larson, R.C. (1974) "A Hypercube Queuing Model for Facility Location and Redistricting in Urban Emergency Services," *Computers and Operations Research*, 1, 67-95.

[738] Larson, R.C. (1975) "Approximating the Performance of Urban Emergency Service Systems," *Operations Research*, 845-868.

[739] Larson, R.C. (1978) "Markov Models of Assignment Sensor AVL Systems", *Transportation Science*, 12, 331-351.

[740] Larson, R.C. and A.R. Odoni (1981) *Urban Operations Research*, Prentice-Hall, Inc., Englewood Cliffs, NJ.

[741] Larson, R. and K. Stevenson (1972) "On Insensitivities in Urban Redistricting and Facility Location," *Operations Research*, 613-618.

[742] Lawrence, S.R. and M.J. Rosenblatt (1992) "Introducing International Issues Into Operations Management Curricula," *Production and Operations Management*, 1, 103-117.

[743] Lawson, C.L. (1965) "The Smallest Covering Cone Or Sphere," *SIAM Review*, 7, 415-417.

[744] Leamer, E.E. (1968) "Locational Equilibria," *Journal of Regional Science*, 8, 229-242.

[745] Lederer, P.J. (1986) "Duolopy Competition in Networks," *Annals of Operations Research*, 6, 99-109.

[746] Lederer, P.J. and J.-F. Thisse (1990) "Competitive Location on Networks Under Delivered Pricing," *Operations Research Letters*, 9, 147-153.

[747] Lee, D.T. (1986) "Geometric Location Problems and Their Complexity," *Proceedings of the 12th Conference on Mathematical Foundations of Computer Science*, Lecture Notes in Computer Science, Springer, Berlin, 233.

[748] Lee, D.T., T.H. Chen and C.D. Yang (1990) "Shortest Rectilinear Paths Among Weighted Obstacles," *Proceedings of the 6th Annual ACM Symposium on Computational Geometry (Berkeley, California)*, ACM, NY.

[749] Lee, D.T. and Y.F. Wu (1986) "Geometric Complexity of Some Location Problems," *Algorithmica*, 1, 193-211.

[750] Lee, H.L. and M.A. Cohen (1985) "Equilibrium Analysis of Disaggregate Facility Choice Systems Subject to Congestion-Elastic Demand," *Operations Research*, 33, 293-311.

[751] Lefebvre, O., C. Michelot and F. Plastria (1991) "Sufficient Conditions for Coincidence in Minisum Multifacility Location Problems with a General Metric," *Operations Research*, 39, 437-442.

[752] Lefebvre, O., C. Michelot and F. Plastria (1992) "Boolean Linear Programming Formulation of the Attraction Tree Detection Problem," *Proceedings of the VI Meeting of the EURO Working Group on Locational Analysis*, J.A. Moreno-Perez, Editor, Universidad De La Laguna, Tenerife, Spain, 123-134.

[753] Lemke, C.E. (1968) "On Complementary Pivot Theory," *Mathematics of the Decision Sciences*, G.B. Dantzig and A.F. Veinott, Editors.

[754] Leonardi, G. (1981) "A Unifying Framework for Public Facility Location Problems – Part 1: a Critical Overview and Some Unsolved Problems," *Environment and Planning A*, 13, 1001-28.

[755] Leonardi, G. and R. Tadei (1984) "Random Utility Demand Models and Service Location," *Regional Science and Urban Economics*, 14, 399-431.

[756] Leong, G.K., D.L. Snyder and P.T. Ward (1990) "Research in the Process and Content of Manufacturing Strategy," *Omega*, 18, 109-122.

[757] Lerner, A. and H. Singer (1937) "Some Notes on Duopoly and Spatial Competition," *Journal of Political Economy*, 39, 145-186.

[758] Levine, A. (1986) "A Patrol Problem," *Mathematical Magazine*, 59, 159-166.

[759] Levitt, T. (1983) "The Globalization of Markets," *Harvard Business Review*, 92-102.

[760] Lin, S. (1965) "Computer Solutions of the Traveling Salesman Problem," *Bell System Tech. J.*, 2245-2269.

[761] Lindner-Dutton, L., R. Batta and M.H. Karwan (1991) "Equitable Sequencing of a Given Set of Hazardous Materials Shipments," *Transportation Science*, 25, 124-137.

[762] List, G. (1993) "Siting Emergency Response Teams: Tradeoffs Among Response Time, Risk, Risk Equity and Cost," in Transportation Of Hazardous Materials, L. Moses and D. Lindstrom, Editors, Kluwer Academic Publishers, Boston, 117-134.

[763] List, G. and M. Abkowitz (1988) "Towards Improved Hazardous Materials Flow Data," *Journal of Hazardous Materials*, 17, 287-304.

[764] List, G. and M. Abkowitz (1986) "Estimates of Current Hazardous Material Flow Patterns," *Transportation Quarterly*, 40, 483-502.

[765] List, G., M. Abkowitz and E. Page (1986) "Information Sources for Flow Analyses of Hazardous Materials," *Transportation Research Record*, 1063, 15-21.

[766] List, G. and P.B. Mirchandani (1991) "An Integrated Network/Planar Multiobjective Model for Routing and Siting for Hazardous Materials and Wastes," *Transportation Science*, 25, 146-156.

[767] List, G. and M.A. Turnquist (1994) "Routing and Emergency Response Team Siting for High-level Radioactive Waste Shipments," Technical Report, Department of Civil Engineering, Rensselaer Polytechnic Institute.

[768] List, G.F., P.B. Mirchandani, M.A. Turnquist and K.G. Zografos (1991) "Modeling and Analysis for Hazardous Materials Transportation: Risk Analysis, Routing/scheduling and Facility Location," *Transportation Science*, 25, 100-114.

[769] Lo, H.K. and M.R. McCord (1991) "Value of Ocean Current Information for Strategic Routing," *European Journal of Operational Research*, 55, 124-135.

[770] Lösch, A. (1954) *the Economics of Location*, Yale University Press, New Haven.

[771] Louveaux, F.V. (1993) "Stochastic Location Analysis," *Location Science*, 1, 127-154.

[772] Louviere, J. (1984) "Using Discrete Choice Experiments and Multinomial Logit Choice Models to Forecast Trial in Competitive Retail Environments: a Fast Food Restaurant Illustration," *Journal of Retailing*, 60, 81-107.

[773] Louveaux, F.V., M. Labbe and J.-F. Thisse (1989) "Facility Location Analysis: Theory and Applications," *Annals of Operations Research*, 18, J.C. Baltzer AG, Editors, Scientific Publishing Company, Basel, Switzerland.

[774] Love, R.F. (1972) "A Computational Procedure for Optimally Locating a Facility with Respect to Several Rectangular Regions," *Journal of Regional Science*, 12, 233-242.

[775] Love, R.F. (1974) "The Dual of a Hyperbolic Approximation to the Generalized Constraint Multifacility Location Problem with ℓ_p Distances," *Management Science*, 21, 22-33.

[776] Love, R.F. (1976) "One Dimensional Facility Location-allocation Using Dynamic Programming," *Management Science*, 22, 614-617.

[777] Love, R.F. (1994) "BQ735 Logistics Programs," School of Business, McMaster University, Hamilton, Ontario, Canada.

[778] Love, R.F. and P.D. Dowling (1985) "Optimal Weighted ℓ_p Norm Parameters for Facilities Layout Distance Characterizations," *Management Science*, 31, 200-206.

[779] Love, R.F. and P.D. Dowling (1989) "A New Bounding Method for Single Facility Location Models," *Annals of Operations Research*, 18, 103-112.

[780] Love, R.F. and P.D. Dowling (1989) "A Generalized Bounding Method for Multifacility Location Models," *Operations Research*, 37, 653-657.

[781] Love, R.F. and S.A. Kraemer (1973) "A Dual Decomposition Method for Minimizing Transportation Costs in Multifacility Location Problems," *Transportation Science*, 7, 297-316.

[782] Love, R.F. and J.G. Morris (1972) "Modelling Inter-city Road Distances by Mathematical Functions," *Operational Research Quarterly*, 23, 61-71.

[783] Love, R.F. and J.G. Morris (1975) "A Computational Procedure for the Exact Solution of Location-allocation Problems with Rectangular Distances," *Naval Research Logistics Quarterly*, 22, 441-453.

[784] Love, R.F. and J.G. Morris (1979) "Mathematical Models of Road Travel Distances," *Management Science*, 25, 130-139.

[785] Love, R.F. and J.G. Morris (1988) "On Estimating Road Distances by Mathematical Functions," *European Journal of Operational Research*, 36, 251-253.

[786] Love, R.F., J.G. Morris and G.O. Wesolowsky (1988) *Facilities Location: Models and Methods*, North Holland, NY.

[787] Love, R.F., W.G. Truscott and J.H. Walker (1985) "Terminal Location Problem: a Case Study Supporting the Status Quo," *Journal of the Operational Research Society*, 36, 131-136.

[788] Love, R.F. and J.H. Walker (1994) "An Empirical Comparison of Block and Round Norms for Modelling Actual Distances," *Location Science*, 2, 21-43.

[789] Love, R.F., G.O. Wesolowsky and S.A. Kraemer (1973) "A Multifacility Minimax Location Method for Euclidean Distances," *International Journal of Production Research*, 11, 32-40.

[790] Luenberger, D.L. (1984) *Linear and Non-Linear Programming*, Addison-Wesley, Reading, Massachusetts.

[791] Mahajan, V., E. Miller and S. Sharma (1984) "An Empirical Comparison of Awareness Forecasting Models of New Product Introduction," *Marketing Science*, 3, 179-197.

[792] Mahoney, J. (1991) "A Location-Allocation Perspective on the Hazardous Waste Management Problem: an Application to the Electroplating Industry in New England," Ph.D. Dissertation, Department of Geography, Boston University.

[793] Maimon, O. (1986) "The Variance Equity Measure in Locational Decision Theory," *Annals of Operations Research*, 6, 147-160.

[794] Maimon, O. (1988) "An Algorithm for the Lorenz Measure in Locational Decisions on Trees," *Journal of Algorithms*, 9, 583-596.

[795] Mairs, T.G., G.W. Wakefield, E.L. Johnson and K. Spielberg (1978) "On a Production and Distribution Problem," *Management Science*, 24, 1622-1630.

[796] Mamer, J. and R.D. McBride (1993) "Solving Multicommodity Flow Problems with a Primal Embedded Network Simplex Algorithm, Working Paper, Anderson Graduate School of Management, May 20.

[797] Mandell, M.B. (1991) "Modeling Effectiveness - Equity Tradeoffs in Public Service Delivery Systems," *Management Science*, 37, 467-482.

[798] Maranas, C.D. and C.A. Floudas (1994) "A Global Optimization Method for Weber's Problem with Attraction and Repulsion," W.W. Hager, D.W. Hearn, and P.M. Pardalos, Editors, *Large Scale Optimization: State of the Art*, Dordrecht, Kluwer, 259-293.

[799] Maranzana, F.E. (1964) "On the Location of Supply Points to Minimize Transport Costs," *Operational Research Quarterly*, 15, 261-270.

[800] Marglin, S. (1967) *Public Investment Criteria*, Cambridge, MA, MIT Press.

[801] Marianov, V. and C. ReVelle (1991) "The Standard Response Fire Protection Siting Problem," *INFOR*, 116-129.

[802] Marianov, V, and C. ReVelle (1992) "The Capacitated Standard Response Fire Protection Siting Problem: Deterministic and Probabilistic Models," *Annals of Operations Research*, 303-322.

[803] Marianov, V. and C. ReVelle (1992) "A Probabilistic Fire Protection Siting Model with Joint Vehicle Reliability Requirements," *Papers in Regional Science*, 217-241.

[804] Marianov, V. and C. ReVelle (1994) "The Queuing Probabilistic Location Set Covering Problem and Some Extensions," *Socio-Economic Planning Sciences*, 28, 167-178.

[805] Marsh, M. and D. Schilling (1991) "A Comparison of Equity Measures in Facility Siting Decisions," 244-249, R.D. Reid, Editor, *Proceedings of the First International Conference of the Decision Sciences Institute*, Brussels, Belgium.

[806] Marsh, M. and D. Schilling (1994) "Equity Measurement in Facility Location Analysis: a Review and Framework," *European Journal of Operational Research*, 74, 1-17.

[807] Martini, H. (1994) "Minsum k-Flats of Finite Point Sets in R^d," *Studies in Locational Analysis*, 7, 123-129.

[808] Martins, E.Q.V. (1984) "On Multicriteria Shortest Paths," *European Journal of Operational Research*, 16, 236-245.

[809] Marucheck, A.S. and A.A. Aly (1981) "An Efficient Algorithm for the Location-allocation Problem with Rectangular Regions," *Naval Research Logistics Quarterly*, 28, 309-323.

[810] Masuyama, S., T. Ibaraki and T. Hasegawa (1981) "The Computational Complexity of the M-center Problems on the Plane," *the Transactions of the IECE of Japan*, E64, 57-64.

[811] McAllister, D.M. (1976) "Equity and Efficiency in Public Facility Location," *Geographical Analysis*, 8, 47-63.

[812] McCord, M.R. and A.Y. Leu (1995) "Sensitivity of Hazardous Material Routes to Limited Preference Specification," *INFOR*, to appear.

[813] McFadden, D. (1974) "Conditional Logit Analysis of Quantitative Choice Behavior," *Frontiers in Economics*, P. Zarembkar, Editor, New York: Academic Press.

[814] McKelvey, R.D. (1986) "Covering, Dominance, and Institution Free Properties of Social Choice," *Journal of Political Science*, 30, 283-314.

[815] McKelvey, R.D. and R.E. Wendell (1976) "Voting Equilibria in Multidimensional Choice Spaces," *Mathematics of Operations Research*, 1, 144-158.

[816] Megiddo, N. (1978) "On the Parametric Nonlinear Complementarity Problem," *Mathematical Programming Study*, 7, 142-150.

[817] Megiddo, N. (1983) "Linear Time Algorithms for Linear Programming in R^3 and Related Problems," *SIAM Journal on Computing*, 12, 759-776.

[818] Megiddo, N. (1983) "The Weighted Euclidean 1-center Problem," *Mathematics of Operations Research*, 8, 498-504.

[819] Megiddo, N. and K.J. Supowit (1984) "On the Complexity of Some Common Geometric Location Problems," *SIAM Journal on Computing*, 13, 182-196.

[820] Megiddo, N. and A. Tamir (1982) "On the Complexity of Locating Linear Facilities in the Plane," *Operations Research Letters*, 1, 194-197.

[821] Megiddo, N., E. Zemel and S.L. Hakimi (1983) "The Maximum Coverage Location Problems," *SIAM Journal of Algebraic and Discrete Methods*, 4, 253-261.

[822] Mehrez, A., Z. Sinuany-Stern and A. Stulman (1985) "A Single Facility Location Problem with a Weighted Maximin-minimax Rectilinear Distance," *Computers and Operations Research*, 12, 51-60.

[823] Mehrez, A., Z. Sinuany-Stern and A. Stulman (1986) "An Enhancement of the Drezner-Wesolowsky Algorithm for Single-facility Location with Maximin of Rectilinear Distance," *Journal of the Operational Research Society*, 37, 971-977.

[824] Mehrez, A. and A. Stulman (1982) "The Maximal Covering Location Problem with Facility Placement on the Entire Plane," *Journal of Regional Science*, 22, 361-365.

[825] Mehrez, A. and A. Stulman (1984) "The Facility Location Problem When the Underlying Distribution Is Either Dominated Or Dominating," *Zeitschrift Für Operations Research*, 28, B157-B161.

[826] Mehrez, A. and A. Stulman (1984) "An Extended Continuous Maximal Covering Location Problem with Facility Placement," *Computers and Operations Research*, 11, 19-23.

[827] Melachrinoudis, E. (1985) "Determining an Optimal Location for an Undesirable Facility in a Workroom Environment," *Applications of Mathematical Models*, 9, 365-369.

[828] Melachrinoudis, E. (1988) "An Efficient Computational Procedure for the Rectilinear Maximin Location Problem," *Transportation Science*, 22, 217-223.

[829] Melachrinoudis, E. and T.P. Cullinane (1985) "Locating an Undesirable Facility Within a Geographical Region Using the Maximin Criterion," *Journal of Regional Science*, 25, 115-127.

[830] Melachrinoudis, E. and T.P. Cullinane (1986) "Locating an Obnoxious Facility Within a Polygonal Region," *Annals of Operations Research*, 6, 137-145.

[831] Melachrinoudis, E. and T.P. Cullinane (1986) "Locating an Undesirable Facility with a Minimax Criterion," *European Journal of Operational Research*, 24, 239-246.

[832] Melzak, Z.A. (1961) "On the Problem of Steiner," *Canadian Mathematical Bulletin*, 4, 143-148.

[833] Mercer, A. and G.K. Rand (1977) "A Structure for Distribution Studies," *European Journal of Operational Research*, 1, 161-168.

[834] Merkhofer, M.W. and R.L. Keeney (1987) "A Multiattribute Utility Analysis of Alternative Sites for the Disposal of Nuclear Waste," *Risk Analysis*, 7, 173-194.

[835] Miaou, S.P. and S.M. Chin (1991) " Computing K-Shortest Path for Spent Nuclear Fuel Highway Transportation," *European Journal Of Operational Research*, 53, 64-80.

[836] Michelot, C. (1986) "Un Algorithme Pour Résoudre Une Famille De Problemes De Localisation Multisources (in French)," *Modelisation Mathematique Et Analyse Numerique*, 20, 341-353.

[837] Michelot, C. (1987) "Localization in Multifacility Location Theory," *European Journal of Operational Research*, 31, 177-184.

[838] Michelot, C. (1987) "Properties of a Class of Location-allocation Problems," *Cahiers Du Centre D'Etudes En Recherche Opérationnelle De L'ULB*, 29, 105-114.

[839] Michelot, C. (1993) "The Mathematics of Continuous Location," *Isolde VI Survey Papers, Studies in Locational Analysis*, 5, 59-83.

[840] Michelot, C. and O. Lefebvre (1987) "A Primal-dual Algorithm for the Fermat-Weber Problem Involving Mixed Gauges," *Mathematical Programming*, 39, 319-335.

[841] Michelot, C. and F. Plastria (1991) "Attraction Tree Detection Is NP-complete," Rapport Du CERMSEM 9107, Université De Paris I Sorbonne-Panthéon, Paris, France.

[842] Microanalytics (1994) TRUCKSTOPS2 Tutorial Documentation, 2300 Clarenden Blvd., Arlington, Virginia, and 1986 Queen St. E., Toronto, Ontario.

[843] Miehle, W. (1958) "Link Length Minimization in Networks," *Operations Research*, 6, 232-243.

[844] Miller, J.G. and A.V. Roth (1994) "A Taxonomy of Manufacturing Strategies," *Management Science*, 40, 285-304.

[845] Miller, T., T. Friesz and R. Tobin (1992) "Heuristic Algorithms for Delivered Price Spatially Competitive Network Facility Location Problems," *Annals of Operations Research*, 34, 177-202.

[846] Miller, T., R. Tobin and T. Friez (1991) "Stackelberg Games on a Network with Cournot-nash Oligopolistic Competitors," *Journal of Regional Science*, 31, 435-454.

[847] Min, H. (1987) "A Multiobjective Retail Service Location Model for Fastfood Restaurants," *Omega*, 15, 429-41.

[848] Min, H. (1992) "Dynamic Location of Multinational Firms: a Chanceconstrained Goal Programming Approach," Technical Report, Northeastern University.

[849] Minieka, E. (1983) "Anticenters and Antimedians of a Network," *Networks*, 13, 359-364.

[850] Minieka, E. (1970) "The M-Center Problem," *SIAM Review*, 12, 138-139.

[851] Minieka, E. (1977) "The Centers and Medians of a Graph," *Operationa Research*, 25, 641-650.

[852] Minieka, E. (1980) "Conditional Centers and Medians of a Graph," *Networks*, 10, 265-272.

[853] Mirchandani, P.B. and R.L. Francis (1990) *Discrete Location Theory*, Editors, John Wiley and Sons, New York.

[854] Mirchandani, P.B. and A. Oudjit (1980) "Localizing 2-Medians on Probabilistic and Deterministic Tree Networks," *Networks*, 10, 329-350.

[855] Mirchandani, P.B. and A. Oujdit (1982) "Probabilistic Demand and Costs in Facility Location Problems," *Environment and Planning A*, 14, 917-32.

[856] Mirchandani, P.B., R. Rebello and A. Agnetis (1995) "The Inspection Station Location Problem in Hazardous Material Transportation: Some Heuristics and Bounds," *INFOR*, to appear.

[857] Misterek, S., R. Schroeder and K. Bates. (1992) "The Nature of the Link Between Manufacturing Strategy and Organizational Culture," C.A. Voss, Editor, *Manufacturing Strategy - Process and Content*, Chapman & Hall, 331-352.

[858] Mitchell, J.S.B. and C.H. Papadimitriou (1991) "The Weighted Region Problem: Finding Shortest Paths Through a Weighted Planar Subdivision," *Journal of the ACM*, 38, 18-73.

[859] Mittal, A.K. and V. Palsule (1984) "Facilities Location with Ring Radial Distances," *IIE Transactions*, 16, 59-64.

[860] Mladineo, R. (1986) "An Algorithm for Finding the Global Maximum of a Multimodal, Multivariate Function," *Mathematical Programming* 34, 188-200.

[861] Mohring, H. and H.F. Jr. Williamson (1969) "Scale and 'industrial Reorganisation' Economics of Transport Improvements," *Journal of Transport Economics and Policy*, 3, 251-271.

[862] Moon, I.D. and S.S. Chaudhry (1984) "An Analysis of Network Location Problems with Distance Constraints," *Management Science*, 30, 290-307.

[863] Moon, S. (1989) "Application of Generalized Benders Decomposition to a Nonlinear Distribution System Design Problem," *Naval Research Logistics*, 36, 283-295.

[864] Moore, G. and C. ReVelle (1982) "The Hierarchical Service Location Problem," *Management Science*, 775.

[865] Moore, L. (1988) "Stated Preference Analysis and New Store Location," *Store Choice, Store Location and Market Analysis*, N. Wrigley, Editor, NY, Routledge.

[866] Morell, D. (1984) "Sizing and the Politics of Equity," *Hazardous Waste*, 1, 555-571.

[867] Morrill, R.L. and J. Symons (1977) "Efficiency and Inequality Aspects of Optimum Location," *Geographical Analysis*, 9, 215-225.

[868] Morris, J.G. (1973) "A Linear Programming Approach to the Solution of Constrained Multifacility Minimax Location Where Distances Are Rectangular," *Operational Research Quarterly*, 24, 419-435.

[869] Morris, J.G. (1978) "On the Extent to Which Certain Fixed-charge Depot Location Problems Can Be Solved by LP," *Journal of the Operational Research Society*, 29, 71-76.

[870] Morris, J.G. (1981) "Convergence of the Weiszfeld Algorithm for Weber Problems Using a Generalized "distance" Function," *Operations Research*, 29, 37-48.

[871] Morris, J.G. and J.P. Norback (1983) "Linear Facility Location - Solving Extensions of the Basic Problem," *European Journal of Operational Research*, 12, 90-94.

[872] Moses, L.N. (1958) "Location and the Theory of Production," *Quarterly Journal of Economics*, 72, 259-272.

[873] Mosler, K.C. (1987) *Continuous Location of Transportation Networks*, Springer Verlag, Berlin.

[874] Muller, J.C. (1979) "La Cartographie D'une Métrique Non Euclidienne: Les Distances-temps," *L'Espace Géographique*, 3, 215-227.

[875] Muller, J.C. (1982) "Non Euclidean Geographic Spaces: Mapping Functional Distances," *Geographical Analysis*, 14, 189-203.

[876] Mulligan, G. (1983) "Consumer Demand with Multipurpose Shopping Behavior," *Geographical Analysis*, 15, 76-80.

[877] Mulligan, G.F. (1991) "Equality Measures and Facility Location," *Papers in Regional Science: the Journal of the RSAI*, 70, 345-365.

[878] Mulligan, G.F. and T.J. Fik (1994) "Price and Location Conjectures in Medium- and Long-run Spatial Competition Models," *Journal of Regional Science*, 34, 179-198.

[879] Mumphrey, A.J. Jr., J.E. Seley and J. Wolpert (1971) "A Decision Model for Locating Controversial Facilities," *Journal of the American Institute of Planners*, 37, 397-402.

[880]] Munoz, J. (1988) "Competitive Location of Two New Facilities with Rectilinear Distances," B. Pelegrin, Editor, *Proceedings of IIIrd Meeting of the Euro Working Group on Locational Analysis*, Sevilla, Spain.

[881] Murtagh, B.A. and S.R. Niwattisyawong (1982) "An Efficient Method for the Multi-depot Location Allocation Problem," *Journal of the Operational Research Society*, 33, 629-634.

[882] Nagurney, A. (1993) *Network Economics: a Variational Inequality Approach*, Kluwer Academic Publishers, Boston.

[883] Nair, K.P.K. and P. Chandrasekaran (1971) "Optimal Location of a Single Service Center of Certain Types," *Naval Research Logistics Quarterly*, 18, 503-510.

[884] Nakanishi, M. and L.G. Cooper (1974) "Parameter Estimate for Multiplicative Interactive Choice Model: Least Squares Approach," *Journal of Marketing Research*, 11, 303-11.

[885] Neebe, A.W. and B.M. Khumawala (1981) "An Improved Algorithm for the Multi-commodity Location Problem," *Journal of the Operational Research Society*, 32, 143-169.

[886] Neebe, A.W. and M.R. Rao (1983) "An Algorithm for the Fixed Charge Assignment of Users to Sources Problem," *Journal of the Operational Research Society*, 34, 1107-1113.

[887] Nemhauser, G.L. and L.A. Wolsey (1988) *Integer and Cominatorial Optimization*, John Wiley, NY.

[888] Nemhauser, G.L., L.A. Wolsey and M.L. Fisher (1978) "An Analysis of the Approximations for Maximizing Submodular Set Functions," *Mathematical Programming*, 14, 265-294.

[889] Newell, G.F. (1973) "Scheduling, Location, Transportation and Continuum Mechanics; Some Simple Approximations to Optimization Problems," *SIAM Journal of Applied Mathematics*, 25, 346-360.

[890] Nguyen, V.H. and J.J. Strodiot (1992) "Computing a Global Optimal Solution to a Design Centering Problem," *Mathematical Programming*, 53, 111-123.

[891] Nijkamp, P. and J. Paelink (1973) "A Solution Method for Neo-classical Location Problems," *Regional and Urban Economics*, 3, 383-410.

[892] Odoni, A. (1987) "Stochastic Facility Location Problems," G. Andreatta, F. Mason and P. Serafini, Editors, *Stochastics in Combinatorial Optimization*, World Scientific, Singapore.

[893] Odoni, A.R. and G. Sadiq (1982) "Two Planar Facility Location Problems with High-speed Corridors and Continuous Demand," WP, Massachusetts Institute of Technology.

[894] Ohsawa, Y. and A. Suzuki (1987) "Location-allocation Problem of Multiperson Facility," *Journal of the Operations Research Society of Japan*, 30, 368-395.

[895] Ohya, T., M. Iri and K. Murota (1984) "Improvements of the Incremental Method for the Voronoi Diagram with Computational Comparisons of Various Algorithms," *Journal of the Operations Research Society of Japan*, 27, 306-336.

[896] Okabe, A. and M. Aoyagi (1991) "Existence of Equilibrium Configurations of Competitive Firms on an Infinite Two-dimensional Space," *Journal of Urban Economics*, 29, 349-370.

[897] Okabe, A., B. Boots and K. Sugihara (1992) *Spatial Tesselations. Concepts and Applications of Voronoi Diagrams*, John Wiley & Sons, Chichester.

[898] Okabe, A. and A. Suzuki (1987) "Stability of Spatial Competition for a Large Number of Firms on a Bounded Two-dimensional Space," *Environment and Planning A*, 16, 107-114.

[899] Okabe, A. and A. Suzuki (1992) *Mathematical Methods for Locational Optimization Problems*, Asakura Shoten, Tokyo.

[900] O'Kelly, M. (1981) "A Model of the Demand for Retail Facilities Incorporating Multistop, Multipurpose Trips," *Geographical Analysis*, 13, 134-48.

[901] O'Kelly, M.E. (1986) "The Location of Interacting Hub Facilities," *Transportation Science*, 20, 92-106.

[902] O'Kelly, M. (1987) "Spatial Interaction Based Location Allocation Models," in *Spatial Analysis and Location Allocation Models*, A. Ghosh and G. Rushton, Editors, New York, Van Nostrand Reinhold.

[903] O'Kelly, M.E. (1992) "A Clustering Approach to the Planar Hub Location Problem," *Annals of Operations Research*, 40, 339-353.

[904] Osleeb, J.P. and S.J. Ratick (1986) "Locational Decisions: Methodology and Applications," *Annals of Operations Research*, J.C. Baltzer AG, Editor, Scientific Publishing Company, Basel, Switzerland, 6.

[905] Ostresh, L.M. Jr. (1973) "Twain - Exact Solution to the Two Source Location Allocation Problem," G. Rushton, M.F. Goodchild and L.M.

Ostresh, Editors, *Computer Programs for Location Allocation Problems*, Dept. Geography, University of Iowa, Iowa City, IA, 15-28.

[906] Ostresh, L.M. Jr. (1973) "MULTI - Exact Solutions to the m-center Location-allocation Problem," G. Rushton, M.F. Goodchild and L.M. Ostresh Jr., Editors, *Computer Programs for Location-allocation Problems*, Monograph Number 6, Department of Geography, University of Iowa, Iowa City, IA, 29-53.

[907] Ostresh, L.M. Jr. (1973) *an Investigation of the Multiple Location-allocation Problem*, Ph.D. Thesis, University of Iowa, Iowa City, IA.

[908] Ostresh, L.M. Jr. (1975) "An Efficient Algorithm for Solving the Two-center Location-allocation Problem," *Journal of Regional Science*, 15, 209-216.

[909] Ostresh, L.M. Jr. (1977) "The Multifacility Location Problem: Applications and Descent Theorems," *Journal of Regional Science*, 17, 409-419.

[910] Ostresh, L.M. Jr. (1978) "On the Convergence of a Class of Iterative Methods for Solving the Weber Location Problem," *Operations Research*, 26, 597-609.

[911] Overton, M.L. (1983) "A Quadratically Convergent Method for Minimizing a Sum of Euclidean Norms," *Mathematical Programming*, 27, 34-63.

[912] Owen, G. and L.S. Shapley (1989) "Optimal Location of Candidates in Ideological Space," *International Journal of Game Theory*, 18, 339-356.

[913] Papadimitriou, C.H. and K. Steiglitz (1982) *Combinatorial Algorithms and Complexity*, Prentice-Hall, NJ.

[914] Papageorgiou, G.J. (1980) "Social Values and Social Justice," *Economic Geography*, 56, 110-119.

[915] Papini, P.L. (1983) "Minimal and Closest Points Nonexpansive and Quasi-nonexpansive Retractions in Real Banach Spaces," P.M. Gruber and J.M. Wills, Editors, *Convexity and Its Applications*, Birkhäuser Verlag, Basel, Switzerland.

[916] Pardalos, P.M., J.H. Glick and J.B. Rosen (1987) "Global Minimization of Indefinite Quadratic Problems," *Computing*, 39, 281-291.

[917] Parker, B.R. and V. Srinivasan (1976) "A Consumer Preference Approach to the Planning of Rural Primary Health-Care Facilities," *Operations Research*, 24, 991-1029.

[918] Parlar M. (1994) "Single Facility Location with Region-dependent Distance Metrics," *International Journal of System Science*, 25, 513-525.

[919] Pasquil F., (1961) Atmospheric dispersion modelling. *Meteorology Magazine*, 90, 33-49.

[920] Patel, M.H. and A.J. Horowitz (1994) "Optimal Routing of Hazardous Materials Considering Risk of Spill," *Transportation Research B*.

[921] Pearn, J. and P.K. Ho (1990) "Public Facilities Location Under Elastic Demand," *Transportation Science*, 24, 117-136.

[922] Peeters, D. (1990) "The Profit-maximizing Weber Problem," F. Orban and J.P. Rasson, Editors, *Proceedings of the Fifth Meeting of the EURO Working Group on Locational Analysis*, Sept. 3-6, 1990, Namur, Belgium, 125-136.

[923] Peeters, D. and J.-F. Thisse (1990) "Spatial Price Policies and the Location of the Firm," M. Chatterji and R.E. Kuenne, Editors, *New Frontiers in Regional Science: Essays in Honour of Walter Isard, Volume 1*, MacMillan, London, 57-74.

[924] Peeters, D. and I. Thomas (1994) "The Effect of the Spatial Structure on the p-median Results," *Transportation Science*, Forthcoming.

[925] Pelegrin, B. (1990) "Generation of 2-approximate Partitions for Some p-center Problems," F. Orban and J.P. Rasson, Editors, *Proceedings Of the Fifth Meeting of the EURO Working Group on Locational Analysis*, Sept. 3-6, 1990, Namur, Belgium, 137-149.

[926] Pelegrin, B. (1991) "The p-center Problem in R^n with Weighted Tchebycheff Norms," *Belgian Journal of Operations Research Statistics and Computer Science*, 31, 49-62.

[927] Pelegrin, B. and L. Canovas (1994) "Primal and Dual Algorithms for the Minimum Covering ℓ_p-hypersphere Problem," *Studies in Locational Analysis*, 7 153-169.

[928] Pelegrin, B. and F.R. Fernandez (1988) "Determination of Efficient Points in Multiple Objective Location Problems," *Naval Research Logistics*, 35, 697-705.

[929] Pelegrin, B. and F.R. Fernandez (1989) "Determination of Efficient Solutions for Point-objective Locational Decision Problems," *Annals of Operations Research*, 18, 93-102.

[930] Pelegrin, B., C. Michelot and F. Plastria (1985) "On the Uniqueness of Optimal Solutions in Continuous Location Theory," *European Journal of Operational Research*, 20, 327-331.

[931] Perkins H.C. (1974) *Air Pollution*, McGraw-Hill, New York.

[932] Perko A. (1986) "Implementation of Algorithms for K Shortest Loopless Paths," *Networks*, 16, 149-160.

[933] Perreur, J. (1989) "L'evolution Des Representations De La Distance Et L'amenagement Du Territoire," *Revue D'Economie Regionale Et Urbaine*, 1, 5-32.

[934] Perreur, J. and J.-F. Thisse (1974) "Une Application De La Métrique Circum-radiale àl'étude Des Déplacements Urbains," *Revue Economique*, 25, 298-315.

[935] Perreur, J. and J.-F. Thisse (1974) "Central Metrics and Optimal Location," *Journal of Regional Science*, 14, 411-421.

[936] Pesamosca, G. (1991) "A Fixed Point Algorithm for Solving the Euclidean Multifacility Location Problem in a Tree," *Applied Mathematics and Computation*, 44, 105-120.

[937] Pesamosca, G. (1991) "On the Optimality Conditions for the Euclidean Multifacility Location Problem in a Tree," *Rendiconto Dell'Academia Delle Science Fisiche E Matematiche (Napoli, Italy)*, 58, 65-78.

[938] Pesamosca, G. (1991) "On the Analytic Solution of the 3-point Weber Problem," *Rendiconti Di Matematica (Roma, Italy) Serie VII*, 11, 39-45.

[939] Peterson, E.L. (1976) "Geometric Programming," *SIAM Review*, 18, 1-52.

[940] Phelps, R.R. (1957) "Convex Sets and Nearest Points," *Proceedings of the American Mathematical Society*, 8, 790-797.

[941] Phelps, R.R. (1958) "Convex Sets and Nearest Points II," *Proceedings of the American Mathematical Society*, 9, 867-873.

[942] Phlips, L. (1983) *the Economics of Price Discrimination*, Cambridge University Press, Cambridge, UK.

[943] Piessens, R., E. De Doncker-Kapenga, C.W. Uberhuber and D.K. Kahaner (1983) *QUADPACK a Subroutine Package for Automatic Integration*, Springer-Verlag, Berlin.

[944] Pijawka, D., S. Foote, and A. Soesilo (1985) "Risk Assessment of Transporting Hazardous Material: Route Analysis and Hazard Management," *Transportation Research Record*, 1020, 1-6.

[945] Pirkul, H., S. Narashimhan and P. De (1987) "Firm Expansion Through Franchising: a Model and Solution Procedure," *Decision Sciences*, 18, 631-45.

[946] Piyavskii, S.A. (1972) "An Algorithm for Finding the Absolute Extremum of a Function," *USSR Computational Mathematics and Mathematical Physics*, 12, 57-67, (Zh. Vȳchisl Mat. Fiz. 12, 4, 888-896).

[947] Plassard, F. (1977) *Les Autoroutes Et Le Développement Régional* Economica, Paris.

[948] Plastria, F. (1983) "Continuous Location Problems and Cutting Plane Algorithms," Ph.D. Thesis, Departement Wiskunde, Vrije Universiteit Brussel, Brussels, Belgium.

[949] Plastria, F. (1983) "Points Efficaces En Localisation Continue," *Cahiers Du Centre De Recherche Opérationelle De L'Université Libre De Bruxelles*, 25, 329-332.

[950] Plastria, F. (1983) "A Note on Fixed Point Optimality Criteria for the Location Problem with Arbitrary Norms," *Journal of the Operational Research Society*, 34, 164-165.

[951] Plastria, F. (1984) "Localization in Single Facility Location," *European Journal of Operational Research*, 18, 215-219.

[952] Plastria, F. (1987) "Solving General Continuous Single Facility Location Problems by Cutting Planes," *European Journal of Operational Research*, 29, 98-110.

[953] Plastria, F. (1989) "The Continuous P-centre Sum Problem," *Proceedings of the 4th Meeting, EURO Working Group on Locational Analysis*, Chios, Greece, J. Karkazis & T.B. Boffey, Editors, 29-38.

[954] Plastria, F. (1991) "Optimal Gridpositioning Or Single Facility Location on the Torus," *Revue D'Automatique D'Informatique Et De Recherche Operationelle - Oper. Res.*, 25, 19-29.

[955] Plastria, F. (1991) "The Effect of Majority in Fermat-Weber Problems with Attraction and Repulsion in Pseudometric Spaces," *Yugoslav Journal of Operations Research*, 1, 141-146.

[956] Plastria, F. (1992) "A Majority Theorem for Fermat-Weber Problems in Quasimetric Spaces with Applications to Semidirected Networks," *Proceedings of the VI Meeting of the EURO Working Group on Locational Analysis*, J.A. Moreno-Perez, Editor, Universidad De La Laguna, Tenerife, Spain, 153-165.

[957] Plastria, F. (1992) "GBSSS, the Generalized Big Square Small Square Method for Planar Single Facility Location," *European Journal of Operational Research*, 62, 163-174.

[958] Plastria, F. (1992) "When Facilities Coincide: Exact Optimality Conditions in Multifacility Location," *Journal of Mathematical Analysis and Applications*, 169, 476-498.

[959] Plastria, F. (1992) "On Destination Optimality in Asymetric Distance Fermat-Weber Problems," *Annals of Operations Research*, 40, 355-369.

[960] Plastria, F. (1993) "Continuous Location Anno 92: a Progress Report," *Isolde VI Survey Papers, Studies in Locational Analysis*, 5, 85-127.

[961] Plastria, F. (1994) "Fully Geometric Solutions to Some Planar Minimax Location Problems," *Studies in Locational Analysis*, 7, 171-183.

[962] Platzman, L. and J. Bartholdi (1989) "Spacefilling Curves and the Planar Traveling Salesman Problem," *Journal of the ACM*, 36, 719-737.

[963] Plott, C.R. (1967) "A Notion of Equilibrium and Its Possibility Under Majority Rule," *American Economic Review*, 57, 787-806.

[964] Pomper, C.L. (1976) *International Investment Planning: an Integrated Approach*, North-Holland.

[965] Ponsard, C. (1983) *a History of Spatial Economic Theory*, Springer-Verlag, Berlin.

[966] Pooley, J. (1994) "Integrated Production and Distribution Facility Planning At Ault Foods," *Interfaces*, 24, 113-121.

[967] Pooley, J. (1994) Private Communication.

[968] Porter, M.E. (1985) *Competitive Advantage*, Free Press-Macmillan, NY.

[969] Porter, M.E., Editor, (1986) *Competition in Global Industries*, Harvard Business School Press.

[970] Porter, M.E. (1990) *the Competitive Advantage of Nations*, the Macmillan Press.

[971] Powell, M.J.D. (1969) "A Method for Nonlinear Constraints in Minimization Problems," *Optimization*, R. Fletcher, Editor, Academic Press.

[972] Preparata, F.P. and M.I. Shamos (1985) *Computational Geometry - an Introduction*, Springer Verlag.

[973] Prescott, E.C. and Visscher, M. (1977) "Sequential Location Among Firms with Foresight," *Bell Journal of Economics*, 8, 378-393.

[974] Primrose, P.L. and R. Leonard (1991) "Selecting Technology for Investment in Flexible Manufacturing," *International Journal of Flexible Manufacturing Systems*, 4, 51-77.

[975] Prosperi, D.C. and H.J. Schuller (1976) "An Alternate Method to Identify Rules of Spatial Choice," *Geographical Perspectives*, 38, 33-38.

[976] Psaraftis, H.N., G.G. Tharakan and A. Ceder (1986) "Optimal Response to Oil Spills: the Strategic Decision Case," *Operations Research*, 34, 203-217.

[977] Puerto, J. and F.R. Fernandez (1994) "Multicriteria Decisions in Location," *Studies in Locational Analysis*, 7, 185-199.

[978] Purdy, G. (1993) "The Measurement of Risk from Transporting Dangerous Goods by Road and Rail, in Transportation of Dangerous Goods: Assessing the Risks," F. Saccomanno and K. Cassidy, Editors, Institute for Risk Research, University of Waterloo, 19-46.

[979] Rado, F. (1988) "The Euclidean Multifacility Location Problem," *Operations Research*, 36, 485-492.

[980] Rahman, M and M. Kuby (1995) "A Multiobjective Model for Locating Solid Waste Transfer Stations Using an Empirical Distance-based Opposition Function," *INFOR*.

[981] Rahman, M., A.E. Radwan, J. Upchurch and M. Kuby (1992) "Modeling Spatial Impacts of Siting a NIMBY Facility," *Transportation Research Record* 1359, 133-140.

[982] Rao, R.C. and D.P. Rutenberg (1979) "Preempting an Alert Rival: Strategic Timing of the First Plant by Analysis of Sophisticated Rivalry," *Bell Journal of Economics*, 10, 412-28.

[983] Ratick, S.J. and A.L. White (1988) "A Risk-Sharing Model for Locating Noxious Facilities," *Environment and Planning, B: Planning and Design*, 15, 165-179.

[984] Rautman, C.A., R.A. Reid and E.E. Ryder (1993) "Scheduling the Disposal of Nuclear Waste Material in a Geologic Repository Using the Transportation Model," *Operations Research*, 41, 459-469.

[985] Rawls, J. (1971) *a Theory of Justice*, Harvard University Press, Cambridge, Massachusetts.

[986] Read, J.A. (1994) "Quantifying Risk," Proceedings, 1st DND Symposium/Workshop on Risk Evaluation and Assessment, 13-19.

[987] Recker, W. and H. Schuler (1981) "Destination Choice and Processing Spatial Information: Some Empirical Tests with Alternative Constructs," *Economic Geography*, 57, 373-83.

[988] Recker, W.W. and L.P. Kostyniuk (1978) "Factors Influencing Destination Choice for Urban Grocery Trips," *Transportation*, 7, 19-33.

[989] Reif, J.H. and J.A. Storer (1985) "Shortest Paths in Euclidean Space with Polyhedral Obstacles," Techn.Rep.CS-85-121, Computer Science Dept., Brandeis University, Waltham, Massachusetts.

[990] Reilly, W.J. (1931)*the Law of Retail Gravitation*, Knickerbocker Press, NY.

[991] Repede, J. and J. Bernardo (1994) "Developing and Validating a Decision Support System for Locating Emergency Medical Vehicles in Louisville, Kentucky," *European Journal of Operational Research*, 75, 567-581.

[992] ReVelle, C. (1986) "The Maximum Capture Or Sphere of Influence Problem: Hotelling Revisited on a Network," *Journal of Regional Science*, 26, 343-357.

[993] ReVelle, C. (1989) "Review, Extension and Prediction in Emergency Service Siting Models," *European Journal of Operational Research*, 58-69.

[994] ReVelle, C.S. (1993) "Facility Siting and Integer-friendly Programming," *European Journal of Operational Research*, 65, 147-158.

[995] ReVelle, C., D. Bigman, D. Schilling, J. Cohen and R. Church (1977) "Facility Location: a Review of Context-free and EMS Models," *Health Services Research*, 129-146.

[996] Revelle, C., J. Cohon and D. Shobrys (1991) "Simultaneous Siting and Routing in the Disposal of Hazardous Wastes," *Transportation Science*, 25, 138-145.

[997] ReVelle, C. and K. Hogan (1988) "A Reliability-constrained Siting Model with Local Estimates of Busy Fractions," *Environment and Planning B: Planning and Design*, 143-152.

[998] ReVelle, C. and K. Hogan (1989) "The Maximum Reliability Location Problem and α-Reliable p-center Problem: Derivatives of the Probabilistic Location Set Covering Problem," *Annals of Operations Research*, 155-174.

[999] ReVelle, C. and K. Hogan (1989) "The Maximum Availability Location Problem," *Transportation Science*, 192-200.

[1000] ReVelle, C. and V. Marianov (1991) "A Probabilistic FLEET Model with Individual Vehicle Reliability Requirements," *European Journal of Operational Research*, 93-105.

[1001] Revelle, C., D. Marks and J.C. Liebman (1970) "An Analysis of Private and Public Sector Location Models," *Management Science*, 16, 692-99.

[1002] ReVelle, C. and D. Serra (1991) "The Maximum Capture Problem Including Relocation," *Information and Operations Research*, 29, 130-138.

[1003] ReVelle, C. and S. Snyder "Integrating Emergency Services: the Fire and Ambulance Siting Techniques," *Socio-Economic Planning Sciences*, Forthcoming.

[1004] ReVelle, C.S. and R.W. Swain (1970) "Central Facilities Location," *Geographical Analysis*, 2, 30-42.

[1005] ReVelle, C., C. Toregas and L. Falkson (1976) "Applications of the Location Set Covering Problem," *Geographical Analysis*, 8, 65.

[1006] Richardson, J. (1992) "Nash-Efficient Siting of Hazardous Facilities," *Socio-Economic Planning Sciences*, 26, 191-202.

[1007] Riley, R.C. (1973) *Industrial Geography*, Chatto and Windus, London.

[1008] ROADNET, (1994) Roadnet Technologies Inc., 10540 York Road, Huntvalley, Maryland, 21030.

[1009] Ro, H. and D. Tcha (1984) "A Branch-and-bound Algorithm for the Two-level Uncapacitated Facility Location Problem with Some Side Constraints," *European Journal of Operational Research*, 18, 349-358.

[1010] Robbins, J.C. (1981) "Routing Hazardous Materials Shipments," Ph.D. Dissertation, Indiana University, Bloomington, Indiana.

[1011] Robinson, E.P., L. Gao and S.D. Muggenborg (1993) "Designing an Integrated Distribution System At DowBrands, Inc.," *Interfaces*, 23, 107-117.

[1012] Rockafellar, R.T. (1970) *Convex Analysis*, Princeton University Press.

[1013] Rodriguez-Ramos, W.E. and R.L. Francis (1983) "Single Crane Location Optimization," *Journal of Construction Engineering and Management*, 109, 387-397.

[1014] Rogers, D.F., R.D. Plante, R.T. Wong and J.R. Evans (1991) "Aggregation and Disaggregation Techniques and Methodology in Optimization," *Operations Research*, 3, 94, 553-582.

[1015] Rosen, J.B. and G.L. Xue (1992) "Computational Comparison of Two Algorithms for the Euclidean Single Facility Location Problems," *ORSA Journal of Computing*, 3, 207-212.

[1016] Rosen, J.B. and G.L. Xue (1992) "On the Convergence of Miehle's Algorithm for the Euclidean Multifacility Location Problem," *Operations Research*, 40, 188-191.

[1017] Rosen, J.B. and G.L. Xue (1992) "A Globally Convergent Algorithm for the Euclidean Multifacility Location Problem," *Acta Mathematica Applicata Sinica - English Series - Yingyong Shuxue Xuebao*, 4, 357-366.

[1018] Rosen, J.B. and G.L. Xue (1993) "On the Convergence of a Hyperboloid Approximation Procedure for the Perturbed Euclidean Multifacility Location Problem," *Operations Research*, 41, 1164-1171.

[1019] Rosing, K.E. (1991) "Towards the Solution of the (Generalized) Multi-Weber Problem," *Environment and Planning*, B, 18, 347-360.

[1020] Rosing, K.E. (1992) "An Optimal Method for Solving the (generalized) Multi-Weber Problem," *European Journal of Operational Research*, 58, 414-426.

[1021] Rosing, K.E. (1992) "The Optimal Locations of Steam Generators in Large Heavy Oil Fields," *American Journal of Mathematical and Management Sciences*, 12, 19-42.

[1022] Rosing, K.E. and B. Harris (1992) "Algorithmic and Technical Improvements: Optimal Solutions to the (generalized) Multi-Weber Problem," *Papers in Regional Science: the Journal of the RSAI*, 71, 331-352.

[1023] Rosing, K.E. and J.J. Van Dijk (1989) "On the Correct Number of Regions in Regionalisation Structures," *Environment and Planning B*, 16, 469-481.

[1024] Rossi, F. and I. Gavioli (1987) "Aspects of Heuristics Methods in the Probabilistic Traveling Salesman Problem," G. Andreatta, F. Mason and P. Serafini, Editors, *Stochastics in Combinatorial Optimization*, World Scientific, Singapore.

[1025] Rote, G. and G. Woeginger (1989) "Geometric Clusterings," WP B-89-04, Fachbereich Mathematik, Serie B Informatik, Freie Universität Berlin, W. Germany.

[1026] Rothschild, R. (1976) "A Note on the Effect of Sequential Entry on Choice and Location," *Journal of Industrial Economics*, 24, 313-320.

[1027] Roy, B. and D. Bouyssou (1986) "Comparison of Two Decision-aid Models Applied to a Nuclear Power Plant Siting Example," *European Journal of Operational Research*, 25, 200-215.

[1028] Rubinstein, R.Y. and G. Samorodnitsky (1986) "Optimal Coverage of Convex Regions," *Journal of Optimization Theory and Applications*, 51, 321-343.

[1029] Rushton, G. (1969) "Analyzing Spatial Behavior by Revealed Space Preference," *Annals Association of American Geographers*, 59, 391-400.

[1030] Rushton, G. and J.C. Thill (1989) "The Effect of Distance Metric on the Degree of Spatial Competition Between Firms," *Environment and Planning A*, 21, 499-507.

[1031] Saccomanno, F.F. and B. Allen (1988) "Locating Emergency Response Capability for Dangerous Goods Incidents on a Road Network," *Transportation Research Record*, 1193, 1-9.

[1032] Saccomanno, F.F. and A. Chan (1985) "Economic Evaluation of Routing Strategies for Hazardous Road Shipments," *Transportation Research Record*, 1020, 12-18.

[1033] Saccomanno, F.F. and J.H. Shortreed (1993) "Hazardous Material Transport Risks: Societal and Individual Perspectives," *Journal of Transportation Engineering*, 119, 177-188.

[1034] Saccomanno, F.F., J.H. Shortreed, M.V. Aerde and J. Higgs (1990) "Comparison of Risk Measures for the Transport of Dangerous Commodities by Truck and Rail," *Transportation Research Record*, 1245, 1-13.

[1035] Sakashita, N. (1967) "Production Function, Demand Function and Location Theory of the Firm," *Papers of the Regional Science Association*, 20, 109-122.

[1036] Sarin, R.K. (1993) "Models of Equity in Public Risk," *Psychological Perspectives on Justice: Theory and Applications*, B. Mellers and J. Baron, Editors, Cambridge, University Press.

[1037] Sarin, R.K. (1985) "Measuring Equity in Public Risk," *Operations Research*, 33, 210-217.

[1038] Savas, E.S. (1978) "On Inequality in Providing Public Services," *Management Science*, 24, 800-808.

[1039] Scanlon, R.D. and E.J. Cantilli (1985) "Assessing the Risk and Safety in the Transportation of Hazardous Materials," *Transportation Research Record*, 1020, 6-11.

[1040] Scarf, H. (1973) *the Computation of Economic Equilibrium*, Yale University Press, New Haven, CT.

[1041] Schaefer, M.K. and A.P. Hurter (1974) "An Algorithm for the Solution of a Location Problem with Metric Constraints," *Naval Research Logistics Quarterly*, 21, 625-636.

[1042] Schärlig, A. (1973) *Où Construire L'usine? La Localisation Optimale D'une Activité Industrielle Dans La Pratique*, Paris, Dunod.

[1043] Scherer, F.M. (1967) "Research and Development Resource Allocation Under Rivalry," *the Quarterly Journal of Economics*, 81, 359-94.

[1044] Schilling, D. (1980) "Dynamic Location Modelling for Public Sector Facilities: a Multi-Criteria Approach," *Decision Sciences*, 11, 714-725.

[1045] Schilling, D.A. (1982) "Strategic Facility Planning: the Analysis of Options," *Decision Sciences*, 13, 1-14.

[1046] Schilling, D., D. Elzinga, J. Cohon, R. Church and C. ReVelle (1979) "The TEAM/FLEET Models for Simultaneous Facility and Equipment Siting," *Transportation Science*, 167.

[1047] Schilling, D.A., J. Vaidyanathan and R. Barkhi (1993) "A Review of Covering Problems in Facility Location," *Location Science*, 1, 25-55.

[1048] Schmalensee, R. and J.-F. Thisse (1988) "Perceptual Maps and the Optimal Location of New Products: an Integrative Essay," *International Journal of Research in Marketing*, 5, 225-249.

[1049] Schrage, L. (1975) "Implicit Representation of Variable Upper Bounds in Linear Programming," *Mathematical Programming Study*, 4, 118-132.

[1050] Schrage, L. (1991) *User's Manual for LINGO*, LINDO Systems Inc., Chicago, IL.

[1051] Schreiber, P. (1986) "Zur Geschichte Des Sogenannten Steiner-Weber-Problems (Towards the History of the So-called Steiner-Weber Problem) - in German," *Wissenschaftliches Zeitschrift Der Ernst-Moritz-Arndt-Universität, Greifswald, Math.-nat.wiss. Reihe*, 35, 53-58.

[1052] Schuler, H.J. (1981) "Grocery Shopping Choices - Individual Preferences Based on Store Attractiveness and Distance," *Environment and Behavior*, 13, 331-347.

[1053] Schuler, H.J. and D.C. Prosperi (1977) "A Conjoint Measurement Model of Consumer Spatial Behavior," *Regional Science Perspectives*, 7, 122-134.

[1054] Schwartz, S.I., R.A. McBride and R.L. Powell (1989) "Models for Aiding Hazardous Waste Facility Siting Decisions," *Journal of Environmental Systems*, 18, 97-122.

[1055] Scorer R.S. (1968) *Air Pollution*, Pergamon, Oxford.

[1056] Scotchmer, S. and J.-F. Thisse (1992) "Space and Competition - a Puzzle," *Annals of Regional Science*, 26, 269-286.

[1057] Scott, A.J. (1971) "Dynamic Location Allocation Systems: Some Basic Planning Strategies," *Environment and Planning A*, 3, 73-82.

[1058] Scott, C.H., B. Murtagh and E. Sirri (1985) "Solution of Constrained Minimax Location Problems Via Conjugate Duality," *New Zealand Operational Research*, 13, 61-67.

[1059] Sen, A. (1973) *on Economic Inequality*, Clarendon Press, Oxford, England.

[1060] Serra, D., V. Marianov and C. ReVelle (1992) "The Hierarchical Maximum Capture Problem," *European Journal of Operational Research*, 62, 3.

[1061] Serra, D. and C. ReVelle (1994) "The Maximum Capture Problem with Uncertainty," *Economics Working Paper 74*, Universitat Pompeu Fabra.

[1062] Serra, D. and C. ReVelle (1994) "Market Capture by Two Competitors: the Preemptive Capture Problem," *Journal of Regional Science*, 34, 549-561.

[1063] Shamos, M.I. and D. Hoey (1975) "Closest Point Problems," *Proceedings of the 16th Annual Symposium on Foundations of Computer Science*, 151-162.

[1064] Shapiro, J.F., V.M. Singhal and S.N. Wagner (1993) "Optimizing the Value Chain," *Interfaces*, 23, 102-117.

[1065] Sheppard, E.S. (1974) "A Conceptual Framework for Dynamic Ocation-Allocation Analysis," *Environment and Planning A*, 7, 143-57.

[1066] Sherali, H.D. and W.P. Adams (1990) "A Hierarchy of Relaxations Between the Continuous and Convex Hull Representations for Zero-one Programming Problems," *SIAM Journal on Discrete Mathematics*, 3, 411-430.

[1067] Sherali, H.D. and S.I. Kim (1992) "Variational Problems for Determining Optimal Paths of a Moving Facility," *Transportation Science*, 4, 330-345.

[1068] Sherali, H.D., S. Ramachandran and S. Kim (1994) "A Localization and Reformulation Discrete Programming Approach for the Rectilinear Distance Location-Allocation Problem," *Discrete Applied Mathematics*, 49, 357-378.

[1069] Sherali, H.D. and C.M. Shetty (1977) "The Rectilinear Distance Location-allocation Problem," *AIIE Transactions*, 9, 136-143.

[1070] Sherali, H.D. and C.H. Tuncbilek (1992) "A Squared Euclidean Distance Location-allocation Problem," *Naval Research Logictics*, 39, 447-469.

[1071] Sherali, H.D. and C.H. Tuncbilek (1992) "A Global Optimization Algorithm for Polynomial Programming Problems Using a Reformulation-linearization Technique," *Journal of Global Optimization*, 2, 101-112.

[1072] Sheth, J. and G. Eshghi, (1989) *Global Operations Perspectives*, Editors, South-Western.

[1073] Shetty, C.M. and H.D. Sherali (1980) "Rectilinear Distance Location-allocation Problem: a Simplex Based Algorithm," *Extremal Methods and System Analysis, Lecture Notes in Economics and Mathematical Systems*, Springer, NY, 174, 442-464.

[1074] Shier, D.R. (1977) "A Min-Max Theorem for P-Center Problems on a Tree," *Transportation Science*, 11, 243-252.

[1075] Shiode, S. and H. Ishii (1991) "A Single Facility Stochastic Location Problem Under A-distance," *Annals of Operations Research*, 31, 469-478.

[1076] Shobrys, D. (1981) "A Model for the Selection of Shipping Routes and Storage Locations for a Hazardous Substance," Ph.D. Dissertation, Johns Hopkins University, Baltimore.

[1077] Simchi-Levi, D. (1991) "The Capacitated Traveling Salesmen Location Problem," *Transportation Science*, 25, 2-18.

[1078] Simchi-Levi, D. (1992) "Hierarchical Planning for Probabilistic Distribution Systems in Euclidean Spaces," *Management Science*, 38, 198-211.

[1079] Simchi-Levi, D. and O. Berman (1987) "Heuristics and Bounds for the Travelling Salesman Location Problem on the Plane," *Operations Research Letters*, 6, 243-248.

[1080] Simchi-Levi, D. and O. Berman (1988) "A Heuristic Algorithm for the Traveling Salesman Location Problem on Networks," *Operations Research*, 36, 478-484.

[1081] Simpson, P. (1969) "On Defining Areas of Voter Choice: Professor Tullock on Stable Voting," *Quarterly Journal of Economics*, 83, 478-490.

[1082] Sivakumar, R.A. (1992) "Transportation of Hazardous Materials: a New Modeling Perspective," Ph.D. Dissertation, Department of Industrial Engineering, State University of New York At Buffalo, Buffalo, New York.

[1083] Sivakumar, R.A., R. Batta and M.H. Karwan (1995) "A Multiple Route Conditional Risk Model for Transporting Hazardous Materials," *INFOR*, to appear.

[1084] Sivakumar, R.A., R. Batta and M.H. Karwan (1993) "A Network-based Model for Transporting Extremely Hazardous Materials," *Operations Research Letters*, 13, 85-93.

[1085] Skinner, W. (1992) "Missing the Links in Manufacturing Strategy," C. Voss, Editor, *Manufacturing Strategy - Process and Content*, Chapman & Hall, 13-25.

[1086] Slater, P. (1975) "Maximin Facility Location," *Journal of Research National Bureau of Standards*, 79B, 107-115.

[1087] Slater, P.J. (1978) "Structure of the K-Centra in a Tree," *Proceedings of the 9th Southeast Conference on Combinatorics, Graph Theory and Computing*, 663-670.

[1088] Smallwood, R.D. (1965) "Minimal Detection Station Placement," *Operations Research*, 13, 632-646.

[1089] Smith, S.A. (1979) "Estimating Service Territory Size," *Management Science*, 25, 301-311.

[1090] Smithies, A. (1941) "Optimum Location in Spatial Competition," *Journal of Political Economy*, 49, 423-439.

[1091] Soland, R.M. (1971) "An Algorithm for Separable Nonconvex Programming Problems II: Nonconvex Constraints," *Management Science*, 17, 759-773.

[1092] Soland, R.M. (1974) "Optimal Facility Location with Concave Costs," *Operations Research*, 22, 373-382.

[1093] Späth, H. (1977) "Computational Experiences with the Exchange Method, Applied to Four Commonly Used Partitioning Cluster Analysis Criteria," *European Journal of Operational Research*, 1, 23-31.

[1094] Späth, H. (1978) "Explizite Lösung Des Dreidimensionales Minimax Standortproblems in Der City-block Distanz," *Zeitschrift Für Operations Research*, 22, 229-237.

[1095] Späth, H. (1981) "The Minisum Location Problem for the Jaccard Metric," *OR Spektrum*, 3, 91-94.

[1096] Spencer, A.H. (1980) "Cognition and Shopping Choice: a Multidimensional Scaling Approach," *Environment and Planning A*, 12, 1235-1251.

[1097] Stassen, R.E. and R.A. Mittelstaedt (1994) "Territory Encroachment in Maturing Franchise Systems," *the Journal of Marketing Channels*. Forthcoming.

[1098] Steele, J. (1981) "Complete Convergence of Short Paths and Karp's Algorithm for the TSP," *Mathematics of Operations Research*, 6, 374-378.

[1099] Steele, J. (1981) "Subadditive Euclidean Functionals and Nonlinear Growth in Geometric Probability," *Annals of Probability*, 9, 365-376.

[1100] Stidham, S. (1971) *Stochastic Design Models for Location and Allocation of Public Service Facilities*, Department of Environmental Systems Engineering, Cornell University. Report to the New York State Science and Technology Foundation.

[1101] Stone, R.E. (1991) "Some Average Distance Results," *Transportation Science*, 25, 83-90.

[1102] Storbeck, J. (1982) "Slack, Natural Slack and Location Covering," *Socio-Economic Planning Sciences*, 99.

[1103] Stough, R.R. and J. Hoffman (1986) "Assessing the Risk of Hazardous Materials Flows: Implications for Incidence Response and Enforcement Training," *Transportation Research Record*, 1063, 27-36.

[1104] Stowers, C.L. and U.S. Palekar (1993) "Location Models with Routing Considerations for a Single Obnoxious Facility," *Transportation Science*, 27, 350-362.

[1105] Strom M. (1976) "Atmospheric Dispersion," in A. Stern (ed) *Air Pollution*, Vol.1, McGraw-Hill, New York.

[1106] Sugihara, K. (1993) "Approximation of Generalized Diagrams by Ordinary Voronoi Diagrams," *Graphical Models and Image Processing*, 55, 522-531.

[1107] Sugihara, K. and M. Iri (1989) *VORONOI2 Reference Manual — Topology-oriented Version for the Incremental Method for Constructing Voronoi Diagrams*, Research Memorandum, RMI 89-04, Department of Mathematical Engineering and Information Physics, Faculty of Engineering, University of Tokyo.

[1108] Sugihara, K. and M. Iri (1994) "A Robust Topology-oriented Incremental Algorithm for Voronoi Diagrams," *International Journal of Computational Geometry & Applications*, 4, 179-228.

[1109] Suzuki, A. and Z. Drezner (1994) "On the p-center Problem in a Square," Submitted for Publication.

[1110] Suzuki, A. and M. Iri (1986) "Approximation of a Tessellation of the Plane by a Voronoi Diagram," *Journal of the Operations Research Society of Japan*, 29, 69-96.

[1111] Suzuki, A. and M. Iri (1986) "A Heuristic Method for the Euclidean Steiner Problems as a Geographical Optimization Problem," *Asia-Pacific Journal of Operational Research*, 3, 109-122.

[1112] Suzuki, T. (1987) "Optimum Locational Patterns of Bus-stops for Many-to-one Travel Demand," *Papers of the Annual Conference of the City Planning Institute of Japan*, 22, 247-252[in Japanese].

[1113] Suzuki, T., Y. Asami and A. Okabe (1991) "Sequential Allocation of Public Facilities in One- and Two-dimensional Space: Comparison of Several Policies," *Journal of Mathematical Programming Series* **B**, 52, 125-146.

[1114] Swain, R. (1971) *a Decomposition Algorithm for a Class of Facility Location Problems*, Ph.D. Dissertation, Cornell University, Ithaca, NY.

[1115] Swain, R. (1974) "A Parametric Decomposition Algorithm for the Solution of Uncapacitated Location Problems," *Management Science*, 21, 189-198.

[1116] Sylvester, J.J. (1857) "A Question in the Geometry of Situation," *Quarterly Journal of Pure and Applied Mathematics*, 1, 79.

[1117] Sylvester, J.J. (1860) "On Poncelet's Approximate Linear Valuation of Surd Forms," *London, Edinburgh and Dublin Philosophical Magazine & Journal of Science*, 20, 203-222.

[1118] Takeda, S. (1985) "On Geographical Optimization and Dynamic Facility Location Problems," Unpublished Master's Thesis, Department of Mathematical Engineering and Information Physics, University of Tokyo [in Japanese].

[1119] Tansel, B.C., R.L. Francis and T.J. Lowe (1983) "Location on Networks: a Survey, Parts I and II," *Management Science*, 29, 482-497 and 498-511.

[1120] Tapiero, C.S. (1971) "Transportation Location Allocation Problem Over Time," *Journal of Regional Science*, 11, 377-384.

[1121] Tapiero, C.S. (1974) "The Demand and Utilization of Recreational Facilities," *Regional Science and Urban Economics*, 4, 173-83.

[1122] Tcha D. and B. Lee (1984) "A Branch-and-bound Algorithm for the Multilevel Uncapacitated Facility Location Problem," *European Journal of Operational Research*, 18, 35-43.

[1123] Teitz, M.B. and P. Bart (1968) "Heuristic Methods for Estimating the Generalized Vertex Median of a Weighted Graph," *Operations Research*, 16, 955-961.

[1124] Tellier, L.N. (1985) *Economie Spatiale: Rationalité économique De L'espace Habité*, Gaetan Morin, Chicoutoumi, Quebec, Canada.

[1125] Tellier, L.N. (1987) "Points D'attraction, Points De Répulsion, Centre Et Périphérie, in Association De Science Régionale De Langue FranÇaise," *Espace Et Périphérie*. Lisbon: Laboratorio Nacional De Engenharia Civil.

[1126] Tellier, L.N. and B. Polanski (1989) "The Weber Problem: Frequency and Different Solution Types and Extension to Repulsive Forces and Dynamic Processes," *Journal of Regional Science*, 29, 387-405.

[1127] Thill, J. and I. Thomas (1987) "Towards Conceptualizing Trip-Chaining Behavior: a Review," *Geographical Analysis*, 19, 1-17.

[1128] Thill, J.C. and G. Rushton (1992) "Demand Sensitivity to Space-price Competition with Manhattan and Euclidean Representations of Distance," *Annals of Operations Research*, 40, 381-401.

[1129] Thisse, J.-F. and Y.Y. Papageorgiou (1981) "Reconciliation of Transportation Costs and Amenities as Location Factors in the Theory of the Firm," *Geographical Analysis*, 13, 189-195.

[1130] Thisse, J.-F. and J. Perreur (1977) "Relations Between the Point of Maximum Total Profit and the Point of Minimum Total Transportation Cost: a Restatement," *Journal of Regional Science*, 17, 227-234.

[1131] Thisse, J.-F., J.E. Ward and R.E. Wendell (1984) "Some Properties of Location Problems with Block and Round Norms," *Operations Research*, 32, 1309-1327.

[1132] Thisse, J.-F. and H.G. Zoller (1983) *Locational Analysis of Public Facilities*, North Holland.

[1133] Tietz, M.B. and P. Bart (1968) "Heuristic Methods for Estimating the Generalized Vertex Median of a Weighted Graph," *Operations Research*, 16, 955-965.

[1134] Timmermans, H. (1982) "Consumer Choice of Shopping Centre: an Information Integration Approach," *Regional Studies*, 16, 171-182.

[1135] Timmermans, H. (1988) "Multipurpose Trips and Individual Choice Behaviour: an Analysis Using Experimental Design Data," in *Behavioural Modelling in Geography and Planning*, Reginald G. Golledge and Harry Timmermans, Editors, Croom Helm, London.

[1136] Ting, S.S. (1984) "A Linear-Time Algorithm for Maxisum Facility Location on Tree Networks," *Transportation Science*, 18, 76-84.

[1137] Tobin, R. and T. Friesz (1986) "Spatial Competition Facility Location Models: Definition, Formulation and Solution Approach," *Annals of Operations Research*, 6, 49-74.

[1138] Todd, M.J. (1982) "An Implementation of the Simplex Method for Linear Programming Problems with Variable Upper Bounds," *Mathematical Programming*, 23, 34-49.

[1139] Toregas, C.R. and C. ReVelle (1972) "Optimal Location Under Time Or Distance Constraints," *Papers of the Regional Science Association*, 28, 133-144.

[1140] Toregas, C. and C. ReVelle, (1973), "Binary Logic Solutions to a Class of Location Problems," *Geographical Analysis*, 145-155.

[1141] Toregas, C., R. Swain, C. ReVelle and L. Bergman (1971) "The Location of Emergency Service Facilities," *Operations Research*, 19, 1363-1373.

[1142] Toussaint, G.T. (1983) "Computing Largest Empty Circles with Location Constraints," *International Journal of Computer and Information Sciences*, 12, 347-358.

[1143] Tovey, C.A. (1992) "A Polynomial-time Algorithm for Computing the Yolk in Fixed Dimensions," *Mathematical Programming*, 57, 259-277.

[1144] Truelove, M. (1993) "Measurement of Spatial Equity," *Environment & Planning C*, 11, 19-34.

[1145] Tsai, W.H., M.S. Chern and T.M. Lin (1991) "An Algorithm for Determining Whether m Given Demand Points Are on a Hemisphere Or Not," *Transportation Science*, 25, 91-97.

[1146] Turner D.B. (1970) *Workbook of Atmospheric Dispersion Estimates*, Office of Air Program, Publication AP-26, US EPA.

[1147] Turnquist, M.A. (1987) "Routes, Schedules, and Risks in Transporting Hazardous Materials," in Strategic Planning in Energy and Natural Resources, 289-302, B. Lev, J.A. Bloom, A.S. Gleit, F.H. Murphy and C.A. Shoemaker, Editors, North Holland, Amsterdam.

[1148] Turnquist M.A. and K. G. Zografos (1991) Special Issue of *Transportation Science*, 25, No. 2, Devoted to Hazardous Materials Transportation.

[1149] Tuy, H. (1990) "On Polyhedral Annexation Method for Concave Minimization," *Functional Analysis, Optimization and Mathematical Economics*, Lev J. Leifman and J.B. Rosen, Editors, Oxford University Press, 248-260.

[1150] Tuy, H. and F.A. Al-Khayyal (1991) "A Class of Global Optimization Problems Solvable by Sequential Unconstrained Convex Minimization," *Journal of Global Optimization*, 141-151.

[1151] Tuy, H. and F.A. Al-Khayyal (1992) "Global Optimisation of a Nonconvex Single Facility Location Problem by Sequential Unconstrained Convex Minimisation," *Journal of Global Optimisation*, 2, 61-71.

[1152] Tuy, H. (1994) D.C., "Optimization: Theory, Methods, and Algorithms," *Handbook on Global Optimization*, Dordrecht, Kluwer.

[1153] UNEP (1990) "The Basel Convention for the Control of Transboundary Movements of Hazardous Waste and Their Disposal," United Nations Environmental Programme Environmental Law Library No. 2.

[1154] Vaish, H. and C.M. Shetty (1976) "The Bilinear Programming Problem," *Naval Research Logistics Quaterly*, 23, 303-309.

[1155] Van Ackere, A. (1992) "Competing Against a Public Service," Working Paper, London Business School.

[1156] Vandell, K.D. and C.C. Carter (1993) "Retail Store Location and Market Analysis: a Review of Research," *Journal of Real Estate Literature*, 1, 13-45.

[1157] Van Dijk, J.J. (1990) "Managerial Theories of the Firm in Locational Analysis," *Proceedings of the 4th Meeting, EURO Working Group on Locational Analysis*, Chios, Greece, J. Karkazis & T.B. Boffey, Editors, 141-155.

[1158] Van Dijk, J.J. and K.E. Rosing (1989) "Vloeibaar Aardgas Op De Maasvlakte - (Liquid Natural Gas on the Maasplain, in Dutch)," *ESB - Economische En Sociale Berichten*, 6-9, 877-879.

[1159] Van Roy, T.J. (1989) "Multi-level Production and Distribution Planning with Transportation Fleet Optimization," *Management Science*, 35, 1443-1453.

[1160] Van Roy, T.J. and L.A. Wolsey (1987) "Solving Mixed Integer Programming Problems Using Automatic Reformulation," *Operations Research*, 35, 45-57.

[1161] Ventura, J.A. and Z.A. Dung (1993) "Algorithms for Computerized Inspection of Rectangular and Square Shapes," *European Journal of Operational Research*, 68, 265-277.

[1162] Ventura, J.A. and S. Yeralan (1989) "The Minimax Center Estimation Problem for Automated Roundness Inspection," *European Journal of Operational Research*, 41, 64-72.

[1163] Verter, V. and M.C. Dincer (1992) "An Integrated Evaluation of Facility Location, Capacity Acquisition and Technology Selection for Designing Global Manufacturing Strategies," *European Journal of Operational Research*, 60, 1-18.

[1164] Verter, V and E. Erkut (1995) "Hazardous Materials Logistics: an Annotated Bibliography," *Operations Research and Environmental Management*, A. Haurie and C. Carraro, Editors, Kluwer Publishing Company.

[1165] Viegas, J. and P. Hansen (1985) "Finding Shortest Paths in the Plane in the Presence of Barriers to Travel (for Any L_p Norm)," *European Journal of Operationnal Research*, 20, 373-381.

[1166] Vijay, J. (1985) "An Algorithm for the P-center Problem in the Plane," *Transportation Science*, 19, 235-245.

[1167] Von Hohenbalken, B. and D.S. West (1984) "Manhattan Versus Euclid: Market Areas Computed and Compared," *Regional Science and Urban Economics*, 14, 19-35.

[1168] Wagner, J.L. and L.M. Falkson (1975) "The Optimal Nodal Location of Public Facilities with Price-Sensitive Demand," *Geographical Analysis*, 7, 69-83.

[1169] Walker, J.H. (1991) "An Empirical Study of Round and Block Norms for Modelling Actual Distances," Ph.D. Thesis, McMaster University, Hamilton, Ontario, Canada.

[1170] Walker, W. (1974) "Using the Set-Covering Problem to Assign Fire Companies to Fire Houses," *Operations Research*, 22, 275-277.

[1171] Ward, J.E. and R.E. Wendell (1980) "A New Norm for Measuring Distance Which Yields Linear Location Problems," *Operations Research*, 28, 836-844.

[1172] Ward, J.E. and R.E. Wendell (1983) "Characterizing Efficient Points in Location Problems Under the One-infinity Norm," J.-F. Thisse and H.G. Zoller, Editors, *Locational Analysis of Public Facilities*, North Holland, Amsterdam, Netherlands.

[1173] Ward, J.E. and R.E. Wendell (1985) "Using Block Norms for Location Modelling," *Operations Research*, 33, 1074-1090.

[1174] Warszawski, A. (1973) "Multi-dimensional Location Problems," *Operational Research Quarterly*, 24, 165-179.

[1175] Waters, N.M., S.C. Wirasinghe, A. Babalola and K.E.D. Marion (1986) "Location of Bus Garages," *Journal of Advanced Transportation*, 20, 133-150.

[1176] Watson, G.A. (1983) "An Algorithm for the Single Facility Location Problem Using the Jaccard Metric," *SIAM Journal on Science and Statistical Computing*, 4, 748-756.

[1177] Watson-Gandy, C.D.T. (1982) "Heuristic Procedures for the M-partial Cover Problem on a Plane," *European Journal of Operational Research*, 11, 149-157.

[1178] Watson-Gandy, C.D.T. (1984) "The Multifacility Min-max Weber Problem," *European Journal of Operational Research*, 18, 44-50.

[1179] Watson-Gandy, C.D.T. (1985) "The Solution of Distance Constrained Mini-sum Location Problems," *Operations Research*, 33, 784-802.

[1180] Webber, M.J. (1979) *Information Theory and Urban Spatial Structure*, Croom Helm.

[1181] Weber, A. (1909) *ÜBer Den Standort Der Industrien, 1. Teil: Reine Theorie Des Standortes*, Tübingen, Germany. English Translation: *on the Location of Industries*, University of Chicago Press, Chicago, IL, 1929. (English Translation by C. J. Friedeich (1957), Theory cf the Location Of Industries, Chicago University Press, Chicago.)

[1182] Webster, S. and E.P. Robinson (1995) "Analytical Results for Two-echelon Distribution Network Design," *Location Science*, to appear.

[1183] Webster, S. and A. Gupta (1995) "The General Optimal Market Area Model with Uncertain and Nonstationary Demand," *Location Science*, to appear.

[1184] Weigkricht, E. and K. Fedra (1995) "Decision Support Systems for Dangerous Goods Transportation," *INFOR*.

[1185] Weisbrod, G.E., R.J. Parcells and C. Kern (1984) "A Disaggregate Model for Predicting Shopping Area Market Attraction," *Journal of Retailing*, 60, 65-83.

[1186] Weiszfeld, E. (1937) "Sur Le Point Pour Lequel La Somme Des Distances De N Points Donnés Est Minimum," *Tohoku Mathematical Journal*, 43, 355-386.

[1187] Wendell, R.E. and A.P. Hurter (1973) "Location Theory - Dominance and Convexity," *Operations Research*, 21, 314-320.

[1188] Wendell, R.E., A.P. Hurter and T.J. Lowe (1977) "Efficient Points in Location Problems," *AIIE Transactions*, 9, 238-246.

[1189] Wendell, R.E. and R.D. McKelvey (1981) "New Perspectives in Competitive Location Theory," *European Journal of Operational Research*, 6, 174-182.

[1190] Wendell, R.E. and S.J. Thorson (1974) "Some Generalizations of Social Decisions Under Majority Rule," *Econometrica*, 42, 893-912.

[1191] Wesolowsky, G.O. (1972) "Rectangular Distance Location Under the Minimax Optimality Criterion," *Transportation Science*, 6, 103-113.

[1192] Wesolowsky, G.O. (1977) "Probabilistic Weights in the One-Dimensional Facility Location Problem," *Management Science*, 24, 224-229.

[1193] Wesolowsky, G.O. (1983) "Location Problems on a Sphere," *Regional Science and Urban Economics*, 12, 495-508.

[1194] Wesolowsky, G.O. (1993) "The Weber Problem: History and Perspectives," *Location Science*, 1, 5-23.

[1195] Wesolowsky, G.O. and R.F. Love (1971) "Location of Facilities with Rectangular Distances Among Point and Area Destinations," *Naval Research Logistics Quarterly*, 18, 83-90.

[1196] Wesolowsky, G.O. and R.F. Love (1972) "A Nonlinear Approximation Method for Solving a Generalized Rectangular Distances Weber Problem," *Management Science*, 18, 656-663.

[1197] West, D.S. (1985) "Testing for Market Preemption Using Sequential Location Data," *Bell Journal of Economics*, 16, 129-43.

[1198] White, A.L. and S.J. Ratick (1989) "Risk, Compensation, and Regional Equity in Locating Hazardous Facilities," *Papers of the Regional Science Association*, 67, 29-42.

[1199] White, J.A. (1971) "A Quadratic Facility Location Problem," *AIIE Transactions*, 3, 156-157.

[1200] White, J. and K. Case (1974) "On Covering Problems and the Central Facility Location Problem," *Geographical Analysis*, 281.

[1201] Whittaker, J. (1991) "A Reappraisal of Probabilistic Risk Analysis," *Engineering Management Journal*, 3, 13-16.

[1202] Widmayer, P., Y.F. Wu and C.K. Wong (1987) "On Some Distance Problems in Fixed Orientations," *SIAM Journal on Control*, 16, 728-746.

[1203] Wijeratne, A.B. (1990) "Routing and Scheduling Decisions in the Management of Hazardous Material Shipments," Ph.D. Dissertation, Cornell University, Ithaca, NY.

[1204] Wijeratne, A.B., M.A. Turnquist and P.B. Mirchandani (1993) "Multiobjective Routing of Hazardous Materials in Stochastic Networks," *European Journal of Operational Research*, 65, 33-43.

[1205] Wijeratne, A. and S.C. Wirasinghe (1986) "Estimation of the Number of Fire Stations and Their Allocated Areas to Minimize Fire Service and Property Damage Costs," *Civil Engineering Systems*, 3, 2-6.

[1206] Williams, H.C.W.L. (1977) "On the Formation of Travel Demand Models and Economic Evaluation Measures of User Benefit," *Environment and Planning A*, 9, 285-344.

[1207] Williams, H.C.W.L. and M.L. Senior (1977) "A Retail Location Model with Overlapping Market Areas: Hotelling Revisited," *Urban Studies*, 14, 203-215.

[1208] Wilson, A.G. (1970) *Entropy in Urban and Regional Modelling*, Pion, London.

[1209] Wilson, A.G. (1976) "Retailers' Profits and Consumers' Welfare in a Spatial Interaction Shopping Model," *Theory and Practice in Regional Science*, I. Masser, Editor, Pion, London, 42-59.

[1210] Winston W.L. (1994) *Operations Research: Applications and Algorithms* 3rd Edition, Duxbury Press, Belmont, California.

[1211] Winter, P. and J. Macgregor Smith (1991) "Steiner Minimal Trees for Three Points with One Convex Polygonal Obstacle," *Annals of Operations Research*, 33.

[1212] Wirasinghe, S.C. and N.M. Waters (1983) "An Approximate Procedure for Determining the Number, Capacities, and Locations of Solid Waste Transfer-stations in an Urban Region," *European Journal of Operational Research*, 12, 105-111.

[1213] Wittink, D.R. and P. Cattin (1989) "Commercial Use of Conjoint Analysis: an Update," *Journal of Marketing*, 53, 91-96.

[1214] Witzgall, C.J. (1964) "Optimal Location of a Central Facility: Mathematical Models and Concepts," National Bureau of Standards Report 8388, US Department of Commerce, National Bureau of Standards, Washington, DC.

[1215] Witzgall, C.J. (1965) "On Convex Metrics," *Journal of Research of the National Bureau of Standards (Sec B)*, 69B, 175-177.

[1216] Wolf, U. (1976) *Location Problems on Tree Network*, Master's Thesis, Tel-Aviv University.

[1217] Wright, J. (1993) "Transport Canada Dangerous Goods Accident Costing Study and Model," *Transportation of Dangerous Goods: Assessing the Risks*, 601-616, F. Saccomanno and K. Casidy, Editors, Institute for Risk Research, Univ. of Waterloo.

[1218] Wrigley, N. (1988) *Store Choice, Store Location and Market Analysis*, London, Routledge.

[1219] Wyman, M. and M. Kuby (1993) "A Multiobjective Location-Allocation Model for Assessing Toxic Waste Processing Technologies," *Studies in Location Science*, 4, 193-196.

[1220] Wyman, M. and M. Kuby (1995) "Turning the Tables: Using Location Science to Choose Technology Rather Than Allowing Technology to Constrain Location," *INFOR*.

[1221] Wyman M.M. and M. Kuby (1995) "Proactive Optimization of Toxic Waste Transportation, Location and Technology," *Location Science*, to appear.

[1222] Xue, G.L. (1994) "A Globally Convergent Algorithm for Facility Location on a Sphere," *Computers and Mathematics with Applications*, 27, 37-50.

[1223] Yamani, A., M.J. Hodgson and L.A. Martin-Vega (1990) "Single Aircraft Mid-air Refueling Using Spherical Distances," *Operations Research*, 38, 792-800.

[1224] Yeralan, S. and J.A. Ventura (1988) "Computerized Roundness Inspection," *International Journal of Production Research*, 26, 1921-1935.

[1225] Yu, J.C. and C.A. Judd (1985) "Cost Effectiveness Analysis of Transportation Strategies for Nuclear Waste Repository Sites," *Transportation Research Record*, 1020, 23-28.

[1226] Zangwill, W.I. (1969) *Nonlinear Programming: a Unified Approach*, Prentice-Hall, Englewood Cliffs, NJ.

[1227] Zeller, R.E., D.D. Achabal and L.E. Brown (1980) "Market Penetration and Locational Conflict in Franchise Systems," *Decision Sciences*, 11, 58-80.

[1228] Zhang, S. (1991) "Stochastic Queue Location Problems, Ph.D. Thesis, Erasmus Universiteit Rotterdam, Rotterdam, the Netherlands.

[1229] Zografos, K.G. and C.F. Davis (1989) "Multiobjective Programming Approach for Routing Hazardous Materials," *Journal of Transportation Engineering*, 115, 661-673.

[1230] Zografos, K.G. and S. Samara (1989) "A Combined Location-Routing Model for Hazardous Waste Transportation and Disposal," *Transportation Research Record*, 1245, 52-59.